# Nonlinear Water Waves

International Union of Theoretical and Applied Mechanics

K. Horikawa, H. Maruo (Eds.)

# Nonlinear Water Waves

IUTAM Symposium, Tokyo/Japan
August 25-28, 1987

Springer-Verlag
Berlin Heidelberg New York
London Paris Tokyo

Prof. Dr. K. Horikawa
Dept. of Civil Engineering
University of Tokyo
7-3-1 Hongo, Bunkyo-ku
Tokyo 113
Japan

Prof. Dr. H. Maruo
Dept. of Naval Architecture
and Ocean Engineering
Yokohama National University
156 Tokiwadai, Hodogaya-ku
Yokohama 240
Japan

**ISBN-13:978-3-642-83333-5 e-ISBN-13:978-3-642-83331-1**
**DOI: 10.1007/978-3-642-83331-1**

Library of Congress Cataloging in Publication Data
Nonlinear water waves, IUTAM Symposium, Tokyo/Japan, August 25-28, 1987.
"Papers presented at the Symposium on Non-linear Water Wavevs"--Pref.
1. Water waves--Congresses.
2. Nonlinear waves--Congresses.
3. Nonlinear theories--Congresses.
I. Horikawa, Kiyoshi.
II. Maruo, H. (Hajime),
III. Symposium on Non-linear Water Waves (1987 : Tokyo/Japan)
IV. International Union of Theoretical and Applied Mechanics.
TC172.N66 1988 627'.042 88-3057
ISBN-13:978-3-642-83333-5 (U.S.)


Softcover reprint of the hardcover 1st edition 1988

# General Information

## International Scientific Committee

H. Maruo, Co-Chairman
K. Kirchgaessner
J. N. Newman
P. G. Saffman
L. van Wijngaarden
K. Horikawa, Co-Chairman
M. S. Longuet-Higgins
M. Roseau
S. Ya. Sekerzh-Zenkovich

## Local Organizing Committee

H. Maruo, Co-Chaiman
H. Hasimoto
I. Imai
H. Mitsuyasu
T. Sawaragi
I. Tani
Y. Yamamoto
K. Horikawa, Co-Chirman
M. Hattori
T. Inui
S. Motora
N. Shuto
T. Tatsumi
R. Yamazaki
M. Hino
K. Kawada
S. Nakamura
S. Takezawa
Y. Tsuchiya

## Executive Committee

H. Maruo, Co-Chairman
M. Ikehata, Secretary
T. Hirayama
T. Kinoshita
M. Mizuguchi
A. Watanabe
K. Horikawa, Co-Chairman
M. Isobe, Secretary
T. Izumiya
H. Maeda
H. Nishimura
H. Kajitani
N. Mimura
T. Shibayama

## Supporting Organizations

International Union of Theoretical
and Applied Mechanics (IUTAM)
Commemorative Association for the Japan World Exposition(1970)
Inoue Foundation for Science
Japan Society for the Promotion of Science
The Kajima Foundation

## Cooperating Scientific Organizations

Science Council of Japan
Japan Society of Civil Engineers
Society of Naval Architects of Japan
Architectural Institute of Japan
Japan Institute of Navigation
Japan Society of Applied Physics
Japan Society of Fluid Mechanics
Mathematical Society of Japan
Meteorological Society of Japan
Physical Society of Japan
The Flow Visualization Society of Japan
The Japan Society of Mechanical Engineers
The Japan Society for Aeronautical and Space Sciences
The Japanese Society of Irrigation, Drainage
and Reclamation Engineering
The Oceanographical Society of Japan

# Preface

Non-linear features of surface waves in water have drawn attention of scientists and engineers in connexion with numerous problems in practical fields such as coastal engineering, naval architecture and ocean engineering in recent years. However, difficulties in the mathematical analysis of the problems have prevented the development in the theoretical elucidation of the phenomena. Thanks to the remarkable progress of the high speed computer, the computational method is now widely available, and it is expected that theoretical findings will be able to provide useful information for the various engineering fields.

The Symposium on Non-linear Water Waves was proposed to the General Assembly of The International Union of Theoretical and Applied Mechanics (IUTAM) at Lyngby in 1984 and was decided to be held in Tokyo during the period of 1986-1987. Then the IUTAM Bureau appointed the International Scientific Committee constituted by nine members. The Local Organizing Committee was established in order to take responsibility for preparation and management of the Symposium. The Symposium has been held on 25-28 August 1987 at Sanjo Conference Hall of the University of Tokyo. The University of Tokyo and the Yokohama National University have become the local hosts of the Symposium.

The Symposium has intended to provide a wide scope of analytical and numerical methods as well as experimental studies for the analysis of the non-linear phenomena related to water waves in various fields of science and engineering, such as applied mathematics, mechanics, civil engineering, naval architecture and ocean engineering at the present stage. Since the emphasis is especially put on the analysis of non-linearity of water waves, the scope of the presented papers includes following topics:

1) Theoretical and experimental studies of non-linear water waves,
2) Non-linear instability and deformation of water waves,
3) Wave breaking,
4) Non-linear wave-current interaction,
5) Non-linear water waves around structures and ships,
6) Wave-body interaction, and
7) Non-linear internal waves.

Fifty four papers have been presented at the Symposium including three keynote lectures. Over one hundred participants from fourteen countries have taken an active part in the Symposium. It is believed that the Symposium has been able to provide valuable information for the progress in various fields from both science and engineering, which are relevant to the topics discussed in the Symposium, and to give the participants rare opportunity to discuss their common interest with specialists from different fields.

Science Council of Japan and other fourteen domestic societies supported the Symposium, and financial support was generously provided by IUTAM, Commemorative Association for the Japan World Exposition (1970), Inoue Foundation for Science, Japan Society for the Promotion of Science, and the Kajima Science Foundation.

All affairs concerning the preparation and management of the Symposium and the editorial work of the proceedings have been carried out by the Exective Committee members. We would like to express our deepest appreciation for their great efforts to accomplish this valuable Symposium.

Kiyoshi Horikawa
Hajime Maruo

February 1988

# Participants

| | |
|---|---|
| Akylas, T.R. | MIT, Room 3-360, Cambridge, MA 02139, USA |
| Aranha, J.A.P. | Dinav, IPT, Cidade Universitaria, CP. 7141, Sao Paulo, S.P. BRAZIL |
| Asano, T. | Dept. of Civil Eng., Kyoto Univ., Yoshida-Honmachi, Sakyo-ku, Kyoto 606, JAPAN |
| Banner, M.L. | School of Math., Univ. of New South Wales, P.O. BOX 1, Kensington, N.S.W. 2033, AUSTRALIA |
| Barthel, V. | Hydraulics Lab., Nat. Res. Council, Montreal Rd. M-32, Ottawa, CANADA K1A OR6 |
| Bertram, V. | Univ. of Hamburg, IfS, Laemmersieth 90, 2000 Hamburg 60, FRG |
| Bryant, P.J. | Math. Dept., Univ. of Canterbury, Christchurch, NEW ZEALAND |
| Choi, H.S. | Dept. of Naval Arch., Seoul Nat. Univ., 56-1, Shinrim 2-Dong, Kwanak-ku, Seoul 151, KOREA |
| Cointe, R. | Inst. Francais du Petrole, B.P. 311, 92506 Rueil Malmaison, FRANCE |
| Dern, J.C. | Bassin d'Essais des Carenes, 8 Boulevard Victor 75732, Paris Cedex 15, FRANCE |
| Donelan, M.A. | Canada Cent. for Inland Waters, P.O. Box 5050, Burlington, Ontario, CANADA, L7R 4A6 |
| Eggers, K. | Horstlooge 16, D-2000 Hamburg 67, FRG |
| Faltinsen, O.M. | Div. of Marine Hyd., Norwegian Inst. of Tech., N-7034 Trondheim-NTN, NORWAY |
| Fenton, J. D. | Dept. of Civil Eng., Univ. of Auckland, Private Bag, Auckland, NEW ZEALAND |
| Ferrant, P. | Lab. Hyd. Navale-ENSM, 1 rue de la Noe, 44072 Nantes Cedex, FRANCE |
| Funakoshi, M. | Res. Inst. for Applied Mech., Kyushu Univ., Kasuga-Koen 6-1, Kasuga, Fukuoka, JAPAN |
| Germain, J.P. | Inst. de Mecanique de Grenoble, Domaine Universitaire, B.P.No. 68, 38402 S'Martin d'Heres Cedex, FRANCE |
| Grabitz, G. | Max-Planck Inst., Bunsenstr. 10, D-34 Göttingen, FRG |

| | |
|---|---|
| Grimshaw, R. | School of Math., Univ. of New South Wales, P.O. Box 1, Kensington, N.S.W. 2033, AUSTRALIA |
| Hansen, J.B. | ISVA, Technical Univ. of Denmark, Build. 115, DK-2800, Lyngby, DENMARK |
| Hasimoto, H. | Dept. of Mechanical Eng., Hosei Univ., Kajino, Koganei 184, JAPAN |
| Hattori, M. | Dept. of Civil Eng., Chuo Univ., Kasuga 1-13-27, Bunkyo-ku, Tokyo 112, JAPAN |
| Hermans, A.J. | Delft Univ. of Tech., Fac. of Math. and Informatics Jnlianalaan 132, 2628 BL Delft, THE NETHERLANDS |
| Hino, M. | Dept. of Civil Eng., Tokyo Inst. of Tech., 2-12-1, O-okayama, Meguro-ku, Tokyo 152, JAPAN |
| Hirayama, T. | Dept. of Naval Arch. and Ocean Eng., Yokohama Nat. Univ., Tokiwadai, Hodogaya-ku, Yokohama 240, JAPAN |
| Hong, G. | East China Tech. Univ., 1 Xikang Road, Nanjing, CHINA |
| Horikawa, K. | Dept. of Civil Eng., Univ. of Tokyo, 7-3-1 Hongo, Bunkyo-ku, Tokyo 113, JAPAN |
| Hui, W.H. | Dept. of Applied Math., Univ. of Waterloo, Waterloo, Ontario, CANADA N2L 3G1 |
| Huijsmans, R.H.M. | Maritime Res. Inst., 2 Haagsteeg, Wageningen, THE NETHERLANDS |
| Hwang, J.H. | Seoul Nat. Univ., San 56-1, Sinrim 2-Dong, Kwanak-ku Seoul 151, KOREA |
| Hwung, H.H. | Hyd. & Ocean Eng. Grad. School, Nat. Cheng Kung Univ. Tainan, TAIWAN, CHINA (70101) |
| Ikehata, M. | Dept. of Naval Arch. & Ocean Eng., Yokohama Nat. Univ., 156 Tokiwadai, Hodogaya-ku, Yokohama 240, JAPAN |
| Imai, I. | 1-11-5 Suido, Bunkyo-ku, Tokyo 112, JAPAN |
| Inoue, Y. | Dept. of Naval Arch. & Ocean Eng., Yokohama Nat. Univ., 156 Tokiwadai, Hodogaya-ku, Yokohama 240, JAPAN |
| Isobe, M. | Dept. of Civil Eng., Univ. of Tokyo, 7-3-1 Hongo, Bunkyo-ku, Tokyo 113, JAPAN |

| | |
|---|---|
| Iwata, K. | Dept. of Civil Eng., Nagoya Univ., Furu-cho, Chikusa-ku, Nagoya 464, JAPAN |
| Izumiya, T. | Dept. of Civil Eng., Univ. of Tokyo, 7-3-1 Hongo, Bunkyo-ku, Tokyo 113, JAPAN |
| Jami, A. | LNH(EDF) Chatou, 7 Ave de Gommonvilliers 91430 IGNY, FRANCE |
| Janssen, P. | ECMWF, Shinfield Park, Reading RG2 9AX, UK |
| Kajitani, H. | Dept. of Naval Arch., Univ. of Tokyo, Hongo 7-3-1, Bunkyo-ku, Tokyo 113, JAPAN |
| Kakutani, T. | Dept. of Mech. Eng., Osaka Univ., 1-1 Machikaneyamacho, Toyonaka, Osaka 560, JAPAN |
| Kambe, T. | Dept. of Physics, Univ. of Tokyo, Hongo, Bunkyo-ku, Tokyo 113, JAPAN |
| Kawahara, T. | Dept. of Physics, Kyoto Univ., Kyoto 606, JAPAN |
| Kim, S. | Inst. of Machinery and Metals, P.O. Box 1, Daeduk Science Town, Chungnam 300-01, KOREA |
| Kim, Y.J. | Dept. of Naval Arch., Nat. Fisheries Univ. of Pusan, Daeyeon-dong Nam-ku, Pusan 608, KOREA |
| Kimura, A. | Fac. of Eng., Tottori Univ., Koyama Minami 4-101, Tottori, JAPAN |
| Kinoshita, T. | Inst. of Industrial Sci., Univ. of Tokyo, 7-22-1 Roppongi, Minato-ku, Tokyo 106, JAPAN |
| Kioka, W. | Nagoya Inst. of Tech., Gokiso-cho, Showa-ku, Nagoya 466, JAPAN |
| Kirby, J.T. | Coastal & Ocean. Eng., Univ. of Florida, 336 Weil Hall, Gainesville, FL 32611, USA |
| Kodama, Y. | Dept. of Physics, Nagoya Univ., Chikusa-ku, Nagoya 464, JAPAN |
| Kyozuka, Y. | Res. Inst. for Applied Mech., Kyushu Univ., 6-1, Kasuga-Koen, Kasuga 816, JAPAN |
| Lee, C.M. | Pohang Inst. Sci. Tech., P.O.BOX 125, Pohang 680, KOREA |
| Lee, S.J. | Dept. of Naval Arch., Chungnam Nat. Univ., Daejeon, Chungnam 300-31, KOREA |

| | |
|---|---|
| Lichter, S. | Aerospace & Mech. Eng. Dept., Univ. of Arizona, Tucson, Arizona 85721, USA |
| McCarthy, J.H. | Office of Naval Res., 7-23-17 Roppongi, Minato-ku, Tokyo 106, JAPAN |
| Maeda, H. | Inst. of Industrial Sci., Univ. of Tokyo, 7-22-1, Roppongi, Minato-ku, Tokyo 106, JAPAN |
| Marthinsen, T. | Saga Petroleum, P.O.BOX 9, N-1322 Høvik, NORWAY |
| Maruo, H. | Dept. of Naval Arch. & Ocean Eng., Yokohama Nat. Univ., Tokiwadai 156, Hodogaya-ku, Yokohama 240, JAPAN |
| Masuya, T. | Nat. Res. Inst. Fisheries Eng., Ebidai, Hasaki-machi Kashima-gun, Ibaraki 314-04, JAPAN |
| Matsui, T. | Dept. of Arch., Nagoya Univ., Furo-cho, Chikusa-ku, Nagoya 464, JAPAN |
| Mei, C.C. | Dept. of Civil Eng., MIT, Cambridge, MA 02139, USA |
| Miao, G. | Shanghai Jiaotong Univ., 1954 Hwashan Road Shanghai, CHINA |
| Miloh, T. | Ocean Eng., Univ. of California Santa Barbara, CA 93106, USA |
| Mimura, N. | Dept. of Civil Eng., Ibaraki Univ., 4-12-1 Nakanarusawa, Hitachi, Ibaraki 316, JAPAN |
| Mitsuyasu, H. | Inst. of Applied Mech., Kyushu Univ., 6-1 Kasugakoen, Kasuga 816, JAPAN |
| Miyata, H. | Dept. of Naval Arch., Univ. of Tokyo, 3-1 Hongo 7, Bunkyo-ku, Tokyo 113, Japan |
| Miyata, M. | Geophys. Inst., Univ. of Tokyo, 2-11-16 Yayoi, Bunkyo-ku, Tokyo 113, JAPAN |
| Mizuguchi, M. | Dept. of Civil Eng., Chuo Univ., Kasuga 1-13-27, Bunkyo-ku, Tokyo 112, JAPAN |
| Mori, K. | Hiroshima Univ., Saijo, Higashi-Hiroshima 724, JAPAN |
| Motora, S. | Found. of Shipbuilding Ad., 1-15-16 Toranomon, Minato-ku, Tokyo 105, JAPAN |
| Nadaoka, K. | Dept. of Civil Eng., Tokyo Inst. of Tech., 2-12-1 O-okayama, Meguro-ku, Tokyo 152, JAPAN |

| | |
|---|---|
| Naito, S. | Fac. of Eng., Osaka Univ., 2-1 Yamadaoka, Suita, Osaka 565, JAPAN |
| Nakamura, S. | Kochi Nat. College of Tech., Monobe, Nankoku City, Kochi 783, JAPAN |
| Nakatake, K. | Kyushu Univ., 2-20-3 Takamiya, Minami-ku, Fukuoka 815, JAPAN |
| Nishimura, H. | Inst. of Eng. Mech., Univ. of Tsukuba, Sakura, Niihari, Ibaraki 305, JAPAN |
| Ogiwara, S. | Res. Inst., IHI Co. Ltd., 1 Shin-nakahara, Isogo-ku, Yokohama 235, JAPAN |
| Okamura, M. | Res. Inst. for Applied Mech., Kyushu Univ., Kasuga, Fukuoka 816, JAPAN |
| Palm, E. | Inst. of Math., Univ. of Oslo, P.B.1053, Blindern, 0316 Oslo 3, NORWAY |
| Papanikolaou, A. | Nat. Tech. Univ. of Athens, 42, Patission Str., 10682 Athens, GREECE |
| Peregrine, D.H. | School of Mathematics, Bristol Univ., University Walk, Bristol BS8 1TW, UK |
| Prüser, H.H. | Univ. of Hannover, Callinstrasse 32, 3000 Hannover 1, FRG |
| Pullin, D.I. | Dept. of Mech. Eng., Univ. of Queensland, St. Lucia 4067, AUSTRALIA |
| Roberts, A.J. | Dept. of Applied Math., Adelaide Univ., G.P.O. BOX 498, Adelaide, South Australia, 5001, AUSTRALIA |
| Roseau, M. | Univ. Pierre et Marie Curie, 144 Avenue du General Leclerc, 92330 Sceaux, FRANCE |
| Saffman, P.G. | California Inst. Tech., 217-50 Firestone, Pasdena, CA 91125, USA |
| Sawaragi, T. | Dept. of Civil Eng., Osaka Univ., Yamada-oka, Suita-shi 565, JAPAN |
| Shemer, L. | Dept. of Fluid Mech., Tel-Aviv Univ., Tel-Aviv 69978, ISRAEL |
| Shibayama, T. | Dept. of Civil Eng., Yokohama Nat. Univ., 156 Tokiwadai, Hodogaya-ku, Yokohama 240, JAPAN |
| Sugimoto, N. | Dept. of Mech. Eng., Osaka Univ., Toyonaka, Osaka 560, JAPAN |

| | |
|---|---|
| Suzuki, Katsuo | Nat. Defense Academy, Hashirimizu, Yokosuka 239, JAPAN |
| Suzuki, Kazuo | Dept. of Naval Arch., Yokohama Nat. Univ., 156 Tokiwadai, Hodogaya-ku, Yokohama 240, JAPAN |
| Takagi, K. | Fac. of Eng., Osaka Univ., 2-1 Yamadaoka, Suita, Osaka 565, JAPAN |
| Takaishi, Y. | Ship Res. Inst., 6-38-1 Shinkawa, Mitaka, Tokyo 181, JAPAN |
| Takaki, M. | Dept. of Naval Arch. & Ocean Eng., Hiroshima Univ., Saijo, Higashi-Hiroshima 724, JAPAN |
| Takezawa, S. | Dept. of Naval Arch. & Ocean Eng., Yokohama Nat. Univ., 156 Tokiwadai, Hodogaya-ku, Yokohama 240, JAPAN |
| Tanaka, M. | Fac. of Eng., Gifu Univ., 1-1 Yanagido, Gifu 501-11, JAPAN |
| Taniuchi, T. | Dept. of Physics, Nagoya Univ., Chikusa, Nagoya 464, JAPAN |
| Tatsumi, T. | Hiroshima Inst. of Tech., 725 Miyake, Itsukaichi, Saeki-ku, Hiroshima 731-51, JAPAN |
| Terao, Y. | Tokai Univ., 3-20-1 Orido, Shimizu, Shizuoka 424, JAPAN |
| Tomita, H. | Ship Res. Inst., 6-38-1 Shinkawa, Mitaka, Tokyo 181, JAPAN |
| Tsai, C.P. | Dept. of Civil Eng., Nat. Chung Hsing Univ., Taichung, TAIWAN, CHINA (40227) |
| Tsuchiya, Y. | Disas. Prev. Res. Inst., Kyoto Univ., Uji, Kyoto 611, JAPAN |
| Tulin, M.P. | Ocean Eng. Lab., Univ. of California S.B., Santa Barbara, CA 93106, USA |
| Umeda, N. | Nat. Res. Inst. Fisheries Eng., Ebidai, Hasaki-machi, Kashima-gun, Ibaraki 314-04, JAPAN |
| Watanabe, A. | Dept. of Civil Eng., Univ. of Tokyo, 7-3-1 Hongo, Bunkyo-ku, Tokyo 113, JAPAN |
| Wijngaarden, L.V. | Univ. of Twente, P.O. Box 217, 7500 AE Enschede, THE NETHERLANDS |

| | |
|---|---|
| Wu, D.M. | Harbin Ship Eng. Inst., Harbin, CHINA |
| Wu, T.Y. | California Inst. of Tech., 104-44 Pasadena, CA 91125, USA |
| Yamamoto, Y. | Dept. of Industrial Mechanical Eng., Tokyo Denki Univ., Ishizaka, Hatoyama-machi, Saitama 350-03, JAPAN |
| Yamashita, T. | Disas. Prev. Res. Inst., Kyoto Univ., Uji, Kyoto 611, JAPAN |
| Yamazaki, R. | 4-19-31 Wakamiya, Higashi-ku, Fukuoka 813, JAPAN |
| Yasuda, T. | Dept. of Civil Eng., Gifu Univ., 1-1 Yanagido, Gifu 501-11, JAPAN |
| Zhao, R. | Div. of Marine Hyd., Norwegian Inst. of Tech., N-7034 Trondheim-NTH, NORWAY |

# Contents

**Poster session 1: Nonlinear water waves**

Chairman: Mei, C.C.



**Poster session 2: Nonlinear wave-body interaction**

Chairman: van Wijngaarden, L.

# Keynote Lecture

# Nonlinear Effects in Water Wave Diffraction

CHIANG C. MEI

Department of Civil Engineering
Massachusetts Instutute of Technololgy
Cambridge, Mass, U.S.A.

## Summary

We present a survey of nonlinear effects on water waves interacting with a structure or topography with nontrivial geometry. Examples will be discussed for narrow-banded incident waves whose average wavelength is comparable to the sea depth. In particular the role played by the diffraction of short waves on the excitation of long-period oscillations will be highlighted.

## Introduction

Great strides have been made in recent decades in the basic theory of nonlinear dynamics of waves in an unbounded sea. In engineering and coastal oceanography it is important to assess the interaction of moderate and large waves with bodies such as oil rigs, ships, breakwaters, complex coastlines, or submarine bathymetry. While intensive efforts are being devoted to the numerical calculation of diffraction of steep waves, many difficulties remain to be resolved. In this paper we present several examples involving weakly nonlinear waves.

To cast our examples in perspective, we categorize existing works on the diffraction of weakly nonlinear waves as follows.

i. Intermediate or great depth. ($kh \geq O(1)$).

i.a. Short fetch. The spatial domain of interest is within a few wavelengths from the body, i.e., $kx = O(1)$ (e.g., second order wave forces due to steady periodic waves).

i.b. Intermediate fetch. Let $\epsilon$ be the measure of wave steepness $kA$ or slow modulation. The spatial and time ranges of interest are $kx, \omega t = O(\epsilon^{-1})$ (e.g., slow

K. Horikawa, H. Maruo (Eds.)
Nonlinear Water Waves
IUTAM Symposium Tokyo/Japan 1987

drift oscillations and long wave radiation).

i.c. Long fetch. The spatial and time ranges of interest are $kx, \omega t = O(\epsilon^{-2})$ (e.g., forward scattering by a slender body or mild bathymetry).

ii. Shallow water ($kh << 1$).

ii.a. One-dimensional scattering (e.g., weak reflection by gradual decrease of depth, strong reflection by an abrupt step, total reflection by a wall).

ii.b. Two-dimensional scattering (e.g., Mach reflection at glancing incidence, resonance in a harbor.)

In this survey most of our examples will be chosen from class i.b, namely intermediate depth and fetch. The incident sea is assumed to be narrow-banded swell. Specifically we shall discuss the slow drift motion of a floating body, the trapping of long waves on a wide ridge attacked by short swell, the penetration of long internal waves behind a breakwater attacked by short swell, and the radiation of long waves over periodic bars. The long waves in all four cases are the consequence of second order nonlinearity. We then discuss a few examples from class i.c.

## Slow Sway of a Floating Body

In a narrow banded sea, the motion of a floating body has a low frequency component due to second order forcing at frequencies much below that of the indident sea. Mooring cables with natural frequencies which are low due either to weak elastic constants or to their great length (as in tension leg platforms) can be resonated, causing severe strains. The problem contains at least two distinct time scales. For unidirectional groups of progressive waves it is well known that at the second order $O(\epsilon^2)$ slow variations in time are accompanied by comparably slow variations in space. Physically, there are long waves of set-down propagating at the group velocity of the short waves. Agnon and Mei [1] have treated the two dimensional sway of a rectangular block sliding on the seabed in beam seas. They point out that in addition to the long waves locked to the incident, transmitted and reflected wave groups, there are also long waves radiated at the shallow water wave speed $\sqrt{gh}$. Agnon, Choi and Mei [4] have further extended the analysis to a rectangular cylinder of finite draft swaying in the free surface. It can be shown that for bodies of size $a$ with $ka = O(1)$ and moderately weak mooring with the elastic constant $K \equiv \epsilon K_1 = O(\epsilon \rho g h)$, the displacement amplitude $X$ of slow sway is at most comparable to the incident wave amplitude in order of magnitude, i.e., $kX = O(\epsilon)$. For very weak mooring with the elastic constant $K \equiv \epsilon^2 K_2 = O(\epsilon^2 \rho g h)$, the slow sway can be comparable to the wavelength, $kX = O(1)$.

The approximate analysis is expedited by separating the flow field into near and far fields of the body. In the near field, the only relevant length scale is $1/k$, while two time scales exist, $1/\omega$ and $1/\epsilon\omega$. Therefore two time coordinates are introduced: $t$ and $t_1 = \epsilon t$, and the potential can be expanded as in the following series,

$$\Phi = \sum_{n=1} \epsilon^n \sum_{m=-n}^{n} \psi_{nm}(x, z, t_1) e^{-im\omega t} \tag{1}$$

At the first order the first harmonic $\psi_{11}$ represents the short waves. The associated diffraction problem is solved by conventional means in the linearized diffraction theory. In particular the reflection and transmission coefficients $R$ and $T$ are found. The first order, zeroth harmonic $\psi_{10}$ represents the long period oscillations in the near field. If the gap between the keel and the sea bottom is not small, $\psi_{10}$ can be shown to be uniform in the near field. At the second order, the slow motion potential $\psi_{20}$ is harmonic in x and z and governed by inhomogeneous Neuman conditions.

In the far field, long waves are expected, hence it is necessary to add the slow spatial coordinate $x_1 = \epsilon x$. We assume the far field expansions:

$$\Phi = \sum_{n=1} \epsilon^n \sum_{m=-n}^{n} \phi_{nm}(x, z, x_1, t_1) e^{-im\omega t} \tag{2}$$

The long wave potential $\phi_{10}$ is found to be governed by the inhomogeneous hyperbolic equation:

$$\begin{aligned} \frac{\partial^2 \phi_{10}}{\partial t_1{}^2} - gh \frac{\partial^2 \phi_{10}}{\partial x_1{}^2} &= \alpha \frac{\partial}{\partial t_1} |A(t_1 - x_1/C_g)|^2 + \\ &\quad \frac{\partial}{\partial t_1} |R\, A(t_1 + x_1/C_g)|^2, \quad \text{if} \quad x_1 < 0; \\ &= \alpha \frac{\partial}{\partial t_1} |T\, A(t_1 - x_1/C_g)|^2, \quad \text{if} \quad x_1 > 0. \end{aligned} \tag{3}$$

where $\alpha$ is some constant. The inhomogeneous solution is of the form

$$\begin{aligned} \phi_{10}^I(t_1 - x_1/C_g) + \phi_{10}^R(t_1 + x_1/C_g), \qquad & x_1 < 0, \\ \phi_{10}^T(t_1 - x_1/C_g), \qquad & x_1 > 0. \end{aligned} \tag{4}$$

The total long wave potential must include the homogeneous solution

$$\begin{aligned} &\phi_{10}^{+}(t_1 + x_1/\sqrt{gh}), \qquad x_1 < 0, \\ &\phi_{10}^{-}(t_1 - x_1/\sqrt{gh}), \qquad x_1 > 0 \end{aligned} \tag{5}$$

which represents free waves propagating at the long wave speed $\sqrt{gh}$. These free waves are determined by matching the near and far fields on both sides of the body. In this fashion, both the long period pressure on the body and the long waves radiated to infinities are found.

If the gap beneath the keel is small, however, the slow potential along the keel changes considerably. This is associated with the so-called blockage effect. Whether the gap is small or not a uniformed valid ordinary differential equation can be derived for the slow sway amplitude $X$

$$\begin{aligned} \frac{d^2X_{10}}{dt_1{}^2} + \frac{K_1}{2\rho g\sqrt{gh}}\frac{dX_{10}}{dt_1} + \frac{K_1}{2\rho k c_1}X_{10} = \\ - \left(\frac{C_g}{\sqrt{gh}}|R|^2 + 1 - |R|^2\right)\frac{C_g}{gh - C_g^2}\frac{\rho S}{2h}\frac{\partial |A|^2}{\partial t_1} \\ + \frac{C_g}{C}\frac{g}{2hc_1}|RA|^2 \end{aligned} \tag{6}$$

with the blockage coefficient $c_1$ playing the role of the apparent mass.

Sample analytical results are shown in Figure 1 for three types of incident wave envelopes: b) the front of a sinusoidally modulated envelope, c) the front of a uniform wave train, and d) a wave packet. Extensions for slow sways of large amplitude have also been made.

## Trapping of Long Waves on a Wide Ridge in Short Swells

When a sinusoidal wave train propagates obliquely from shallow water towards increasing depth, its direction turns away from the normal to depth contours. If the initial incidence is sufficiently oblique, complete reflection can occur. Near a long submarine ridge the depth increases in both directions away from the axis, hence sinusoidal waves can be trapped along the ridge. For a given ridge profile, the eigen frequencies and wavenumbers of the trapped modes must obey certain dispersion relation. Since no wave can exist outside the ridge with the same frequency and the same wavenumber component in the direction parallel to the ridge, those trapped on the ridge cannot be excited by steady incident waves through any linearized mechanism.

Persistent wind waves of short periods however can enter the ridge. The first order process is easily predicted by the linearized theory of shoaling and diffraction. At the second order the long waves traveling at the group velocity can enter the ridge and excite the trapped waves.

Agnon and Mei [2] have considered a wide rectangular ridge of depth $h'$ and width $2L$ in a sea of depth $h$. As a slight modification of §2, we divide the flow region into strips parallel to the ridge axis, which is taken to be the $y$ axis. Near the left and right edges of the ridge we define two near fields of the widths $k(|x| - L) = O(1)$. There are three far fields, one to the left and one to the right of the ridge, and one between the two edges. In these far fields the horizontal scales are $1/k$ and $1/\epsilon k$. The first order short waves $\phi_{11}$ in the far fields are easily found and the reflection and transmission coefficients $R$ and $T$ are calculated. The first order long waves in the three far fields are governed by wave equations similar to (1) where we must change

$$\frac{\partial^2}{\partial {x_1}^2} \quad \text{to} \quad \frac{\partial^2}{\partial {x_1}^2} + \frac{\partial^2}{\partial {y_1}^2} = \nabla_1^2 \tag{7}$$

The righthand side of (3) must be modified slightly. Now the forcing terms have sinusoidal components of the structure

$$\exp[\pm 2i\Omega(t_1 \pm \mu x_1 - \nu y_1)] \tag{8}$$

If

$$\frac{1}{\sqrt{gh}} < \nu < \frac{1}{\sqrt{gh'}} \tag{9}$$

free long waves can be forced on the ridge, but they must be exponentially attenuating away from the ridge. The amplitude of the free long wave is again determined by matching the near and far fields.

Figure 2 shows the time history of trapped long waves excited by an obliquely incident wave train whose envelope is a packet with a carrier frequency equal to the resonant frequency of a trapped mode. An interesting feature is that as the packet advances along the ridge, a stationary observer sees a steady reverberation of long periods in the trail of the disappearing packet.

Penetration of Long Internal Waves behind a Barrier in Short Surface Waves

It is well known that two trains of surface waves with slightly different frequencies can interact nonlinearly to resonate internal waves in an open sea. Agnon and Mei

[3] have recently considered the effects of a thin and long barrier perpendicular to the incident waves. In the horizontal plane the barrier coincides with the positive $y$ axis. A train of sinusoidally modulated surface waves advances in the positive $x$ direction and is scattered by the barrier. From the ray approximation for monocromatic waves we expect the first quadrant of the horizontal plane to be a shadow free of short waves, the second quadrant to have incident and normally reflected waves of equal amplitudes, and the third and fourth quadrants to be illuminated by the incident waves only. Correction to the ray approximation is needed in the parabolic boundary layers near the $x$ axis, resulting in diffraction fringes in the boundary layer. Because of the slow modulation, nonlinearity now provides a second order forcing at twice the modulational frequency. The first effect of this forcing is to induce set-down waves traveling with the envelope of each wave train at the group velocity. But these waves cannot ensure continuity on the long scale across the edge rays i.e., the positive and negative $x$ axis. It is therefore necessary to add free long waves away from the x axis and also consider the long period motions in the near field of the $x$ axis. The free waves satisfy the equation

$$\left(\frac{\partial^2}{\partial {t_1}^2} - U^2\nabla_1^2\right)\left(\frac{\partial^2}{\partial {t_1}^2} - U'^2\nabla_1^2\right)\begin{pmatrix}\phi'^F_{10}\\ \phi^F_{10}\end{pmatrix} = 0. \tag{10}$$

where primed (unprimed) quantities refer to the upper(lower) layer. $U$ and $U'$ are respectively the phase velocities of the internal and surface mode of the long period $2\Omega$. Now $U$ and $U'$ are in general different from the group velocity $C_g$ of the short surface wave. Therefore with the same wavenumber component along $x$ as the group locked waves, the $y$ component of the free long waves (surface and internal modes) are different from zero. It can be shown that under some conditions the $y$ wavenumber component of the free internal wave of frequency $2\Omega$ is real while that of the free surface wave of the same frequency is imaginary. This means that the free long internal waves are inclined with respect to the x axis. Thus while the lee of the breakwater is a shadow for short waves, the long internal waves can enter it without hindrance. This penetration phenomenon is most prominant when the internal long waves are not resonated by the interacting surface waves through the usual mechanism.

Figure 4 gives sample results of the long internal wave amplitude.

<u>Radiation of Long Waves over Periodic Bars and Bragg Scattering</u>

Experiments in a long wave tank by Heathershaw [7] have shown that a long train of sinusoidal bars on the seabed can trigger strong reflection if the Bragg resonance condition is satisfied, even if the bars have small amplitude. Mei [10] has extended a theory of Davies and Heathershaw [5] and treated the first order effects by a theory valid in the neighborhood of resonance. In particular it is shown that, over the bars, the envelopes of the incident and reflected waves $A$ and $B$ satisfy a pair of coupled linear hyperbolic equations. The coupling coefficient $\Omega_o$ is proportional to the bar amplitude and also defines a cut-off frequency. If the incident wave is detuned from the resonance frequency by an amount below the cutoff frequency, resonance is strong. In particular a patch of bars can cause complete reflection if the patch has a large enough width.

Again at the second order long waves are forced. For constant mean depth and normal incidence the long waves are governed by

$$\frac{\partial^2 \bar{\zeta}_{20}}{\partial t_1{}^2} - gh\frac{\partial^2 \bar{\zeta}_{20}}{\partial x_1{}^2} = \frac{1}{\rho}\frac{\partial^2 S}{\partial x_1{}^2} + \frac{g\Omega_o}{C}\frac{\partial}{\partial x_1}(iA^* B + *) \tag{11}$$

where $S$ is the radiation stress due to the incident and reflected envelopes. The last term is proportional to the local bar amplitude. Thus there are also locked long waves which accompany the incident, reflected and transmitted short wave envelopes over the bars and on both sides of the bar patch. Furthermore free long waves must be radiated over and away from the bars. In particular, over a wide patch of bars, the short wave can be totally reflected while the free long waves can penetrate deeply the transmission side. Thus the bar patch becomes a nonlinear filter and can change the sea spectrum in the coastal zone.

Figure 5 gives an example that on the transmission side of a wide patch of parallel bars short waves are prevented while long waves can be quite noticeable. This suggests that long-period beats can occur on a beach in the absense of surfs. Experiments have confirmed these features (Hara and Mei, [6]).

## Forward Scattering and the Nonlinear Parabolic Approximation

It is well known from the classical theory of steady Stokes waves that the third order nonlinearity can affect the phase of the first order envelope after a fetch of the order $kx = O(\epsilon^{-2})$. Modern theories on the instability have further revealed for unidirectional waves that a complex process of evolution may take place if sideband disturbances are added to a uniform Stokes wave train. Thus if one is interested in the competing effects of diffraction and nonlinearity on the first order envelope, one must consider a large scatterer of the dimension at least $O(1/\epsilon^2 k)$.

For arbitrarily shaped bathymetry or structure all kinds of length scales can be present in all horizontal directions; the associated mathematical task is difficult. It is therefore reasonable to attempt first those cases for which the linearized diffraction problem can be dealt with by simple analytical means. The method of parabolic approximation, introduced in optics long ago, has been an effective tool for a broad class of linear diffraction problems such as glancing incidence by a slender body and edge diffraction. In its simplest form the approximate equation is

$$2ik\frac{\partial A}{\partial x} + \frac{\partial^2 A}{\partial y^2} = 0 \tag{12}$$

Being parabolic, the boundary value problem of diffraction becomes an initial value problem which can be solved numerically by marching forward in $x$. Partly for this numerical advantage its applications have been greatly extended in recent years to underwater acoustics and coastal oceanography. The basic requirement for it to apply is that the waves must be essentially propagating forward. This is the case with head sea incidence on a slender body which is much longer than the wavelength and with a seabed of depth changing slowly over a very long distance. Typically the diffraction effects appear as modulations of the envelope. The modulation is slow in the lateral direction and still slower in the longitudinal direction.

When nonlinearity is included, the governing equation for constant depth and steady waves becomes

$$2\frac{\partial A}{\partial x_2} - \frac{i}{k}\frac{\partial^2 A}{\partial {y_1}^2} + iK'|A|^2 A = 0 \tag{13}$$

where $x_2 = \epsilon^2 x, y_1 = \epsilon y$ and

$$K' = k^3 \frac{C}{C_g} \frac{\cosh 4kh + 8 - 2\tanh^2 kh}{8\sinh^4 kh} \tag{14}$$

This equation is a special case of the transient, two dimensional evolution equation for constant depth. On the wall of a breakwater slightly slanted with respect to the $x$ direction the no flux boundary condition gives

$$\frac{\partial A}{\partial y_1} = \frac{ikA}{\epsilon}\frac{\partial Y_B}{\partial x_2} \tag{15}$$

and the "initial condition" at the tip of the breakwater is

$$A = A_o \tag{16}$$

On this basis Yue and Mei [16] find that nonlinearity reduces the amplitude along the wall, and there is a wedge region adjacent to the wall in which the envelope is flat and the crests are normal to the wall. This sudden bending of crests resembles the Mach stem phenomenon observed in shallow water. A similar problem of wave focussing by a subsurface lens has been examined by Stamnes *et al* [15] both theoretically and experimentally. A wedge of flat envelope is also found in the focal region. The gross features of the Mach stem can be viewed as a shock wave. Further insights of the jump has been given by Peregrine [13].

There have been two kinds of extensions to these works. One is the combination of diffraction and refraction by Stokes waves over a large scale bathymetry. The cubic Schrodinger equation (13) now must have space-dependent coefficients. Kirby and Dalrymple [8] have considered a circular shoal with depth $h(\epsilon^2 r)$ while Liu and Tsay [10] have treated a channel with a semicircular shoal connecting two different but constant depths. In both examples a cusped caustic is expected from the linearized theory. Nonlinear calculations again show a reduction of amplitude near the caustic and the feature of Mach stem is apparent. The second extension is the diffraction of short waves in a current when both are directed along the axis of a large slender body fixed in the free surface (Mei and Lo, [12]). Here refraction is caused by the variation of current strength near the body.

Because of the sideband instability of Stokes waves diffraction of nonlinear waves is an inherently transient phenomenon. Some exploratory calculations have been made by Yue [17] for a wave packet incident on the slowly converging banks in an otherwise straight channel. The water depth is taken to be infinitely deep. Figure 6 shows the evolution of a wave packet along the transmission branch after it passes the transition. If the depth is finite, one should further anticipate long waves propagating both along and opposite to these envelopes. These long waves must be solved in conjunction with the wave envelope. The numerical task is greater but should be worthwhile.

In closing, one may mention the problem of resonance of long-period oscillations in a harbor by narrow-banded short waves. The linear mechanism of excitation by long tsunamis is well developed, and there have also been studies of nonlinear modifications (Rogers and Mei [14], Lepelletier [9]). But seiches large enough to strain the mooring lines of ships in berthing have been reported in several harbors when there is no tsunami. A quantitative theory will therefore be very valuable to coastal engineering.

## 7. Acknowledgement

Financial support for the work done by the author and his collaborators has been received from the Office of Naval Research (Contracts N000 14 80 C 531 with the Fluid Mechanics Program and N00014 83K 0550 with the Ocean Engineering Program) and the National Science Foundation (Grant MEA 8210649). The first three topics described here are based on a recent thesis by Y. Agnon.

## References

1. Agnon, Y. and Mei, C. C.: Slow drift motion of a two dimensional block in beam seas. J. Fluid Mech. 151 (1985) 279-294.

2. Agnon, Y., and Mei, C. C.: Trapping and resonance of long shelf waves due to incident short wave groups. (1987 submitted).

3. Agnon, Y., and Mei, C. C.: Excitation of long internal waves by groups of short surface waves incident on a breakwater. (1987 submitted).

4. Agnon. Y., Choi, H. S. and Mei, C. C.: Slow drift of a floating cylinder in narrow-banded seas (1987 submitted).

5. Davies, A. G. and Heathershaw, A. D.: Surface wave propagation over sinusoidally varying topography. J. Fluid Mech. 144 (1984) 419-433.

6. Hara, T. and Mei, C. C.: Bragg scattering of surface waves by periodic bathymetry: theory and experiment. J. Fluid Mech. 174 (1987) 46-126.

7. Heathershaw, A. D.: Seabed wave resonance and sandbar growth. Nature, 296 (1982) 343-345.

8. Kirby, J. T., and Dalrymple, R. A.: A parabolic equation for the combined refraction diffraction of Stokes waves gy mildly varying topography. J. Fluid Mech. 136 (1983) 453-467.

9. Lepelletier, T. G.: Tsunami-harbor oscillations induced by nonlinear transient long waves. Report No. KH-R-41, Keck Laboratory, Calif. Inst. Tech. 1980.

10. Liu, P. L-F. and Tsay, T. K.: Refraction diffraction model for weakly nonlinear water waves. J. Fluid Mech. 141 (1984) 265-274.

11. Mei, C. C.: Resonant reflection of surface water waves by periodic sandbars. J. Fluid Mech. 152 (1985) 315-335.

12. Mei, C. C. and Lo, E.: Nonlinear development of head seas along a slender body. Proc. Symp. Naval Hydrody. 15 (1985) 137-147. Hamburg.

13. Peregrine, D. H.: Wave jumps and caustics in the propagation of finite amplitude water waves . J. Fluid Mech. 136 (1983) 35-452.

14. Rogers, S. R. and Mei, C. C.: Nonlinear resonant excitation of a long and narrow bay. J. Fluid Mech. 88 (1977) 161-180.

15. Stamnes, J. J., Løvhaugen, O., Spjelkavik, B., Mei, C. C., Lo, E. and Yue, D. K. P.: Nonlinear focusing of surface waves by a lens, theory and experiment. J. Fluid Mech. 135 (1983) 71-94.

16. Yue, D. K. P.: Numerical study of Stokes wave diffraction at grazing incidence. Sc.D. thesis, Department of Civil Engineering, Mass. Inst. Tech. 1980.

17. Yue, D. K. P., and Mei, C. C.: Forward scattering of Stokes waves by a thin wedge. J. Fluid Mech. 99 (1980) 33-52.

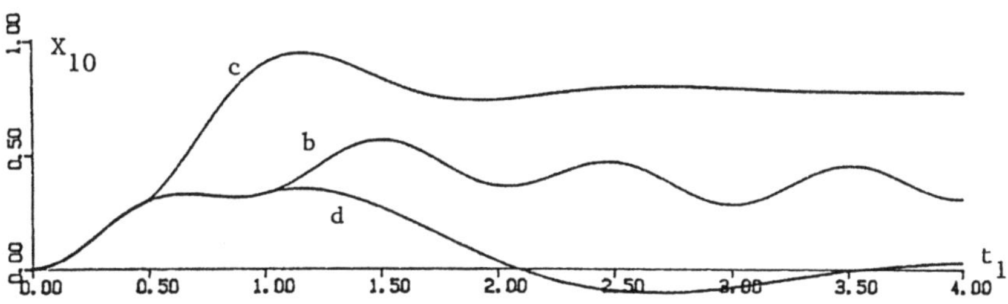


Figure 1 : Slow sway vs time

Figure 2a : Incident envelope

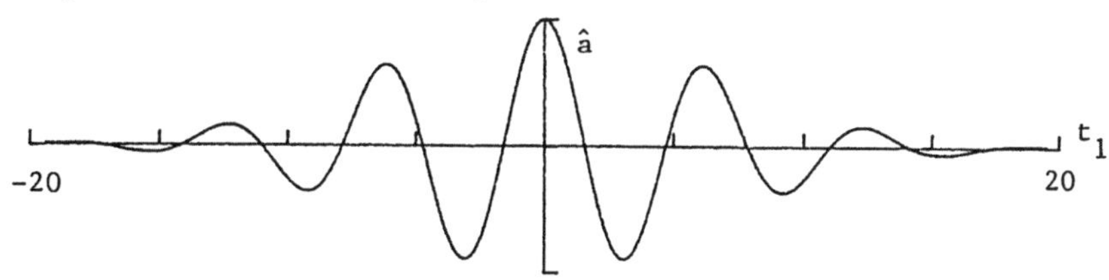


Figure 2b : Typical response on the shelf

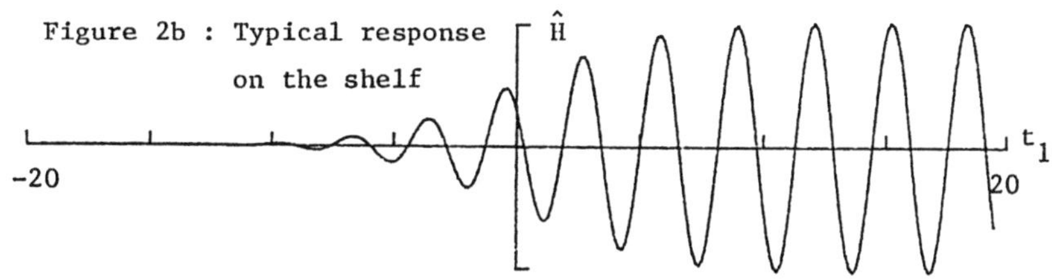

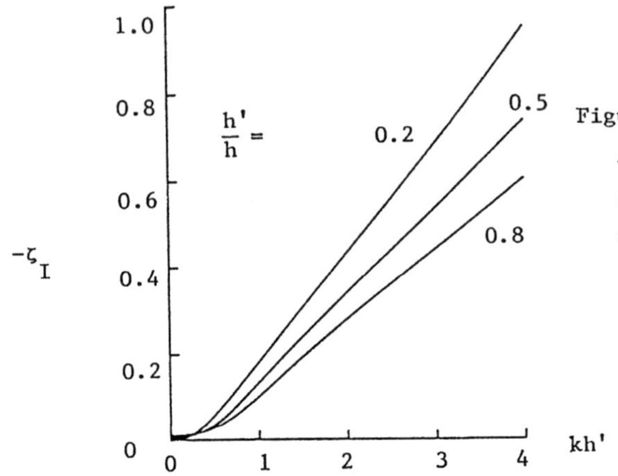


Figure 3 : Amplitude of long internal waves

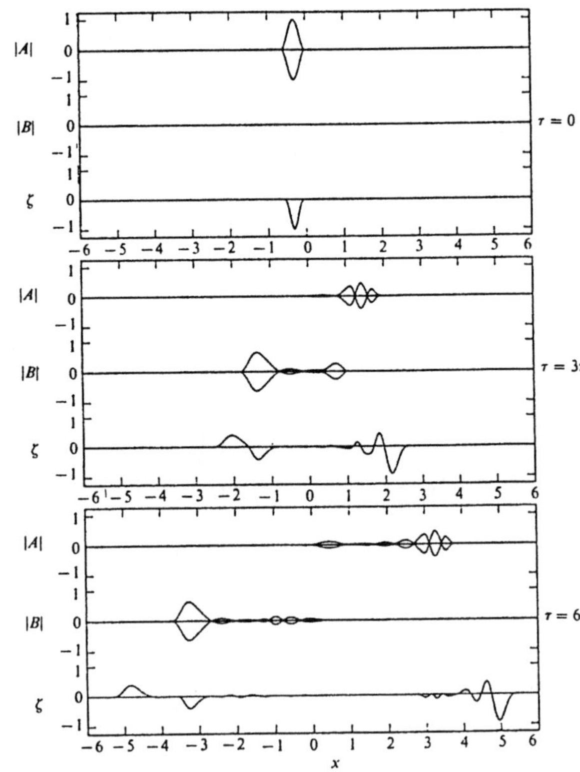


Figure 4 : Scattering by bars in $0<x<1$. $\zeta$ refers to long waves

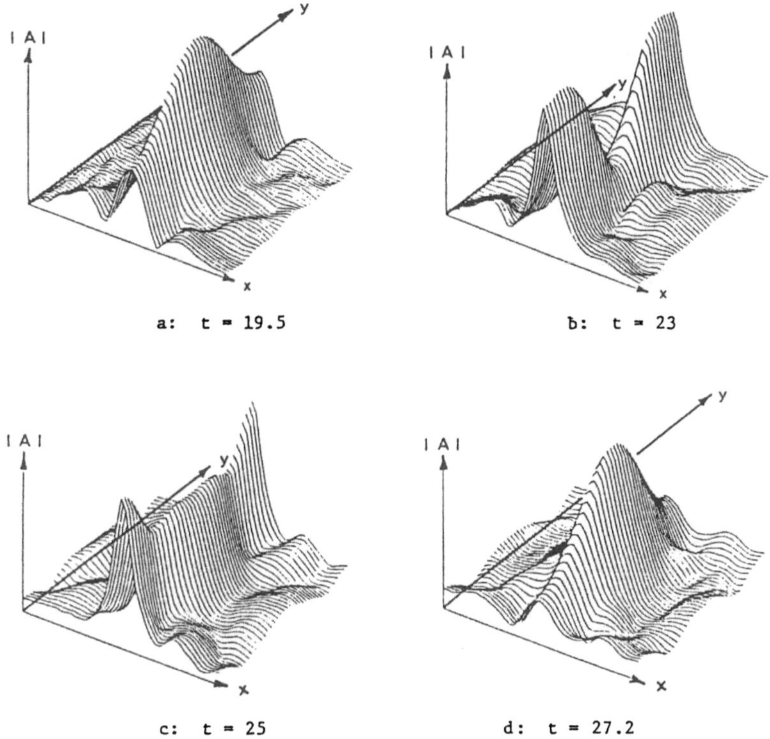


Figure 5 : Two-dimensional variations of |A| in the uniform channel at selected times.

# Recent Developments in the Modelling of Unsteady and Breaking Water Waves

D.H. PEREGRINE

School of Mathematics
University of Bristol, Bristol BS8 1TW, United Kingdom.

## Introduction

Steep unsteady water waves were first successfully modelled by Longuet-Higgins and Cokelet (1976). A boundary integral method for solving Laplace's equation for irrotational flow was used together with time marching of the position and velocity potential of surface particles using the free surface boundary conditions. Most subsequent work uses boundary integral formulations. Marker and cell (MAC) methods have also been used to a lesser extent; however they do not appear to be as successful in describing wave overturning and are not described here.

## Boundary integral methods

Boundary integral methods may be implemented in a number of ways since there is a considerable choice of (i) boundary integral (ii) type and accuracy of discretization and (iii) method of marching in time. No concensus has emerged as to the best method, for example at the recent 2nd Workshop on Water Waves and Floating Bodies (Evans, 1987, see also Evans & Newman, 1987) seven speakers described aspects of, or results from, boundary integral methods all of which were different.

Perhaps the best discussion of boundary integrals in this context is that of Baker et al (1982). The main varieties are those using a logarithmic Green's function like Longuet-Higgins and Cokelet (1976) or those derived from Cauchy's integral theorem for complex functions, such as used by Vinje and Brevik (1981). There is also a choice between a potential/stream function equation or an equation for velocity components, or dipoles, as used

K. Horikawa, H. Maruo (Eds.)
Nonlinear Water Waves
IUTAM Symposium Tokyo/Japan 1987

by Baker et al (1982) and Dold and Peregrine (1986a). These latter methods permit iterative solution of the resulting matrix equations.

In discretization there are two philosophies. A simple approximation with constant, linear or quadratic elements is easy to implement. This approach is that of Vinje and Brevik (1981) and this is followed by many others of which the work of Lin et al (1984) is perhaps the most noteworthy recent example. On the other hand Longuet-Higgins and Cokelet (1976) use high-order approximations, a route most recently followed by Dold and Peregrine (1986a) where at least 11 point approximations are made and contribute to the high efficiency of their method.

The time marching of these computations involves the integration of a set of ordinary differential equations for each discretization point. Many of the standard methods are used.

A problem with many implementations of the boundary integral method is the development of a zig-zag instability. This is often suppressed by smoothing. Dommermuth et al (1987) follow the example of vortex sheet solvers and redistribute surface discretization points. Vinje and Brevik (1981) and Dold and Peregrine (1986a) usually have less problems of this type. In complex wave motions, such as are now being studied, computational efficiency can be improved by judicious point redistribution.

## The Initiation of Wave Breaking

The major advantage of the boundary integral method is that it can describe the initial overturning stage of wave breaking. Thus, from the beginning its results have been examined to try to understand the initiation of the breaking process. Using some of the results of Longuet-Higgins and Cokelet's (1976) computations Peregrine et al (1980) were able to describe important features of the breaking (further examples are given in New et al, 1985). The most remarkable feature is the strong accelerations that act on water rising into the jet; though with hindsight it is easy to understand them. I have been pleasantly surprised at the analysis stimulated by the computational results which has lead through the work of New (1983) and Longuet-Higgins (1980) to Greenhow's (1983) approximate analytical solutions with several of the features of a plunging breaker.

Although the start of breaking is well described, there is still no good explanation for it. Some people describe it as an instability, but I question this. There are water-wave instabilities which lead to breaking, but breaking only occurs after the instabilities have evolved into a nonlinear phase. Examples where this evolution towards breaking is studied include Longuet-Higgins and Cokelet (1978) for subharmonic alternate-wave instabilities: Dold and Peregrine (1986b) give a preliminary account of their study of the modulational Benjamin-Feir study. In this case waves break as they pass through a strong modulation and grow beyond maximum steepness. In one case a wave with a crest very close to the Stokes $120^{\circ}$ corner solution was found. Tanaka et al (1987) find that the Tanaka instability of high solitary waves may or may not lead to breaking. If breaking occurs then its development is on a shorter time scale than the instability. If breaking does not occur there is a smooth transition to the lower solitary wave of equal energy.

Some understanding of breaking may come from considering very steep standing waves. Studies by Saffman & Yuen (1979), Srokosz (1981) and our work at Bristol show a variety of behaviour. McIver and Peregrine (1981) showed that the tip of a jet did behave as Longuet-Higgins (1972) free-fall parabola solution. Recent computations of steep solitary waves meeting (M. Lewy, private communication) give a fine example of this type of flow. Pugh (1985) found that Longuet-Higgins's (1973) simple log(sec x) approximation to a maximum steepness standing wave gave an almost periodic solution for 6 periods and then developed a singularity when "falling". This singularity is the first we have found which appears to be a property of the free surface rather than self intersection and is clearly resolved. On the other hand problems of numerical resolution as surfaces become very sharp are very common. Longuet-Higgins's (1983) theory, and his intriguing experiments on axisymmetric standing waves should be noted.

These gave rise to remarkable high speed vertical jets.

The approach to a completely different kind of breaking has recently been modelled by Hansen (1987). In calculation for very viscous flow down a wall moving into a pool of liquid, Hansen finds steady solutions are limited by an overhanging profile on which there is a stagnation point. It thus seems likely that wave overturning and breaking of a free surface is possible even in the absence of inertia effects.

## The Generation of Breakers

Most computations to date have forced waves to break by a variety of means including, pressure forcing, "over-energetic" initial conditions, such as a large sine wave, or a steady wave shape and surface potential appropriate to deeper water. In one sense these are like real water waves in that breaking usually occurs because of some disturbance, such as wind, shoaling water, convergence of wave energy, or motion of a rigid surface. In all these cases there is no simple way of relating the results to laboratory or prototype waves.

One of the aims of most modellers is to develop a "numerical wave-tank", which can be used like experimental tanks. Experimental tanks are usually used for testing models of structures such as ships, oil rigs and breakwaters. Dommermuth et al (1987) have made comparisons between a tank and a numerical model, including the wave paddle, for an experiment in which waves are focussed to form a breaker with acceptable results.

A recent example of our work at Bristol with a submerged breakwater gave unexpected flows. A solitary wave meeting a semicircular cylinder on the bed is illustrated in figure 1a. As the wave encounters the cylinder a new crest grows in front of it. Then, as the wave passes a trough forms close to the cylinder. Figure 1b shows the subsequent evolution toward breaking in a *backward* direction. In some circumstances the front of the wave breaks, in others waves break both ways(!) and sometimes there is no breaking. The backward breaking is best understood by noting that flow over a cylinder leads to the generation of waves. These waves give the wave drag

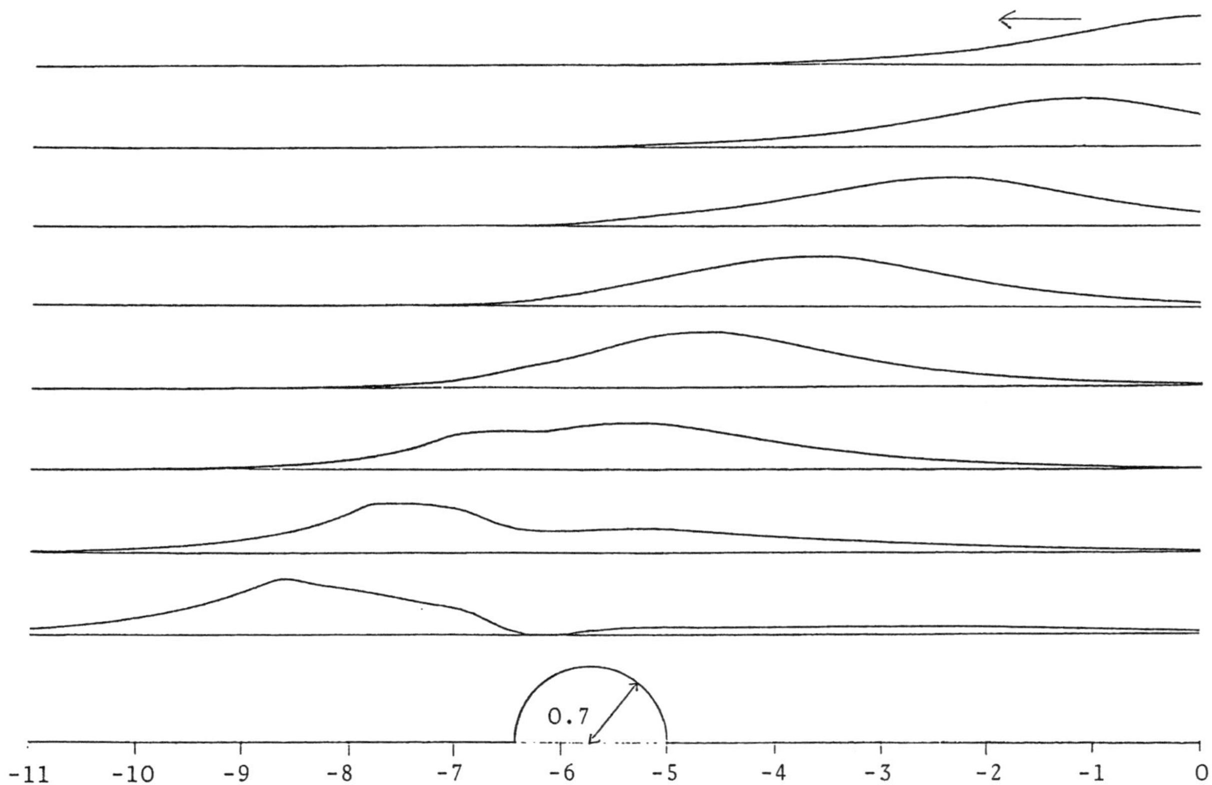


Fig.1a Successive surface profiles, with no vertical exaggeration, showing a solitary wave of amplitude 0.462 times the depth, h, meeting a semi-circular bump on the bed with radius 0.7 h, at times 0, 1, 2, 3, 4, 5, 6 and 7 $(h/g)^{\frac{1}{2}}$. (Courtesy of M. Cooker).

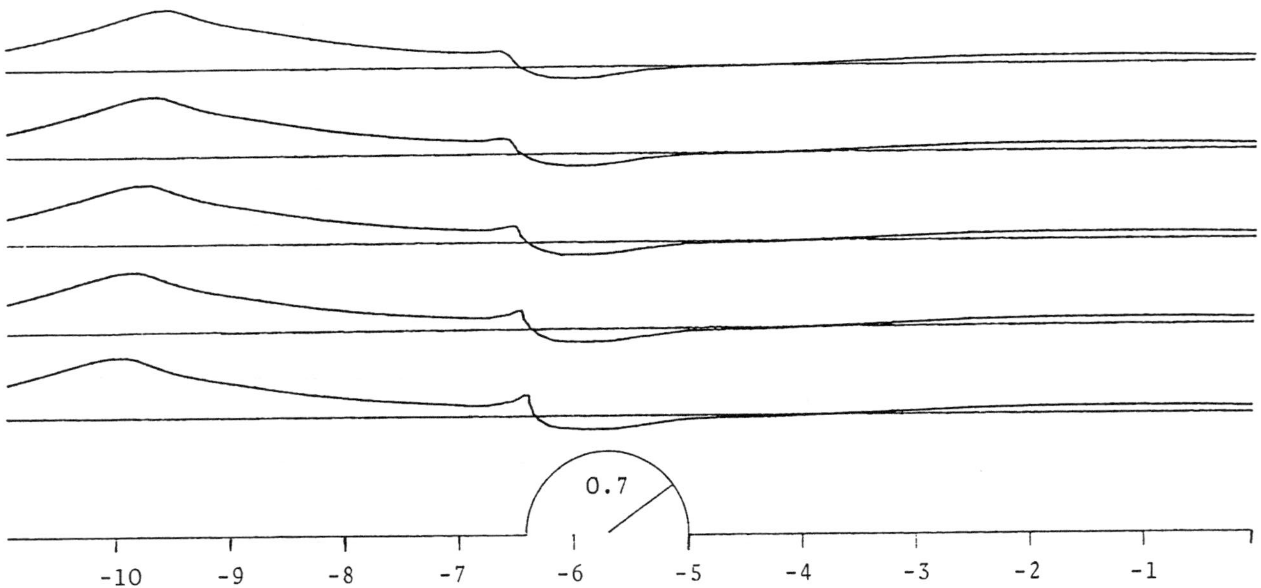


Fig. 1b Successive surface profiles, with no vertical exaggeration, showing a solitary wave of amplitude 0.462 times the depth, h, meeting a semi-circular bump on the bed with radius 0.7 h, at times 8, 8.1, 8.2, 8.3 and 8.4 $(h/g)^{\frac{1}{2}}$. (Courtesy of M. Cooker).

on the cylinder. Here we see an unsteady flow and the initial growth of one wave, which could become a wave train if the flow were sustained.

The program used in the figures is a development by Cooker and Lewy of one described by Dold and Peregrine (1986a) which includes a suitable conformal transformation of the bed. It is equally suited to obstacles of different shapes: figure 3 shows a large solitary wave encountering a large shoal-like elliptical obstacle. The computations for figures 1 and 2 took about an hour on a VAX 750 multi-user system.

The line along which the free surface meets a rigid boundary can cause difficulties in modelling especially if the rigid boundary is moving, as illustrated by Greenhow et al (1982). Some implementations of the boundary integral method have satisfactorily described this point, Lin et al (1984) give a good discussion with nice examples. Greenhow (1987) has continued with examples of wedge entry and cylinder entry to and exit from the water.

## Further Stages of Breaking

There is no effective deterministic model of the splashing stage of breaking, but once large-scale organized motions, described by Peregrine (1983) and Basco (1985), have degenerated into turbulence the wave may become quasi-steady as a spilling breaker or bore. Madsen and Svendsen (1983) and Sevendsen and Madsen (1984) have made a promising start modelling bores with a "turbulent layer" model. Tulin and Cointe (1986) make progress in describing spilling breakers. They use a model in which the "roller" is carried by the wave and the strong turbulent mixing described by Peregrine and Svendsen (1978) is incorporated with a mixing layer and a thin turbulent wake. Using results from Duncan's (1981, 1983) measurements, Tulin and Cointe are able to give explanations of several aspects of breaking waves.

Unsteady aspects of spilling breakers are described by Banner (this volume) in a relatively simple model.

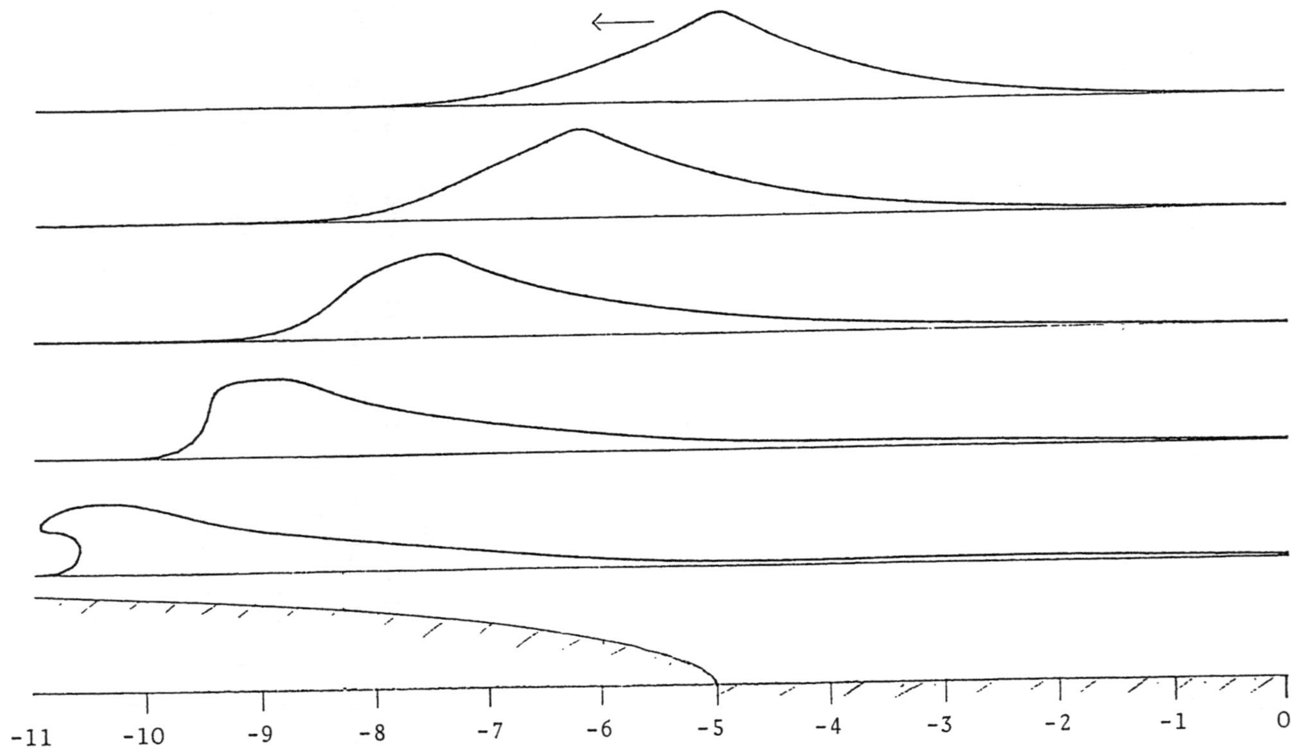


Fig.2. A solitary wave of amplitude 0.804h meeting an elliptical shoal of height 0.9h and half length 10h. Surface profiles are given for times 4, 5, 6, 7 and 8 $(h/g)^{\frac{1}{2}}$ since the start of the computation. (Courtesy of M. Cooker).

References


1. Baker, G.R., Meiron, D.I. and Orzag, S.A.: Generalised vortex methods for free-surface flow problems. J. Fluid Mech. 123 (1982) 477-501.

2. Basco, D.R.: A qualitative description of wave breaking. J. Waterway, Port, Coastal & Ocean Engng., Proc. A.S.C.E. 111 (1985) 171-188.

3. Dold, J.W. and Peregrine, D.H.: An efficient boundary-integral method for steep unsteady water waves. Numerical methods for fluid dynamics II, eds: K.W. Morton and M.J. Baines, Clarendon Press, Oxford (1986a) 671-679.

4. Dold, J.W. and Peregrine, D.H.: Water wave modulation, Proc. 20th Internat. Conf. Coastal Engng., Taipei, A.S.C.E. (1986b).

5. Dommermuth, D.G., Yue, D.K.P., Rapp, R.J., Chan, F.S. and Melville, W.K.: Deep-water breaking waves: a comparison between potential theory and experiments. Submitted for publication (1987).

6. Duncan, J.H.: An experimental investigation of breaking waves produced by a towed hydrofoil. Proc. Roy. Soc. Lond. A 377 (1981) 331-348.

7. Duncan, J.H.: The breaking and non-breaking wave resistance of a two-dimensional hydrofoil. J. Fluid Mech. 126 (1983) 507-520.

8. Evans, D.V. and Newman, J.N.: Report on First International Workshop on Water Waves and Floating Bodies. J. Fluid Mech. 174 (1987) 521-528.

9. Evans, D.V. (ed.): Proceedings of Second International Workshop on Water Waves and Floating Bodies, Bristol University, School of Mathematics (1987).

10. Greenhow, M.: Free-surface flows related to breaking waves. J. Fluid Mech. 134 (1983) 259-275.

11. Greenhow, M.: Water entry and exit of a horizontal cylinder, in Evans (1987).

12. Greenhow, M., Vinje, T., Brevik, P. and Taylor, J.: A theoretical and experimental study of the capsize of Salter's duck in extreme waves. J. Fluid Mech. 118 (1982) 221-239.

13. Hansen, E.B.: Stokes flow down a wall into an infinite pool. J. Fluid Mech. 178 (1987) 243-256.

14. Lin, W.-M., Newman, J.N. and Yue, D.K.: Non-linear forced motions of floating bodies. 15th Symp. Naval Hydrodynamics, Hamburg (1984).

15. Longuet-Higgins, M.S.: A class of exact, time-dependent, free-surface flows. J. Fluid Mech. 55 (1972) 529-543.

16. Longuet-Higgins, M.S.: On the form of the highest progressive and standing waves in deep water. Proc. Roy. Soc. Lond. A, 331 (1973) 445-456.

17. Longuet-Higgins, M.S.: On the forming of sharp corners at a free surface. Proc. Roy. Soc. London. A, 371 (1980) 453-478.

18. Longuet-Higgins, M.S.: Bubbles, breaking waves and hyperbolic jets at a free surface. J. Fluid Mech. 127 (1983) 103-121.

19. Longuet-Higgins, M.S. and Cokelet, E.D.: The deformation of steep surface waves on water. I. A numerical method of computation. Proc. Roy. Soc. Lond. A, 350 (1976) 1-26.

20. Longuet-Higgins, M.S. and Cokelet, E.D.: The deformation of steep surface waves on water. II. Growth of normal-mode instabilities. Proc. Roy. Soc. Lond. A, 364 (1978) 1-28.

21. McIver, P. and Peregrine, D.H.: Comparison of numerical and analytical results for waves that are starting to break. Int. Sympos. on Hydrodynamics in Ocean Engng., Trondheim, Norway (1981) 203-215.

22. Madsen, P.A. and Svendsen, I.A.: Turbulent bores and hydraulic jumps. J. Fluid Mech. 129 (1982) 1-25.

23. New, A.L.: A class of elliptical free-surface flows. J. fluid Mech. 130 (1983) 219-239.

24. New, A.L., McIver, P. and Peregrine, D.H.: Computations of breaking waves. J. Fluid Mech. 150 (1985) 233-251.

25. Peregrine, D.H.: Breaking waves on beaches. Ann. Rev. Fluid Mech. 15 (1983) 149-178.

26. Peregrine, D.H. and Svendsen, I.A.: Spilling breakers, bores and hydraulic jumps. Proc. 16th Coastal Engng. Conf. A.S.C.E., Hamburg 1 (1978) 540-550.

27. Peregrine, D.H., Cokelet, E.D. and McIver, P.: The fluid mechanics of waves approaching breaking. Proc. 17th Conf. Coastal Engng., A.S.C.E. 1 (1980) 512-528.

28. Pugh, D.: The computation of velocity field beneath steep unsteady water-waves. M.Sc. dissertation, Bristol University (1985).

29. Saffman, P.G. and Yuen, H.C.: A note on numerical computations of large amplitude standing waves. J. Fluid Mech. 95 (1979) 705-715.

30. Srokosz, M.A.: Breaking effects in standing and reflected waves in "Hydrodynamics in ocean engineering", Sympos. Trondheim, Norwegian Inst. Tech. (1981) 183-202.

31. Svendsen, I.A. and Madsen, P.A.: A turbulent bore on a beach. J. Fluid Mech. 148 (1984) 73-96.

32. Tanaka, M., Dold, J.W., Lewy, M. and Peregrine, D.H.: Instability and breaking of a solitary wave. J. Fluid Mech. To appear.

33. Tulin, M.P. and Cointe, R.: A theory of spilling breakers. 16th Sympos. Naval Hydrodynamics, Berkeley, National Acad. Press, Washington, D.C. (1986).

34. Vinje, T. and Brevik, P.: Numerical simulation of breaking waves. Adv. Water Resources 4 (1981) 77-82.

# Second Order Nonlinear Interactions Between Waves and Low Frequency Body Motion

Odd M. Faltinsen

Division of Marine Hydrodynamics,
Norwegian Institute of Technology,
Trondheim, Norway.

Abstract

Second order wave structure interaction problems are discussed in detail. Weak points in the traditional way of analyzing slowdrift motion is discussed. A consistent theory to second order in wave amplitude and first order in slowdrift velocity for the slowdrift motion of a structure is presented. The interaction between the waves and the local quasi-steady flow due to the slowdrift velocity is incorporated. A new numerical procedure to solve the first and second order problem is briefly presented.

## Introduction

The normal way to solve non-linear wave-structure problems in ship and offshore hydrodynamics is to use perturbation analysis with the wave amplitude as a small parameter. Potential theory is assumed and the problem is solved to second order in incident wave amplitude. The solution of the second order problem results in mean forces, difference frequency and sum frequency forces. Difference frequency forces can excite resonance oscillations in surge, sway and yaw of moored structures and in heave, roll and pitch of floating structures with low water plane area. Sum frequency forces can excite resonance oscillations in heave, pitch and roll of tension leg platforms. Perturbation solutions obviously fail when the waves are close to breaking or structural parts move in and out of the water with large amplitude of motion. They have also limited applicability when viscous forces matter.

K. Horikawa, H. Maruo (Eds.)
Nonlinear Water Waves
IUTAM Symposium Tokyo/Japan 1987

Second order solutions have been extensively studied in the literature. Ogilvie(1983) has given a good state of the art report on the subject. One popular second order problem is the slow drift problem. The traditional way to set up the equation of slowdrift motion is to estimate second order excitation forces without the effect of slowdrift motion and to use a linear response model with mass, damping and spring terms. In the simplest damping model the damping is set empirically as a fraction of critical damping. A more rational way is to include wave drift damping (Wichers(1982)) and damping due to skin friction and eddy making. The presence of wave drift damping can be explained by interpreting the slowly varying velocity as a quasi-steady speed of the structure, and by realizing that mean wave forces are functions of the structural speed. The traditional way of hydrodynamically analyzing slowdrift motion has weak points. The necessity of introducing wave drift damping indicates that the slowdrift motion should be hydrodynamically analyzed in combination with the slowdrift excitation forces. This is done in the main text and it is shown that additional terms to Wichers' wave drift damping term occur.

Triantafyllou(1982) has pointed out that the first order motions cannot be decoupled from the slowdrift translatory and angular motions. The slowly varying angular motions cause a quasi-steady change in wave heading, and the slowdrift motions create a slowly varying frequency of encounter effect between the waves and the structure. The latter effect may be particularly important when a lightly damped roll resonance oscillation is excited. This may cause important memory effects in the roll motion.

Resonance roll motion causes also other difficulties in the analysis. Non-linear viscous effects are significant in the estimation of roll at resonance. This implies that a linear first order roll motion does not exist at resonance. It is shown in the main text that this causes inaccuracies in the prediction of second order hydrodynamic forces.

An aim of the analysis must be to accurately predict extreme values of slowdrift motion in irregular sea. Naess(1986) has studied this problem from a statistical point of wiev. An input to his numerical study was the second order transfer functions for slowdrift excitation forces. The damping was simply set equal to 3% of the critical damping. An important finding in Naess' work was that use of Newman's(1974) approximation for the slowdrift excitation forces gave extreme values within 20% from the more elaborate procedure based on second order transfer functions. Newman's approximation involves only calculation of drift forces in regular waves and is far less time consuming than a consistent calculation of second order transfer function. A major aim in this report is to obtain an accurate and consistent hydrodynamic theory for slowdrift motion. We will concentrate on two-dimensional problems, but the procedure is possible to generalize to three dimensions. The sum frequency problem is not dealt with in details. One reason is that there are many similarities mathematically with the difference frequency problem.

Theory of Slowdrift Motion

Consider an infinitely long horizontal cylinder that is moored in longcrested irregular beam sea waves in deep water. There are two time scales. One is high frequency and associated with the incident waves. The other one is slowly varying and associated with the horizontal resonance motion of the moored cylinder. The problem is two-dimensional in the cross-sectional plane and we will choose a coordinate system (y,z) which is fixed with respect to the slowdrift horizontal motion of the section. The (y,z) coordinate system coincides with the inertial system (y,z) when the structure is at rest. z=0 is in the mean free surface and positive z-axis is upwards (see Fig.1). Mathematically we can write

$$\bar{y} = \int_0^t U(t)dt + y \tag{1}$$

$$\bar{z} = z \tag{2}$$

Here U(t) is the slowly varying horizontal velocity of the section and t is the time variable.

Consider the fluid to be incompressible and the analyzed fluid motion irrotational so that there exists a velocity potential $\phi$ which satisfies Laplace equation. The solution will be written as a series expansion in a parameter $\delta$ characterizing the smallness of the wave amplitudes, i.e. we write

$$\phi = \phi_S + \phi_1 + \phi_2 + \dots\dots\dots \tag{3}$$

where $\phi_S$ is independent of $\delta$, $\phi_1$ is linear with respect to $\delta$, $\phi_2$ is proportional to $\delta^2$ etc..

Let the incident waves propagate along the positive y-axis. The incident wave potential correct to $0(\delta)$ is written as

$$\phi^I = \sum_{j=1}^{N} \frac{gA_j}{\omega_o^j} e^{i(\omega_j t - k_j y - \epsilon_j)} e^{k_j z} \tag{4}$$

Here g is the acceleration of gravity, $\omega_j$ the circular frequency of oscillation and $k_j$ the wave number of wave component no. j. $\omega_j$ are Kj connected through the relations

$$\omega_j t = \omega_o^j t - k_j \int_o^t U(t)\, dt \tag{5}$$

$$k_j = \frac{(\omega_o^j)^2}{g} \tag{6}$$

Here $\omega_o^j$ is the circular frequency of oscillations of the waves in the inertial system(y,z).

Equation (5) expresses what we in the introduction referred to as a slowly varying frequency of encounter effect and that the first order motions depend on the slowdrift motion. The phase angles $\epsilon_j$ may be considered as random phase angles and the amplitudes $A_j$ of the wave components may be determined by a wave spectrum $S(\omega_0)$ characterizing the sea state.

We will assume U is small and solve the problem correct to O(U). We should note that $\partial U/\partial t = o(U)$. This means that the $\phi_s$ problem will be solved as a quasi-steady problem, and the (y,z) coordinate system is not considered to be an accelerated coordinate system. We will in the end of the analysis introduce the effect of $\partial U/\partial t$.

$\phi_s$-problem

The steady velocity potential $\phi_s$ satisfies the rigid free surface condition ($\partial\phi_s/\partial z=0$ on z=0) and a zero normal velocity condition on the body.
We can write

$$\phi_s = - Uy + \phi_B + \phi_\Gamma \tag{7}$$

Here $\phi_\Gamma$ is the velocity potential due to the shed vorticity. It satisfies the same body and free surface condition as $\phi_s$.

To evaluate the order of magnitude of $\phi_\Gamma$ we will consider a small KC-number flow. Here $KC=U_{max} T_N/b$, where $T_N$ is the natural sway period and b is the beam. In practical cases KC is less than 10. We will consider the asymptotic case $KC \to 0$. A Brown-Michael solution(1955) is used to study the flow locally around a separation point. The separated flow is approximated by a single vortex. In the far-field from the local flow this is a plausible representation when $KC \to 0$ and diffusion is neglected. But we should note that vorticity can reach the far-field even for small KC-numbers. During different phases of the body oscillations vorticity of different signs are generated. Vortices of nearly the same magnitude and opposite sign may interact and convect to the far-field. But we should note that the combined effect of two far-field vortices of the same magnitude and opposite sign which stay close to each other, has a small effect on the body pressure.

Anyway we want to keep our mathematical model of $\phi_\Gamma$ simple at this stage. The intention is to decide if we have to include $\phi_\Gamma$. If we denote the circulation of the single vortex by $\Gamma_0$ and its distance relative to the separation point by $|z_0|$ we find from the Kutta condition that

$$\Gamma_0 = 0(U\,|z_0|^{1/\lambda}) \tag{8}$$

Here $\lambda=2-\delta/\pi$, where $\delta$ is the interior angle of the body at the separation point. From the zero-force condition on the vortex and the cut adjoining the vortex with the separation point we find that $\Gamma_0=0(U^{2\lambda/(2\lambda-1)})$. For a rectangular cross-section it implies that $\Gamma_0=0(U^{3/2})$, for a cross-section with bilge keels it implies that $\Gamma_0=0(U^{4/3})$ and for a cross-section With no sharp corners it implies that $\Gamma_0=0(U^2)$.

In the vicinity of the vortex we may set $0(\phi_\Gamma)=0(\Gamma_0)$. At some distance from the vortex we may set $0(\phi_\Gamma)=0(\Gamma_0|z_0|^{1/\lambda})=0(U^{(2\lambda+1)/(2\lambda-1)})$. This is due to the combined behaviour of the vortex and its "image" inside the body. This implies that for a rectangular cross-section $0(\phi_\Gamma)=0(U^2)$. For a cross-section with no sharp corners it implies that $0(\phi_\Gamma)=0(U^3)$. For a cross-section with bilge keel it implies that $0(\phi_\Gamma)=o(U5/3)$.The consequence of this discussion is that we can neglect the influence of the shed vorticity as long as we solve the problem correct to order U and the slowdrift horizontal motion is a small Keulegan-Carpenter number flow. The $\phi_B$ -problem can be solved by a standard method.

<u>$\phi_1$ and $\phi_2$-problems</u>

Formulation and solution of the $\phi_1$ and $\phi_2$ problem is quite lengthy to describe in detail. In this context we are limited to give a more general description of the $\phi_1$ and $\phi_2$ problem.

By using the dynamic and kinematic free surface condition it is possible to derive the following free surface conditions for the $\phi_1$ and $\phi_2$ problem

$$\frac{\partial^2\phi_i}{\partial t^2} + 2\frac{\partial\phi_s}{\partial y}\frac{\partial^2\phi_i}{\partial y\partial t} + \frac{\partial^2\phi_s}{\partial y^2}\frac{\partial\phi_i}{\partial t} + g\frac{\partial\phi_i}{\partial z} = h_i(y,t) \quad (9)$$

on $z = 0$

Here $h_1(y,t)=0$ and $h_2(y,t)$ depends on $\phi_s$ and $\phi_1$. Far away from the body this becomes

$$\frac{\partial^2\phi_i}{\partial t^2} - 2U\frac{\partial^2\phi_i}{\partial y\partial t} + g\frac{\partial\phi_i}{\partial z} = h_i(y,t) \quad (10)$$

on $z = 0$

Equation (10) resembles the classical linear free surface condition with forward speed when i=1. It should, however, be noted that terms of $O(U^2)$ are neglected. Close to the body the free surface condition is quite different from the far-field expression. In popular terms one may say that the waves "ride" on the local steady flow. The body boundary condition for $\phi_1$ can be found in Newman(1978). The body boundary condition for $\phi_2$ depends on $\phi_1$ and $\phi_s$.

The radiation conditions are needed in addition. A numerical procedure to solve the first and second order problem will be briefly presented. In the first order problem this is based on using a multipole expansion in the domain where equation (10) applies, together with Green's second identity for the whole fluid domain. The multipoles satisfy equation (10) with $h_i=0$ and have the singularity point inside the body. For a free-surface piercing body the origin of the coordinate system is chosen as the singularity point. In Green's second identity the source and dipole functions for infinite fluid domain are used. Very far from the body the velocity potential generated by the body and satisfying equation (10) represents radiating wave solutions. There is only one wave system on each side of the body. The wave length is different on the two sides. If we interpret the slowdrift velocity as an incident current, the wave length is shortest on the side where the current is opposite the wave propagation direction.

The second order potential can in principle be solved as the 1. order problem. A difference would be that we add a particular solution satisfying the inhomogeneous free surface condition (10). The difference frequency part of the solution decays slowly with depth when two frequency components are close to each other. This means that finite depth effect may be important. The solution procedure that we have outlined can be generalized to incorporate finite depth effects.

From the solution of the first order problem we can find added mass and damping coefficients and linear wave excitation loads. By means of the equations of the first order motions we can find the linear motions. But it should be noted that the equations of motions depend parametrically on the horizontal slowdrift motion.

The second order horizontal force $F_2$ follows from integrating the pressure correct to second order over the wetted body surface.

We shall concentrate on the difference frequency problem and write the part of $F_2$ that is of interest in the slowdrift problem as

$$
\begin{aligned}
F_2^{SV} = {} & \sum_{j=1}^{N} \sum_{k=1}^{N} A_j A_k \Big[ T_{jk}^{c} \cos\{(\omega_k - \omega_j)t - (\epsilon_k - \epsilon_j)\} \\
& \quad + T_{jk}^{s} \sin\{(\omega_k - \omega_j)t - (\epsilon_k - \epsilon_j)\} \Big]_{U=0} \\
& + U \Bigg[ \sum_{j=1}^{N} \sum_{k=1}^{N} A_j A_k \frac{\partial}{\partial U} \Big[ T_{jk}^{c} \cos\{(\omega_k - \omega_j)t - (\epsilon_k - \epsilon_j)\} \\
& \quad + T_{jk}^{s} \sin\{(\omega_k - \omega_j)t - (\epsilon_k - \epsilon_j)\} \\
& \quad + V_{jk}^{c} \cos\{(\omega_k + \omega_j)t - (\epsilon_k + \epsilon_j)\} \\
& \quad + V_{jk}^{s} \sin\{(\omega_k + \omega_j)t - (\epsilon_k + \epsilon_j)\} \Big] \Bigg]_{U=0}
\end{aligned}
\qquad (11)
$$

The first term can be interpreted as the slowdrift excitation force while the second term as the slowdrift damping term. The slowdrift excitation force is calculated by setting U=0 in the first order problem. The damping term is evaluated by calculating $F_2$ for some small positive and negative constant velocity $\pm|U|$ and deducting the two results. $T^c_{jk}$, $T^s_{jk}$, $\partial T^c_{jk}/\partial U$, $\partial T^s_{jk}/\partial U$, $\partial V^c_{jk}/\partial U$ and $\partial V^s_{jk}/\partial U$ are independent of $A_j$, $A_k$, $\epsilon_j$ and $\epsilon_k$ and can be interpreted as second order transfer function when $j \neq k$. We will follow Newman's(1974) definition and set $T^c_{jk} = T^c_{kj}$, $T^s_{jk} = -T^s_{kj}$, $\partial T^c_{jk}/\partial U = \partial T^c_{kj}/\partial U$, $\partial T^s_{jk}/\partial U = -\partial T^s_{kj}/\partial U$, $\partial V^c_{jk}/\partial U = \partial V^c_{kj}/\partial U$, $\partial V^s_{jk}/\partial U = \partial V^s_{kj}/\partial U$.

The damping term can be simplified by using equivalent linearization. An equivalent damping coefficient can be written as

$$B_{22} = -\sum_{j=1}^{N} A_j^2 \left.\frac{\partial T^c_{jj}}{\partial U}\right|_{U=0} \tag{12}$$

This agrees with Wichers'(1982) expression. But from equation (11) we note that there are additional time dependent damping terms.

Let us now formulate the equations of slowdrift motions $x_2$. The restoring force due to the mooring system will be denoted $-C_{22}x_2$. We have in the hydrodynamic analysis treated $U \equiv dx_2/dt$ as a steady slowdrift velocity. The consequence of this is that we have not derived an added mass force $A_{22}\, dU/dt$. But we know that this term is of equal importance as the mass term $M\, d^2x_2/dt^2$. We can introduce the added mass force in a simple way without consequences for the boundary problem for $\Phi_1$ and $\Phi_2$. We know that $\Phi_s$ is a lower order term than $\Phi_1$ and $\Phi_2$. By requiring that $\partial\Phi_s/\partial t$ causes forces of the same order of magnitude as the second order force, we avoid taking into account effects of $dU/dt$ in the $\phi_1$ and $\phi_2$-problem. The unsteady effects of $\phi_s$ can easiest be taken care of in the non-accelerated coordinate system $(y,z)$. The result is an added mass force $-A_{22}\, d^2x_2/dt^2$ on the body, where $A_{22}$ is added mass in sway at zero frequency. We can now write up the following equation system

$$(M + A_{22}) \frac{d^2x_2}{dt^2} + B_{22}^{SD} \frac{dx_2}{dt} + C_{22}x_2 = F_2^{SD} \tag{13}$$

Here $F_2^{SD} - B_{22}^{SD}\, U$ is equal to equation (11).

Since we have neglected all term of $O(U^2)$ in our analysis there are no drag term in equation (13). In reality the drag damping may be important, in particular for low sea states. A consistent introduction of the drag term would mean that we have to solve the $\phi_1$ and $\phi_2$-problem to $O(U^2)$. This matter will not be pursued here. A skin friction term due to laminar boundary layer flow is of $O(U)$ and could consistently be introduced into the equation of motion. But the term is expected to be of small practical importance.

Numerical Results

A computer program based on the theoretical procedure presented above has been derived. The numerical results show that it is important that the first order waves "ride" on the local quasi-steady flow. The numerical solution breaks down when we use condition (10) for a free surface-piercing body. The reason is that equation (10) is based on U being a first order approximation of the steady flow everywhere on the free surface. This cannot be true at the intersection points between the free surface and the body. We have investigated when we can neglect terms of $O(U^2)$ in the free surface condition. It seems to be a good approximation to keep only terms of $O(U)$ when $\tau = \omega_{je}U/g < \sim 0.1$.

In Fig. 2 are presented numerical results for drift forces in regular waves on a circular cylinder with axis in the mean free surface. The cylinder is free to oscillate in heave, sway and roll. The metacentric height is 0.127 times the beam, the roll radius of gyration is 0.25 times the beam, the z-coordinate of the centre of gravity is -0.125 times the beam. The current velocity is zero. There is a significant influence of viscous roll damping on the roll in the vicinity of roll resonance. No theoretical methods were used in estimating the viscous roll damping. It was simply set as a factor times the roll wave damping. The main purpose was to illustrate the qualitative influence.

In order to predict realistic roll angles at resonance viscous effects have to be introduced. Generally speaking this means a term proportional to the square of the roll velocity. In the perturbation scheme presented here it is not clear how viscous roll damping can be taken care of in a consistant way.

In Fig.3 are shown results for drift forces in regular waves and current incident on a circular cylinder with axis in the mean free surface. The cylinder is free to oscillate in heave and sway, but was restrained from oscillating in roll. The results show that the influence of current velocity on the mean wave forces is of practical importance. The drift force increases with increasing current velocity in the incident wave propagation direction.

From the results in Fig.3 it is possible to find wave drift damping (see Eq.12). A high accuracy is needed in the theoretical prediction of mean wave force as a function of U to satisfactorily predict wave drift force damping. The results are presented in Fig.4. Nakamura et al.(1986) experimental results are also shown in Fig.4. The agreement between theory and experiments is not completely satisfactory. For instance the theoretical wave drift damping is practically zero for small frequencies, while the experimental damping has a finite value. This difference is likely to be due to viscous effects. The results show that wave drift damping is of practical importance for slowdrift sway motion of a moored structure. But viscous effects cannot be completely disregarded in general. A consistent theoretical evaluation of viscous effects is beyond present state of the art.

Second order transfer functions for slowdrift excitation forces and slow drift damping have also been evaluated. It was found that the force and damping due to the second order effects are sensitive to small inaccuracies in the first order potential. This will be discussed in more detail in a future publication. Evaluation of a complete set of second order transfer functions is time consuming. Having in mindalso influence of viscous effects it is questionable to recommend calculations of a complete set of second order transfer functions for practical evaluation of slowdrift oscillations of a moored structure in irregular waves.

## Conclusion

Second order wave structure interaction problems are discussed in detail. Weak points in the traditional way of analyzing slowdrift motion of a moored structure is discussed. A consistent theory to second order in wave amplitude and first order in slowdrift velocity for the slowdrift motion of a moored two-dimensional structure in irregular sea is presented. The interaction between the waves and the local quasi-steady flow due to the slowdrift velocity is incorporated.It is shown that:

a. The effect of separated flow can be neglected to first order in slow drift velocity.
b. The first order motions are dependent on the second order motions.
c. It is important that the first order waves "ride" on the local quasi-steady flow.
d. The presence of resonance roll oscillations create theoretical difficulties due to non-linear viscous damping effects.

e. Current has to be incorporated in practical predictions of mean wave drift forces on marine structures.

f. Wave drift damping is of significant importance for slow-drift sway motion of a moored structure. But viscous effects cannot be neglected in practical evaluations of slowdrift damping.

g. Time dependent slow drift damping terms due to second order non-linear potential flow effects cannot a priori be neglected.

h. The second order force is sensitive to small inaccuracies in the first order potential.

ACKNOWLEDGEMENT

The work done by Rong Zhao in developing the computer programs is highly appreciated.

REFERENCES

Brown, C.E. and Michael, W.H. (1955). Nat. Adv. Comm. Aero techn. Note 3430.

Nakamura, S. Saito, K. and Takagi, M. (1986): "On the Increased Damping of a Moored Body During Low Frequency Motions in Waves", OMAE '86 , Tokyo.

Newman, J.N. (1974): "Second Order, Slowly Varying Forces on Vessels in Irregular Waves", International Symposium on the Dynamics of Marine Vehicles and Structures in Waves, University College, London, Publ. by the Inst. of Mech. Eng.

Newman, J.N. (1978): "The Theory of Ship Motions", Advances in Applied Mechanics, Volume 18.

Naess, A. (1986): "On the Statistical Analysis of Slow-drift forces and motions of floating offshore Structures", Proc. 3rd OMAE Conf., Tokyo.

Ogilvie, T.F.(1983): "Second Order Hydrodynamic Effects on Ocean Platforms", Ship and platform Motions, International Workshop at University of California, Berkeley.

Triantafyllou,A.W. (1982): "A Consistent Hydrody-namic Theory for Moored and Positioned Vessels", Journal of Ship Research 26,pp. 97-105.

Wichers, J.E.W. (1982): "On the Low-frequency surge motions of vessels moored in high seas", 14th Annual OTC,Houston, Texas.

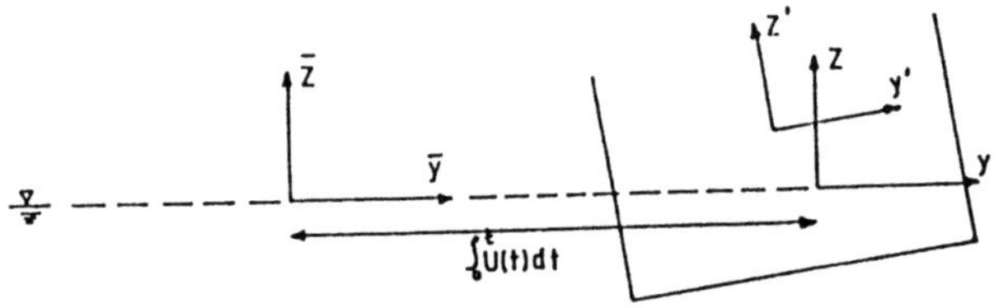

Fig. 1 Coordinate systems

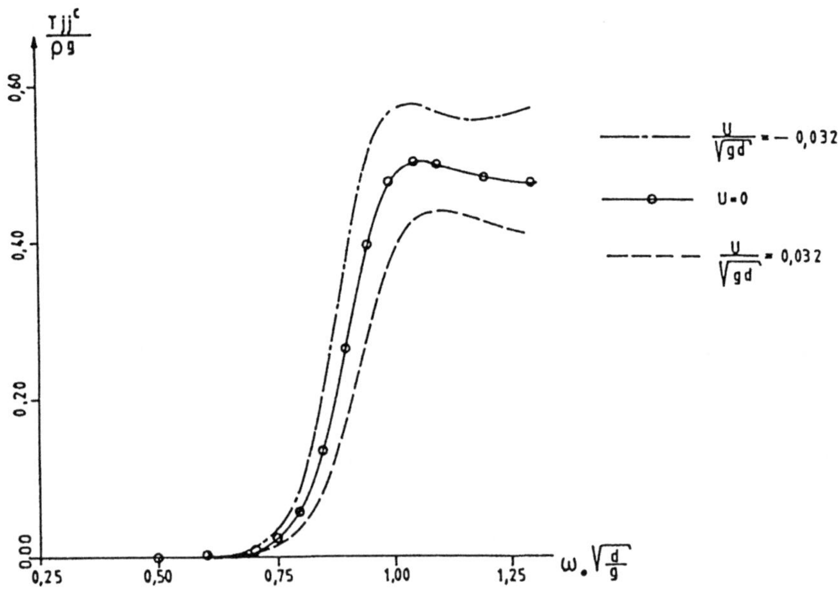

Fig. 2 Influence of viscous roll damping on drift forces for a circular cylinder with axis in the mean free surface. U = 0

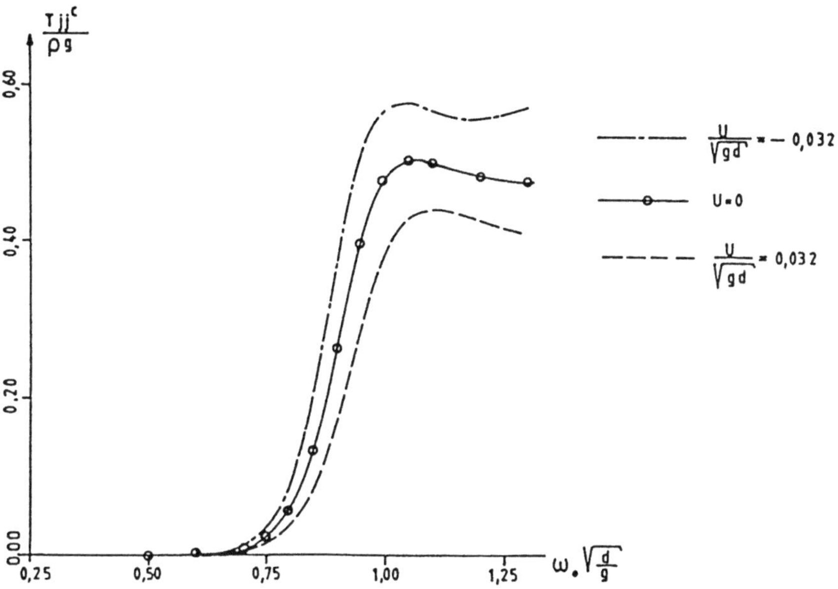


Fig. 3 Drift force per unit length on a circular cylinder with axis in the mean free surface. Influence of current velocity. The cylinder is restrained from oscillating in roll. Positive current direction along negative y- axis. Incident wave propagation direction along positive y-axis.

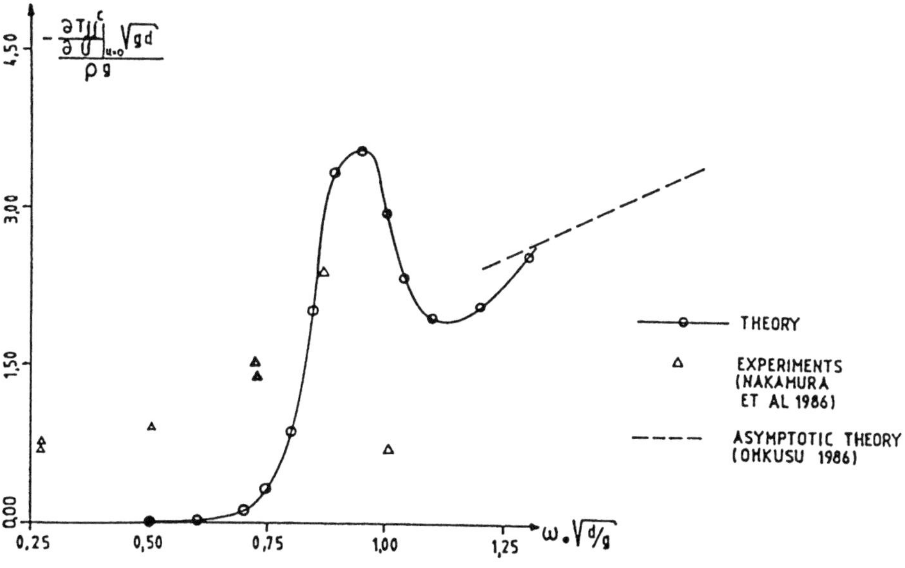


Fig. 4 Wave drift damping per unit length for a circular cylinder in the mean free surface. The cylinder is restrained from oscillating in roll.

# Session 1

# Evolution of Nonlinear Water Waves

# Evolutional Behaviour and Instability of a Single Wave Packet

Takao YAMASHITA and Yoshito TSUCHIYA

Disaster Prevention Research Institute
Kyoto University
Uji, Kyoto, Japan

Summary

Propagation of a single wave packet on a gently sloping bottom of 1/50 has been investigated experimentally over a wide range of maximum carrier wave steepness in order to elaborate on the wave packet instability. An instability diagram was derived as a function of wave steepness $ak$ and the ratio of the water depth to the wave lengthness of maximum carrier wave $kh$ (we will call it shallowness). Our conclusions indicate that even highly nonlinear carrier waves form an envelope propagating on a slope.

Introduction

We have often observed the wave-packet-like groupiness in wave data recorded even in the shallow water near the breaker zone. However, the combined effects of shoaling, diffraction, refraction and breaking on the variable wave propagation depth create a difficult scenario through which the coastal engineer must attempt to understand the process by which waves propagate in the shallow seas. Judging from the abundance of nearshore field wave data which reflects the groupiness property, we have come to conjecture that the wave packet transformation in shallow seas plays an important role in defining nearshore wave irregularity. Furthermore, the long period components of nearshore waves which are closley associated with sediment transport are also expected to be directly related to the wave packet transformation. For these reasons we have initiated an investigation into the wave packet behaviour in shallow seas.

The investigation into the nonlinear resonant interaction originated by Benjamin & Feir(1) shows by their experimental work that Stokes waves are unstable with respect to side-band disturbances. This was confirmed and extended theoretically by Whitham(2), Lighthill(3), Benny & Roskes(4), Chu & Mei(5) and Davey & Stewartson(6). Further theoretical advances for deep water waves have been

K. Horikawa, H. Maruo (Eds.)
Nonlinear Water Waves
IUTAM Symposium Tokyo/Japan 1987

made by Zakharov & Shabat(7) resulting in a widely applicable equation which showed an elegant way to investigate the mechanism of weakly nonlinear resonant interaction of ocean waves. Extensions of the Cubic Schrödinger Equation (CSE) have been done to water waves on a sloping bottom (4),(5),(6). Tsuchiya et. al.(8) also derived the CSE in the characteristic coordinates of the eikonal equation in order to introduce reff-action effects. Recent interest in the wave packet propagation seems to gather on the treatment of the higher-order nonlinear Schrödinger equation which includes the effects of variable water depth. We, here, performed wave flume experiments of a single wave packet instability on a gently sloping bottom of 1/50 over a wide range of maximum carrier wave steepness.

## Cubic Schrödinger Equation on a gently sloping bottom

Assuming a very mild bottom slope and no energy leak between wave ray we can obtain the CSE with bottom slope effects.(8) The 2-D version of this CSE is :

$$iA_{\xi_2} + \alpha A_{\tau\tau} + \beta |A|^2 A + \gamma A = 0 \tag{1}$$

where

$$\alpha = \frac{1}{2c_g^3}\omega_{kk} = \frac{1}{2\omega c_g} + \frac{\omega h \cosh^2 kh}{c_g^3 k \sinh 2kh} - \frac{h \sinh kh}{c_g^2 \cosh kh} \tag{2}$$

$$\beta = -\frac{\omega k^2(\cosh 4kh + 8 - 2\tanh^2 kh)}{16 c_g \sinh^4 kh} + \frac{\omega(2\omega \cosh^2 kh + k c_g)^2}{2 c_g \sinh^2 2kh (gh - c_g^2)} \tag{3}$$

$$\gamma = i\left\{\frac{k(\sinh 2kh + 2kh)\sinh kh - 4k^3 h^2 \cosh kh}{(\sinh 2kh + 2kh)^2 \sinh kh}\right\} h_{\xi_2} \tag{4}$$

In the above equations, $k$ : wave number, $h$ : water depth, $c_g$: group velocity, $\omega$: frequency, $\xi_2 = x_2$ and $\tau = \varepsilon^{-1}\int c_g^{-1}\,dx_2 - t_1$ .
From the numerical tests of this equation, it was rccognized that (i) initial envelope soliton solution propagates on a slope in a stable form, (ii) shrunk initial soliton deforms in radiation, (iii) for extended initial soliton recurrence or fission appears, as shown in Fig.1.
Here, the deformation parameter of envelope soliton $\delta$ is defined as,

$$A_1 = a_0 \operatorname{sech}\left\{\delta\tau\sqrt{\frac{\beta}{2\alpha}}\right\} \exp\left(\frac{i\beta a_0^2 \xi_2}{2}\right) \tag{5}$$

where $a_0$ is the amplitude of maximum carrier wave and $\delta > 1$ depicts shrinking, $\delta < 1$ extending, of the initial condition respectively.

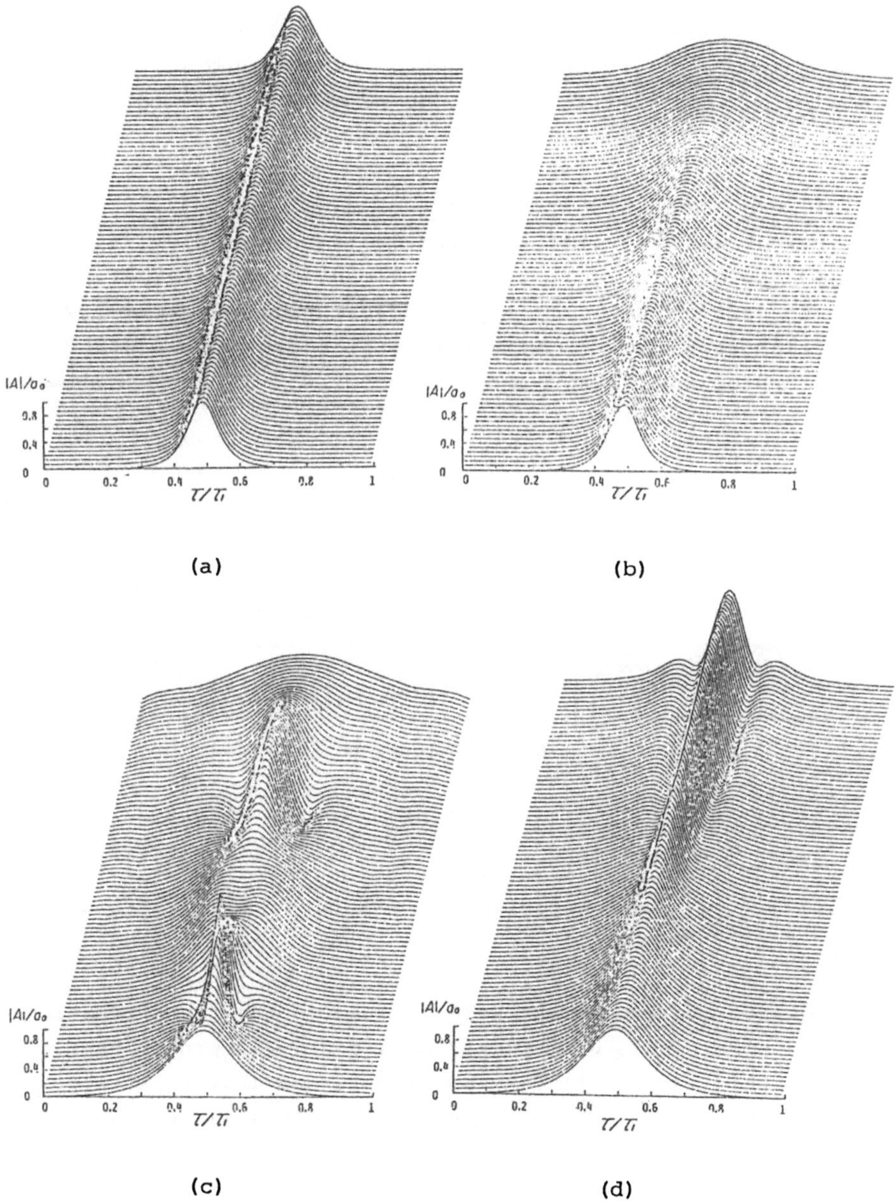


Fig. 1. Numerical solutions of the CSE; (a) initial condition is soliton solution(stable), (b) shrunk initial soliton(radiation), (c) extended initial soliton(recurrence), (d) extended initial soliton(fission), $\tau_1$ is the time when $A<0.01a_o$

As the CSE is symmetric with respect to the spatial coordinate, only the symmetrically changing solution is obtained if the initial condition is symmetric. It might be necessary to make clear the applicability of the CSE by flume experiments.

Wave flume experiment for instability of envelope soliton

The experiment was performed by using the 2-D wave flume, 0.7m wide, 1/50 uniformly sloping bottom. A mini-computer system was employed to control wave generator paddle motion and to analyze wave profiles measured by 6 wave gauges of capacitance type. The positions of wave gauges and outline of wave flume are shown in Fig.2.

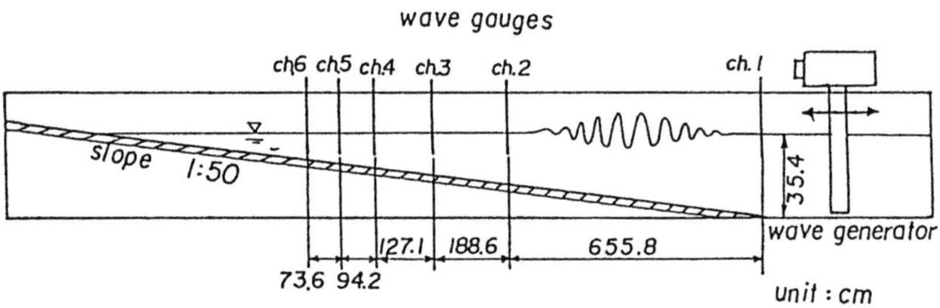


Fig.2 Experimental set-up

The objects of the experiment were to observe the transformation properties of a single wave packet on a gently sloping bottom and to clarify the instability diagram of extended initial envelope solitons with respect to wave steepness $ak$ and shallowness $kh$. With the numerical results of obtained CSE, it has been recognized that wave packets given by the envelope soliton solution propagate on a constant bottom in a stable form, while the shrunk envelope solitons always deform into radiation. The most interesting phenomenon can be observed in the extended envelope cases. It is well-known in the previous numerical studies that the long-time evolution of an envelope soliton with small disturbance results in fission or recurrence, and the higher nonlinearity of carrier waves makes the wave packet skewed(forward leaning) after a short distance.

It is also interesting to examine the deformation of the wave packet in the shallow water where depth uniformly changes across the turning point of the CSE. We performed a wave packet shoaling experiment on a 1/50 uniform slope, by prescribing various extended symmetric initial wave packets of the envelope solitons. Figs. 3 and 4 are the typical deformation patterns of wave packets and the obtained instability/shoaling diagram. Three types of deformation patterns were observed, i.e., (i) stable symmetric, (ii) skewed (forward leaning) and (iii) skewed plus fission (splitting into solitons asymmetrically), as shown in Fig. 3. These can be clearly classified by the plot of $ak$ vs $kh$ (Fig. 4) which represent the nonlinearity of maximum carrier wave and shallowness, in which $U_r$ indicates Ursell's parameter defined by $U_r \fallingdotseq (2\pi)^2 ak/(kh)^3$. Some examples of experimental wave packet profiles are exhibited in Appendix.

In Fig. 4, the solid circles indicate that the wave packets resulted in skewed & fission and the half-solid circles are those cases in which forward-leaning was observed. The stable propagations were observed in the initial conditions as indicated by white circles. However, no symmetric fissions shown in Fig. 1(c) appeared. Moreover, solid lines in this figure depict the changes in $ak$ of stably propagating packets and broken lines are those of unstable wave packets. The dotted lines show the breaking of several carrier waves.

Throughout the experiments, we did not observed the symmetric fission of wave packet, which is much different from the numerical results of the CSE. As the usual nonlinear Schrodinger equation with third order accuracy in the wave steepness $ak$, which governs the phase of motion called the slow modulation of wave packet, has the symmetric soliton solution, the experimental results above are beyond the region of applicability of the CSE.

Furthermore, wave packets initiated by soliton solution in the range of $ak<0.2$ propagate stably on the sloping bottom crossing the point of $ak = 0.2$ as well as the turning point.

Then it can be concluded that the instability of wave packets strikingly depends on the initial condition. The well-posed ini-

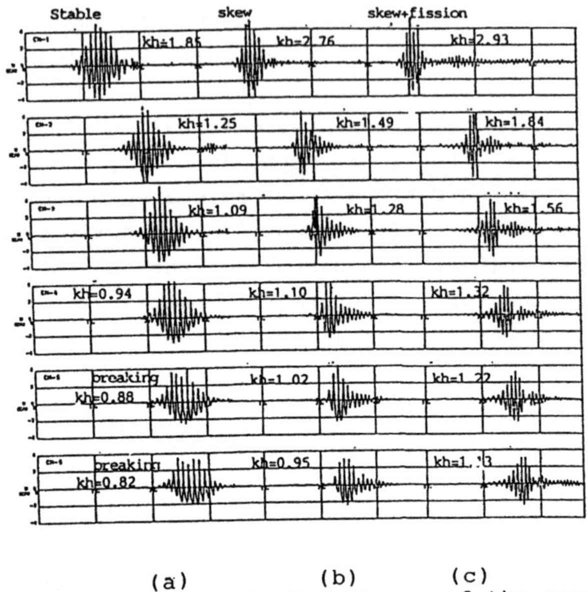


Fig. 3. Three typical patterns of the experimental profiles (a) stable, (b) skewed, (c) skewed+fission

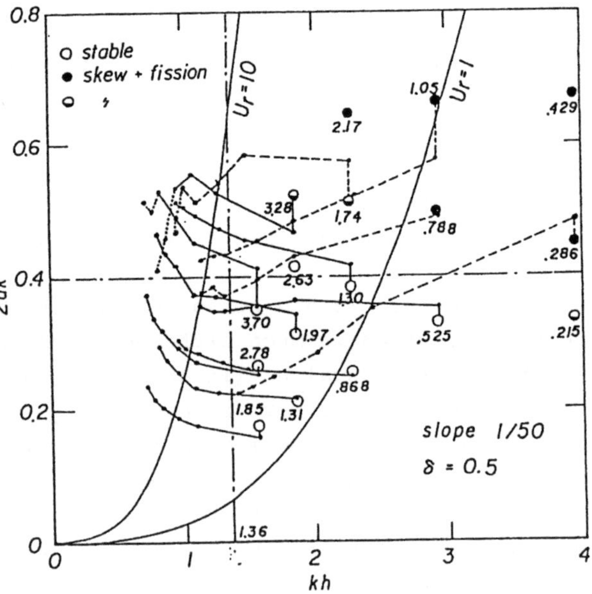


Fig. 4. Stability-diagram of the extended initial solitons (numbers under circles indicate Ursell's parameters of the maximum carrier waves)

tial envelope soliton solutions propagate stably on the sloping bottom through $ak = 0.2$ . This implies that there is a possibility that higher nonlinear carrier waves make a wave packet.

## Conclusions

From the experimental considerations on wave packet deformation and instabilities on a gently sloping bottom, we conclude that: (i) The CSE works of the slow modulation of wave packet in the region of $ak < 0.2$ (ii) The shoaling wave packet propagates on the sloping bottom in symmetric form in the realm of $ak < 0.2$. (iii) There is a possibility that higher nonlinear carrier waves of $ak > 0.2$ form an envelope soliton solution.
From engineering view points, we strongly need the higher-order nonlinear Schrodinger equation to describe the wave packet propagation and transformation in the coastal seas.

## References


1. Benjamin, T.B. and Feir, J.E.; The disintegration of water wavetrains on deep water. J. Fluid Mech., 27, 1967, pp.417-430.

2. Whitham, G.B.; Nonlinear dispersion of water waves. J. Fluid Mech., 27, 1967, pp.399-412.

3. Lighthill, M.J.; Some special cases by the Whitham theory. Proc. Roy. Soc. London, Ser.A, 299, 1967, pp.28-53.

4. Benny, D.J. and Roskes, G.J.: Wave instabilities. Studies Appl. Math., 48, 1969, pp.377-385.

5. Chu, V.C. and Mei, C.C.: The nonlinear evolution of Stokes waves in deep water. J. Fluid Mech., 47, 1971, pp.337-352.

6. Davey, A and Stewartson, K.: On the three-dimensional packets of surface waves. Proc. Roy. Soc. London, Ser. A, 338, 1974, pp. 101-110.

7. Zakharov, V.E.: Exact theory of two- dimensional self-focusing and one-dimensional self modulation of waves in nonlinear media. Sov. Phys., JETP, 34, 1972, pp.62-69.

8. Tsuchiya, Y.; Yasuda, T.; Yamashita, T. and Shibano, S.: Modulation and wavegrouping of nonlinear waves. Proc. 32nd Japanese Conf. Coas. Eng., 1985, pp.179-183 (in Japanese).

## Appendix

Two examples of transformations of experimental wave packets, whose initial conditions are extended envelope solitons, are shown in the figures below. It can be recognized that wave packet propagates in a stable form on the sloping bottom through the point $ak = 0.2$ up to the breaking point.

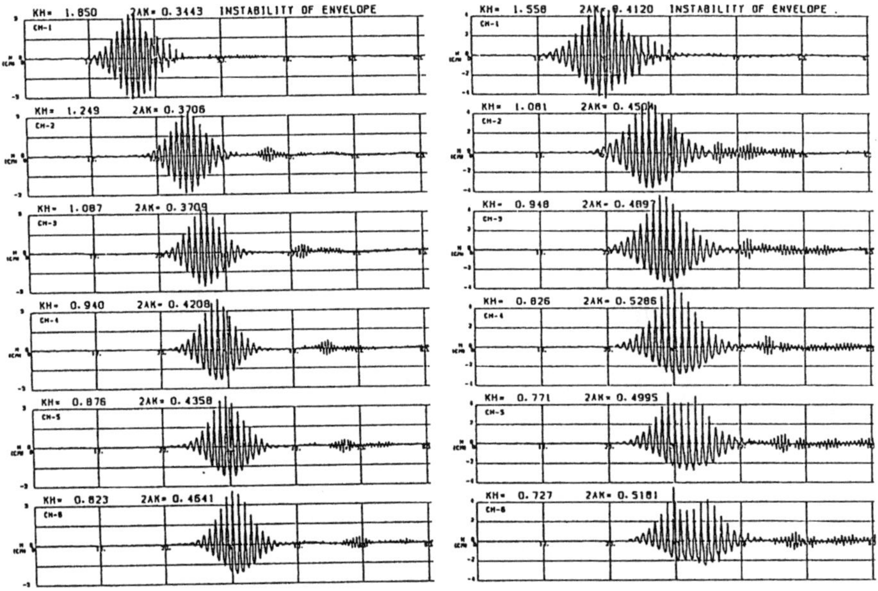


(Parameters in the figures indicate as; 2AK=2*ak*, KH=*kh*)

# On the Initial Evolution of Gravity-Capillary Waves

P.A.E.M. Janssen

KNMI, The Netherlands
ECMWF, United Kingdom

## 1. INTRODUCTION

Currently there is much interest in the properties of wind-generated gravity-capillary waves because of promising remote sensing techniques such as the SAR, that can provide information about surface waves with a wavelength of the order of 4-40 cm, i.e. in the gravity-capillary range. It was thus found that in shallow waters these waves experience a strong modulation in the presence of a non-uniform current and a varying bottom topography. (see e.g. van Gastel 1987).

In this paper we discuss the initial evolution of wind-generated gravity-capillary waves by means of a dynamical model that includes the effect of wind input, viscous dissipation and three wave interactions. Our starting point is a gravity- capillary wave field in isolation that, as was shown by Zakharov (1968) constitutes a Hamiltonian system and the dynamics of the waves are determined by Hamilton's equations. This Hamiltonian system is perturbed by including the "small" effects of wind input and viscous dissipation.

## 2. PROMINENT ROLE OF SECOND HARMONIC RESONANCE

We shall concentrate on the phenomenon of second harmonic resonance. Second harmonic resonance occurs if the phase speed of the free wave at wave number $k_1$ equals the phase speed of the free wave at wave number $k_2=2k_1$. For pure gravity-capillary waves with dispersion relation

$$\omega = (gk+Tk^3)^{\frac{1}{2}} \tag{1}$$

where g is the acceleration of gravity and T is surface tension, the condition for second harmonic resonance is satisfied for the wave number

$$k_1 = \left(\frac{g}{2T}\right)^{\frac{1}{2}} \tag{2}$$

K. Horikawa, H. Maruo (Eds.)
Nonlinear Water Waves
IUTAM Symposium Tokyo/Japan 1987

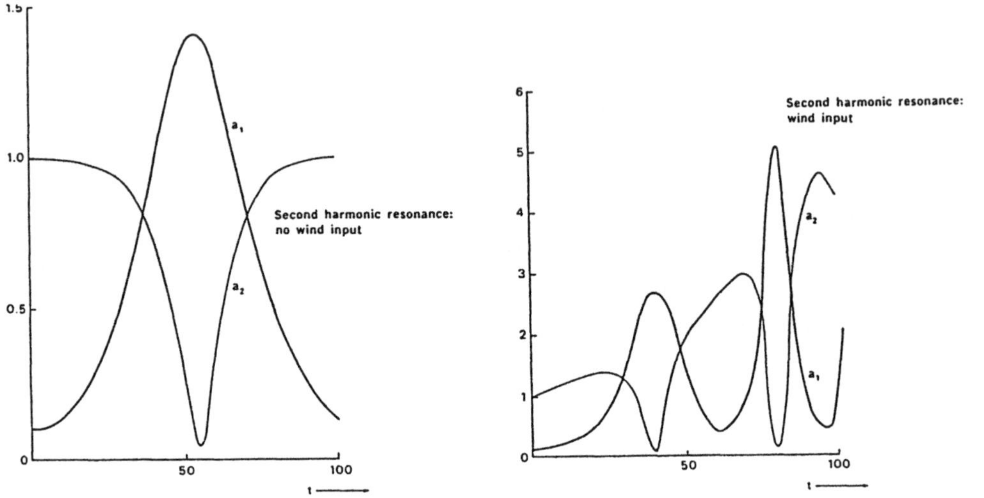


Fig. 1 Resolution in time of waves enjoying a strong interaction due to second harmonic resonance.

In case of second harmonic resonance a strong non- linear interaction between the free wave at wave number $k_1$ and its second harmonic is to be expected. Simmons (1969) conjectured that if gravity-capillary waves are generated by wind near the wave number $k=k_2$ the effect of second harmonic resonance is to transfer energy from the second harmonic to the first harmonic. This is, however, not always true as was pointed out by e.g. Chen and Saffman (1979) and Janssen (1986). No energy transfer was found from second to first harmonic unless the wave amplitude of the second harmonic exceeded a certain amount that depends on the wave number mismatch. If a is the amplitude of the wave at wave number k this threshold condition is given by

$$a > \frac{1}{3k} \left| k^2 - k_2^2 \right| \frac{T}{g} \; ; \qquad (3)$$

this condition holds true for $k \simeq k_2$ and in the absence of wind. For k far away from $k_2$, see Chen and Saffman (1980), and for the effects of wind input and shear in the water current see Janssen (1986).

The physical picture that now emerges is as follows. If the growth rate due to wind has a maximum around the second harmonic wave number $k_2$ then the second harmonic will be generated first. This continues to happen until the wave amplitude exceeds the threshold (3) then, depending on the position of the maximum of the wind input there will be a gradual or a sudden shift of the peak of the spectrum. If this maximum is close to $k_2$, condition (3) will be satisfied quite soon giving a rather gradual shift. If, however, the maximum is far from $k_2$ it takes quite a long time before the energy transfer condition (3) is met. But, when now (3) is satisfied, the amplitudes of the first and second harmonic have become appreciable so a sudden transfer of energy from second to first harmonic now occurs.

We illustrate this physical picture with two examples. First of all, we integrated the perturbed Hamiltonian system for the special case that only the waves obeying the condition for second harmonic resonance are present. The results are given in Fig.1, where we have plotted the amplitude of the first harmonic ($a_1$) and the second harmonic ($a_2$) as a function of time t. Clearly this figure illustrates the strong interplay between the first and second harmonic. If the growth rate $\gamma$ by wind vanishes the nonlinear interaction is periodic in time (recurrence). For finite growth rate the variables $a_1$ and $a_2$ are not periodic in time t, but instead, the variables $a_1e^{-\gamma t}$ and $a_2e^{-\gamma t}$ are periodic on the time scale $\tau = (e^{\gamma t}-1)/\gamma$.

A more complete discussion of this special case is given in Janssen (1986) where also good agreement between this simple theory and an experiment of Choi (1977) is pointed out (see also Kawai (1979).

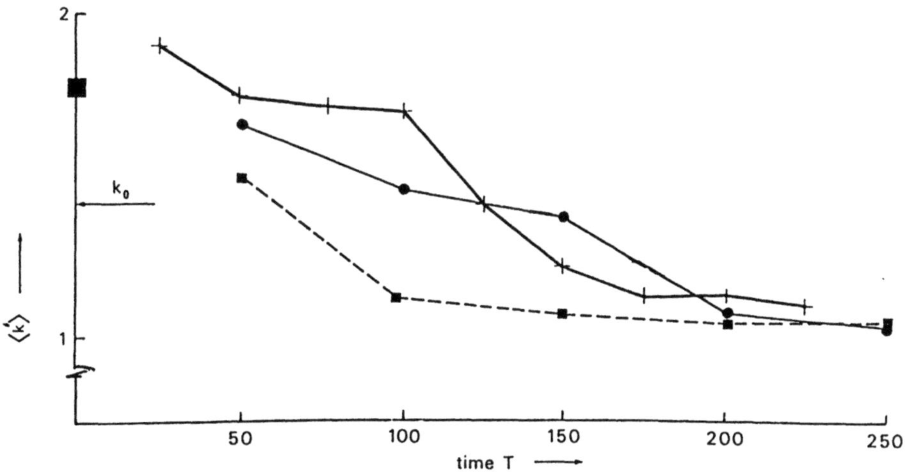


Fig. 2 Evolution of average wave number

Second harmonic resonance is just a special degenerate case of three wave interactions. As in nature a continuum of waves is present that obeys the condition for three-wave interactions, the question now is whether second harmonic resonance plays such a prominent role. In order to answer this question we have numerically solved the perturbed Hamiltonian system for 50 waves, including the waves enjoying second harmonic resonance. We have chosen the mesh width in wave number space in such a way that the waves with the largest wave number are heavily damped because of viscous dissipation. Therefore the inertial subrange and a good part of the viscous range is included in the simulation.

We performed three experiments for $u_* = 17$ (I), 21 (II) and 24 cm/s (III). Experiment I has maximum growth rate at a wave number below $k_2$, which is the one for second harmonic resonance, experiment II has maximum growth rate just at wave number $k_2$, while experiment III has maximum growth rate above $k_2$. Figure 2 shows the evolution of the dimensionless average wave number $\langle k' \rangle = \langle k \rangle \sqrt{T/g}$ as a function of dimensionless time $T = g^{3/4}t/T^{1/4}$. Here, time T = 250 roughly corresponds to 150 wave periods. The figure confirms the physical picture we sketched above, namely that in experiments I and II (■ and ●, resp.) a rather slow migration of the average wave number occurs whereas in experiment III (+) a sudden shift in average wave number of a factor of two is found. It should be noted that during the length of the simulation runs no recurrence was now found.

## CONCLUSIONS

A rather sudden migration of the average wave number of the spectrum is found provided the peak wave number is larger than the wave number $k_2$ for second harmonic resonance and provided the wave height corresponding to this peak is sufficiently large. This is in qualitative agreement with wave tank experiments of Choi (1977) and Kawai (1979).

Remarkably, even in the stage of the initial wavelets (a term introduced by Kawai (1979)) the effect of nonlinear three-wave interactions may be comparable to the effect of wind. For a more complete discussion of this point see Janssen (1987). There is no need to emphasize then that the determination of the growth rate of the waves by wind through an analysis of the time series of the surface elevation of the waves might be in error by a factor of two. Therefore, an alternative technique for the determination of energy transfer from the wind to the gravity-capillary waves has to be found.

The author is pleased to acknowledge numerous discussions with Gerbrand Komen and Philip Saffman.

## References

CHEN, B. and SAFFMAN, P.G. (1979). Stud.Appl.Math. Vol. 60, pp.183-210.

CHEN, B. and SAFFMAN, P.G. (1980). Stud.Appl.Math. Vol. 62, pp.95-111.

CHOI, I. (1977). Thesis, Université d'Aix Marseille.

VAN GASTEL, K. (1987). Ph.D. Thesis, University of Utrecht.

JANSSEN, P.A.E.M. (1986). J.F.M., Vol. 172, pp.531-546.

JANSSEN, P.A.E.M. (1987). J.F.M., Vol. 184, pp.581-597.

KAWAI, S. (1979). J.F.M., Vol. 93, pp.661-703.

SIMMONS, W.F. (1969). Proc.Roy.Soc. Lond. A309, pp.551-575.

ZAKHAROV, V.E. (1968). J.Appl.Mech.Tech.Phys., Vol. 2, 190-194.

# Session 2

# Nonlinear Shallow Water Waves

# Nonlinear Forced Water Waves in a Shallow Channel near a Cut-off Frequency

T. R. AKYLAS and YORGOS D. KANTZIOS

Department of Mechanical Engineering
Massachusetts Institute of Technology
Cambridge, Massachusetts 02139, U.S.A.

*Abstract*

A theoretical study is presented of the free-surface flow induced by a wavemaker performing torsional oscillations about a vertical axis in a shallow channel near a cut-off frequency. It is shown, through a perturbation analysis using characteristic variables, that the nonlinear response is governed by a forced Kadomtsev-Petviashvili (KP) equation with periodic boundary conditions across the channel; this nonlinear initial-boundary-value problem is investigated analytically and numerically. When surface-tension effects are negligible, so that dispersion in the KP is negative, it is found that the nonlinear response reaches a steady state and exhibits jump phenomena, typical of nonlinear forced oscillations. On the other hand, in the high-surface-tension regime, when dispersion in the KP is positive, there is no sign of a steady state and the response undergoes a recurrence phenomenon.

*Introduction*

A semi-infinite channel of width $W$ contains water of undisturbed (constant) depth $h$. At the closed end of the channel, a rigid wavemaker, performing torsional oscillations with frequency $\omega_0$, generates standing waves of wavelength $2W$ across the channel. According to linearized water-wave theory, these waves are propagating along the channel only if the driving frequency $\omega_0$ exceeds the first cut-off frequency $\omega_c$,

$$\omega_0 > \omega_c = \omega\left(k = \frac{\pi}{W}\right) , \tag{1}$$

where $\omega = \omega(k)$ is the linear water-wave dispersion relation; otherwise, if $\omega_0 < \omega_c$, the wave field is evanescent along the channel. In the neighborhood of $\omega_0 = \omega_c$, the streamwise component of group velocity is very small, so that energy cannot propagate away from the wavemaker, and a resonance phenomenon occurs. Close to resonance conditions, the linearized water-wave theory fails and finite-amplitude effects become important, however small the oscillation amplitude of the wavemaker. For water of moderate or infinite depth ($h/W \geq O(1)$), earlier work [1, 2] has shown that, close to resonance, the nonlinear response remains sinusoidal across the channel

K. Horikawa, H. Maruo (Eds.)
Nonlinear Water Waves
IUTAM Symposium Tokyo/Japan 1987

while the wave evolution along the channel is governed by a one-dimensional nonlinear Schödinger equation. Numerical solutions are in reasonable agreement with experiments for moderate-amplitude waves [3].

In this paper, we examine the nonlinear forced response near cut-off conditions for a shallow channel ($h/W << 1$). In this case, because the wave motion is non-dispersive to leading order, the nonlinear response is not sinusoidal across the channel, although the motion imposed by the wavemaker is sinusoidal; the waves are affected by nonlinearity both along and across the channel and are described by a Kadomtsev-Petviashvili (KP) equation, subject to boundary conditions at the wavemaker and initial conditions. This linear initial-boundary-value problem is investigated analytically and numerically using a combination of pseudospectral and finite-difference methods. When surface-tension effects are negligible, so that dispersion in the KP is negative, it is found that the nonlinear response reaches a steady state and exhibits jump phenomena, typical of nonlinear forced oscillations. On the other hand, in the high-surface-tension regime, when dispersion in the KP is positive, there is no sign of a steady state and the response undergoes a recurrence phenomenon.

*Evolution Equation Near Cut-off Conditions*

A schematic of the problem is shown in Fig. 1. The motion of the wavemaker imposes a specified velocity distribution at $x = 0$:

$$\phi_x = \frac{ac_o}{h} \cos \frac{\pi}{W} e^{i\omega_0 t} + c.c. \quad (x = 0,\ 0 \le z \le W) , \tag{2}$$

where $\phi$ is the velocity potential, $a$ is a typical wave amplitude, $c_0$ is the long-wave speed, and $\omega_0$ is assumed to be close to $\omega_c$. Adopting dimensionless variables appropriate in the shallow-water limit, and using asymptotic expansions in the small parameter $\epsilon = a/h$ and $\mu = h/W$, which are measures of nonlinearity and dispersion, respectively, is is found that close to resonance the free-surface elevation $\eta$ (made dimensionless with $a$) is given by [4]

$$\eta = \frac{1}{\epsilon^{1/3}} [u(\xi; X, T) + u(\varsigma; X, T)] \tag{3}$$

where

$$\xi = t + \lambda T + z , \qquad \varsigma = t + \lambda T - z \tag{4}$$

$$\omega_0 = \pi(1 + \lambda\epsilon^{2/3}) , \qquad \lambda = O(1) \tag{5}$$

and $u$, which is a periodic function of $\xi$ with period 2, satisfies the initial-boundary value problem:

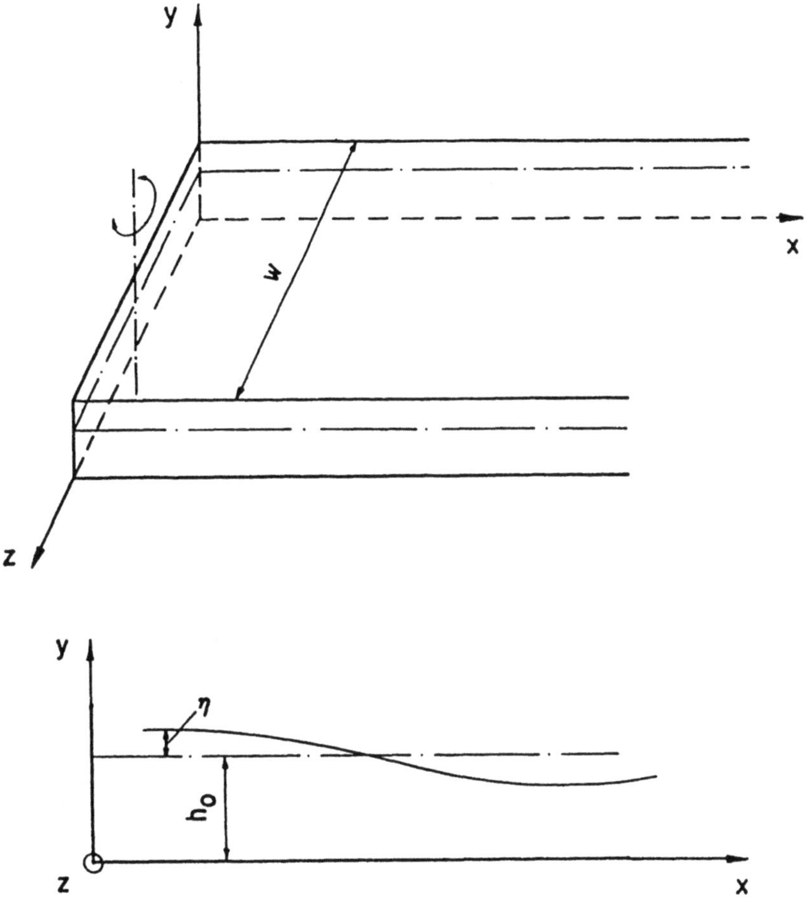


Figure 1. Semi-infinite channel with wavemaker.

$$u_{T\xi} + \lambda u_{\xi\xi} - \frac{3}{4}(u^2)_{\xi\xi} - \frac{1}{2}\left(\frac{1}{3} - \gamma\right) u_{\xi\xi\xi\xi} - \frac{1}{2} u_{XX} = 0 \qquad (0 \le \xi \le 2, X > 0, T > 0) \tag{6}$$

$$u_X = \frac{\pi}{2} \sin \pi\xi \qquad (X = 0, T > 0, 0 \le \xi \le 2) \tag{7}$$

$$u(\xi; X, T = 0) = 0 \qquad (0 \le \xi \le 2, X > 0, T > 0)\ . \tag{8}$$

Here $X, T$ are slow streamwise and time scales, $X = \epsilon^{1/3}x$, $T = \epsilon^{2/3}t$, and $\gamma = \sigma/\rho g h^2$, $\sigma$ being the surface tension, $g$ the gravitational acceleration and $\rho$ the fluid density. Note that $\xi, \varsigma$ in (4) are the characteristic variables of the classical wave equation; in view of the periodicity of $u$ in $\xi, \varsigma$, expression (3) represents a standing (but not sinusoidal in general) wave across the channel, which is slowly modulated along the channel and in time. The initial-boundary-value problem (6)-(8) is solved numerically by the pseudospectral (FFT) method in $\xi$ and finite differences in $X, T$.

*Results*

Numerical calculations were first carried out assuming $\gamma = 0$. In this case, linear theory predicts evanescent waves for $\lambda < -\pi^2/6$ and propagating waves for $\lambda > -\pi^2/6$. Exactly at $\lambda = -\pi^2/6$ linear theory predicts no steady state (the response grows like $T^{1/2}$). The results of nonlinear theory are summarized in Fig. 2 in terms of the steady-state power input as a function of $\lambda$. Note that at $\lambda = -\pi^2/6$ an evanescent nonlinear steady state is reached whereas at $\lambda \approx -1.4716$ a jump occurs in the response. This response is analogous to the response of a deep-water channel near a cut-off frequency, where nonlinear steady state is reached [1].

We also considered the case $\gamma > 0$, in particular $\gamma = 2/3$, in which case the linear cut-off occurs at $\lambda = \pi^2/6$. Here it is found that the nonlinear response does not reach steady state and the power input remains oscillatory for $T$ large. Closer examination of the solution reveals that, for excitation frequencies above the linear cut-off, the response ultimately consists of a series of individual groups, as the generated unsteady disturbance propagates away from the wavemaker. Exactly at linear cut-off, only one such group is generated, while below linear cut-off the response is entirely localized close to the wavemaker but remains unsteady. Further details concerning the methods of analysis and the results can be found in Kantzios & Akylas [4].

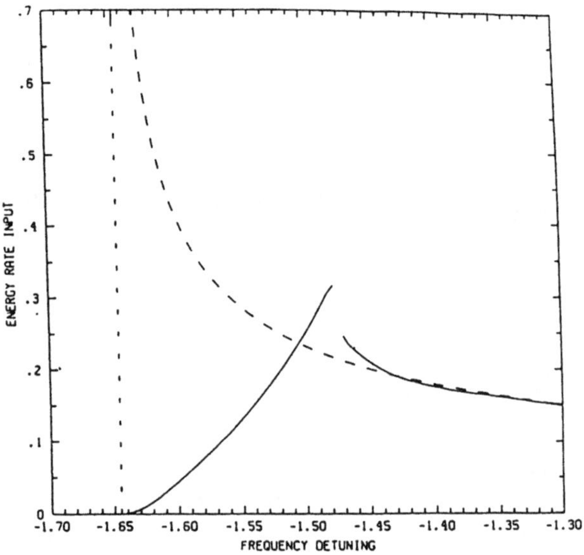


Figure 2. Steady-state power input as a function of frequency detuning $\lambda$ for $\gamma = 0$.

*References*


1. Aranha, J.A.; Yue, D.K.P.; Mei, C.C.: Nonlinear waves near a cut-off frequency in an acoustic duct - a numerical study. *J. Fluid Mech.* **121** (1982) 465–485.
2. Barnard, B.J.S.; Mahony, J.J.; Pritchard, W.G.: The excitation of surface waves near a cut-off frequency. *Phil. Trans. R. Soc. Lond.* **A 286** (1977) 87–123.
3. Kit, E.; Miloh, T.; Shemer, L.: Nonlinear wavemaker in a channel operating at cut-off frequencies: theory and experiment. *Proc. 16th Symp. Naval Hydrodyn.* 1986.
4. Kantzios, Y.D.; Akylas, T.R.: Long nonlinear water waves in a channel near a cut-off frequency. *Stud. Appl. Math.* to appear.

# Precursor Solitons Generated by Three-dimensional Disturbances Moving in a Channel[1]

D. M. WU
Harbin Shipbuilding Engineering Institute, Harbin CHINA
T. Y. WU
California Institute of Technology, Pasadena, California USA

## Introduction

We investigate the phenomenon of upstream-propagating solitons generated periodically by three-dimensional forcing disturbances moving steadily with a transcritical velocity in a nonlinear dispersive medium tranversely bounded. The two-dimensional case of this phenomenon was described numerically first for forced motions in shallow water by Wu & Wu (1982). Experimentally, this phenomenon was also found by Huang et al.(1982) for ship models moving in a towing tank, and more recently by Lee (1985) for a single layer of homogeneous water and by Zhu (1986) for a layer of stratified liquid with a two- dimensional cambered topography moving along the horizontal floor of the water layer. In principle, such a phenomenon can occur in all soliton-bearing systems of various physical and biological contexts. A review of the literature can be found in Wu (1987).

Recent attempts to extend two-dimensional theoretical models to a three-dimensional formulation include the works by Mei (1986) who introduced an inhomogeneous KdV equation to analyze a slender ship hull, by Ertekin et al. (1986) who used the Green-Naghdi directed-sheet model, and by Akylas & Katsis (1987) who employed the Kadomtsev-Pet-viashivili (KP) equation. While these models all predict periodic generation of upstream- advancing two-dimensional solitons by three-dimensional disturbances, their predicted three- dimensional wave patterns behind the disturbance appear to have considerable differences.

[1]This work was jointly sponsored by the U.S. Office of Naval Research and the National Science Foundation. One of us, DMW acknowledges the support by the China State Shipbuilding Corporation.

K. Horikawa, H. Maruo (Eds.)
Nonlinear Water Waves
IUTAM Symposium Tokyo/Japan 1987

This study examines in particular the three-dimensional aspects of forcing distributions, including the three- dimensionality of the pattern of the trailing waves generated by the forcing function. For theoretical calculation we apply the generalized Boussinesq (GB) model introduced by Wu (1979, 1981) to obtain numerical results. These results are analyzed to assess the theoretical prediction of precursor solitons and their period of generation, to enlighten the physical mechanism, and to make relevant comparisons.

## The Generalized Boussinesq Model and Numerical Method

The weakly-nonlinear, weakly-dispersive and weakly-forced three- dimensional long waves are modelled here by the GB model:

$$\varsigma_t + (h+\varsigma)(\phi_{xx}+\phi_{yy}) + \varsigma_x\phi_x + \varsigma_y\phi_y = 0, \tag{1}$$

$$\phi_t + \frac{1}{2}(\phi_x^2+\phi_y^2) + g\varsigma - \frac{h^2}{3}(\phi_{xxt}+\phi_{yyt}) = -p_a/\rho, \tag{2}$$

where the subscripts $x$, $y$, and $t$ denote partial differentiation, $p_a(x,y,t)$ is a free-surface pressure given as a function of two horizontal coordinates $(x,y)$ and time $t$, acting over the undetermined free surface $z=\varsigma(x,y,t)$ of a layer of water of constant density $\rho$. The water was initially at rest under gravity (of acceleration $g$ in the negative $z$- direction) occupying a layer of uniform depth $h$. Further, $\phi$ is the water-layer depth-averaged velocity potential of the fluid so that the $x$- and $y$-components of the layer-averaged flow velocities are $u=\partial\phi/\partial x$, $v=\partial\phi/\partial y$ . This model is related to the Boussinesq family by being characterized by the parameters:

$$\alpha = a/h \ll 1, \quad \epsilon = (h/\lambda)^2 \ll 1, \quad \text{and} \quad \alpha = O(\epsilon), \tag{3}$$

$$|\nabla p_a|/\rho g, \quad |\partial p_a/\partial t|/\rho g c_o \le O(\alpha\epsilon^{\frac{1}{2}}), \tag{4}$$

where $c_o=\sqrt{gh}$. With such forcing functions, (1) is exact while (2) has an error of $O(\alpha\epsilon,\epsilon^2)$ relative to its leading term.

The present initial-boundary value problem considers specifically (as a typical case) the free-surface pressure which for $t<0$ was at hydrostatic equilibrium with the fluid system at rest, with the forcing distribution (symmetrical about $x=0$ and $y=0$) given by

$$p_a(x,y) = P_m P_1(x) P_2(y) \tag{5}$$

$$\begin{aligned} P_1(x) &= \frac{1}{2}\{1+\sin[(L-x)\pi/2\ell]\} \quad (|x-L| \le \ell) \\ &= 1 \qquad\qquad (0 \le x \le L-\ell), \\ P_2(y) &= \frac{1}{2}\{1+\sin[(B-y)\pi/2b]\} \quad (|y-B| \le b) \\ &= 1 \qquad\qquad (0 \le y \le B-b), \end{aligned}$$

and $P_1(-x) = P_1(x)$, $P_2(-y) = P_2(y)$ and $P_a(x,y)$ vanishes for $|x| > L+\ell$, or $|y| > B+b$. Impulsively after $t = 0$, $P_a(x,y)$ moves with a steady transcritical velocity $U$ in the negative $x$-direction, maintaining the same distribution as (5) in the force frame, corresponding to the depth Froude number $F = U/c_o$ . Henceforth, the length, velocity and time will be scaled by h, $c_o$ and $h/c_o$, while the pressure, by $\rho gh$. The channel is bounded by side walls at $y = \pm W$ ($W > B+b$) and is unbounded in the $x$-direction. Occurrence of this phenomenon of resonance with transcritical forcing has been noted (see also Huang et al., 1982; Mei, 1986) to depend critically on the presence of side walls, which introduce the boundary condition $\partial\phi/\partial y = 0$ ($y = \pm W$).

The generation and evolution of the wave field is associated with a wave-making resistance which in coefficient form is

$$C_{DW} = \frac{D}{(\rho g h_o LB)} = \frac{-1}{\rho g h_o LB}\int\int p_a(x,y,t)\frac{\partial\varsigma}{\partial x}dxdy. \tag{6}$$

The numerical method employed for computation of the above set of basic equations is based on original version established by Wu & Wu (1982) but is further extended here. It is based on the modified Euler's predictor-corrector algorithm in advancing time and on the central difference formalism for the space derivatives. A mixed implicit-explicit scheme is adopted for the time forward difference computation of $\varsigma$ and $\phi$, the implicit part being incorporated in order to achieve the desired numerical stability and accuracy with relatively large time step $\Delta t$ (of order 0.1).

## Results

The salient features of the numerical results within a transcritical range are exhibited by the following specific case:

$$F = 1,\ P_m = 0.25,\ L = 1,\ B = 0.5,\ \ell = 0.4, \tag{7}$$

$$b = 0.4,\ \Delta x = \Delta y = 0.4,\ \Delta t = 0.2,\ W = 2,$$

where $\Delta x$, $\Delta y$, and $\Delta t$ are the grid sizes in space and time.

As shown in figure 1, generation of the upstream-advancing two- dimensional solitons (spanning across the water tank) and the down-stream extending three- dimensional waves behind the forcing region (the forcing distribution is depicted by the hydrostatical response, shown with rectangular contours at time $t = 0$) can be clearly seen from the time sequence of computer plots of the free-surface elevation in perspective. The process of growth and eventual breaking away of each forward advancing wave from the forcing agency to become a free soliton is also quantitatively indicated by the corresponding result of the wave resistance coefficient shown in figure 3. As the coefficient $C_{DW}$ reaches its first maximum at $t = 34.4$ and its second at $t = 80.4$, we see from the wave elevation graphics that the first precursor soliton is just emerging at $t = 36$, and at $t = 80$ the second is about to break away. Thus using the period of $C_{DW}$ variation as the standard, we have for the case of (7) the soliton generation period as $T_S = 46$.

The trailing wave system has a conspicuously three- dimensional wave pattern, as shown in figure 1 and in figure 2 for the free-surface relief contours (in which the hydrostatical response is superimposed in dashed lines). Its configuration will be examined in detail to reveal both the depth effect and the side-wall effect. The latter is especially important in exerting influence on the generation period of precursor solitons as well as on having the trailing wave system modified in amplitude and phase distributions. An expository study will be made on comparisons between three-dimensional and the related two-dimensional cases.

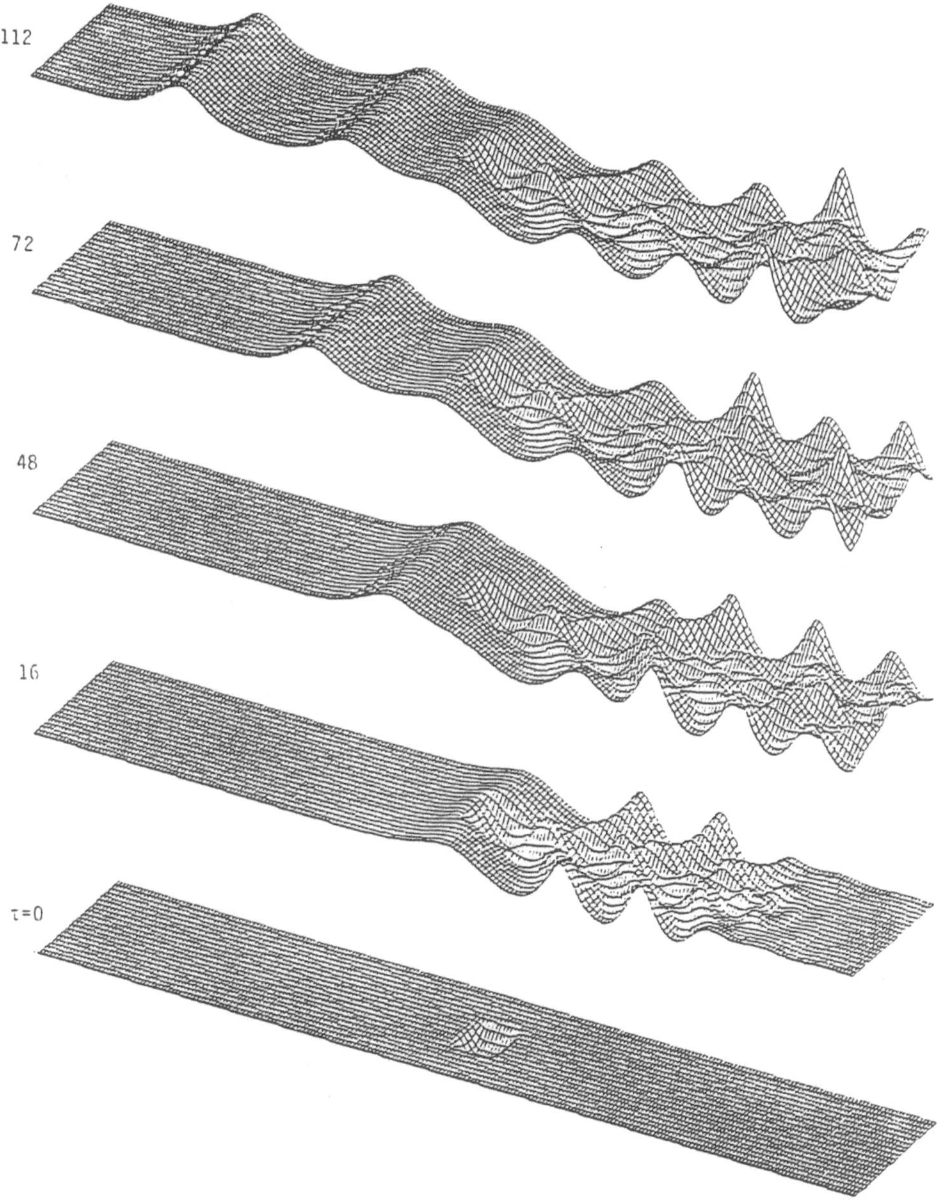


Figure 1. Time sequence of computer plots of the perspective free-suface elevation generated by a pressure distribution given by eq.(5) (and depicted by the hydrostatical response at time $t = 0$), moving in a water tank with the critical velocity $U = \sqrt{gh} = 1$ towards the left, but shown here in the body frame.

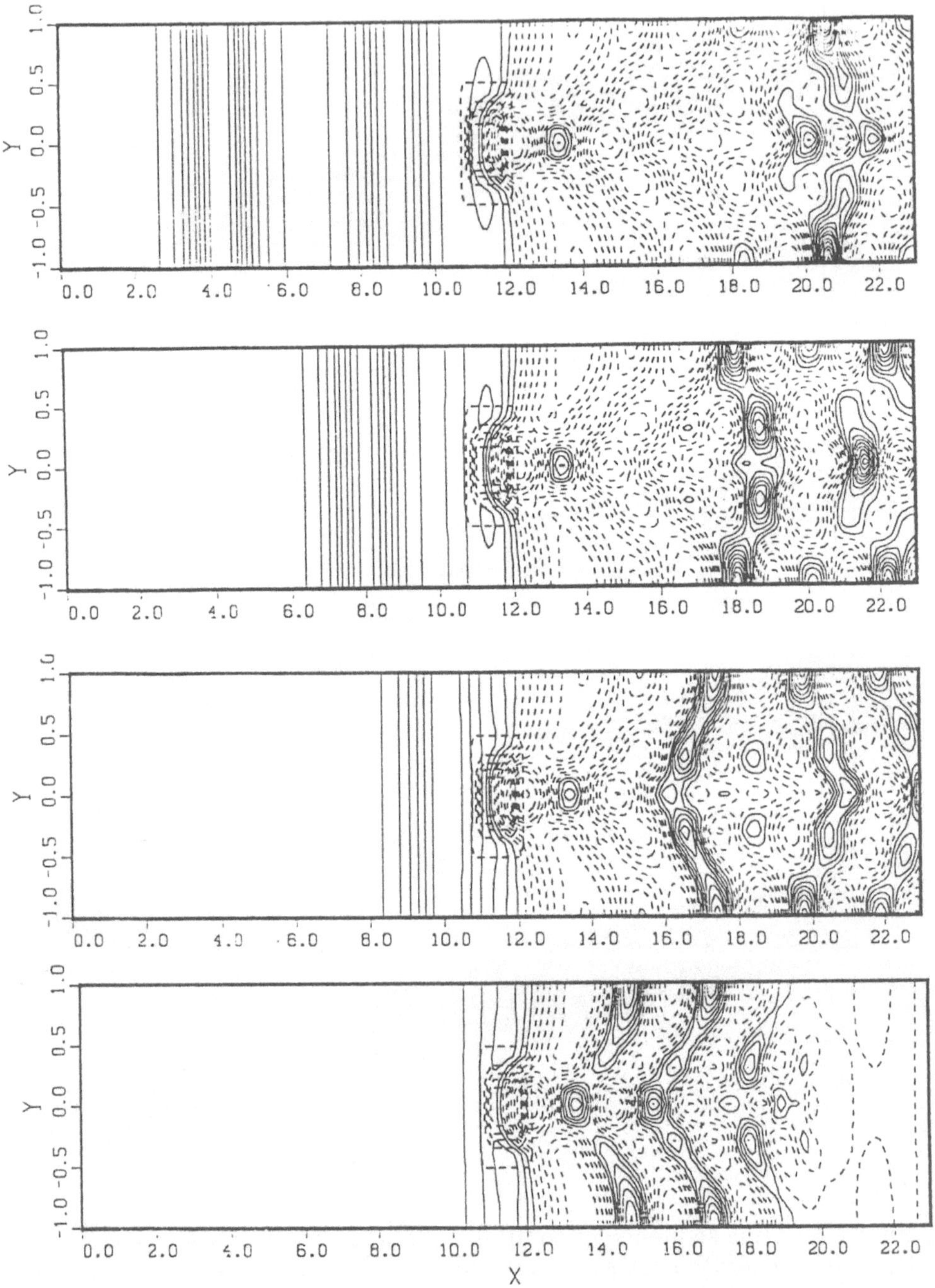


Figure 2. The free surface relief contours corresponding to the wave field shown in figure 1 for $t = 16$, 48, 72, and 112. Constant values of $\varsigma$ are shown with solid (or dashed) lines for positive (or negative) $\varsigma$, level changes being 0.05 between consecutive contours.

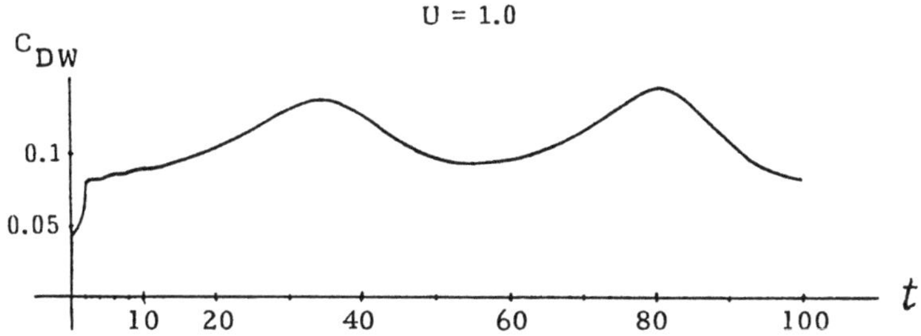


Figure 3. Variation of the wave-resistance coefficent $C_{DW}$ with time $t$ for the case corresponding to that shown in figure 1.

## References

Akylas, T.R. & Katsis, C. 1987 Manuscript

Ertekin, R.C., Webster, W.C. & Wehausen, J.V. 1986 J. Fluid Mech. **169** 275-292.

Huang, D.D., Sibul, O.J., Webster, W.C., Wehausen, J.V., Wu, D.M. & Wu, T.Y. 1982 Proc. Conf. Behavior of Ships in Restricted Waters, II 26. Varna: Bulgarian Shiphydrodynamics Centre.

Lee, S.J. 1985 Ph.D. Thesis, Calif. Inst. Tech, Pasadena, Calif. USA

Mei, C.C. 1986 J. Fluid Mech. **162**, 53-67.

Wu, D.M. & Wu, T.Y. 1982 Proc. 14th Symp. on Naval Hydrodynamics, 103-125. National Academy of Sciences, Washington, D.C. USA

Wu, T.Y. 1981 ASCE J. Engng. Mech. Div.**107**, 501-522.

Wu, T.Y. 1987 J.Fluid Mech. (To appear.)

Zhu, J.L. 1986 Ph.D. Thesis, Calif. Inst. Tech. Pasadena, Calif, USA

# Asymptotic Behavior of a Shallow-Water Soliton Reflected at a Sloping Beach

N. SUGIMOTO and T. KAKUTANI

Department of Mechanical Engineering
Faculty of Engineering Science
Osaka University
Toyonaka, Osaka 560
Japan

## Summary

This paper deals with a long-time asymptotic behavior of a shallow water soliton reflected at a sloping beach. Based on the edge-layer theory, it is shown that the boundary-value problem for the Boussinesq equation under the "reduced" boundary condition is simplified to an "initial value" problem for the Korteweg-de Vries equation in the form of the spatial evolution of the reflected wave. Solving it numerically, the asymptotic behavior is demonstrated and compared with the one which evolves from the "initial value" obtained by solving the full nearshore behavior by the boundary element method.

## Introduction

When a shallow-water soliton is reflected by a sloping beach, how will it evolve eventually? Since the soliton must be deformed by reflection, its resulting long-time behavior is an interesting problem to ask. Up to the present, however, this question does not seem to be clearly answered.

The present report considers this problem based on the edge-layer theory developed in a previous paper [1]. The theory assumes that when a soliton is incident upon a "steep" beach whose horizontal depth is small compared with a characteristic wavelength, the effect of the beach is confined within its vicinity called an "edge-layer", so that the assumption of shallow-water is again valid outside this narrow region.

In the edge-layer, it is assumed that the vertical acceleration of fluid becomes as significant as the horizontal one, while the nonlinearity due to shoaling is still negligible because

K. Horikawa, H. Maruo (Eds.)
Nonlinear Water Waves
IUTAM Symposium Tokyo/Japan 1987

the beach is too steep to accumulate the nonlinearity. Applying the matched asymptotic expansion method, one can derive a "reduced" boundary condition relevant to the Boussinesq equation, which can now be solved without paying any attention to a complicated nearshore behavior.

The Boussinesq equation under the boundary condition thus derived has been solved over a short time span after reflection [2]. We are concerned here with a long-time asymptotic behavior of a reflected soliton. For this purpose, it is more convenient to simplify the Boussinesq equation by the uni-directional Korteweg-de Vries (K-dV) equation focusing our attention on the reflected wave. It is shown that the boundary-value problem for the Boussinesq equation is reduced to an "initial value" problem for the K-dV equation in the form of spatial evolution. Then the "reduced" boundary condition gives a differential equation to determine this "initial value". By solving the K-dV equation numerically, the asymptotic behavior of the reflected soliton is investigated. The result is compared, as a check, with the one which evolves from the "initial value" obtained by solving the full nearshore behavior, such as the surging movement along the beach surface, by the boundary element method [2].

## Method of Analysis

Let a semi-infinite water-layer of uniform depth h be bounded by a plane beach inclined by $\theta$ to the horizontal. Imagine a shallow-water soliton be incident upon the beach and reflected back seaward. We assume that the bi-directional propagation of the soliton is described by the Boussinesq equation for the velocity potential $f(x,t)$:

$$f_{tt}-f_{xx}-\frac{\beta}{3}f_{xxtt}=-\alpha(f_t f_{xx}+2f_x f_{xt})+O(\alpha^2,\alpha\beta,\beta^2), \qquad (1)$$

where x and t denote the space and time variables normalized, respectively, by a characteristic wavelength $\ell$ and a characteristic time $\ell/(gh)^{\frac{1}{2}}$ (g is the gravity constant), while f is normalized by $ag\ell/(gh)^{\frac{1}{2}}$ (a is a characteristic elevation). Suffixes imply the partial differentiation. Two dimensionless small parameters $\alpha$ (=a/h) and $\beta$ (=$(h/\ell)^2$ ) measure, respectively, the weakness of nonlinearity and that of dispersion ($\alpha \sim \beta \ll 1$). In terms of f, the surface elevation $\eta$ , normalized by a, is given by

$$\eta = -f_t - \frac{\alpha}{2} f_x^2 + \frac{\beta}{2} f_{ttt} + O(\alpha\beta, \beta^2). \tag{2}$$

According to the edge-layer theory, the "reduced" boundary condition, which reflects the presence of the "steep" beach, is given by

$$f_x = \mu f_{xx} - \frac{\mu^2}{2} f_{xxx} + \frac{\mu^3}{6} f_{xxxx} + O(\mu^5, \beta) \quad \text{at } x=0, \tag{3}$$

with $\mu = \beta^{\frac{1}{2}} \cot\theta$ . It should be remarked here that the condition (3) can be imposed at x = 0, although the actual beach extends over the region $-\mu < x < 0$ on the left side of the uniform depth region x>0. This is, indeed, the very result of the edge-layer theory.

In order to pursue a long-time behavior of the reflected soliton, it is more convenient and efficient to simplify the Boussinesq equation by the K-dV equation. To single out the unidirectional propagation, f is assumed to be in a form of superposition of the incident wave $F^-$ and the reflected wave $F^+$ within the lowest order of approximation with respect to $\alpha$ and $\beta$ [3]:

$$f = F^+(\sigma^+, \tau) + F^-(\sigma^-, \tau) + O(\alpha, \beta), \tag{4}$$

where $\sigma^{\pm}$ (=$t \mp x$) ($\pm$ signs correspond vertically) measure the time in each moving frame and $\tau$ (=$\alpha x$) is the "long-space" variable. Introducing (4) into (1) and neglecting terms of $O(\alpha, \beta)$, two K-dV equations are derived for the surface elevation $\eta^-$ of the incident wave and $\eta^+$ of the reflected wave, respectively:

$$\frac{\partial\eta^{\pm}}{\partial\tau} \mp \frac{3}{2}\eta^{\pm}\frac{\partial\eta^{\pm}}{\partial\sigma^{\pm}} \mp \frac{\beta}{6\alpha}\frac{\partial^3\eta^{\pm}}{\partial\sigma^{\pm 3}} = 0, \tag{5}$$

where $\eta^{\pm} = -F^{\pm}_{\sigma^{\pm}} + O(\alpha, \beta)$.

On the other hand, the boundary condition (3) is also approximated in accordance with (4). Substitution of (4) into (3) yields the following differential equation between $\eta^+$ and $\eta^-$ with respect to t:

$$L_\mu \eta^+ = L_{-\mu} \eta^-, \tag{6a}$$

with the operator $L_\mu$ defined by

$$L_\mu = \frac{\mu^3}{6}\frac{d^3}{dt^3} + \frac{\mu^2}{2}\frac{d^2}{dt^2} + \mu\frac{d}{dt} + 1, \tag{6b}$$

where note that $\sigma^+=\sigma^-=t$ at $x=0$.

For the given boundary-value due to the incident soliton:

$$\eta^- = \mathrm{sech}^2[D(x+\lambda t-C)] \quad \text{at } x=0, \tag{7}$$

where $\lambda = 1+\alpha/2$, $D=[3\alpha/(4\beta\lambda^2)]^{1/2}$, and C is the position of the peak at t=0, the solution of (6) provides the boundary value of the reflected wave $\eta^+$ at x=0. If (6) with (7) is expressed in terms of a new variable $\chi \equiv Dt$, it is seen that $\eta^+$ depends on $\mu$ only through the combination $\mu D \doteqdot (3^{\frac{1}{2}}/2)\alpha^{\frac{1}{2}}\cot\theta$ . Thus the reflected wave is solely determined by the parameter $\alpha^{\frac{1}{2}}\cot\theta \equiv \mu_\alpha$ , which is nothing but the one that Pedersen and Gjevik claimed as the crucial parameter to determine the run-up height at the beach [4]. But it should be remembered that since the ratio $\alpha/\beta$ is fixed as the Ursell parameter $U_r$, the relation between $\mu_\alpha$ and $\mu$ is simply given by $\mu_\alpha = U_r^{\frac{1}{2}}\mu$.

Since the boundary x=0 corresponds to $\tau=0$, the solution $\eta^+$ to (6) with (7) plays a role of the "initial value" of the K-dV equation for the reflected wave. Using the well-known explicit finite difference scheme [5], this "initial-value" problem is solved numerically to pursue a long-time asymptotic behavior of the reflected soliton.

Results

Numerical computations are carried out for some typical values of the beach angle $\theta$ and the nonlinear parameter $\alpha$. The other parameter $\beta$ is chosen with the Ursell parameter $U_r$ fixed as $U_r$=9.25 (for the reason of this choice, see the reference [2]).

The results of computation are displayed in Figs. 1 and 2 for

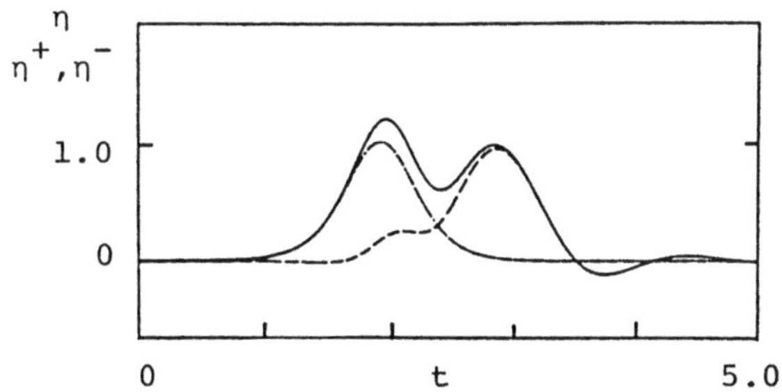


Fig. 1. Boundary-values of the surface elevation.
———— ; total wave $\eta$,
— - — - — ; incident wave $\eta^-$,
- - - - - - ; reflected wave $\eta^+$.

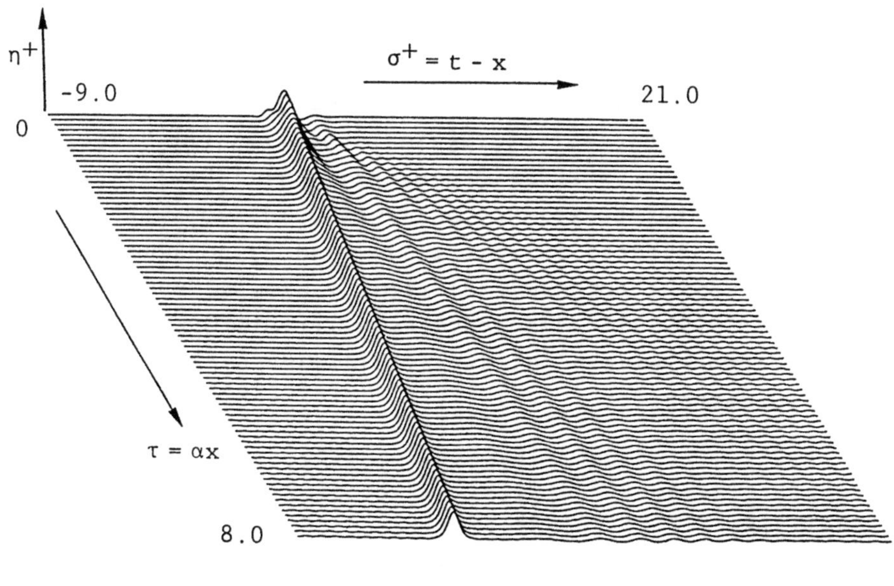


Fig. 2. Long-time evolution of the reflected soliton.

an example with $\theta = \pi/15$ and $\alpha = 0.07$. Figure 1 shows the boundary-value of the surface elevation at x=0. The solid line represents the total elevation $\eta$, while the chain and broken lines represent, respectively, the elevation of the incident wave $\eta^-$ and that of the reflected wave $\eta^+$. It is seen that the reflected wave lags behind the incident wave, which explains the time interval elapsed during reflection at the "actual" beach. It is also observed that the waveform of the reflected wave is considerably distorted from the sech$^2$-function, but the peak of the reflected wave is only slightly less than that of the incident one (94%).

The long-time evolution of $\eta^+$ from the boundary-value is shown in Fig. 2. The reflected wave evolves into one soliton accompanied by an oscillating tail. During the early times up to $\tau = 2\sim3$, the "initial" waveform appears to evolve into a series of solitons. For the times later, however, only one sharp soliton emerges, leaving the oscillating tail behind. At $\tau = 8$, the peak of the surviving soliton is slightly less than the incident one (95%). As is well known, the tail decays eventually but very slowly. Thus the reflected wave reshapes its soliton-form asymptotically with small amplitude decay, so that the resultant effect of the "steep" beach yields only a slight reduction of the soliton amplitude. This may be understood as follows. Noting that the boundary condition (3) implies approximately $f_x=0$ at $x=-\mu$ by Taylor's theorem, the "steep" beach behaves effectively as if it were replaced by a vertical beach or wall indented to $x=-\mu$. In the case of vertical wall, the soliton is subjected only to a phase shift without any amplitude variation. But since (3) is of a form of truncated Taylor series at $O(\mu^4)$, the difference from the vertical wall distorts the reflected wave $\eta^+$ at x=0, which develops in a very complicated manner for $\tau \lesssim 2\sim3$.

## Discussion

To check the above results (though not a perfectly independent check as will be seen below), the parallel calculations combined with the boundary element method (BEM simply) are also made. Although the BEM solves the Laplace equation under the full nonlinear boundary conditions, the large amount of calculations restricts us from taking the shallow-water region wide enough to pursue a long-time behavior. The BEM seems to be useful, however, if it is combined with the idea of the edge-layer theory. That is, calculate first the surface elevation at x=0 by the BEM, and decompose it into the incident and ref-

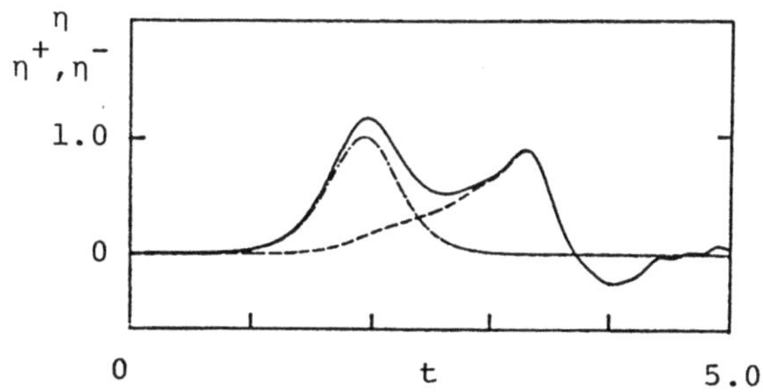


Fig. 3. Boundary-values of the surface elevation obtained by calculating the nearshore behavior by the BEM.
———— ; total wave $\eta$,
— - — - — ; incident wave $\eta^-$,
- - - - - - ; reflected wave $\eta^+$.

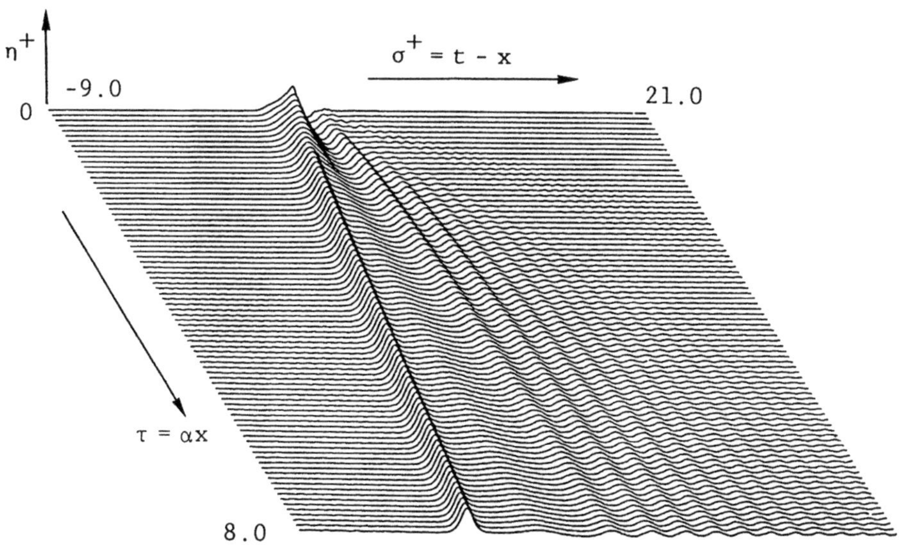


Fig. 4. Long-time evolution of the reflected wave from the "initial value" calculated by the BEM shown in Fig.3.

lected waves, then solve the K-dV equation for the reflected wave. An advantage of this method lies in taking account of a surging movement rigorously along the actual beach surface, which is beyond the scope of the edge-layer theory. Figure 3 shows the boundary-values at x=0 calculated by the BEM. It is seen that because the surging movement lengthens substantially the horizontal beach depth, the arrival of the peak of the reflected wave is delayed compared with the one shown in Fig. 1 and that the waveform is again distorted remarkably. Using these data as the "initial value", the long-time behavior of the reflected soliton is shown in Fig. 4. It is confirmed that the reflected wave evolves again into one sharp soliton with an oscillating tail, and the peak of the surviving soliton recovers 89% of the incident soliton, which is slightly smaller than that obtained in the preceding section. It is thus concluded from this analysis that the edge-layer theory provides a fairly good result.

The authors acknowledge the instrumental support of the computer graphics from Daikin Comtech. Inc., Osaka, Japan.

## References

1. Sugimoto, N.; Kakutani, T.: Reflection of a shallow-water soliton. Part 1. Edge-layer theory for shallow-water waves. J. Fluid Mech. 146 (1984) 369-382.

2. Sugimoto, N.; Kusaka, Y.; Kakutani, T.: Reflection of a shallow-water soliton. Part 2. Numerical evaluation. J. Fluid Mech. 178 (1987) 99-117.

3. Miles, J. W.: Obliquely interacting solitary waves. J. Fluid Mech. 79 (1977) 157-169.

4. Pedersen, G.; Gjevik, B.: Run-up of solitary waves. J. Fluid Mech. 135 (1983) 283-299.

5. Zabusky, N. J.; Kruskal, M. D.: Interaction of "solitons" in a collisionless plasma and the recurrence of initial state. Phys. Rev. Letts. 15 (1965) 240-243.

# Normal Form and Solitons in the Shallow Water Waves

Yuji KODAMA

*Department of Physics, Nagoya University*
*Chikusa-ku, Nagoya, 464 Japan*

*Department of Mathematics, The Ohio State University*
*Columbus, Ohio 43210 U.S.A.*

## 1. INTRODUCTION

It is well known that the shallow water waves can be described by the KdV equation in the first order approximation of an asymptotic perturbation expansion. The KdV equation admits a family of solitary waves called solitons having the remarkable property that the result of interaction of solitons leaves their shape unaltered, except a phase shift. This elastic interaction is due to the fact that the KdV equation possesses an infinite number of conserved quantities. (This is often to be a definition of the integrability.) However, for the problem related to large time behavior of the solutions or large amplitude solitary waves, one needs to study the effects of the higher order terms which are neglected in the derivation of the KdV equation.

In this report, we consider the higher order approximation of the KdV equation for the shallow water waves, and discuss the interaction of the solitary waves. We first derive the higher order corrections to the KdV equation. Then, we transform the higher order equation into a simpler equation called the normal form, and show that the normal form takes an integrable form up to second order, but not to third order in general. The integrable normal form up to third order is obtained for the special values of the surface tension. Finally, we study the interaction of two solitary waves, and show that the interaction is inelastic and leads to the production of new solitary wave as well as the radiations. The asymptotic theory developed here can be applied to the general weakly dispersive-nonlinear wave systems, and it shows that the interaction property holds commonly for the nonintegrable systems.

K. Horikawa, H. Maruo (Eds.)
Nonlinear Water Waves
IUTAM Symposium Tokyo/Japan 1987

## 2. THE HIGHER ORDER APPROXIMATION OF THE KDV EQUATIONS

We consider the inviscid fluid which is assumed to be both incompressible and irrotational. The basic equations with x and y-axes as the horizontal and vertical directions are given in the following demensionless form [1],

$$\left.\begin{array}{ll}\beta\phi_{xx}+\phi_{yy}=0\,, & \text{for}\quad 0<y<1+\eta(x,t)\,,\\ \phi_y=0\,, & \text{for}\quad y=0\ \text{(bottom)}\,,\\ \left.\begin{array}{l}\phi_t+\frac{1}{2}\alpha\phi_x^2+\frac{\alpha}{2\beta}\phi_y^2+\eta\\ \quad=\tau\beta\eta_{xx}(1+\alpha^2\beta\eta_x^2)^{-3/2}\,,\\ \eta_t+\alpha\eta_x\phi_x=\frac{1}{\beta}\phi_y\,.\end{array}\right\} & \text{for}\quad y=1+\eta(x,t)\ \text{(surface)}\,.\end{array}\right\} \tag{1}$$

where $\phi=\phi(x,y,t)$ and $\eta=\eta(x,t)$ represent the velocity potential and the height of the water surface from the equilibrium. Here the parameters $\alpha=a/h$, $\beta=h^2/\ell^2$ and $\tau=T/(gh^2)$ are the scales of the nonlinearity, the dispersion and the surface tension, respectively (h: depth, $\ell$: wavelength, g: gravitational constant, T: surface tension). In the theory of the KdV equation, we choose the balance $\alpha\sim\beta\sim O(\varepsilon)$ with a small parameter $\varepsilon$. From the first two equations in (1), $\phi$ can be written in

$$\phi(x,y,t) = \cos\left(y\sqrt{\beta}\,\frac{\partial}{\partial x}\right)f(x,t)\,. \tag{2}$$

Defining $w=f_x$, we have the following system of equations for $U=(w,\eta)^t$,

$$\frac{\partial}{\partial t}U + A(U)\frac{\partial}{\partial x}U + B(U) = 0\,, \tag{3}$$

where $A=A_0+A_1$ and $B=B_1+B_2+\cdots$, with $A_n\sim B_n\sim O(\varepsilon^n)$, are given by

$$\left.\begin{array}{l}A_0=\begin{pmatrix}0 & 1\\ 1 & 0\end{pmatrix}\,,\qquad A_1=\alpha\begin{pmatrix}w & 0\\ \eta & w\end{pmatrix}\\ B_1=\begin{pmatrix}\frac{1}{2}\beta(1-2\tau)\eta_{3x}\\ -\frac{1}{6}\beta w_{3x}\end{pmatrix}\qquad\text{and so on.}\end{array}\right\} \tag{4}$$

Here and hereafter we use the notation $\eta_{nx}=\partial^n\eta/\partial x^n$. In (3), we look for a solution corresponding to a unidirectional wave propagating in the positive x-axis, i.e. with the eigenvectors $R_\pm$ of $A_0$, $A_0R_\pm=\pm R_\pm$,

$$U(x,t) = u(x,t)R_+ + f(x,t)R_- \tag{5}$$

where u and f are scalar functions given by $u=w+\eta$ and $f=w-\eta$. Note that the unidirectional approximation implies $f\sim O(\varepsilon)$ and f can be expressed by u and its derivatives only. Then we obtain the following equation of u with an appropriate normalization for (x,t,u),

$$u_t + u_x + \varepsilon X_0^{(0)}(u) + \varepsilon^2 X^{(1)}(u) + \varepsilon^3 X^{(2)}(u) = O(\varepsilon^4) , \tag{6}$$

where the vector fields $X^{(n)}(u)$ are given by

$$\left.\begin{aligned}
X_0^{(0)}(u) &= u_{3x} + 6uu_x , \qquad (\mathrm{KdV}) , \\
X^{(1)}(u) &= a_1^{(1)}u_{5x} + a_2^{(1)}uu_{3x} + a_3^{(1)}u_xu_{2x} + a_4^{(1)}u^2u_x , \\
X^{(2)}(u) &= a_1^{(2)}u_{7x} + a_2^{(2)}uu_{5x} + a_3^{(2)}u_xu_{4x} + a_4^{(2)}u_{2x}u_{3x} \\
&\quad + a_5^{(2)}u^2u_{3x} + a_6^{(2)}uu_xu_{2x} + a_7^{(2)}u_x^3 + a_8^{(2)}u^3u_x .
\end{aligned}\right\} \tag{7}$$

The coefficients $a_\ell^{(n)}$ depend on the value of the surface tension, $a_\ell^{(n)}= a_\ell^{(n)}(\tau)$, e.g. $a_1^{(1)}=(19-30\tau-45\tau^2)/10(1-3\tau)^2$, $a_2^{(1)}=2(5-3\tau)/(1-3\tau)$, $a_3^{(1)}= 2(19-6\tau)/(1-3\tau)$ and $a_4^{(1)}=-3$. It should be noted that the coefficients $a_\ell^{(n)}$ are not determined uniquely due to the freedom of coordinate change, $u\to u+\varepsilon F(u)$, e.g. $F(u)=c_1u^2+c_2u_{2x}$. With regard to this freedom, we give, in the next section, the normal form which is a typical form for the equation (6).

## 3. NORMAL FORM

As was shown in ref.2, (6) can be transformed into an equation called the normal form given by

$$v_t + v_x + \varepsilon X_0^{(0)}(v) + \varepsilon^2 X_0^{(1)}(v) + \varepsilon^3 X_0^{(2)}(v) = O(\varepsilon^4) , \tag{8}$$

where the transformed vector fields $X_0^{(n)}(v)$ are

$$X_0^{(1)}(v) = a_1^{(1)} \partial_x \nabla I_3^{(0)}(v) \ ,$$
$$X_0^{(2)}(v) = a_1^{(2)} \partial_x \nabla I_4^{(0)}(v) + \mu \partial_x R^{(2)}(v) \ . \tag{9}$$

Here $I_\ell^{(0)}[v]$ are the conserved quantities of the KdV equation, i.e. $I_0^{(0)} = \frac{1}{2}\int v dx$, $I_1^{(0)} = \frac{1}{2}\int v^2 dx$, $I_2^{(0)} = \frac{1}{2}\int (2v^3 - v_x^2) dx$, $I_3^{(0)} = \frac{1}{2}\int (v_{2x}^2 - 10vv_x^2 + 5v^4) dx$, etc. (These are generated by the recurrent formula [2].) $\nabla I_\ell^{(0)}(v)$ are their gradient (the functional derivative) with respect to v. $\mu = \mu(a^{(1)}, a^{(2)})$ is the constant given by

$$\mu = \sum_{\ell=1}^{8} r_\ell a_\ell^{(2)} + f(a^{(1)}) \tag{10}$$

with $r = (-560, 170, -60, 24, -8, -9, 24, 3)$ and $f(a^{(1)}) = [10a_1^{(1)}(240a_1^{(1)} - 67a_2^{(1)} + 6a_4^{(1)}) + a_2^{(1)}(4a_2^{(1)} + 30a_3^{(1)} + a_4^{(1)}) - 3a_3^{(1)}(2a_3^{(1)} + a_4^{(1)})]/3$. $R^{(2)}(v)$ is given by $R^{(2)}(v) = -(2vv_{4x} + v_x v_{3x} - 3v_{2x}^2 + 10v^2 v_{2x} + 10vv_x^2 + 5v^4)/100$. The transformation $u = T_\varepsilon(v)$ is given by the Lie transformation [2],

$$u = v + \varepsilon\phi^{(1)}(v) + \varepsilon^2(\phi^{(2)} + \frac{1}{2}\phi^{(1)} \cdot \nabla\phi^{(1)})(v) + O(\varepsilon^3) \ , \tag{11}$$

with the generating functions,

$$\left.\begin{aligned} \phi^{(1)}(v) &= \alpha_1^{(1)} v_{2x} + \alpha_2^{(1)} v^2 + \alpha_3^{(1)} v_x \int_x^\infty v dx \ , \\ \phi^{(2)}(v) &= \alpha_1^{(2)} v_{4x} + \alpha_2^{(2)} vv_{2x} + \alpha_3^{(2)} v_x^2 + \alpha_4^{(2)} v^3 \\ &\quad + \alpha_5^{(2)} v_x \int_x^\infty v^2 dx + \alpha_6^{(2)} X_0^{(0)}(v) \int_x^\infty v dx \ , \end{aligned}\right\} \tag{12}$$

where the constants $\alpha^{(1)} = (\alpha_1^{(1)}, \cdots, \alpha_3^{(1)})$ and $\alpha^{(2)} = (\alpha_1^{(2)}, \cdots, \alpha_6^{(2)})$ are determined by $a^{(1)}$ and $a^{(2)}$ in (6). The normal form defined by (8) with (9) is uniquely determined from (6) (i.e.(8) gives a classification of the equations (6)). Note that the terms $\partial_x \nabla I_\ell^{(0)}(v)$ are the integrable equations called the Lax-hierarchy. Thus (6) can be transformed into an integrable equation up to second order, and when $\mu=0$ corresponding to the values of the surface tension, $\tau=0.164$, $0.377$ and $0.562$, the normal form becomes integrable up to third order. (Note that for $\tau=1/3$, the lowest order dispersion vanishes, and (6) is not valid.) We also note that the normal form (8) has three conserved quantities corresponding to the mass, energy and Hamiltonian, and these conserved quantities, say $J_\ell[v;\varepsilon]$, can be written in terms of these of the KdV equation, i.e.

$$J_\ell[v;\varepsilon] = \sum_{n=0} \varepsilon^n C_\ell^{(n)} I_{\ell+n}^{(0)}[v] \quad \text{for} \quad \ell = 0,1,2 \quad . \tag{13}$$

where the constants $C_\ell^{(n)}$ are determined from the transformation (11). Note that $J_\ell$ in (13) exists only for $\ell$=0,1,2, and if $J_4$ exists up to $\varepsilon^2$ then $\mu$=0 in (9) [2]. From this fact, the normal form admits a solitary wave in the form of the KdV soliton, $v(x,t)=s(x-\lambda t)=2\kappa^2\operatorname{sech}^2\kappa(x-\lambda t)$ (note that $R^{(2)}(s)=0$ in (9)). From the transformation (11), the solitary wave solution of (6) can be written in the form,

$$u(x,t) = \sum_{n=1} \sum_{j=1}^{n} \varepsilon^{n-1}\beta_j^{(n-1)}\kappa^{2n}\operatorname{sech}^{2j}\kappa(x-\lambda t) \quad . \tag{14}$$

This form has been found by Fenton [3] in the study of the greatest solitary waves of the shallow water waves.

## 4. SOLITARY-WAVE INTERACTION

We now study the interaction of the solitary waves of the normal form (8) by employing the method of the inverse scattering transform (IST) [4]. Using the IST method, the conserved quantities (13) can be expressed in terms of the scattering data which give the parameters of the solitary waves, i.e.

$$J_\ell[v;\varepsilon] \propto \left(\sum_{j=1}^{M} \kappa_j^{2\ell+1} - (-1)^\ell\Delta_\ell\right) + o(\Delta_\ell) \tag{15}$$

where $\kappa_j$ ($j=1,\cdots,M$) are the soliton parameters giving the amplitude $2\kappa_j^2$, and $\Delta_\ell \geq 0$ corresponds to the radiation level [5]. As an initial condition, we consider two solitary waves with the amplitudes $2\kappa_\ell^2$ for $\ell$=1,2 ($\kappa_1>\kappa_2$), approaching each other. Then from the conserved quantities (15), we obtain the shift of the soliton parameters, $\Delta\kappa_1$, $\Delta\kappa_2$ and $\Delta\kappa_3$ ($\kappa_3$=0 at initial) for new solitary waves as a result of the interaction. These shifts are due to the emission of the radiation, $\Delta_\ell \neq 0$ in (15), which can be calculated from the perturbation method based on the IST method [4]. The radiations are generated from the nonintegrable term $R^{(2)}(v)$ in (9).

The results are as follows:

i) $\Delta\kappa_1>0$, $\Delta\kappa_2<0$, that is, the amplitude of the larger solitary wave is indreased, and contrary that of the smaller one is decreased by the interaction. The amount of the shift has the property $\Delta\kappa_1<|\Delta\kappa_2|$, i.e. the energy expense of the smaller solitary wave is more than the energy gain of the larger one.

ii) $\Delta\kappa_3>0$; this implies a production of new solitary wave.

The graph in Fig.1 shows the shifts of the soliton parameters $\Delta\kappa_1(t)$ and $\Delta\kappa_2(t)$. (The detail of the calculation is given in [5].)

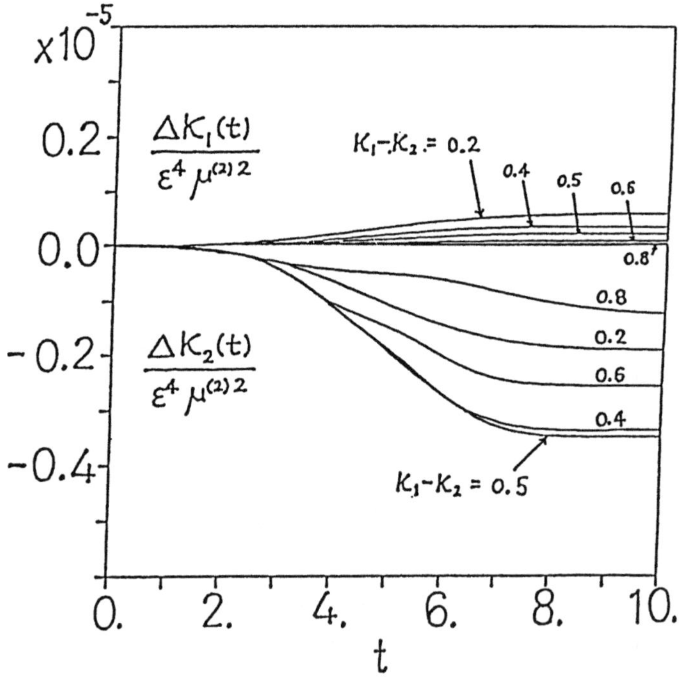


Fig.1 The time evolutions of $\Delta\kappa_1(t)$ and $\Delta\kappa_2(t)$ for $\kappa_1-\kappa_2=0.2$, 0.4, 0.6, and 0.8 ($\kappa_1+\kappa_2=1$).

These results are consistent with the numerical observation done by Bona et al. [6] for the regularized long wave equation. It should be noted that these interaction properties hold commonly for the nonintegrable systems. Fig.2 shows the behavior of two solitary waves through their collision.

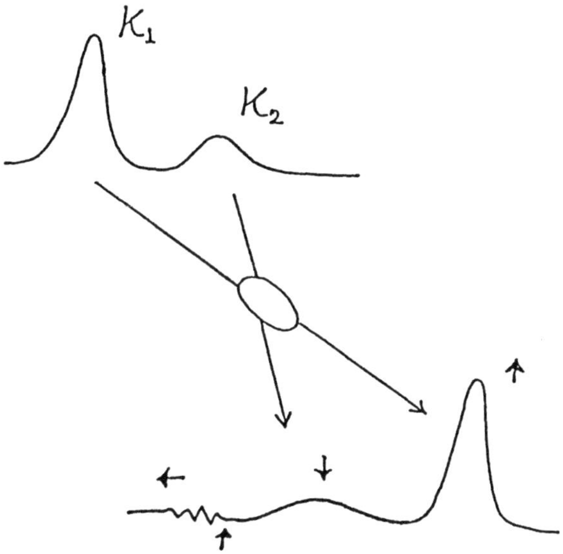


Fig.2 Illustration of Solitary-Wave Interaction

ACKNOWLEDGEMENTS

This research has been supported in part by the NSF grant, No.DMS 8521055.

REFERENCES

1. G.B. Whitham: Linear and Nonlinear Waves (Wiley-Interscience, 1974).
2. Y. Kodama: Physica 16D, (1985) 14; Phys. Lett. 112A, (1985) 193.
3. J.D. Fenton: J. Fluid Mech. 53, (1972) 257.
4. A.C. Newell: Solitons in Mathematics and Physics (SIAM, 1985).
5. Y. Kodama: Normal Form and Solitons, in Topics in Soliton Theory and Exactly Solvable Non-linear Equations, (World Science Publ. Co. 1987); Phys. Lett. 123A (1987) 226.
6. J.L. Bona, W.G. Pritchard and L.R. Scott: Phys. Fluids 23, (1980) 438.

# Energy Distribution of Shallow Water Swell under the Maximum Probability Condition

T. YASUDA and T. MISHIMA
Department of Civil Engineering Gifu University, Gifu

Y. TSUCHIYA
Disaster Prevention Research Institute Kyoto University, Kyoto

Summary

A statistical approach is suggested to determine theoretically the energy distribution of the shallow water swell having a pronounced nonlinearity and finite spectral band-width. The approach is based on two assumptions; they are: i) the swell is approximately representable as a train of random solitons which are governed by the Korteweg-de Vries equation and ii) its initial state satisfies the maximum occurrence probability condition, that is, under this condition the wave energy is assigned to each possible soliton in obedience to the way that is extremely liable to take place. It is revealed that the energy distribution calculated by this approach agrees well with the observed results.

## Introduction

Shallow water swell, having travelled far from their generation area , are generally unidirectional and can be described, approximately, by employing the Korteweg-de Vries (for simplicity KdV) equation under the assumptions that viscosity is negligible and the bottom topography in shallow water is almost planar. It is shown [1,2] that the kinematics of shallow water swell observed at fixed positions can be described accurately and easily by applying the asymptotic multi-soliton solution of the KdV equation, where the swell is represented as a train of many solitons having random amplitudes and phase constants. Since the amplitude of the soliton is invariant, a problem results in determining the initial energy distribution of the swell.

In this study, a statistical approach is attempted in order to determine theoretically the initial energy distribution.

K. Horikawa, H. Maruo (Eds.)
Nonlinear Water Waves
IUTAM Symposium Tokyo/Japan 1987

The results derived through this statistical approach are then compared with field observations.

Energy Distribution

The shallow water swell to be studied here is assumed to be representable as a train of many solitons governed by the KdV equation,

$$\partial_\tau \eta + \frac{3}{4}\partial_\sigma \eta^2 + \frac{1}{6}\partial_\sigma{}^3 \eta = 0 \quad , \tag{1}$$

where

$$\sigma = \varepsilon^{1/2}(x'-t'),\ \tau = \varepsilon^{3/2}t',\ \eta = (1/\varepsilon)\zeta/h,\quad x'=x/h,$$
$$t' = t\sqrt{g/h},\quad \varepsilon = (h/L)^2 \quad , \tag{2}$$

in which primes denote dimensionless variables, $\partial_\tau$ and $\partial_\sigma$ the operators of partial differentiation with respect to $\tau$ and $\sigma$, respectively, x the horizontal coordinate onshore, t the time, h the mean water depth, L the representative wave-length, g the acceleration of gravity and $\zeta$ the water surface displacement around the mean water level.

Here, we carry out the following Galilean transformation in order to separate $\eta$ into components consisting of multi-solitons $\eta_s$ and a constant term $\eta_o$ introduced to satisfy the continuity condition for wave profiles around the mean water level,

$$\eta = \eta_s(\sigma,\tau) - \eta_o,\ \xi = \sigma + \frac{3}{2}\eta_o\tau \ . \tag{3}$$

So that, eq.(1) is rewritten as the following KdV equation,

$$\partial_\tau \eta_s + (3/4)\partial_\xi \eta_s^2 + (1/6)\partial_\xi^3 \eta_s = 0 \quad , \tag{4}$$

$$\eta_o = \int_{\xi_o}^{\xi_x} \eta_s d\xi \quad , \tag{5}$$

where $\xi_o$ is an arbitrary point on $\xi$ coordinates, $\xi_x$ the point distant enough from $\xi_o$ to include many solitons and $(\xi_x - \xi_o)$ is regarded approximately as one period on $\xi$ coordinates.

Observing the swell from a coordinate system moving with the speed $\sqrt{gh}(1-3\varepsilon\eta_o/2)$, and following Gardner[3], we define the Hamiltonian corresponding to eq.(4) as:

$$H = \int_{\xi_0}^{\xi_x} [\frac{1}{4}\eta_s^3 - \frac{1}{12}(\partial_\xi \eta_s)^2] d\xi \quad . \tag{6}$$

The use of eq.(6) gives the following canonical form of the KdV equation,

$$\partial_\tau \eta_s = \partial_\xi (\frac{\delta H}{\delta \eta_s}) \quad , \tag{7}$$

where $\delta / \delta\eta_s$ indicates an operator of functional differentiation.

Since the swell is assumed to consist of N solitons, eq.(6) is rewritten as:

$$H = \sum_{i=1}^{N} (9/20\,\sqrt[3]{2})\, p_i^{5/3} , \tag{8}$$

using the multi-soliton solution of eq.(4), and the following canonical equations of motion are derived for the swell,

$$\frac{dq_i}{d\tau} = \frac{3}{4\sqrt[3]{2}}\, p_i^{2/3} \quad , \quad \frac{dp_i}{d\tau} = 0 \quad , \tag{9}$$

where $q_i$ is the i-th element of the generalized coordinates {q} with N degrees of freedom, and $p_i$ is the i-th element of their conjugate momenta {p} and is related to the amplitude $A_i$ of the i-th soliton non-dimensionalized by h such that

$$p_i = (4\sqrt{3}/9) A_i^{3/2} \quad . \tag{10}$$

The above canonical set states that the trajectory within the phase space spanned on the canonical coordinates {q , p } is determined by the initial representative point alone and the observational state of the swell can be described on the trajectory.

In order to formulate the momentum distribution function by assigning the probability to each possible value of the momentum of soliton, we assume here that the selection of the initial representative point is made under the law of probability and the condition that both values of N and H are conserved. If the law of probability obeys the principle of equal *a priori* probabilities, we can be easily convinced that this selection is carried out in accordance with the condition of the maximum occurrence probability, where N solitons are excited in the way that is most liable to take place, therefore, the momentum distribution function $\rho_N$ can be reduced to the so-called Boltzmann distribution.

We define the occurrence probability S of an initial state as

$$S = -\int \rho_N \log \rho_N \, dq dp \tag{11}$$

and select a representative point specifying the initial state so as to maximize the above occurrence probability S under the following conservation laws of total number N of solitons and the averaged Hamiltonian $\bar{H}$;

$$\int \rho_N dq dp = 1 , \tag{12}$$

$$\int \rho_N H dq dp = \bar{H} , \quad \bar{H} = H/N . \tag{13}$$

As a result, we obtain the following distribution function making the occurrence probability S maximum under both conditions of eqs. (12) and (13),

$$\rho_N = [(\frac{75}{128}\beta^3)^{1/5} \frac{1}{\Gamma(3/5)(\xi_x-\xi_o)}]^N \exp(-\beta H) , \quad \beta = \frac{3}{5}\frac{1}{\bar{H}} , \tag{14}$$

where $\Gamma$ is the Gamma function and $(\xi_x - \xi_o)$ the disturbance length with weak periodicity including N-solitons on the generalized coordinates. The Boltzmann-type distribution function $\rho_1$ in the 2-dimensional phase-space can be derived by carrying out the following 2(N-1)-fold integral with regard to $\rho_N$,

$$\rho_1 = \frac{1}{(3/5)(\xi_x-\xi_o)} (\frac{75}{128}\beta^3)^{1/5} \exp(-\frac{9\beta}{20\sqrt[3]{2}} p^{5/3}) , \tag{15}$$

where subscript i is omitted for simplicity.

The wave energy E is defined by

$$E = \int_{\xi_o}^{\xi_x} \eta^2 d\xi = \sum_{i=1}^{N} p_i \tag{16}$$

and is represented as the summation of the momentum $p_i$ of each soliton, so that $p_i$ can be regarded as the energy quantity and eq.(15) can be used to represent the energy distribution function of the swell.

Further, eq. (15) is rewritten to the following energy distribution function directly applicable to the swell observed temporally over the period of T at a fixed position,

$$R^*(p) = \frac{\beta^*(3p^{2/3}/4\sqrt[3]{2}+1)}{1+\Gamma(3/5)\beta^{*2/5}(128/75)^{1/5}} \exp\left(-\frac{9\beta^*}{20\sqrt[3]{2}}p^{5/3}\right). \tag{17}$$

Here $\beta^*$ can be determined from the temporally observed data through the following equation

$$\frac{H^*}{N^*} = \frac{1+\Gamma(3/5)\beta^{*2/5}(3/5)(128/75)^{1/5}}{\beta^*[1+\Gamma(3/5)\beta^{*2/5}(128/75)^{1/5}]}, \tag{18}$$

in which $N^*$ represents the number of solitons contained within the swell which were observed over the period of T, and $H^*$ the Hamiltonian of the swell.

## Comparisons with observed results

Several tests were performed on both statistical properties of the observed waves and the applicability of the aforementioned assumptions to the observed waves. The wave data were obtained off Ogata coast, facing the Japan Sea, [1] and off Caldera Port in Costa Rica, facing the Pacific Ocean, [4]. The former (DNO 7) was recorded as temporal water surface displacements at the observation point in a depth of 5.7m and at a distance of 250m from the shore. The latter (CR 20) was also recorded temporally in a depth of 18.2m and at a distance of 1.7Km from the shore.

Figure 1 shows the sample distributions of the temporal water surface displacement for the above observed waves (DNO7 & CR20). For comparison purposes the Gaussian distribution is given in this figure as a solid line. Both distributions deviate considerably from the Gaussian and both values of skewness exceed 0.5.

Figure 2 shows the power spectra S(f) for the observed waves. The spectral width parameter $\mu$ is given in this figure. The power spectra exhibit pronounced secondary peaks at frequencies of about twice the main peak frequencies due to

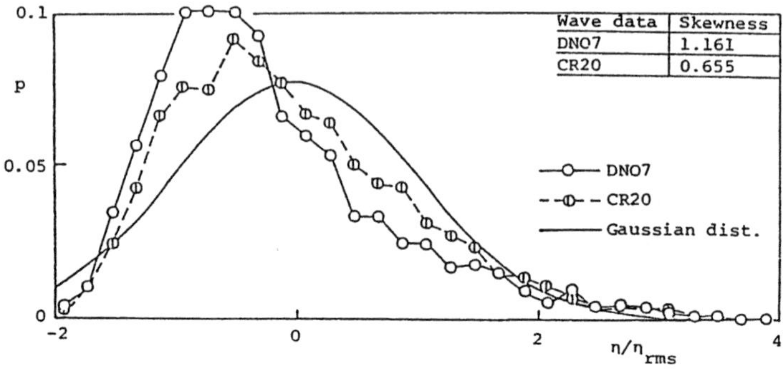


Fig. 1. Distribution of temporal water surface displacement of the observed waves

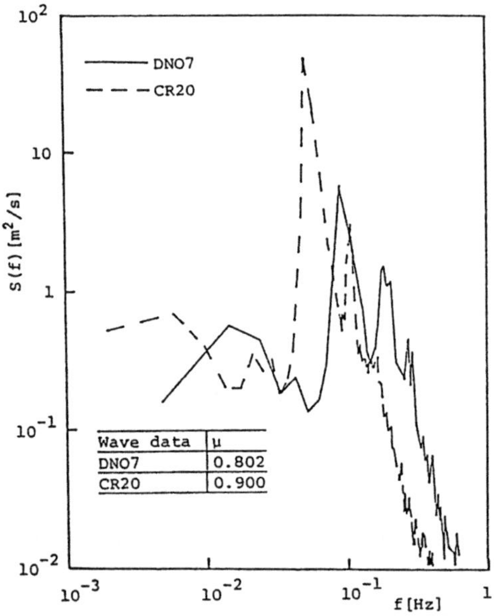


Fig. 2. Power spectra of the observed swell

nonlinear interaction among normal modes. From these results we can conclude that the observed shallow water swell have a pronounced nonlinearity and finite band width.

Figure 3 shows comparisons of wave profiles between the observed waves and the theoretical results calculated by using the N-soliton solution of the KdV equation. It is easily convincing that the swell is representable as a train of

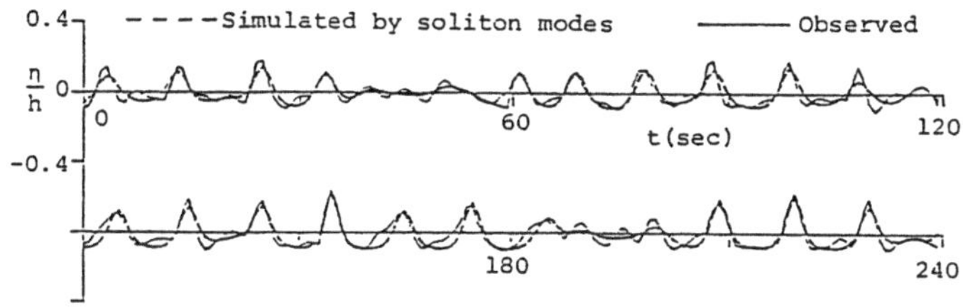


(a) DNO7

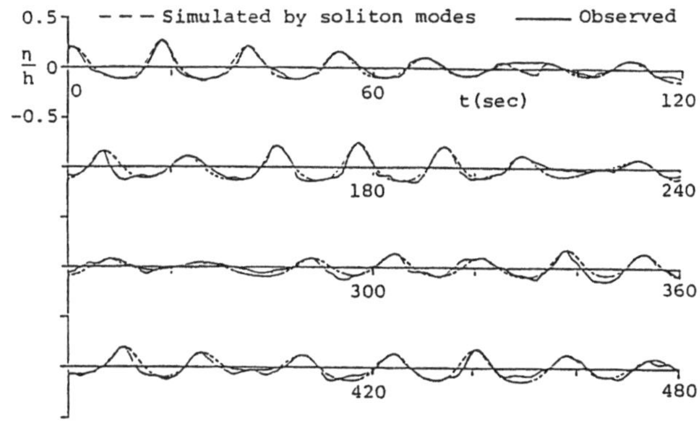


(b) CR20

Fig. 3. Comparisons of temporal wave profiles between observed data and the results simulated by soliton modes

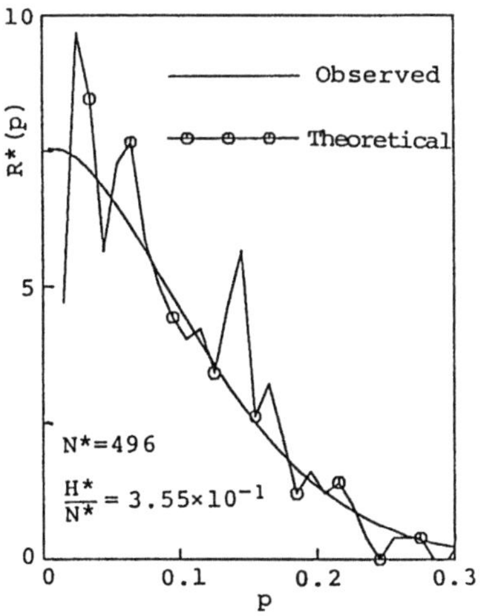


(a) DNO7

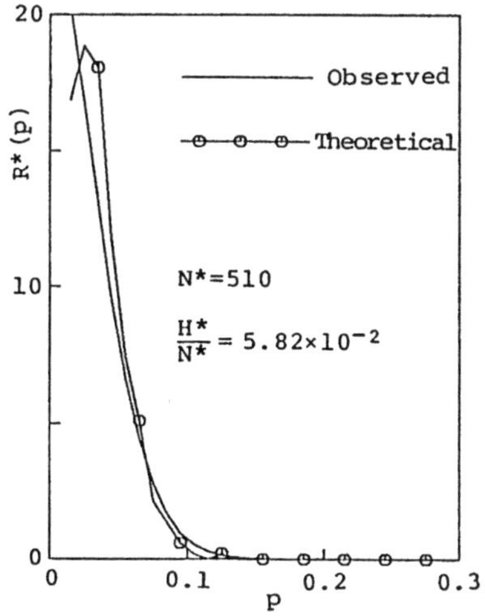


(b) CR20

Fig. 4. Comparisons of the energy distribution between the observed results and the theoretical ones calculated by eq. (17)

solitons, although there is a small discrepancy between both the results.

Figure 4 compares the energy distribution of the observed and theoretical [Eq.(17)] results in the region where the value of p exceeds 0.03. The total number $N^*$ of solitons constituting the temporally observed swell and their Hamiltonian $H^*$ are shown in this figure. It is found that the theoretical energy distribution is very close to the energy distribution of the observed swell. From these comparisons it can be stated that the swell model suggested here does not conflict with the observed results and the energy distribution of the swell could be determined by the maximum probability condition.

Consequently, it is concluded that the shallow water swell can be treated, approximately, as a random train of solitons of which the energy distribution is determined by the maximum probability condition.

## Acknowledgement

The authors wish to express their sincere thanks to Prof. T. Taniuchi of Nagoya Univ. for his valuable advice and to thank Director General Y. Goda of the Port & Harb. Res. Inst. for his kind permission to make use of wave records. This research was supported by the Grant-in-Aid for Scientific Research 61550368 from the Ministry of Education, Science and Culture.

## References

1. Tsuchiya, Y.; Yasuda, T.: A dynamical expression of waves in shallow water, Proc. 19th Int. Conf. on Coastal Eng., 1984, 435-451.

2. Yasuda, T.; Shinoda, S.: Soliton mode representation for kinematics of shallow water swell, in Coastal Hydrodynamics (ed. R. A. Dalrymple, ASCE), 1987, 751-764.

3. Gardner, C. S.: Korteweg-de Vries equation and generalizations. IV., J. Math. Phys. 12 (8), 1971, 1548-1551.

4. Goda, Y.: Analysis of wave grouping and spectra of long-travelled swell, Rep. Port & Harb. Res. Inst., 22 (1), 1983, 3-41.

# Session 3

# Nonlinear Water Waves in a Finite Region

# Experimental and Numerical Study of Long-time Evolution of Standing Waves in a Rectangular Tank

L. SHEMER, E. KIT and T. MILOH

Dept. of Fluid Mechanics, Faculty of Engineering
Tel-Aviv University, Ramat-Aviv 69978, Israel

Summary

Standing waves with their crests parallel to the side walls of a rectangular tank, were generated by a segmented wavemaker. The long-time evolution of these waves was investigated experimentally, and the results of the measurements were compared to the numerical solutions of the appropriate nonlinear Schrödinger equation. An agreement between theory and experiments was obtained when dissipation was taken into account in the model equations.

## I. Introduction

We study resonant waves which appear in a rectangular tank, when excited by a wavemaker having a typical instantaneous length scale $\lambda_n = 2b/n$, where b is the width of the tank and n is an integer representing the mode number. The frequency of forcing should be close to the linear cut-off value given by $\omega_n^2 = 2g\pi/\lambda_n \tanh(2\pi h/\lambda_n)$, h being the mean water depth in the tank. These standing waves have their crests parallel to the tank side walls and are usually referred to as "sloshing waves" (Barnard et al. [1]). Two different steady sloshing wave regimes which may exist in the tank under identical flow conditions were observed in [1]. Aranha et al. [2] presented a derivation of a nonlinear Schrödinger (NLS) equation for the propagation of an acoustic wave in a duct with a wavemaker-like forcing. Their numerical results indicated that for a case which is analogous to deep water in a channel, no steady solution can be obtained. This contradiction between the existing steady experimental results and the numerical predictions triggered our interest in the problem.

In a recent investigation [3] we have presented some experimental results on sloshing waves in the vicinity of the 2nd cut-off frequency, $\omega_2$. These results were compared to the numerical solutions of the NLS equation which was derived for our experimental conditions. This comparison has supported the conclusion in [1] that dissipation has to be incorporated into the model equations in order to obtain qualitative agreement between the experiment and the numerical results. In the present work we investigate in

K. Horikawa, H. Maruo (Eds.)
Nonlinear Water Waves
IUTAM Symposium Tokyo/Japan 1987

greater detail the long time modulation of sloshing waves and draw some conclusions about the validity and the limitations of the theoretical model.

## II. Experimental facility and procedure

The experiments were carried out in a rectangular tank which is 18m long, 1.2m wide and filled to a mean water depth of 0.6m. Waves were generated by a 4-segment paddle-type wavemaker. As it was demonstrated by [1] and [3], the stability of the forcing frequency is of crucial importance for the experiment, since the most interesting nonlinear phenomena are observed in the narrow range of frequencies in the vicinity of the cut-off value $\omega_2$. The wavemaker was therefore operated using the PDP 11/23 minicomputer analog output, which simulated the sine-wave generator and used 1MHz clock of the computer. The relative accuracy of tuning was as good as $3.6 \cdot 10^{-4}$. The sloshing waves of the 2nd mode were obtained by operating all wavemaker segments with identical amplitude and with 180° phase shift between the two inner and two outer segments. The instantaneous surface elevation was measured by 4 conductance-type wave gauges, which could be moved along the tank. This information, together with outputs of position potentiometers representing the instantaneous inclination angle of each segment of the wavemaker, and the forcing signal which served as a reference, were sampled by an A/D converter of the same minicomputer and recorded on magnetic tape for further processing. The general view of the experimental facility and the wave gauges is given in Fig. 1(a). Additional information about the experimental procedure was presented in [3] and [4].

## III. Model equations

The derivation of the governing equations is based on full inviscous nonlinear boundary conditions at the free surface and the Laplace equation (see [3]).The equations are expanded to the 3rd order in a small parameter $\varepsilon$ based on the dimensional stroke of the wavemaker s/b at the mean water level. The calculations based on the actual geometry of our wavemaker when operated at the 2nd sloshing wave mode yield

$$\varepsilon = 0.022 s/b \tag{1}$$

The scaled slow length variable along the tank X and the slow time variable $\tau$ are defined for the case of deep water, which approximately holds for our experimental conditions ($\tanh 2\pi h/\lambda = 0.996$),as

$$X = (2\pi)^{5/2}\varepsilon^{1/2}x/b; \qquad \tau = 2\pi^3\varepsilon\omega_2 t \tag{2}$$

The velocity potential $\phi$ for the sloshing waves of the 2nd mode is given by

$$\phi = \sqrt{\varepsilon b^3 g}\,\cos(2\pi y/\lambda)\,\frac{\cosh 2\pi(z+h)/\lambda}{\cosh(2\pi h/\lambda)}[C(X,\tau)e^{-i\Omega t} + c.c]; \tag{3}$$

with the bottom of the tank at z=-h. The dimensionless complex amplitude of potential C satisfies the following nonlinear governing equation:

$$i\,\frac{\partial C}{\partial \tau} + \frac{\partial^2 C}{\partial X^2} + \lambda C + 2|C|^2 C = 0 \tag{4}$$

where the detuning parameter $\lambda$ represents the ratio of the deviation of the forcing frequency $\Omega$ from the linear cut-off value $\omega_2$ to the amplitude of forcing $\varepsilon$ and is given by

$$\lambda = \frac{4\omega_2(\Omega-\omega_2)}{(2\pi)^4\varepsilon}\,\frac{b}{g} \tag{5}$$

Equation (4) is subject to the boundary condition at the wavemaker

$$\partial C/\partial X|_{X=0} = -i \tag{6}$$

Following [3], we now take into account the dissipation. One can distinguish between the dissipation at the wavemaker, which leads to the modification of the boundary condition (6), and the dissipation at the tank walls which results in an additional term in the governing equation (4):

$$i\,\frac{\partial C}{\partial \tau} + \frac{\partial^2 C}{\partial X^2} + \lambda C + a_1(i-1)C + 2|C|^2 C = 0 \tag{4a}$$

The boundary condition (6) is replaced by

$$\partial C/\partial X|_{X=0} = -i - a_2(1+i)C \tag{6a}$$

In the present work we have chosen the following values of the dissipation coefficients: $a_1$=0.08 and $a_2$=0.7. The criterion for the choice of the dissipation coefficients will be discussed further. The equation (4a) with zero initial condition and the boundary condition (6a) was solved numerically. The numerical scheme was identical to that used by [3].

## IV. Experimental and numerical results

In [3], measurements were performed at the values of $\varepsilon$ ranging from $0.42\cdot10^{-4}$ to $0.9\cdot10^{-4}$. In the present investigation, stronger forcing was used, and the amplitude of the wavemaker corresponded to $\varepsilon=1.29\cdot10^{-4}$ and $\varepsilon=1.56\cdot10^{-4}$. For all amplitudes of forcing, steady wave pattern was observed for low values of the detuning parameter $\lambda$. In this regime, the wave field is basically decaying, and the surface elevation differs notably from zero only in the vicinity of the wavemaker. The decay is not necessarily monotonous, so that the maximum in the wave amplitude may be detached from the wavemaker. This mode may be considered as a trapped soliton, as discussed in [5]. When the frequency of forcing (and thus $\lambda$) is gradually increased, a sharp transition in the wave pattern occurs. The wave field becomes unsteady, and the wave amplitudes and the phases undergo a dramatic change.

At $\lambda$ which exceeds slightly the transitional value, solitons are generated quasi-periodically at the wavemaker and then propagate away with a seemingly constant speed. Such propagating soliton can be seen in Fig. 1(a). The "snapshot" of the numerical solution of equation (4b) with the boundary condition (6b) is presented in Fig. 1(b).

When the frequency of forcing is gradually decreased again, the back transition to the steady state is attained. This transition occurred at lower frequencies, and thus at lower values of $\lambda$, than the transition from steady to unsteady regime for increasing $\lambda$ [3]. The hysteresis loop at $\varepsilon=1.29\cdot10^{-4}$ is shown in Fig. 2. The transition between the steady and the modulated regimes manifests itself in this figure in the jump in the phase of the sloshing wave relative to that of the wavemaker, when measured close to the wavemaker. In the case of the modulated wave regime, the phase angles shown in the figure represent the averaged over long time values. The hysteresis loop covers the values of $\lambda$ between -0.52 and -0.57. These values are close to the average transitional $\lambda$ reported in [3] for weaker forcing. At strong forcing, $\varepsilon=1.56\cdot10^{-4}$, the transition between the two wave patterns was observed at $\lambda\approx-0.6$, and no clear conclusions about the existence of hysteresis could be drawn from the experiments.

The dissipation coefficient along the tank $a_1$ was estimated using the results of [6], where the correction factor suggested in [7] was taken into account. As was noticed in [3], the transition between the steady and the modulated wave regimes is governed mainly by the dissipation at the wavemaker. The value of $a_2$ was chosen so that the transition between these two wave patterns in the numerical calculations and in the experiments occurs at approximately the same value of the detuning parameter $\lambda$.

At sufficiently low frequencies ($\lambda < -0.6$), in both the numerical solutions and in the experiments the steady wave pattern was obtained. The details of this steady wave field were discussed in [3]. In the numerical calculations, the transition to the modulated regime occurred at $\lambda = -0.6$. The evolution of the wave field at four locations along the tank at frequency slightly exceeding the transitional value ($\lambda = -0.58$), is presented in Fig. 3(a). Periodic generation of solitons can be clearly seen. The dimensionless period of modulation estimated from the figure is T=22.6. The experimental results obtained at the forcing frequency f=1.1338 Hz, $\lambda$=-0.52, immediately beyond the transitional region, are presented in Fig. 3(b). The absolute values of the normalized velocity potential $|C|$ in this figure were obtained from the experimentally measured surface elevation amplitudes, using eq. (3).

The results in Figs. 3(a) and 3(b) are shown at identical locations along the tank. The scaling of the vertical coordinate is also identical. The period of modulation in Fig.3(b) is about 402s, corresponding to 456 periods of forcing and giving the dimensionless value of T=22.8, indistinguishable from the numerically obtained modulation period. The general shape of the wave modulation pattern and the measured amplitudes are also very similar in both figures at all locations.

When the forcing frequency is increased, the period of modulation decreases both in experiments and in numerical calculations. The character of modulation also changes; periodic appearance of solitons is gradually replaced by a wave-train-like modulation. The results of numerical calculations at $\lambda=0$ are presented in Fig.4(a), and the experimental results obtained at identical value of $\lambda$ (f=1.1393 Hz) are shown in Fig.4(b). Note that the total time span in this figure is half of that in Fig.3. The period of modulation at this $\lambda$ from the numerical solution is T=6.4, while the experiments give the value of 133s, which corresponds to 151 periods of forcing and to the dimensionless value of T=7.6. While the general shapes of modulation in both figures are very similar, the quantitative agreement in the values of the period of modulation and the amplitudes is less impressive than in the previous figure. The notable disagreement in wave amplitudes in Fig. 4 may be attributed to possible reflection of the slowly propagating modulation wave train from the beach at the far end of the tank. This reflection apparently becomes more essential when the period of modulation and the modulation wave length become shorter.

## V. Concluding remarks

The comparison of the numerical and experimental results presented in this paper allows us to conclude that the theoretical model, which includes dissipation both at the wavemaker and along the tank, provides a very good description of main features of the wave field. The model predicts correctly the existence of two qualitatively different wave patterns, steady and modulated, as a function of a single governing parameter, $\lambda$. More than that, with the dissipation coefficients properly chosen, the numerically obtained values of periods of long-time modulation, wave amplitudes and general modulation pattern, are in close quantitative agreement with the experiment, for the whole range of forcing frequencies and amplitudes. The only major physical feature which the model does not predict at the present stage, is the existence of hysteresis in the transition between the two wave regimes.

Acknowlegement.

This work was supported in part by U.S.-Israel Binational Science Foundation under the grant 85-00343.

References.

1. Barnard, B.J.S., Mahony, J.J. and Pritchard, W.G.: The excitation of surface waves near a cut-off frequency. Phil. Trans. R. Soc. Lond. **A286** (1977) 87-123.

2. Aranha, J.A., Yue, D.K.P. and Mei, C.C.: Nonlinear waves near a cut-off frequency in an acoustic duct - a numerical study. J. Fluid Mech. **121** (1982) 465-485.

3. Kit, E., Shemer, L. and Miloh, T.: Experimental and theoretical investigation of nonlinear sloshing waves in a rectangular channel. J. Fluid Mech. **181** (1987) 265-291.

4. Shemer, L., Kit, E. and Miloh, T.: Measurements of two- and three-dimensional waves in a channel, including the vicinity of cut-off frequencies. Exp. Fluids **5** (1987) 66-72.

5. Miles, J.W.: Note on parametrically excited trapped cross-wave. J. Fluid Mech. **151** (1985) 391-394.

6. Mei, C.C. and Liu, L.F.: The damping of surface gravity waves in a bounded liquid. J. Fluid Mech. **59** (1973) 239-256.

7. Miles, J.W.: Surface-wave damping in a closed basins. Proc. Roy. Soc. Lond. **A297** (1967) 459-475.

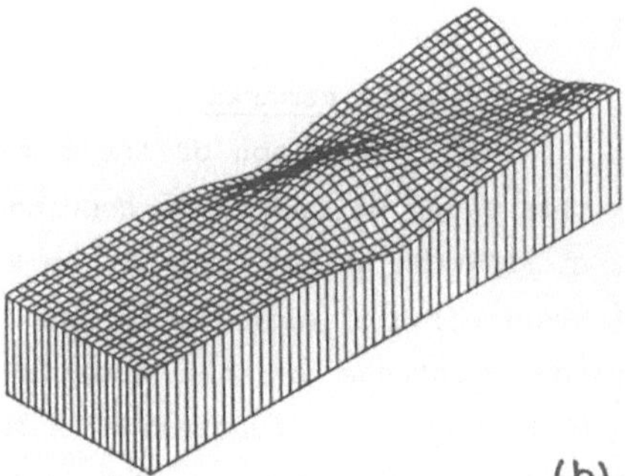

(b)

Figure 1(a) General view of the experimental facility with propagating soliton.

Figure 1(b) Numerical solution of the NLS equation ($\lambda$ = -0.58).

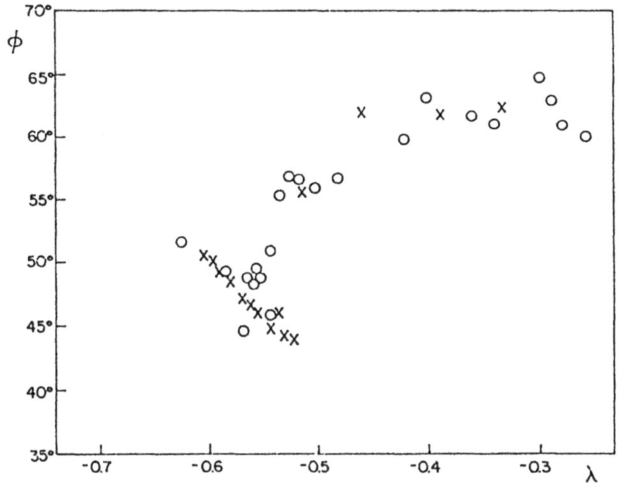


Figure 2 The dependence of the phase angles of the surface elevation at the distance 10cm from the wavemaker on $\lambda$; x, for increasing frequency, O , for decreasing frequency; $\varepsilon = 1.29 \cdot 10^{-4}$.

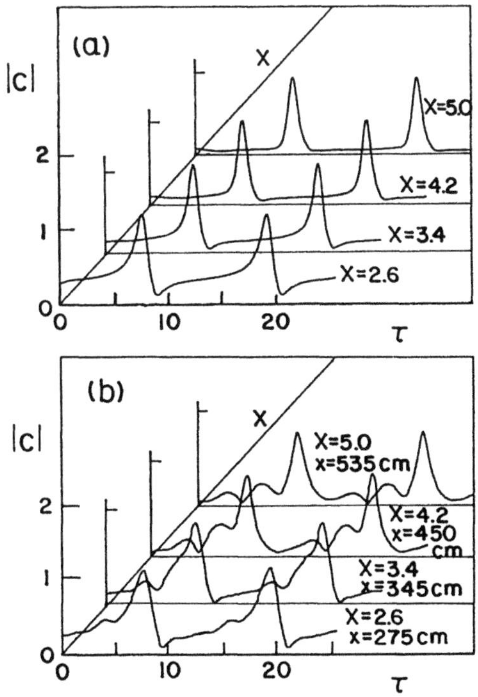


Figure 3 The dependence of the amplitude of the velocity potential on slow time at 4 locations along the tank: (a) numerical calculations, $\lambda = -0.58$; (b) experimental results, $\lambda = -0.51$.

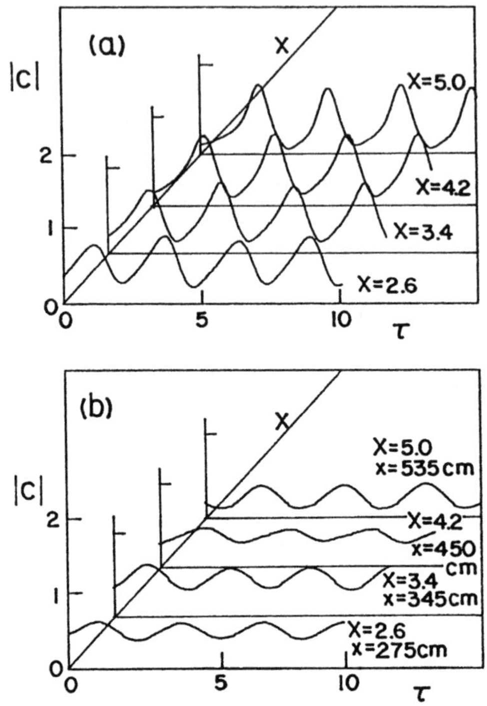


Figure 4 As in previous figure, $\lambda = 0$.

# Nonlinear Waves in Circular Basins

PETER J. BRYANT

Mathematics Department, University of Canterbury,
Christchurch, New Zealand.

## Summary

Waves generated near resonance in bays or harbours, by storms or earthquakes, may rise to large wave heights. This investigation examines waves progressing around a circular basin of finite uniform depth, with particular attention being given to those waves near resonance. Multiple families of free steady waves are associated with each of the depths at which resonance occurs, with different families having different orderings of the wave components composing them. Linear stability calculations indicate that those free waves dominated by resonating wave components are linearly unstable, but calculations of the nonlinear time evolution over many wave periods do not confirm the instability in the weakly unstable examples. When the model is made more realistic by including periodic forcing and wavenumber dependent damping, some of these weakly unstable examples are found to be stable, while others are unstable but evolve slowly in time into periodically modulated waves. Although the occurrence of multiple families of free waves is interesting for theoretical reasons, practical calculations of waves near resonance should include both forcing and damping. For any given depth ratio, forcing amplitude, and forcing period, there may be no, one, or multiple steady progressive waves.

## Introduction

Resonant waves generated in enclosed regions may be the result of external forcing of waves near their natural frequencies by a local source, such as a storm or earthquake, or the result of internal forcing of some wave components by others propagating from a distant source. Internal resonance is investigated first in the absence of external forcing or wave damping, then the model is made more realistic by including external forcing and wave damping.

Internal resonance changes the ordering of the components of a wave. For example, in a circular basin with a depth to

K. Horikawa, H. Maruo (Eds.)
Nonlinear Water Waves
IUTAM Symposium Tokyo/Japan 1987

radius ratio of 0.831, the first harmonic of a wave progressing around the basin forces resonantly the second harmonic of the wave, causing the two components to be of comparable magnitude. The same phenomenon occurs with capillary-gravity waves, where it is known as Wilton's ripples. Miles [1,2] developed a Hamiltonian formulation for the analytical calculation of forced damped waves near resonance in circular basins. Prior assumptions had to be made about the ordering of the wave components, assumptions which fail near resonance. Fully nonlinear calculations on the standing wave system studied by Miles show good agreement with his weakly nonlinear results only when the prior ordering assumptions are valid. The present fully nonlinear method makes no prior ordering assumptions.

Progressive waves propagating around a circular basin of uniform depth are investigated. The water motion outside the boundary layers is taken to be inviscid and irrotational. Following the linear solution, the water surface displacement $\eta(r,\theta,t)$ and the velocity potential $\phi(r,\theta,z,t)$ are both represented by Fourier-Bessel series whose coefficients may have slow time dependence. Damping, due to boundary layer friction, is included as a linear attenuation dependent on wavenumber in the equations for the time evolution of the wave coefficients. Forcing is imposed as a steady circular motion of the centre of the basin with the orientation of the basin remaining unchanged. This forcing motion of the basin drives irrotational progressive waves of the same period as the forcing period.

The shallow water approximation is inappropriate for the calculation of resonant waves because of the modification it makes to wave dispersion, and hence to the occurrence of resonance. Wave calculations based on the shallow water approximation should be supplemented by accurate fully nonlinear calculations whenever resonance occurs.

## Progressive waves and resonance

The Fourier-Bessel series for waves progressing with angular velocity c in the positive $\theta$-direction, in a circular basin of uniform depth to radius d, are

$$\eta = \sum_{m=0}^{\infty} \sum_{n=1}^{\infty} J_m(k_{mn}r)(a_{mn}\cos m(\theta-ct) + \alpha_{mn}\sin m(\theta-ct)), \tag{1}$$

$$\phi = \sum_{m=0}^{\infty} \sum_{n=1}^{\infty} J_m(k_{mn}r) \frac{\cosh k_{mn}(z+d)}{\cosh k_{mn}d} (b_{mn}\sin m(\theta-ct) + \beta_{mn}\cos m(\theta-ct)), \tag{2}$$

where $k_{mn}$ is the nth zero of $J_m'$. The series for $\phi$ satisfies Laplace's equation and the side and bottom boundary conditions, and the wave coefficients have to be chosen so that $\eta$ and $\phi$ satisfy the free surface boundary conditions also. These boundary conditions, in nondimensional form, are

$$\eta_t + \alpha D[\eta] - \phi_z + c\varepsilon'(\eta_r \sin(\theta-ct) + \frac{1}{r}\eta_\theta \cos(\theta-ct)) + \varepsilon(\eta_r\phi_r + \frac{1}{r^2}\eta_\theta\phi_\theta) = 0 \quad \text{on} \quad z = \varepsilon\eta , \tag{3}$$

$$\eta + \phi_t + \alpha D[\phi] - \frac{\varepsilon'}{\varepsilon} c^2 r\cos(\theta-ct) + c\varepsilon'(\phi_r \sin(\theta-ct) + \frac{1}{r}\phi_\theta \cos(\theta-ct)) + \frac{1}{2}\varepsilon(\phi_r^2 + \frac{1}{r^2}\phi_\theta^2 + \phi_z^2) = 0 \quad \text{on} \quad z = \varepsilon\eta . \tag{4}$$

The amplitude parameter $\varepsilon$ is chosen so that the surface displacement $\eta$ has a root mean square value of 1, making constant $\varepsilon$ equivalent to constant potential energy. The linear damping operator D transforms to the square of wavenumber in the equations for the rates of change of the wave coefficients, with $\alpha \ll 1$ being the weak damping coefficient. This empirical relation is chosen as a first approximation to the effects of boundary layer damping. The parameter $\varepsilon'$ is a measure of the amplitude of the conservative forcing motion, the centre of the basin moving on the horizontal circle

$$x = \varepsilon'\cos ct \quad , \quad y = \varepsilon'\sin ct . \tag{5}$$

A first approximation to the wave velocity of a progressive free wave dominated by the (1,1) wave component is given by

$$c = (k_{11}\tanh k_{11}d)^{\frac{1}{2}} . \tag{6}$$

This means that resonance occurs near depth ratios d satisfying

$$m(k_{11}\tanh k_{11}d)^{\frac{1}{2}} = (k_{mn}\tanh k_{mn}d)^{\frac{1}{2}} . \tag{7}$$

All nontrivial solutions of this equation up to m = 4 are listed in table 1. It should be emphasised that the values in table 1 provide only a first approximation to the depth ratios at which resonance occurs, because nonlinear effects, forcing, and damping

TABLE 1

| m | n | d |
|---|---|---|
| 2 | 2 | 0.831 |
| 3 | 2 | 0.279 |
| 3 | 3 | 0.455 |
| 3 | 4 | 0.748 |
| 4 | 2 | 0.158 |
| 4 | 3 | 0.249 |
| 4 | 4 | 0.330 |
| 4 | 5 | 0.423 |
| 4 | 6 | 0.542 |
| 4 | 7 | 0.721 |
| 4 | 8 | 1.204 |

are all ignored, and it is assumed that the wave is dominated by the (1,1) wave component.

When the forcing and damping are both zero ($\alpha = \varepsilon' = 0$), symmetric steady waves are calculated by putting the wave coefficients $a_{mn}$, $b_{mn}$ all constant with $\alpha_{mn}, \beta_{mn}$ all zero. For given values of the amplitude parameter $\varepsilon$ and depth d, trial solutions for the wave coefficients and wave velocity c are substituted into the nonlinear boundary conditions (3) and (4) with $\alpha = \varepsilon' = 0$, and are improved by Newton's method. When the forcing and damping are nonzero, the approach used is to calculate the forcing amplitude $\varepsilon'$ needed to drive a progressive wave of given amplitude ratio $\varepsilon$ and wave velocity c, and to calculate the coefficients for the resulting wave. Forced damped steady waves are nonsymmetric in general, with the wave coefficients $a_{mn}$, $b_{mn}$, $\alpha_{mn}$, $\beta_{mn}$ all constant. For given values of $\varepsilon$,d,$\alpha$, and c, trial solutions for $\varepsilon'$ and the wave coefficients are substituted in the nonlinear boundary conditions (3) and (4), and are improved by Newton's method.

The linear stability of the steady waves is investigated by perturbing all the wave coefficients in the Fourier-Bessel series (1) and (2), substituting in the nonlinear boundary conditions (3) and (4), and linearizing in the perturbations. The series are inverted numerically to obtain a set of linear first order differential equations for the slow time evolution of the perturbations, which are solved numerically by finding the eigenvalues and eigenvectors of the matrix of coefficients. The same technique is followed for the nonlinear time evolution of unsteady waves, except that there is no linearization, and the resulting set of nonlinear first order differential equations is solved numerically by time-stepping.

Symmetric steady free waves

Multiple families of steady waves are associated with the resonant depths listed in table 1. A family of steady waves arising from the resonance between the (1,1) wave component and the (4,7) wave component near the depth ratio d = 0.721, at constant ε = 0.02 and no forcing or damping, is described here. The wave coefficients $a_{11}$, $a_{22}$, $a_{34}$, and $a_{47}$ are plotted as functions of d in figures (1a,b,c,d).

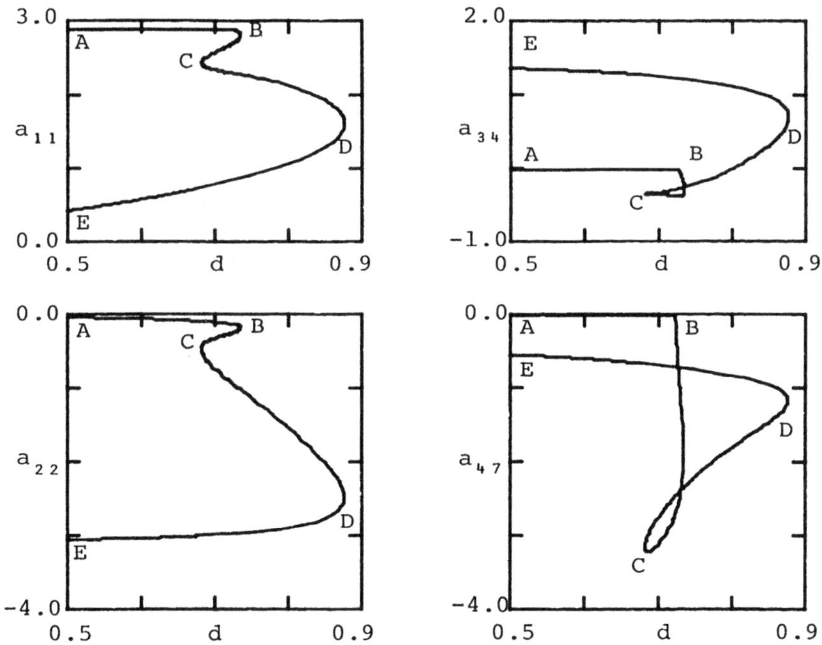


Figure 1. The amplitudes of the dominant wave components of steady free waves in the neighbourhood of the depth ratio d where the (4,7) wave component is resonant.

The wave at the point A in the four figures, at depth ratio 0.5, has an ordering of the wave components consistent with non-resonant interactions between them, and is linearly stable. A pitchfork bifurcation in $a_{47}$ occurs at B, near d = 0.725, with a stable branch on which $a_{47}$ remains near zero, and unstable branches on which $a_{47}$ becomes large and positive or large and negative. The latter branch is illustrated in figure 1(d). Although resonance between the (1,1) and (2,2) wave components is predicted in table 1 to occur near d = 0.831, the nonlinear modification of the linear resonance relation (7) moves it to depth ratios near the point C, where $a_{22}$ begins to increase to large magnitudes. It is followed in turn by the $a_{34}$ wave component near the point D. The normalization of the wave components, corresponding to constant wave energy, means that the $a_{11}$ wave component decreases from B to E.

Linear stability calculations at different points along the curve found stable waves between A and B only, with the instability becoming stronger from B towards E. Fully nonlinear calculations on perturbed waves confirmed the stability of the waves between A and B and the instability of the waves between C and E. However, in the weakly unstable section between B and C, fully nonlinear calculations confirmed neither stability nor instability over the long times for which numerical integration could be made. Wave damping is found to stabilize the weaker unstable wave solutions between B and C, because the damping absorbs energy transferred to the higher wavenumber components by the resonant interactions.

Although waves on the section AB are stable to small perturbations, a large perturbation, such as to the point C, destabilizes an otherwise stable wave. A large perturbation may occur when waves generated by a distant source arrive in a basin, and these waves are able to resonate with an existing stable wave in the basin.

## Forcing and damping

Forced damped steady waves are examined first at fixed amplitude ratio $\varepsilon = 0.02$, depth ratio d = 1, and damping coefficient $\alpha = 10^{-5}$. The forcing amplitude $\varepsilon'$ necessary to drive such a wave is plotted as a function of wave speed c in figure 2(a). The spike in the curve near c = 1.295 is due to external forcing of the (2,2) wave component, which occurs when c is given by

$$mc = (k_{mn} \tanh k_{mn} d)^{\frac{1}{2}} \tag{8}$$

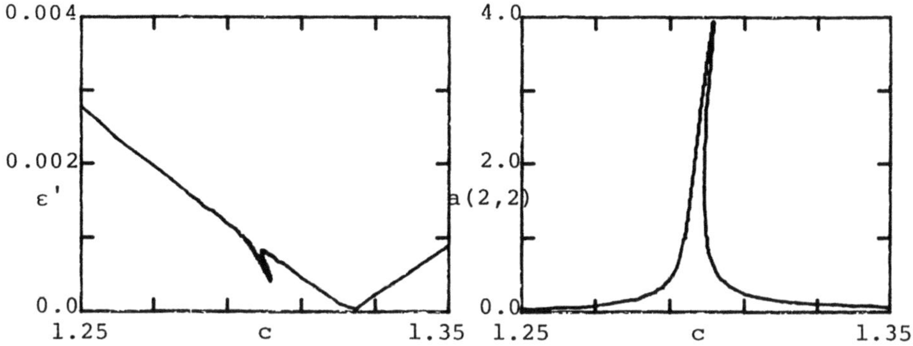


Figure 2. The forcing amplitude ε' and the (2,2) wave amplitude of a forced damped steady wave near resonance.

for d = 1, m = 2, n = 2. The amplitude of the (2,2) wave component defined by

$$a(m,n) = (a_{mn}^2 + \alpha_{mn}^2)^{\frac{1}{2}}$$

for m = 2, n = 2 is plotted in figure 2(b), showing a resonant peak at this same value of c. Equation (8) is satisfied for d = 1, m = 1, n = 1 when c = 1.323, the wave velocity at which resonant forcing of the (1,1) wave component occurs. The forcing amplitude ε' has a minimum of $10^{-4}$ at this value of c, while the phase of the (1,1) wave component changes steeply by almost π, from near zero to near − π.

A linear stability analysis of the family of steady wave solutions sketched in figure 2 finds that the waves are linearly unstable over the peak of the spike only, from c = 1.3012 to the vertical tangent at c = 1.2993. This means that there is at least one stable wave for each value of c, with 2 stable waves for $1.2993 < c < 1.3012$. The interesting property here is that there are linearly stable wave solutions when the resonating (2,2) wave component becomes large. The equivalent wave solution when forcing and damping are zero is weakly unstable according to the linear calculations, although this could not be confirmed by fully nonlinear calculations. It is clear that the weak damping has stabilized the forced wave, or that in real applications where damping is always present, waves with large resonating components may be stable.

The depth ratio d in this example was chosen to be 1 because table 1 shows that there are no significant internal resonances near this value. In the next example, the depth ratio d is chosen to be 0.831 because this is the value at which the major internal resonance between the (1,1) and (2,2) wave components occurs. The values of $\varepsilon$ and $\alpha$ are the same as in the previous example. The results have the same form as those in figure 2, except that the peak in a(2,2) coincides with the minimum in $\varepsilon'$, and the spike in $\varepsilon'$ is very small. A linear stability analysis finds two bands of linear instability for the wave solutions in this example. The first band proves to be stable when fully non-linear time evolution calculations are made, indicating failure there of the Floquet method for calculating linear stability. The second band lies over the peak of the spike down to the vertical tangent, as in the first example, except that here it leads to no stable solutions for $1.2977 < c < 1.3000$. Miles [2] indicates that in the absence of stable steady wave solutions, periodically modulated waves (limit cycle solutions) or chaotically modulated waves can occur. A fully nonlinear time evolution calculation of the wave solution at $c = 1.298$, on the left of the a(2,2) peak, confirms its linear instability, and finds that it tends towards a periodically modulated wave.

The band of values of c for which there are no stable steady wave solutions is so narrow in this example that it probably cannot be realised experimentally. However, Miles [2] showed that this band increases in width when the amplitude parameter is increased or the damping coefficient is decreased. It should be possible to assess the physical relevance of periodically modulated and chaotically modulated waves when more parameter ranges have been calculated. In practical applications, where damping is always present, there may be no, one, or many stable steady progressive wave solutions for a given depth ratio, forcing period and forcing amplitude.

Acknowledgement is made to the San Diego Supercomputer Center, which is sponsored by the National Science Foundation, for the computing time used in this research.

References

1. Miles, John W. Internally resonant surface waves in a circular cylinder. J. Fluid Mech. 149 (1984) 1-14.

2. Miles, John W. Resonantly forced surface waves in a circular cylinder. J. Fluid Mech. 149 (1984) 15-31.

# Numerical Analysis of Wave Motion Using the Lagrangian Description

H. NISHIMURA and S. TAKEWAKA

Institute of Engineering Mechanics
University of Tsukuba

## Summary

The Lagrangian coordinate system has a notable advantage in the description of free surface conditions which most essentially prescribe the dynamics of surface waves, although some difficulties are involved in numerical solution of equations. In this context, two alternative schemes (velocity and pressure formulations) are presented for analyzing nonlinear transformation of water waves. Their usefulness is examined through application to the resonant behavior of sloshing waves as well as to breaking of a solitary wave.

## Lagrangian Equation System

It has been a common practice to use the Eulerian coordinate system in numerical analyses of wave motion. The system fixed to a physical plane, however, requires highly artificial and inaccurate manipulations in the treatment of fluid motions at and near a moving surface. The boundary conditions at the free surface are in fact the most important factors governing wave behavior.

The Lagrangian equations of motion and continuity are written as follows:

$$(Z_{tt}+H)\cdot Z_a=-P_a/\rho, \tag{1}$$

$$(Z_{tt}+H)\cdot Z_b=-P_b/\rho, \tag{2}$$

$$X_aY_b-X_bY_a=J, \tag{3}$$

K. Horikawa, H. Maruo (Eds.)
Nonlinear Water Waves
IUTAM Symposium Tokyo/Japan 1987

where $Z=(X, Y)$ is the position vector at time t, (a, b) is a pair of Lagrangian parameters identifying a fluid particle, H is the acceleration due to external forces, P is the pressure, $\rho$ is the density of the fluid, J is an arbitrary function of a and b, and the subscripts denote the partial differentiations.

Here, for simplicity, let us assume that the water body is initially at rest and define (a, b) as the positional coordinates of a fluid particle (X, Y) at the initial instant. Since the surface particles are conveniently identified by a unique and constant value $b_0$ of the parameter b in this framework, the dynamic condition on the free surface is given as

$$P=0 \qquad \text{on } b=b_0 \tag{4}$$

or, in terms of Z,

$$(Z_{tt}+H)\cdot Z_a=0 \quad \text{on } b=b_0. \tag{5}$$

No kinematic constraint is imposed in the present system fixed to the water body. Kinematic conditions on any curvilinear and/or moving boundaries are automatically satisfied by employing a suitable mesh scheme.

In spite of the notable advantages in treating boundary conditions, this system has been adopted by rather few authors (e. g., Brennen [1], Chan [2], and Fritts and Borris [3]), perhaps, because of the following difficulties:

1) Both the equations of motion and continuity are nonlinear and, in particular, the equation of motion must be integ-twice with respect to time for solution.
2) In order to calculate the position of a water particle, it is desirable to define the discretized values of velocity components at the same mesh points. The mesh scheme of this type often causes the discrepancy between the numbers of discretized equations and variables.

A sufficiently fine mesh has to be employed to maintain the accuracy and stability of computation and the generation and accumulation of residual errors are to be carefully monitored throughout the computation. It is often necessary to ignore a few local conditions to obtain a solvable system of simultaneous equations particularly if the velocity formulation is adopted.

## Velocity Formulation

In this method, the basic equations are written for the horizontal and vertical velocity components, U and V. The pressure terms are eliminated from the equations of motion (1) and (2) on the assumption of irrotationality:

$$U_aX_b-U_bX_a+V_aY_b-V_bY_a=0. \tag{6}$$

Through differentiation with respect to time, the equation of continuity (3) leads to

$$U_aY_b-U_bY_a-V_aX_b+V_bX_a=0. \tag{7}$$

The surface condition is also rewritten in terms of U and V:

$$(U_t+F)X_a+(V_t+G)Y_a=0 \quad \text{on } b=b_0, \tag{8}$$

where F and G are the horizontal and vertical components of the external force acceleration H. The particle position (X, Y) is estimated from an obvious relationship,

$$X_t=U, \qquad Y_t=V. \tag{9}$$

For the application of finite difference method, an orthogonal mesh is provided to cover the a-b plane and all the equations and conditions as well as the variables are discretized. The particle velocity and position are calculated alternately at lagged time-steps. By virtue of this arrangement, the X and Y values at the previous time-step are rationally used in solving the system of Eqs. (6), (7) and (8).

## Pressure Formulation

The principal unknown variable here is the pressure, P. The equations of motion (1) and (2) are rewritten into the following set of equations:

$$X_{tt}=(-Y_b P_a+Y_a P_b)/\rho, \tag{10}$$

$$Y_{tt}=(\ X_b P_a-X_a P_b)/\rho. \tag{11}$$

The pressure distribution is determined in such a way that the particle positions estimated through numerical integration of these equations satisfy the equation of continuity (3) for each time-step. The particle position (X, Y) itself is used as an auxiliary variable and the rate of deformation, $X_a$, etc., are evaluated from the known X and Y values at the current time-step. Note that J remains unity since (a, b) has been defined as the initial particle position.

This formulation is more straightforward compared with the velocity formulation; the additional terms such as damping terms can be readily introduced into the equation system if necessary. The size of coefficient matrices to be handled is roughly half as large as that in the case of velocity formulation, and they are of better characteristics with predominant diagonal components since the basic equation is Poisson-type in principle. A difficulty, however, is found in the specification of appropriate P values at side and bottom boundaries. The problem is more serious if the mesh is arbitrarily shaped.

## Sloshing Waves in a Rectangular Tank

Both the methods described above were applied to analyses of two-dimensional sloshing waves in a rectangular tank. Figure 1 shows the deformation of computational mesh and flow field due to a finite-amplitude oscillation of water body. The calculated variation of surface elevation at a sidewall is plotted in Fig. 2 in comparison with the perturbation solution given by

(a) t = 5.75 T

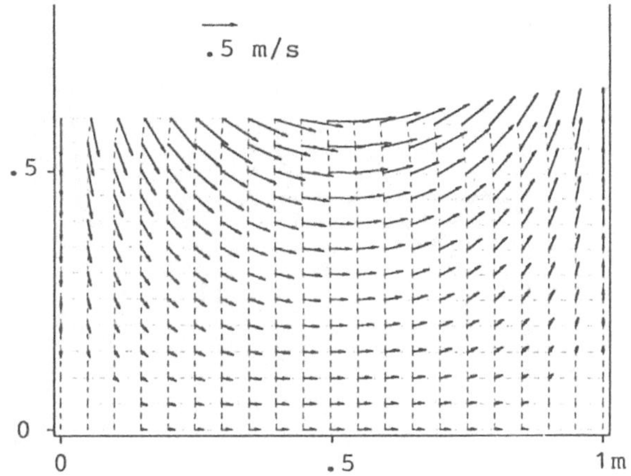


(b) t = 6 T

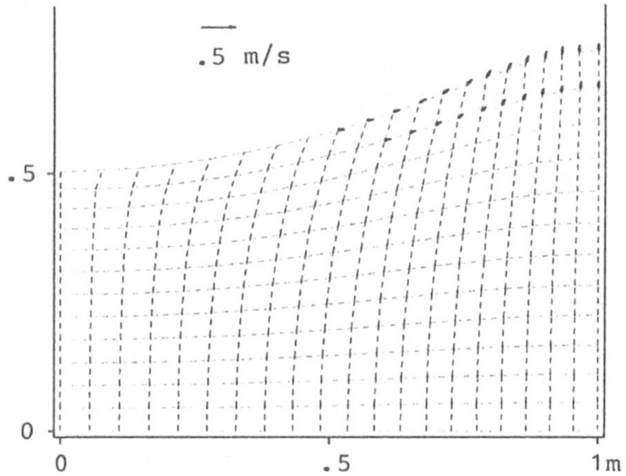


Fig.1. Two-dimensional sloshing of water in a rectangular tank.

Takayama [4].

So far as these waves with moderate amplitudes are concerned, no significant difference is found depending on the method employed. Further deformation of the mesh, however, causes the instability of computation owing to checkerboard split and the accumulation of vorticity. This tendency is stronger if the computation is based on the pressure formulation.

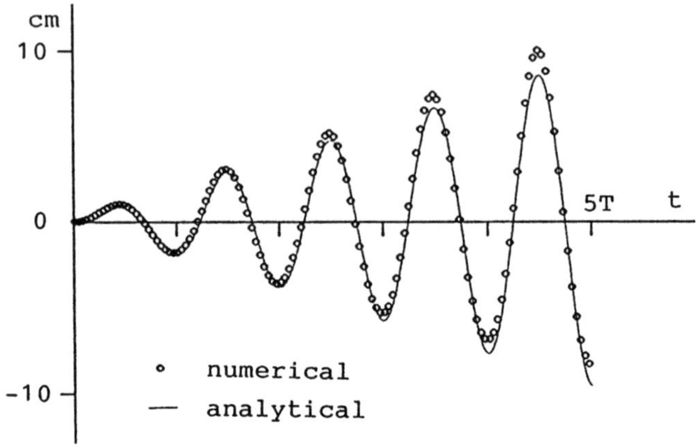


Fig.2. Variation of surface elevation at a sidewall.

## Shoaling of a Solitary Wave

The usefulness of these numerical models was examined also through analyses of a solitary wave in a rectangular basin. The wave was generated by a horizontal move of a sidewall. First, its propagation over a flat bottom and reflection at another sidewall were well simulated. Then shoaling of the solitary wave over a 1/20 slope was treated.

Figure 3 illustrates the changes in surface profile, giving an overall view of the wave transformation. In Fig. 4, the calculated profiles before breaking are plotted together with the experimental results obtained by Nagao [5]. Figure 5 shows a

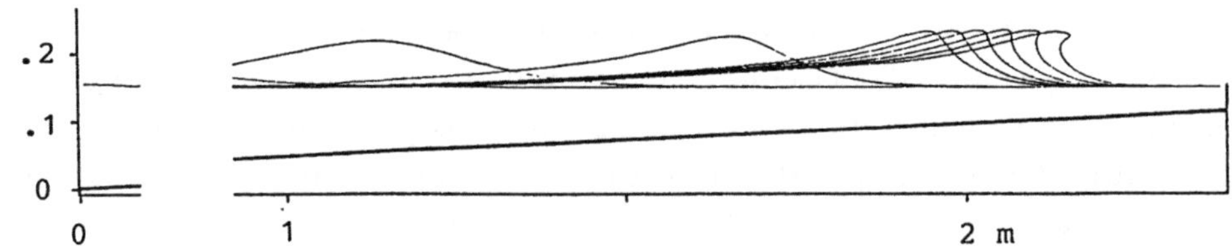


Fig.3. Shoaling of a solitary wave over a 1/20 slope.

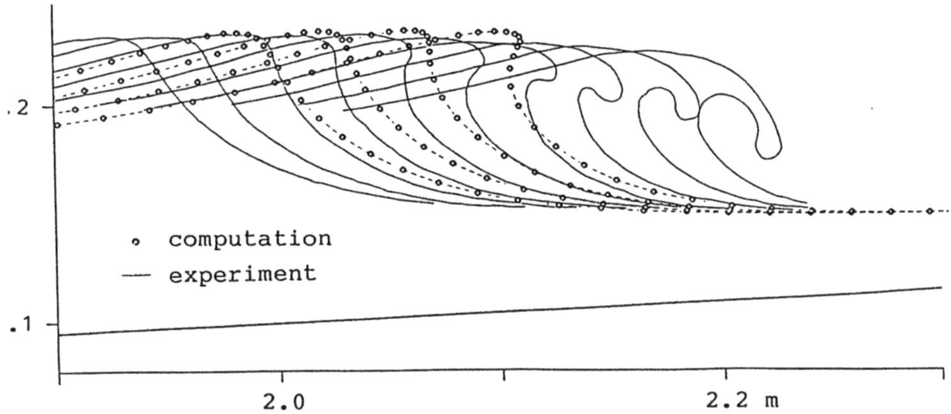


Fig.4. Calculated and observed changes in surface profile before breaking.

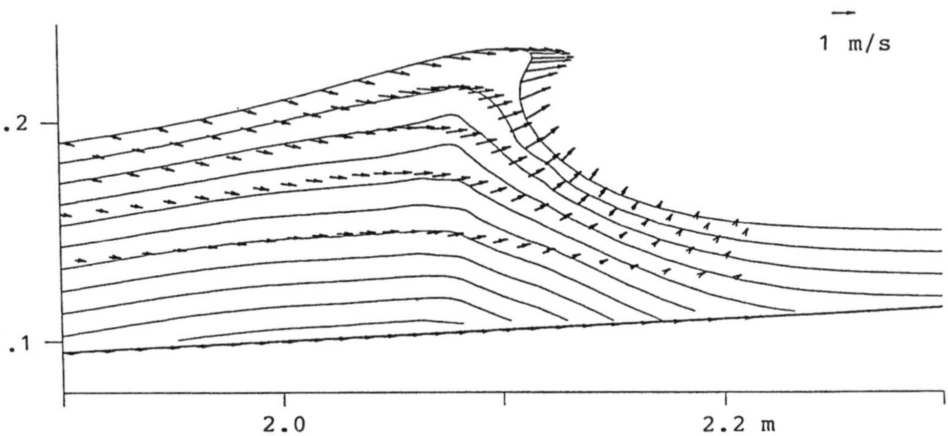


Fig.5. Water particle velocities and isobars.

calculated vector field of water particle velocity and isobars. A region near the wave crest is characterized by high particle velocities and mild pressure gradient, which may be the direct causes of wave breaking.

## Conclusion

The number of mesh points involved in the last example was 455 in total and roughly 5 min. was required for a 1600 step computation by the high-speed computer system M-780/20. The improvement in the efficiency of stepwise iteration will allow more precise analyses, and then the numerical computation based on the Lagrangian description can be a powerful means for studying two-dimensional wave transformations.

## References


1. Brennen, C.: Some numerical solutions of unsteady free surface wave problems using the Lagrangian description of the flow. Lecture Notes in Physics 8 (1970) 403-409.

2. Chan, R. K.-C.: A generalized arbitrary Lagrangian-Eulerian method for incompressible flows with sharp interfaces. J. Comp. Phys. 17 (1975) 311-331.

3. Fritts, M. J. and Borris, J. P.: The Lagrangian solution of transient problems in hydrodynamics using a triangular mesh. J. Comp. Phys. 31 (1979) 173-215.

4. Takayama, T.: Theory of transient fluid waves in a vibrated storage tank, Rep. Port and Harbour Res. Inst. 15-2 (1976) 3-53.

5. Nagao, M.: Study of wave transformation on a sloping beach. Master Thesis, Tohoku Univ. (1986) 121p.

# Transient Approaches of Non-linearities in Naval Hydrodynamics

A. JAMI, G. POT

Laboratoire National d'Hydraulique (EDF)
6, quai Watier 78400 Chatou, France

Summary

A general formulation is given for the solution of the potential flow induced by the free motion of a floating body. Using a variational formulation in the fluid domain a matrix form is given for the problem including the non-linear free surface conditions and the equations of motion for the rigid body. An integral representation formula, exact in the case of linearized free surface conditions is used for the formulation in unbounded domains. Various types of test cases are presented in which a specific algorithm is applied for each type of non-linearity.

## Introduction

Non-linear free surface waves have been lengthly studied ; although the physical phenomena are rather well known, their numerical simulation are still an open problem particularly in the practical applications of naval hydrodynamics. The numerical code named MELINA and developped at ENSTA (*) being robust in application to the linearized case, we tackled the account for non-linearities. In this paper, we present a terse but exhaustive classification for those non-linearities together with the algorithm adapted for each of them.

Let us denote $\Omega(t)$ the fluid domain bounded by a free surface $\Gamma(t)$ , a rigid body with wetted surface $S(t)$ , and eventually a fixed bottom $\mathcal{B}$. In general $\Omega(t)$ is an unbounded domain ; in case where $\Omega$ is bounded, the limiting boundary is denoted $\Sigma$ .

The position of the rigid body is determined by the six-components vectors of initial position $\vec{X}_0$ and displacements $\vec{X}(t)$. The motion is then given by the velocity vector $\vec{V}(t) = \frac{d\vec{X}}{dt}(t)$. (1)

When dealing with a freely floating body, the fundamental law of dynamics reads :

$$\mathbb{M}\,\frac{d\vec{V}}{dt}(t) + \mathbb{B}\,\vec{V}(t) + \mathbb{C}\left[\vec{X}(t) + \vec{X}_0(t)\right] = \vec{F}(t) + (\rho\mathcal{V} - \mu_G)\vec{\zeta} \qquad (2)$$

K. Horikawa, H. Maruo (Eds.)
Nonlinear Water Waves
IUTAM Symposium Tokyo/Japan 1987

where $\mathbb{M}$ is a generalized mass matrix, $\mathbb{B}$ a damping matrix (accounting for viscous effects from experiments), $\mathbb{C}$ is the matrix of restoring coefficients, $\rho$ is the fluid density, $\mathcal{V}$ is the displaced volume of fluid, $\mathcal{M}$ is the total mass of the body, $\vec{z}$ being upward vertical. The forcing term $\vec{F}(t)$ is the sum of given forces $\vec{F}_0(t)$ and the fluid pressure force

$$F_1(t) = \int_{S(t)} [p + \rho g z] \vec{N} \, ds \tag{3}$$

with $g$ the acceleration of gravity and $\vec{N}$ a generalized normal on $S$. Note that given $\vec{X}_0$, $\vec{F}(t)$ and an initial data for $\vec{V}$, the system (1), (2) can be uniquely solved if $\mathbb{M}$ is invertible which we suppose.

Accounting for the classical assumptions for potential flows, the fluid system reads :

$$\left.\begin{array}{lll} \Delta \phi = 0 & \text{on} & \Omega(t) \\ \frac{\partial \phi}{\partial n} - \vec{V}.\vec{N} = 0 & \text{on} & S(t) \\ \left.\begin{array}{l} p = 0 \\ \frac{d}{dt}[z - \eta] = 0 \end{array}\right\} & \text{on} & \Gamma(t) \end{array}\right\} \tag{5}$$

with a constant pressure taken as reference above the free surface $\Gamma(t)$ the elevation of which is the unknown function $\eta$. This system is completed by initial and far field conditions for which we shall suppose the fluid is at rest.

Thanks to Bernoulli's relation the pressure is related to the velocity potential in the closed domain $\Omega(t)$,

$$p/\rho + \frac{1}{2} |\vec{\text{grad}}\, \phi|^2 + \frac{\partial \phi}{\partial t} + g z = 0 \tag{6}$$

and this expression is used both for the dynamic condition on $\Gamma(t)$ in (5) and for the fluid force on the body $S$. Finally the kinematic condition on $\Gamma(t)$ in (5) reads

$$\frac{1}{\vec{n}.\vec{z}} \frac{\partial \eta}{\partial t} - \frac{\partial \phi}{\partial n} = 0 \tag{7}$$

where $\vec{n}$ is the unit normal on $\Gamma(t)$.

Variational formulation for the fluid system

Suppose $\vec{V}$ is a given function of time on $S(t)$ and $\Omega(t)$ bounded by $\Sigma(t)$. Taking $\Psi$ as a test function in a properly chosen Hilbert space $\mathcal{H}$ with dot product $(\cdot, \cdot)$ of real valued functions defined on $\Omega(t)$, the traces of which are square integrable ; we define the operators $A(t)$ and $\delta_c$ where $c$ stands for a boundary, by :

$$\left.\begin{array}{l} (A(t)\,\phi, \Psi) = \int_{\Omega(t)} \vec{\text{grad}}\, \phi \,.\, \vec{\text{grad}}\, \Psi \, d\omega \\ (\delta_c\, \phi, \Psi) = \int_c \phi\, \Psi \, ds \end{array}\right\} \tag{8}$$

A variational formulation of equations (5) then reads :

Find $(\phi,\eta) \in \mathcal{H} \times L^2(\Gamma)$ such that

$$\left.\begin{aligned} &\delta_{\Gamma(t)}\left[\frac{\partial\phi}{\partial t} + g\eta + \frac{1}{2}|\overrightarrow{\text{grad}}\,\phi|^2\right] = 0 \\ &-\delta_{\Gamma(t)}\frac{1}{\vec{n}.\vec{z}}\frac{\partial\eta}{\partial t} + A(t)\,\phi - \vec{V}(t).\,\delta_{S(t)}\vec{N} - \delta_\Sigma\frac{\partial\phi}{\partial n} = 0 \end{aligned}\right\} \quad (9)$$

Note that in the case of a bounded basin, $\Sigma$ is a rigid wall where $\partial\phi/\partial n$ vanishes. In the general case a boundary condition on $\Sigma$ must be expressed. A method has been developed in the case of a linearized free surface condition which uses an integral representation formula [1].

A finite element discretisation in space classically gives, for the operators $A(t)$ and $\delta_c$ defined in (8), matrices of real valued coefficients.

## Linearized free surface formulation

In this case, although the function $\eta$ remains an unknown function of our problem, the boundary $\Gamma$ is fixed on the plane $z=0$ and the quadratic terms in (6) and (7) are dropped. The fluid system is then linear and a Green function (satisfying the so-called linear free surface condition for the potential, the Laplace equation in the fluid domain except at a point $M$ (the rigid body being ignored), and the bottom condition) is defined by :

$$H_M(P;t) = \delta(t)\,G_M(P) + Y(t)\,F_M(P;t) \quad (10)$$

where $\delta$ and $Y$ are respectively the Dirac and Heaviside distributions, $G_M$ is the elementary solution of the Laplace equation at point M and vanishing on the plane $z=0$ and $F_M(P;t)$ is a regular function (see [1]). Thanks to vanishing initial conditions for $\phi$ and $\frac{\partial\phi}{\partial t}$ on the linearized free surface, an integral representation formula is written for $\phi$ such that :

$$\begin{aligned} \phi(M;t) = &\int_{S(t)}\left[\phi(P;t)\frac{\partial G_M}{\partial n_P}(P) - \vec{V}(t).\vec{N}(P)\,G_M(P)\right]ds_P \\ &+ \int_0^t d\tau\int_{S(\tau)}\left[\phi(P;\tau)\frac{\partial F_M}{\partial n_P}(P;t-\tau) - \vec{V}(\tau).\vec{N}(P)\,F_M(P;t-\tau)\right]ds_P \end{aligned} \quad (11)$$

Note that when $M$ does not belong to $S(t)$ even the first integral in the R·H·S of (11) has a regular kernel. The second term reports on the history of the moving body to take into account the waves formed on the unbounded free surface. In the system (9) the normal derivative on $\Sigma$

can then be replaced by :

$$\frac{\partial \phi}{\partial n_M}\Big|_\Sigma (t) = \delta_{S(t)} \mathbb{D}_M \phi(t) - \vec{V}(t) \cdot \delta_{S(t)} \mathbb{G}_M \vec{N} + \int_0^t \Big[ \delta_{S(\tau)} \mathbb{E}_M(t-\tau)\, \phi(\tau) - \vec{V}(\tau)\, \mathbb{F}_M(t-\tau) \vec{N} \Big] d\tau \tag{12}$$

where the space variable $P$ has been dropped and the matrices of coupling coefficients are defined for $M \in \Sigma$, by :

$$\left.\begin{array}{ll} \mathbb{G}_M = \left\{ \frac{\partial G_M}{\partial n_M}(P) \right\} , & \mathbb{D}_M = \frac{\partial}{\partial n_P} \mathbb{G}_M(P) \quad ; \; P \in S(t) \\ \mathbb{F}_M(\tau) = \left\{ \frac{\partial F_M}{\partial n_M}(P;\tau) \right\} , & \mathbb{E}_M(\tau) = \frac{\partial}{\partial n_P} \mathbb{F}_M(P;\tau) ; \; P \in S(\tau) \end{array}\right\} \tag{13}$$

Because of properties of energy conservations for both the linear and non-linear problems in the unbounded domains and the fact that the energy of the fluid system is entirely given by the motion of the body, the boundary condition on $\Sigma$ which is exact for the linear case is a good approximation for the non-linear case. This has yet to be more developed and confirmed by numerical tests.

## A general matrix form for the fluid-body system

We denote $U = {}^t(\phi, \eta, \vec{V}, \vec{X})$ the vector of unknowns of the full system of equations which then reads in matrix form :

$$\mathcal{A}(t) \frac{d}{dt} U(t) + \mathcal{B}(t)\, U(t) + \mathcal{N}\big(U(t);t\big) + \int_0^t \mathcal{C}(t;\tau)\, U(\tau)\, d\tau = \mathcal{F}(t) \tag{14}$$

where the matrices of operators $\mathcal{A}$ , $\mathcal{B}$ and $\mathcal{C}$ can easily be derived from the previous equations. [2] .

As we formulate the problem, non vanishing initial conditions for $U$ can only be given for its third component $\vec{V}(t)$ but even if $\vec{V}(0) = 0$ the system can start a motion if $\vec{X}_0$ does not correspond to an equilibrium position. Note that in the linear case, we have : $\vec{n} . \vec{z} = 1$ , $\mathcal{N}(U(t);t) \ll 1$ the domain being fixed the matrices of operators $\mathcal{A}$ and $\mathcal{B}$ are constant in time and when the motion of the body is known, $\mathcal{C}(t,\tau)$ is a function of $t-\tau$ .

Without going further on the discussion on this complete but approximate solution of the general case, we show here some numerical results each of which putting in light a special aspect of the different non-linearities appearing in this problem and we present the time discretization we adopted.

Bounded fluid domain application

It is necessary to clarify the scheme by the simplest case. We look for the solution of a coupled spring-aquarium (see [3]). Starting at rest from a non-equilibrium position, the oscillations of the aquarium result in a free surface elevation. The problem is set in a bounded domain for which there is no boundary $\mathcal{S}$ and $\Sigma$ is the inner surface of the aquarium.

In case where the motions remain of small amplitude, the linearization applies. The system (14) is then reduced to :

$$\mathcal{A}\,\frac{dU}{dt}(t) + \mathcal{B}\,U(t) = \mathcal{F}(t) \tag{15}$$

for which a theorem of existence and uniqueness is known and the mechanical energy is conserved. Moreover the solution has a modal decomposition.

For a given constant time step $\Delta t$, a second order time discretization of (15) is

$$\frac{2}{\Delta t}\mathcal{A}(U^{n+1} - U^{n}) + \mathcal{B}(U^{n+1} + U^{n}) = \mathcal{F}^{n+1} + \mathcal{F}^{n} \tag{16}$$

It is then easily checked that both the mechanical energy of the system and the volume of fluid are conserved.

In the general case, non-linearities have to be taken into account. We then adopted the following algorithm :

- non-linear terms occurring in the vector $\mathcal{N}$ are calculated at time $t_n$,
- a $\theta$-method is introduced to ensure the stability of the numerical algorithm ; consequently a dissipation of energy is forced by the numerical scheme but which can be controlled in this case of a bounded domain,
- as the fluid domain is varying with time, in the time interval the operators $\mathcal{A}(t)$ and $\mathcal{B}(t)$ do not apply on $U^n$ or $U^{n+1}$. We thus use a two step method which reads :

(i) Find $U^*$ such that :

$$\frac{1}{\Delta t}\mathcal{A}^n(U^* - U^n) + \theta\mathcal{B}^n U^* + (1-\theta)\mathcal{B}^n U^n = \frac{1}{2}(\mathcal{F}^{n+1} + \mathcal{F}^n) - \mathcal{N}^n(U^n) \tag{17}$$

(ii) From $U^*$ compute $\mathcal{A}^*$ and $\mathcal{B}^*$

(iii) Find $U^{n+1}$ such that :

$$\frac{1}{\Delta t}\mathcal{A}^* U^{n+1} - \frac{1}{\Delta t}\mathcal{A}^n U^n + \theta\mathcal{B}^* U^{n+1} + (1-\theta)\mathcal{B}^n U^n = \frac{1}{2}(\mathcal{F}^{n+1} + \mathcal{F}^n) - \mathcal{N}^n(U^n)$$

This method results in solving twice the same system in the linearized case. Numerical tests show that higher order harmonics appear in the free surface elevation in the non-linear case which consequently needs a finer grid.

Unbounded domain application with linearized free surface

This type of approximation can be used when dealing with a fully submerged body even when it undergoes a large amplitude free motion. In this case equation (14) reduces to :

$$\mathcal{A} \frac{dU}{dt}(t) + \mathcal{B}\, U(t) + \int_0^t \mathcal{C}(t;\tau)\, U(\tau)\, d\tau = \mathcal{F}(t) \qquad (18)$$

where $\mathcal{A}$ and $\mathcal{B}$ are constant operators but $\mathcal{C}$ depends on the body displacement $\vec{X}(t)$. When the motion of the body is given this only results in a complete computation of the matrices defined in (13) at each time step ; the test case shown on figure[1] of the forced heave of a submerged sphere shows that higher order harmonics appear in the solution due to this non-linearity (see also [4]).

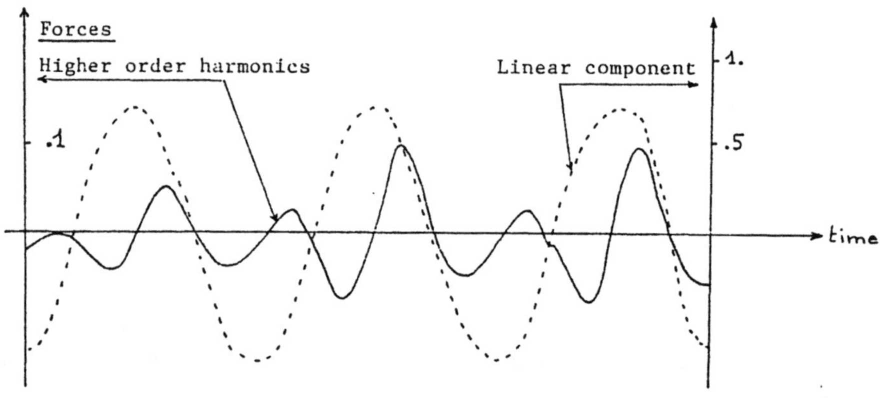


Fig. 1 : Vertical force for the periodic forced heave of a submerged sphere.

A more general test-case has been performed : the free raise of a submerged sphere with density less than the fluid density. As the body position $\vec{X}_0 + \vec{X}(t)$ is not known a two step algorithm has been developed :

- Solve the dynamic equations for the body with fixed hydrodynamic forces at time $t_n$ .

- The solution of the previous system giving a predictor $\vec{X}^*$ for the displacement of the body, compute the updated $\mathscr{C}^*$ operator.
- Solve the full system (18) with $\mathscr{C}^*$.

On the figures 2 and 3 acceleration and hydrodynamic forces are shown. Recall that in the free space the acceleration and linear forces remain constant and the quadratic forces vanish ; an effect of added mass is due to the free surface.

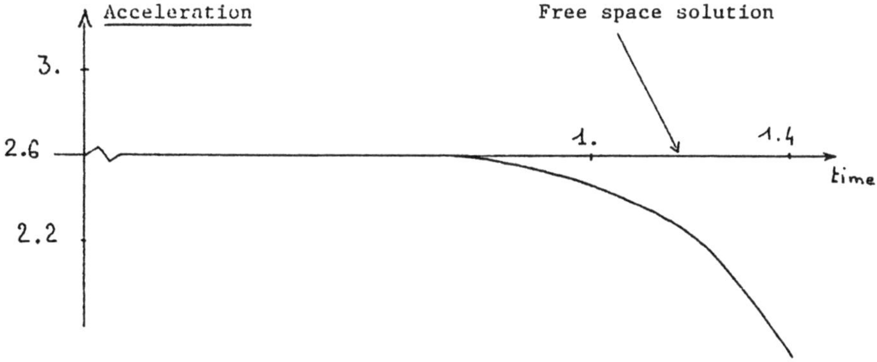


Fig. 2 : Acceleration of a freely raising submerged sphere.

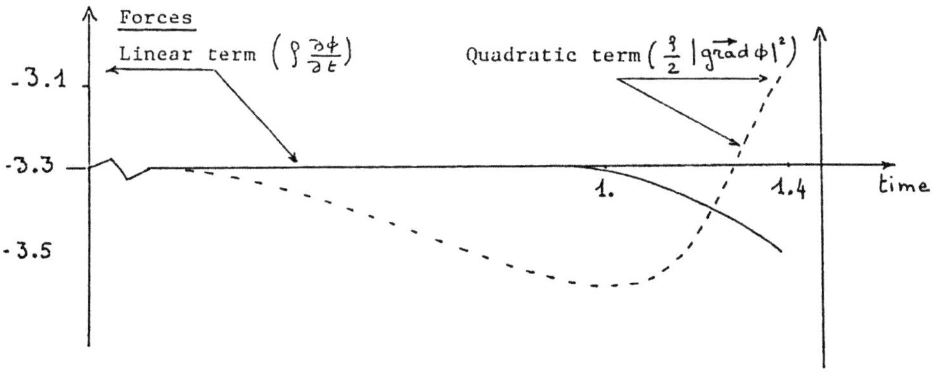


Fig. 3 : Linear and quadratic terms in the hydrodynamic forces.

## Conclusion

Many more numerical examples can be performed among which the forced translation parallel to the free surface of a submerged body. Nevertheless the general formulation is still under development for the numerical code is not yet able to deal with a sufficient number of elements to represent significant non linear behaviour of the free surface.

## References

* Work done under contract DRET/ENSTA n° 1061 (Ministry of Defense).

[1] A. JAMI : ENSTA Report n° 154, 1982.

[2] G. POT : ENSTA Report n° 211, 1986.

[3] L. GELEBART : ENSTA Note, CHN, 1985.

[4] P. FERRANT & A. CLEMENT : IUTAM SYMPOSIUM ON NON-LINEAR WATER-WAVES, TOKYO, 1987.

# The Theoretical Studies on Nonlinear Standing Waves

Y. Y. CHEN

Institute of Harbor and Marine Technology, Taiwan, China

C. P. TSAI

National Chung-Hsing University, Taiwan, China

H. H. HWUNG

National Cheng-Kung University, Taiwan, China

## Summary

A succinct approximation method to the standing waves is proposed in this paper. The solutions are carried out to the leading terms, which are remained in the form of implicit function of the corresponding surface profile. The highest gravity standing wave is performed by this approximation method, and the primary variation of dynamical properties is also discussed in the paper.

## Introduction

The gravity standing waves are produced while two wavetrains of equal wavelength and amplitude are propagating in the opposite direction. Because of the time-dependence characteristics which can not be transformed into steady motion, the analysis of the standing waves is more complicated than the classical Stokes progressive waves.

A number of complicated theoretical analysis in mathematical procedures have been carried out by predecessors, such as double Fourier series expansion (Penney & Price [1]), Stokes' perturbation expansion (Tadjbakhsh & Keller [2], Goda & Kakizaki [3], Tsai & Chen [4]), and conformal mapping method (Grant [5], Schwartz & Whitney [6]). But now, a more succinct approximation method obtained from Lagrange's expansion is proposed in this paper. Detailed descriptions of the characteristics on standing wave are presented, besides, the primary variation of dynamical properties of the standing waves is also discussed.

K. Horikawa, H. Maruo (Eds.)
Nonlinear Water Waves
IUTAM Symposium Tokyo/Japan 1987

## Theoretical Approach

A sketch diagram of the standing wave motion is shown in Fig.1. In the following analysis, the x-axis is taken to be coincident with the mean still water level while the y-axis is chosen upward through the crest, and d is the depth from the mean still water level to the bottom.

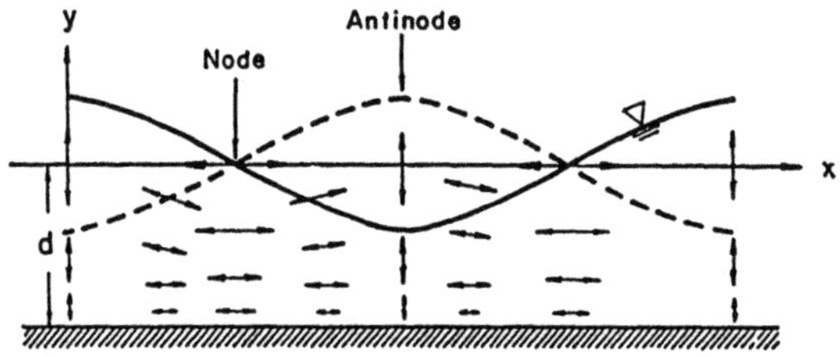


Fig.1. A sketch diagram of the standing wave

Since the flow is to be assumed inviscid, incompressible on a uniform finite depth, the velocity potential $\phi$ for irrotational motion gives rise to the Eulerian water particle velocity components as

$$u = \phi_x, \quad v = \phi_y \tag{1}$$

where the velocity potential $\phi$ statisfies Laplace's equation

$$\nabla^2\phi = \phi_{xx} + \phi_{yy} = 0 \tag{2}$$

The dynamic and kinematic boundary conditions at free surface can be expressed as

$$g\eta + \phi_t + \tfrac{1}{2}(\phi_x^2 + \phi_y^2) = F(t) \quad \text{at} \quad y = \eta \tag{3}$$

$$\phi_y = \eta_t + \eta_x\phi_x \quad \text{at} \quad y = \eta \tag{4}$$

and the boundary conditions at the bottom and wall are

$$\phi_y = 0 \text{ at } y = -d, \text{ and } \phi_x = 0 \text{ at } x = 0 \tag{5}$$

where $\eta = \eta(x,t)$ is the water surface elevation, g is the gravitational acceleration, and the subscripts (x,y and t) designate differential with respect to the Cartesian coordinates and the time. F(t) represents an arbitrary function of time which may be merged in $\phi_t$.

A general solution for $\phi$ which satisfies equations (2) and (5) can be assumed in Fourier series as

$$\phi = \sum_{m=0}^{\infty} \sum_{n=0}^{\infty} C_{mn} \frac{\cosh mk(d+y)}{\cosh mkd} \cos mkx \sin n\sigma t \tag{6}$$

in which m and n are the summation indexes, if (m+n) is odd, then $C_{mn}=0$ (Roberts & Schwartz[7]). To obtain a simple but sufficiently accurate solution for the flow field, the leading term in equation (6) is selected in the following approximate analysis.

Let F(t) be merged in $\phi_t$, then

$$\phi = C_1 \frac{\cosh k(d+y)}{\cosh kd} \cos kx \sin \sigma t + F(t) \tag{7}$$

Substituting equation (7) into (3), and let the wave profile,$\eta$, be kept in the primary implicit terms, then

$$\eta = \frac{1}{g} \{ -C_1 \sigma \frac{\cosh k(d+\eta)}{\cosh kd} \cos kx \cos \sigma t - F'(t)$$

$$- \frac{1}{8} C_1^2 k^2 \frac{\cosh 2kd - \cos 2kx}{\cosh^2 kd} (1-\cos 2\sigma t) \} \tag{8}$$

in which F'(t) denotes the time derivative of F(t). After inserting the derivatives of $\phi$ and $\eta$ from (7) and (8) into (4), we get the following approximate formula,

$$C_1 gk \sinh k(d+\eta) = C_1 \sigma^2 \cosh k(d+\eta) + \text{higher order terms} \tag{9}$$

From the above equation, equation (8), can be transformed into

$$\eta = \frac{1}{g}\{ - \frac{C_1 gk}{\sigma} \frac{\sinh k(d+\eta)}{\cosh kd} \cos kx \cos \sigma t - F't)$$

$$- \frac{1}{8} C_1^2 k^2 \frac{\cosh 2kd - \cos 2kx}{\cosh^2 kd} (1-\cos 2\sigma t) \} \tag{10}$$

and the linear dispersion relationship can be obtained

$$\sigma^2 = gk \tanh kd \tag{11}$$

By Lagrange's expansion on equation (10) to the order $(C_1)^2$, then

$$\eta = \frac{1}{g}\{-\frac{C_1 gk}{\sigma} \tanh kd \cos kx \cos\sigma t - F'(t)$$

$$- \frac{1}{8} C_1^2 k^2 \frac{\cosh 2kd - \cos 2kx}{\cosh^2 kd} (1-\cos 2\sigma t)\}$$

$$+ \frac{1}{4} \frac{C_1^2 k^3}{\sigma^2} \tanh kd (1+\cos 2kx)(1+\cos 2\sigma t) \qquad (12)$$

Because of the x-axis is taken at the mean still water level, then $F'(t)$ can be obtained by the spatial average of equation(12), it is found that

$$F'(t) = \frac{1}{4} \frac{gC_1^2 k^3}{\sigma^2} \tanh kd(1+\cos 2\sigma t)$$

$$- \frac{1}{8} C_1^2 k^2 \frac{\cosh 2kd}{\cosh^2 kd} (1-\cos 2\sigma t) \qquad (13)$$

and substituting equation (13) into (10), then a semi-final expression of $\eta$ is

$$\eta = \frac{1}{g} \{ - \frac{C_1 gk}{\sigma} \frac{\sinh k(d+\eta)}{\cos kd} \cos kx \cos\sigma t$$

$$+ \frac{1}{8} \frac{C_1^2 k^2}{\cosh^2 kd} \cos 2kx(1-\cos 2\sigma t)$$

$$- \frac{1}{4} \frac{gC_1^2 k^3}{\sigma^2} \tanh kd(1+\cos 2\sigma t) \} \qquad (14)$$

When $\sigma t = n\pi$ , where n is an integer, the velocity vanishes. In the situation, the water surface is either at the highest or the lowest position. The crest is at $x=0$, equation (14) gives

$$\eta_o = - \frac{C_1 k}{\sigma} \frac{\sinh k(d+\eta_o)}{\cosh kd} - \frac{1}{2} \frac{C_1 k^3}{\sigma^2} \tanh kd \qquad (15)$$

and the trough is at $x=\pi$, equation (14) gives

$$\eta_\pi = \frac{C_1 k}{\sigma} \frac{\sinh k(d+\eta_\pi)}{\cosh kd} - \frac{1}{2} \frac{C_1^2 k^3}{\sigma^2} \tanh kd \qquad (16)$$

The total wave height of the standing waves is defined as $H=\eta_o - \eta_\pi$, we have

$$H=-C_1\frac{k}{\sigma}\frac{\sinh k(d+\eta_o)+\sinh k(d+\eta_o-H)}{\cosh kd} \tag{17}$$

$C_1$ can be determined

$$C_1=-\frac{H\sigma}{k}\frac{\cosh kd}{\sinh k(d+\eta_o)+\sinh k(d+\eta_o-H)} \tag{18}$$

Therefore, inserting equations (13) and (18) into (7) and (14), the complete solutions of $\phi$ and $\eta$ can be obtained.

Theoretical results

Although all the above results are maintained in an implicit function form, under a set of given conditions of standing waves, that is water depth d, wave height H and wavelength L or period T, the results of the standing wave motion can be easily solved by numerical iteration method. And the related properties of standing waves are discussed as follows.

(i) The case of the highest waves:

When $\sigma t=n\pi$, the slope at arbitrary point on the wave profile can be formulated by using equations (14) and (18), we get

$$\frac{d\eta}{dx}=\frac{-kHG^{-1}\sinh k(d+\eta)\sin kx}{1-kHG^{-1}\cosh k(d+\eta)\cos kx} \tag{19}$$

where $G=\sinh k(d+\eta_o)+\sinh k(d+\eta_o-H)$ (20)

However, a summit will form at the crest of wave profile when wave motion is at the moment of the highest state. Now we set zero by zero in equation (19) at $\eta=\eta_o$, then the maximum wave height can be obtained

$$(kH)_{max}=\frac{G}{\cosh k(d+\eta_o)} \tag{21}$$

here, $\eta_o$ is expressed in equation (15). In the case of deep water, i.e. $d\to\infty$, the maximum steepness can be obtained as

$$\left(\frac{H}{L}\right)_{max}=0.2035 \tag{22}$$

Accordingly, the corresponding wave profile, some of Taylor's[8] results and Penney & Price's[1] results are all plotted in Fig.2. From the comparison shown in this figure, we find that the corresponding profile has only little difference between our approximation and Taylors' results.

After we use L'Hôspital rule in equation (19), the slope in the neighborhood of the summit can be easily found

$$\left(\frac{d\eta}{dx}\right)^2_{\eta=\eta_o} = 1 \tag{23}$$

This result indicates that the crest of the highest standing wave is constructed by a right-angled nodal form.

(ii) The dynamical properties

Based on the above results of $\phi$ and $\eta$ , the average kinetic and potential energy related to one period in any position can be obtained. If the order $(H)^2$ is carried out, then

$$\bar{K} = \frac{1}{8}\rho gH^2G^{-2}(\sinh^2kd - kd\tanh kd\cos 2kx) \tag{24}$$

$$\bar{V} = \frac{1}{8}\rho gH^2G^{-2}\sinh^2kd(1+\cos 2kx) \tag{25}$$

and their computational results are presented in Fig.3. At antinode position, although the potential energy reaches to the maximum, the kinetic energy is not equal to zero because there is contribution by vertical velocity of water particle. And at node position, the potential energy is zero, but the kinetic energy is only contributed by horizontal motion of water particle. Hence, the total average energy related to one period is sinusoidal variation, and the maximum is at antinode; the minimum is at node. As for average kinetic and potential energy related to wavelength, we get

$$\bar{K}_L = \frac{1}{8}\rho gH^2G^{-2}\sinh^2kd(1-\cos 2\sigma t) \tag{26}$$

$$\bar{V}_L = \frac{1}{8}\rho gH^2G^{-2}\sinh^2kd(1+\cos 2\sigma t) \tag{27}$$

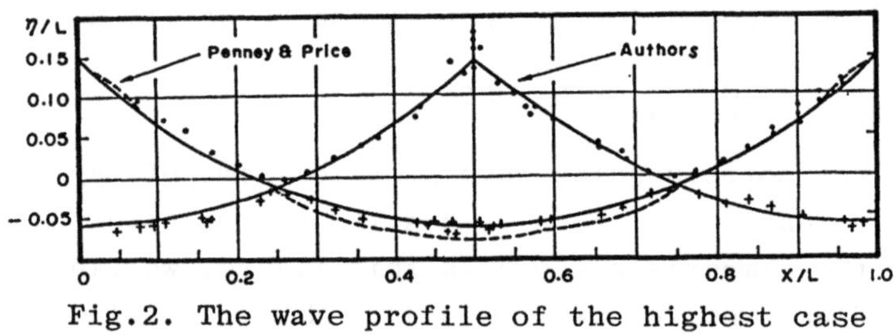


Fig.2. The wave profile of the highest case

And the results are shown in Fig.4. From the computational results, we see the total average energy is a constant, that is, the total energy over one wavelength is conservative because that there is no dissipative or generative force in the wave motion system. The radiation stress of standing waves can be obtained from the formulas by Longuet-Higgins & Stewart [ 9 ], and the result of the average radiation stress related to one period and one wavelength is

$$S_{xx}=\frac{1}{8}\,\rho\, gH^2G^{-2}(\sinh^2 kd+2kd\tanh kd) \tag{28}$$

if the wave is small ampltitude, equation (28) can be simplified to the result of Longuet-Higgins & Stewart [ 9 ]. Figure 5 presents the dimensionless radiation stress related to wave steepness and the relative water depth.

As for the total dynamic pressure of standing wave, it is just equal to the transverse radiation stress, as follows;

$$P=\frac{1}{8}\,\rho gH^2G^{-2}(kd\tanh kd) \tag{29}$$

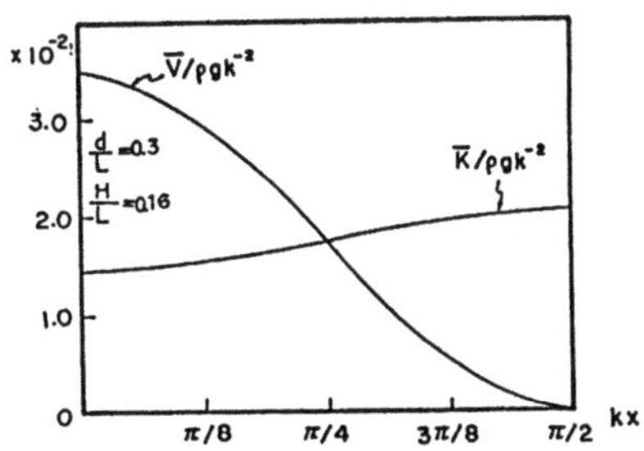


Fig.3 Average energy vs position

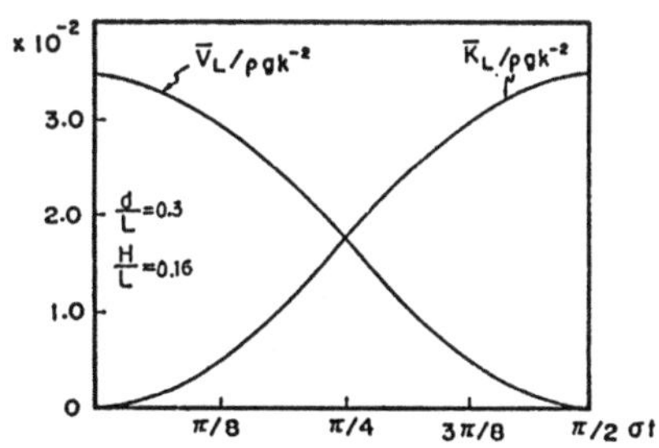


Fig.4 Average energy vs time

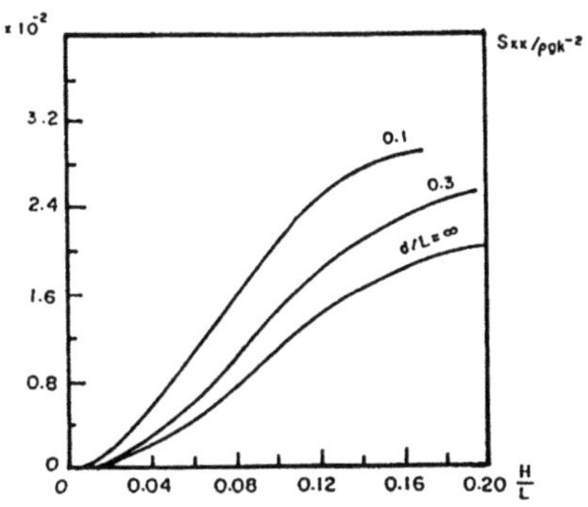


Fig.5 Radiation stress

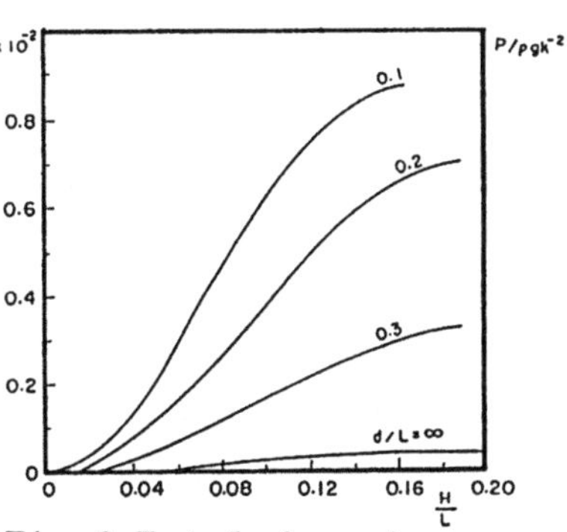


Fig.6 Total dynamic pressure

The theoretical result is shown in Fig.6, which is similar to the variation of radiation stress.

## Conclusions

From the above theoretical approximation, a succinct method was proposed to the analysis of the standing wave motion. The highest standing waves in a uniform water depth can be obtained in this paper, and the maximum wave steepness in deep water is found about 0.2035. The result has only about 10% difference from Taylor's experimental results. Finally, the dynamic properties of standing wave motion are also described in this paper.

## References


1. Penney, W.G.; Price, A.T.: Some gravity wave problems in the motion of perfect liquids. Part II. Finite period stationary gravity waves in a perfect liquid. Phil. Trans. R. Soc. London Ser. A, 244(1952) 255-284.
2. Tadjbakhsh, I.; Keller, J.B.: Standing surface waves of finite amplitude. J.Fluid Mech. 8(1960) 442-451.
3. Goda, Y.; Kakizaki, S.: Study on finite amplitude standing waves and their pressure upon a vertical wall. Rep. Port Harbour Res. Inst., 5(10) (1966) 1-57(in Japanese).
4. Tsai, C.P.; Chen, Y.Y.: Third-order approximation to gravity standing waves by systems perturbation method. Harbour Technol. 2 (1986) 99-120 (in Chinese).
5. Grant, M.A.: Standing Stokes waves of maximum height. J. Fluid Mech., 60 (1973) 593-604.
6. Schwartz, L.W.; Whitney A.K.: A semianalytic solution for nonlinear standing waves in deep water. J. Fluid Mech. 107 (1981) 147-171.
7. Roberts, A.J.; Schwartz, L.W.: The calculation of nonlinear short-crested gravity waves. Phys. Fluid. 26(9) (1983) 2388-2392.
8. Taylor, S.G.: An experimental study of standing waves. Proc. R. Soc. London Ser. A, 218 (1953) 44-59.
9. Longuet-Higgins, M.S.; Stewart, R.W.: Radiation stresses in water waves; a physical discussion with applications. Deep Sea Res., 11 (1964) 529-562.

# Session 4

# Breaking Water Waves

# Critical Condition for the Appearance of Steady Breakers on 2-dimensional Wave Generated by Submerged Foil

Kazu-hiro MORI

Faculty of Engineering
Hiroshima University
Saijo-cho Higashi-Hiroshima 724 Japan

Summary

Assuming the steady breaker of an infant stage as a turbulent free-surface flow, a critical condition for its appearance is derived based on the inviscid instability analysis. Besides the curvature of the wave profile, the velocity on the free-surface appears in the condition; the condition is approximately given by

$$U/M\ \partial M/\partial x - \partial U/\partial x > 0$$

where $M=KU^2-g$, K: curvature of the wave, U: velocity on the free-surface, g: gravity acceleration, x: flow direction. The measured data give an acceptable criterion for the appearance of the breaking.

## Introduction

Steady breakers, observed on the waves generated by a submerged 2-dimensional foil have been studied to make clear their flow mechanism; e.g., Battjes-Sakai [1], Duncan [2], Mori [3] and Tulin-Cointe [4]. Most of the studies are on the characteristics of the breakers. It is, however, also important to find out a critical condition for their appearance not only for practical purposes to control undesirable breaking but for numerical simulations of the breaking.

Critical conditions for their appearance were studied in Mori [3] through measurements focussed on the flows where the breaking was to appear. The breaking was concluded as a free-surface turbulent flow as far as the breakers of infant stage were concerned. The flow mechanism was supposed that the surplus energy accumulated around the wave crest by the increment of the free-surface elevation was dissipated through the turbulence production and the free-surface could eventually maintain itself without any overturning or backward flows.

For such turbulent-oriented breakers, called sub-breaking waves

K. Horikawa, H. Maruo (Eds.)
Nonlinear Water Waves
IUTAM Symposium Tokyo/Japan 1987

there, the commonly used criteria in terms only of the wave slope or the steepness seem not enough. This guess may be supported by a comparison of the limiting wave height of various breaking waves by Schultz et. al. [5] where the steady breakers have appreciably less steepness than the other types of breakers.

In the present paper an attempt is made to introduce a criterion for the sub-breaking waves based on an inviscid instability analysis and examine it by comparing with the measured results.

## Instability Analysis

We introduce a small disturbance to the basic flow whose amplitudes are $o(\varepsilon)$ and assume $\varepsilon \ll 1$. Instability analysis will suggest that the flow may be unstable if the introduced disturbance diverges with time; in other words, sub-breaking waves can appear.

The velocity potential, denoted by $\phi$, satisfies

$$\nabla^2\phi = 0 \tag{1}$$

$$\frac{\partial\phi}{\partial t} + \frac{1}{2}|\nabla\phi|^2 + \frac{p}{\rho} + gz = c_0 \tag{2}$$

where t: time, z: vertical coordinate, $\rho$: density of fluid, p: pressure, g: gravity acceleration and $c_0$: constant. On the free-surface $z=\zeta(x)$, we have

$$p = p_a \tag{3}$$

$$\frac{D}{Dt}(z-\zeta) = 0 \tag{4}$$

where $p_a$: atmospheric pressure.

We introduce a disturbance into the basic flow and write the variables by

$$\phi = \phi_0 + \varepsilon\phi', \quad p = p_0 + \varepsilon p', \quad \zeta = \zeta_0 + \varepsilon\zeta' \tag{5}$$

where the subscript 0 means the basic flow while ' the disturbance.

Taking a curvilinear coordinates system (s,n), where s is along the streamlines of the basic flow and n the outward normal to s, we substitute (5) into (1), (2) and (4). Here the surface n=0 represents the free-surface of the basic flow, and the analysis being limited within a thin layer beneath the free-surface, n can be assumed straight and its metric coefficient unity.

Substituting (5) into (1), we have

$$\nabla^2\phi' = 0 \tag{6}$$

for $\phi'$, and into (2),

$$\frac{1}{2}|\nabla\phi_0|^2 + \frac{p_0}{\rho} + gz + \varepsilon\left\{\frac{\partial\phi'}{\partial t} + \nabla\phi_0\cdot\nabla\phi' + \frac{p'}{\rho}\right\} + o(\varepsilon^2) = c_0 \qquad z=\zeta_0+\varepsilon\zeta' \tag{7}$$

Expanding (7) around $z=\zeta_0$ and taking the leading terms for the introduced disturbance, we have

$$\frac{\partial\phi'}{\partial t} + U\frac{\partial\phi'}{h\partial s} - \frac{1}{\rho}\frac{\partial p_0}{\partial n} n_z\zeta' = 0 \tag{8}$$

where $U=\partial\phi_0/h\partial s$ and h is the metric coefficient along s. In the derivation of (8), p' on $z=\zeta_0$ is given by

$$p'|_{z=\zeta_0} = -\frac{\partial p_0}{\partial n}\delta n|_{z=\zeta_0} + o(\varepsilon) \tag{9}$$

which is also obtained by the Taylor expansion of the pressure around $z=\zeta_0$. The momentum equation in the n-direction gives

$$\frac{1}{\rho}\frac{\partial p_0}{\partial n} n_z = (KU^2 - n_z g)n_z \quad (\equiv M) \tag{10}$$

on the free-surface where $W=\partial\phi_0/\partial n=0$. Here K is the curvature of the basic wave profile, given by $\partial h/h\partial n$, and $n_z$ is the direction cosine of n to z-direction. By virtue of (10), (8) can be rewritten by

$$\frac{\partial\phi'}{\partial t} + U\frac{\partial\phi'}{h\partial s} - M\zeta' = 0 \tag{11}$$

On the other hand, the substitution of (5) into (4) yields the following equation again after the Taylor expansion around $z=\zeta_0$;

$$\frac{\partial\zeta'}{\partial t} + U\frac{\partial\zeta'}{h\partial s} - \frac{1}{n_z}\frac{\partial\phi'}{\partial n} - \left(\frac{\partial W}{\partial n} - \frac{1}{n_z}U\frac{\partial n_z}{h\partial s}\right)\zeta' = 0 \tag{12}$$

From (11) and (12), after eliminating $\zeta'$, we have

$$\frac{\partial^2\phi'}{\partial t^2} + 2U\frac{\partial^2\phi'}{h\partial s\partial t} - L\frac{\partial\phi'}{\partial t} + U\left(\frac{\partial U}{h\partial s} - L\right)\frac{\partial\phi'}{h\partial s} - U^2\frac{\partial^2\phi'}{\partial n^2} - \left(U^2K + \frac{M}{n_z}\right)\frac{\partial\phi'}{\partial n} = 0 \tag{13}$$

where

$$L = \frac{\partial W}{\partial n} - \frac{1}{n_z} U \frac{\partial n_z}{h\partial s} + \frac{U}{M}\frac{\partial M}{h\partial s} \tag{14}$$

Now the problem is to solve (6) under the boundary condition (13) where all the coefficients are provided by the basic flow. We assume the solution to satisfy (6) in the form of

$$\phi' = \phi_1(s,n)e^{ct} \tag{15}$$

Then (13) is written by

$$c^2 + (2U\phi_{1s} - L)c + U(\frac{\partial U}{h\partial s} - L)\phi_{1s} - U^2\phi_{1nn} - (KU^2 + \frac{M}{n_z})\phi_{1n} = 0 \tag{16}$$

and c is given by

$$c = \frac{1}{2}L - U\phi_{1s} \pm \sqrt{D} \tag{17}$$

where

$$D = \frac{1}{4}L^2 + U^2\phi_{1s}^{\ 2} - U\frac{\partial U}{h\partial s}\phi_{1s} + U^2\phi_{1nn} + (U^2K + \frac{M}{n_z})\phi_{1n},$$
$$\phi_{1s} = \frac{\partial\phi_1}{h\partial s}/\phi_1, \quad \phi_{1n} = \frac{\partial\phi_1}{\partial n}/\phi_1, \quad \phi_{1nn} = \frac{\partial^2\phi_1}{\partial n^2}/\phi_1 \tag{18}$$

The condition for this flow to be stable is

$$Re\{c\} \equiv c_r < 0 \tag{19}$$

where Re{ } means the real part of c.

To have more direct insight, we assume $\phi_1(s,n)$ has a wave number of k and is given by

$$\phi_1 \propto e^{k\{i f(s) + g(n)\}} \tag{20}$$

where $i$ is the imaginary unit. Then $\phi_{1s}$ is given by

$$\phi_{1s} \propto ikR \tag{21}$$

In (21), R is a real function and $\phi_{1s}$ can be supposed imaginary. For a large wave number k, which is very common for such flows as sub-breaking waves, D can be approximated by

$$D \sim -k^2\{U^2R^2 + o(\frac{1}{k})\} \quad (22)$$

Therefore $\sqrt{D}$ can be presumed imaginary and the unstable condition, i.e., the critical condition for the appearance of the sub-breaking waves is given by

$$c_r \doteqdot \frac{1}{2}L > 0 \quad (23)$$

Now we evaluate $c_r$ for the measured data to examine the validity of the criterion given by (23). Measurements were carried out for the NACA-0012 wing (chord length:30cm) at two different submerged depth; d=19 cm and 21cm. Velocities were measured by a hot film anemometer 3mm below the free-surface. Precise descriptions and results are found in Mori [3]. In the analysis the assumptions of

$$n_z = 1.0, \quad \frac{\partial}{h\partial s} = \frac{\partial}{\partial x} \quad (24)$$

are used; x: the horizontal axis to the flow direction.

The values of $c_r$ are compared between the two cases; one is the case where the wave is perfectly free from breaking, while the other is that where the wave is fully mature for breakings. Here the former is denoted by NON and the latter by JB. Figures 1 and 2 show the estimated values of $c_r$ for d=19cm ($U_0$=0.726m/s for NON and 0.754m/s for JB) and d=21cm ($U_0$=0.883m/s for NON and 0.923m/s for JB) respectively; where $U_0$: uniform flow velocity, $x_0$: distance from the wave crest in the x-direction.

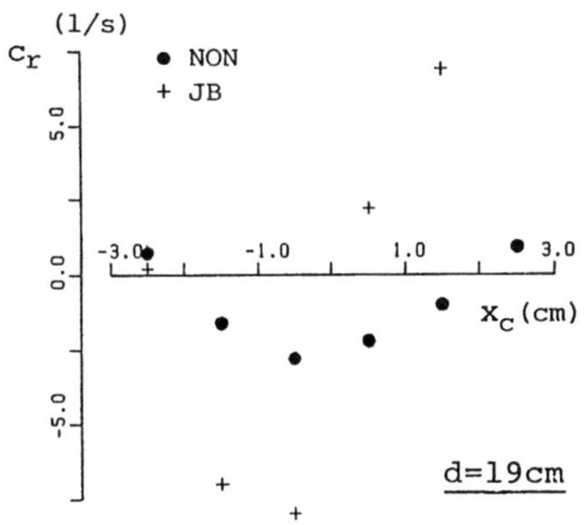


Fig.1 Comparison of $C_r$

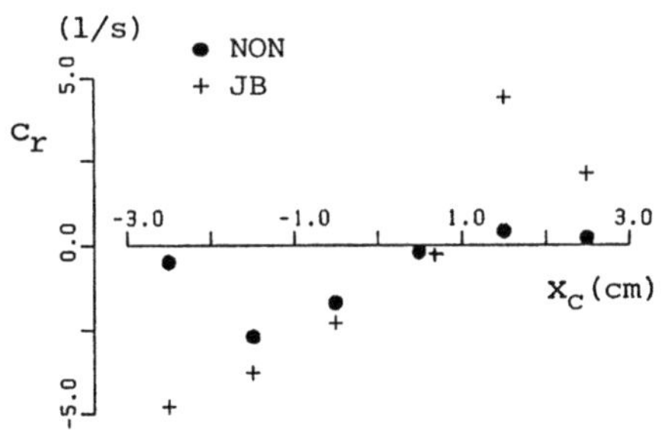


Fig.2 Same as Fig.1

For the both cases of d=19cm and 21cm, $c_r$ of JB changes its sign from the negative to the positive around the wave crest, while that of NON remains almost negative. This may suggest that the flows of JB can become unstable around the wave crest if a disturbance is introduced; in other words, sub-breaking can take place. Contrarily, in cases of NON, unstable flows can never appear even if disturbances are introduced.

Actually the flows of JB were quite unstable; although the wave crest was still round, even by a slight increment of the oncoming flow velocity, the breaking could have been induced. The breaking appeared a little behind the wave crest where $c_r$ has changed its sign and the breaking region moved forward in a moment to cover all around the wave crest.

## Conclusion

Assuming the steady breakers by a submerged wing as a turbulent free-surface flow, a critical condition for their appearance is derived based on the instability analysis. The condition is well acceptable compared with the measured. The criterion is expected to be used in practical designs or in numerical computations for the detection of breaking.

## References

1. Battjes, J. A.; Sakai, T.: Velocity field in a steady breaker. Jour. of Fluid Mech. 111 (1981) 421-437.

2. Duncan, J. H.: The breaking and non-breaking wave resistance of a two-dimensional hydrofoil. Jour. of Fluid Mech. 126 (1983) 507-520.

3. Mori, K.: Sub-Breaking waves and critical condition for their appearance. Jour. of The Soc. of Naval Arch. of Japan 159 (1986) 1-8.

4. Tulin, M. P.; Cointe, R.: A Theory of spilling breakers. Proc. of 16th Sym. on Naval Hydrod. (1986) 93-104.

5. Schultz, W.W.; Ramberg, S. E.; Griffin, O. M.: Steep and breaking deep water waves. Proc. of 16th Sym. on Naval Hydrod. (1986) 106-115.

# Surging Characteristics of Spilling Zones of Quasi-steady Breaking Water Waves

MICHAEL L. BANNER

Department of Applied Mathematics
University of New South Wales
Kensington, N.S.W., Australia

## Summary

Wave breaking on a wide range of length scales is a commonly observed feature of the wind-driven ocean surface. Despite the importance of breaking waves in various interfacial processes of basic and technological significance, our detailed knowledge of the attendant hydrodynamics is relatively sparse, both observationally and theoretically.
This contribution is concerned with a description of the breaking region following crest overturning for the case of spilling breakers. It draws on detailed measurements of basic aspects of quasi-steady breaking waves to examine a simple mathematical model for the perturbation response of the spilling zone. The model predictions are found to compare favorably with observed perturbation response characteristics. The use of the model in estimating the rate of energy dissipation in the wave field associated with spilling breakers is considered.

## Introduction

Several contributions on various aspects of wave breaking have been published in the open literature as well as in recent conference proceedings (eg. Phillips and Hasselmann [1], Toba and Mitsuyasu [2]).These contributions include studies of the instabilities which appear to precede breaking, criteria for the onset of breaking at sea, assessments of the momentum transfer from the wave field due to breaking and of the resultant subsurface currents and turbulence induced by the breaking. In the context of obtaining a firmer understanding of the local details of the hydrodynamics associated with this intermittent, turbulent free surface phenomenon, it is apparent that both direct observation and theoretical modelling are intrinsically difficult. Understandably, previous approaches have examined the somewhat simpler case of quasi-steady realizations of spilling breakers. Longuet-Higgins and Turner [3] modelled a steady spilling region as a developing entraining plume, deducing a similarity solution for

K. Horikawa, H. Maruo (Eds.)
Nonlinear Water Waves
IUTAM Symposium Tokyo/Japan 1987

the downslope flow. Battjes and Sakai [4] and Duncan [5] made observations of quasi-steady spilling breakers. Duncan [5] reported a comprehensive set of observations on the characteristics of spilling breakers produced by a hydrofoil towed steadily in a wave tank at various speeds, inclination angles and depths. Two observations of particular interest in the present context were reported by Duncan [5]: firstly, the spilling region imparts a shearing force along the forward slope equal to the component of its weight in that direction; secondly, the spilling region had a small oscillation in its length with a regular period ranging between four to five times the period of the linear wave with phase speed equal to the hydrofoil speed. This was attributed to forcing by background wave components generated when the hydrofoil was started from rest.

These observations motivated the present study which extends existing knowledge of spilling zones via further detailed measurements of basic aspects of quasi-steady breaking waves and describes a simple mathematical model aimed at describing the perturbation quasi-steady dynamics of the spilling region. Observations of the response of the spilling zone to perturbation from equilibrium are also reported which provide a useful comparison for the model predictions. The use of the model for estimating the rate of energy dissipation associated with oceanic spilling breakers is considered.

## Quasi-steady spilling breaker characteristics

Banner [6] recently reported a very detailed set of geometric and kinematic observations for spilling breakers with a wavelength of 0.4m. For the present purpose of supporting the model to be outlined below, the salient characteristics of the wave flume and typical quasi-steady stationary breaking waves generated by a subsurface hydrofoil in the flume are shown in Figures 1 and 2. In the wave flume, the wavelength and water depth were set by the volume flow and outflow level controls. The hydrofoil section (96mm × 21mm), located at mid-water depth, was orientated to provide a given strength of breaking. Measurements of the surface elevation were made with a standard capacitance wire (0.1mm dia) gauge. The flow velocity data was measured using a single channel laser anemometer with frequency shifting capability.

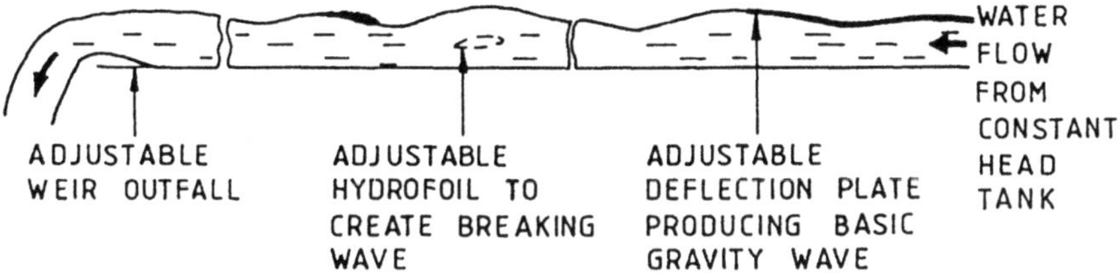


Fig.1. Experimental wave flume configuration: working length 7.3m; channel width 0.225m; water depth 0.22m. The measurement site was halfway along the channel.

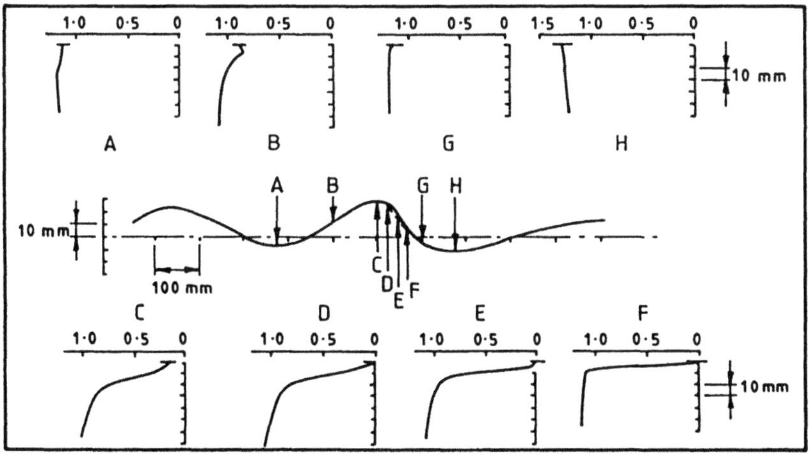


Fig.2. Vertical profiles of the mean velocity (——) (normalised by the linear-phase speed C = 0.85m/s) parallel to the local mean water surface at the various sites indicated along the breaking wave profile (wavelength ~ 0.4m), corresponding to fairly gentle breaking.

The measurements are consistent with the simplified spilling zone parameterisation shown schematically in Figure 3.

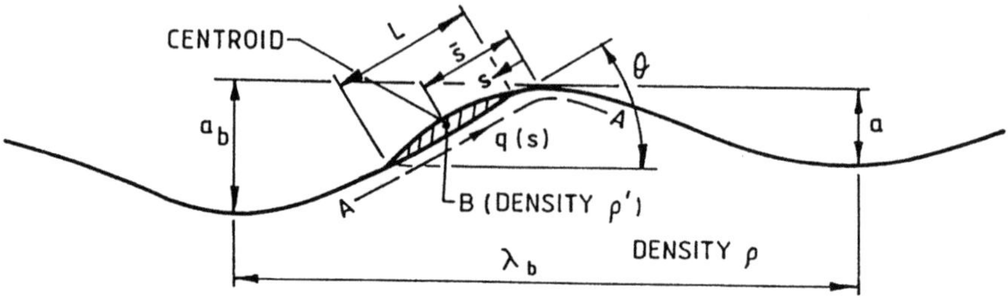


Fig.3. Definition sketch for spilling zone variables.

It should be noted that these detailed observations locate the rear of the mean spilling zone region somewhat downslope of the breaking crest; the latter is not a mean stagnation point as was assumed previously. However, the measurements indicate an approximate equality of the mean spilling zone length L and mean downslope location of the centroid $\bar{s}$ of the mean spilling cross-sectional area B(t). Also, the mean spilling zone thickness $\bar{h}$ is expressed conveniently in terms of an aspect ratio r, where $r = \frac{\bar{h}}{L}$, with $\bar{h} = L^{-1}\int \eta \, ds$ over the spilling region length.

Brief description of model

Several simplifying assumptions have been made to obtain an initial, tractable time-dependent model of the compact spilling region for the quasi-steady breaking wave. These are as follows:

(a) Under equilibrium conditions [suffixed 'e']

(i) beneath the highly sheared transition layer, a streamline AA exists in the underlying flow with an upslope mean velocity distribution $\bar{q}(s)$. At the equilibrium centroid $\bar{s}_e$, $\bar{q}(\bar{s}_e) = q_e = 0(C)$, the linear phase speed of neighbouring unbroken waves. In the absence of more detailed information on the underlying flow field, a basic underlying flow has been assumed on the basis of Bernoulli's equation in which the unsteady term is assumed to be small, yielding

$$\bar{q}(\bar{s}) = \{q_e^2 + D\, g\, \bar{s}\sin\theta - D_e\, g\, \bar{s}_e \sin\theta_e\}^{\frac{1}{2}}$$

where $q_e$ is the equilibrium upslope velocity and $D = 2[1 - \rho'/\rho\; r \cot\theta]$. The second term in D approximates the additional pressure effect of the spilling region.

(ii) the mean tangential shear force $\bar{\tau}_e L_e$ (where $\bar{\tau}$ is the effective mean shear stress) is given by the downslope spilling zone weight $\rho' g\, B_e \sin\theta_e$. (Duncan [5]).

(b) for departures from equilibrium:

(i) the aspect ratio r and inclination θ are prescribed functions of the breaking extent (specified by the downslope position of the centroid $\bar{s}$). Also $\bar{s} \simeq L$ is assumed during excursions from equilibrium, so that the 'toe' location is $(\bar{s} + L/2)$.

(ii) all fluid particles in the mathematical volume B(t) have a common average up/downslope velocity specified by $\dot{\bar{s}}$.

(iii) the bulk shear force counteracting the downslope body force $\rho' g\, B \sin\theta$ is given by $\bar{\tau} L = -\rho' g\, B \sin\theta\; [\dot{\bar{s}} + q(\bar{s})]/q_e$.

According to this formulation, the shear stress is given by an effective eddy viscosity $\rho' g\, B \sin\theta\; \delta/q_e L$ produced by the mean velocity gradient $[\dot{\bar{s}} + q(\bar{s})]/\delta$ where δ is the characteristic depth of the highly sheared transition layer. From the observations, the effective mean shear stress is predominantly turbulent Reynolds stress despite the very high mean shear in the transition layer.

With these assumptions, the equation of motion for the up/downslope motion of the spilling region is $\frac{d}{dt}(\rho' g B \dot{\bar{s}}) = -\bar{\tau} L + \rho' g B \sin\theta$. Using $B = r L^2$, $L \simeq \bar{s}$, the form for $\bar{\tau} L$ in (b) (iii) above and transforming to non-dimensional variables using reference length, time and velocity scales of $\frac{C^2}{g \sin\theta}$, $\frac{C}{g \sin\theta}$ and C respectively, the equation of motion reduces to

$$\frac{d^2u}{dt^2} + K\frac{du}{dt} + K(3u)^{\frac{2}{3}}\left(\gamma^{-1}\sqrt{\alpha + K D (3u)^{\frac{1}{3}}} - 1\right) = 0$$

where $u = \frac{1}{3}\bar{s}^3$, $\alpha = \gamma - D_e \bar{s}_e$, $\gamma = q_e/C$ and $K = \sin\theta/\sin\theta_e$.

This nonlinear ODE was solved numerically for conditions under which the model was a reasonable approximation to realizable perturbations of the spilling zone and used parameters determined from the mean breaking wave configuration data described earlier in this article. The simulation involved calculation of the response to a travelling velocity deficit pulse superimposed on the mean upslope velocity $\bar{q}(s)$. A triangular pulse with duration and magnitude comparable with the experimental values was used in this study for the case of the 0.4m breaking wave. The response characteristics are shown in Figure 4 for conditions typical of the complementary experiments.

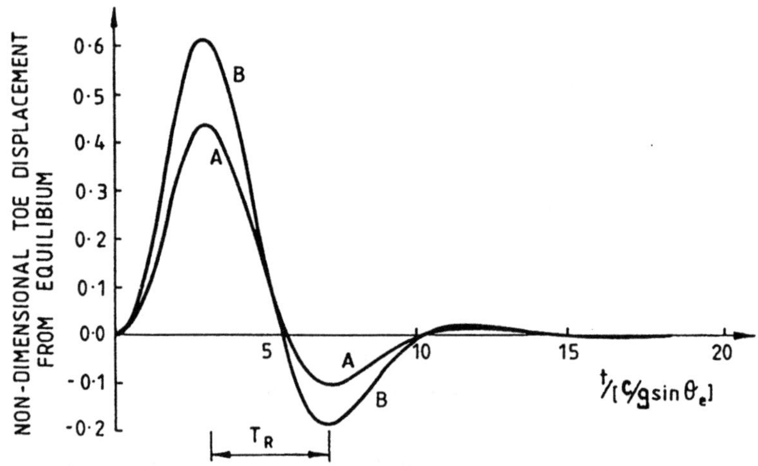


Figure 4. Computed spilling zone temporal response to a travelling triangle pulse deficit of velocity. The conditions match the data in Fig.5. Toe displacement is normalised by its equilibrium value. A and B represent velocity deficits of 0.1C and 0.15C.

## Comparison with observed perturbation response

The method chosen for applying controlled, approximately known perturbations to the spilling zone, while causing only minor geometrical changes to the underlying flow field, was to briefly introduce tangentially and withdraw a small (15 × 15cm) piece of wire gauze at the water surface at the wave crest upstream of the breaking crest. The pulse of velocity deficit induced near the surface was monitored in the trough ahead of the spilling region using the laser anemometer. The major response was observed to be a transient downslope surge of the spilling region, with a small displacement of the crest location. The bulk motion of the spilling zone was recorded by a suitably orientated capacitance wire gauge. The imposed perturbations were found to have a pulse duration of about 1 sec. and a peak velocity deficit of about 0.15C for the fairly gently breaking wave with $\lambda_b \simeq 0.4$m. examined in detail. Figure 5 shows a typical measured spilling zone response. The associated parameters ($\theta = 17°$, $\bar{s}_e = 0.24$, $q_e = 1.15$) were used in the model calculations, which showed a relatively weak dependence on variations of these parameters.

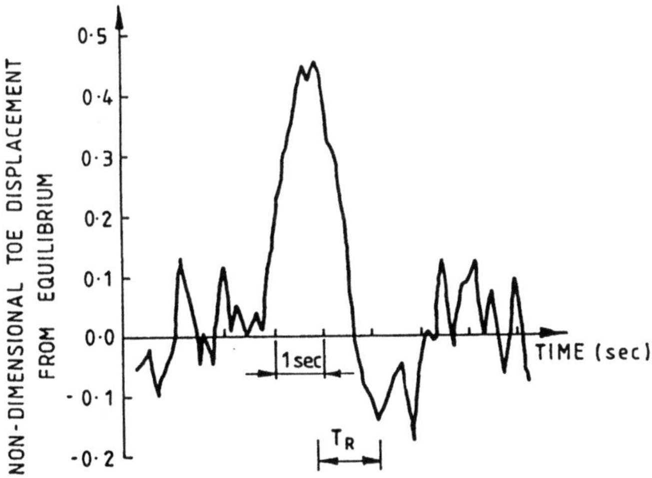


Fig.5. Typical spilling zone response (normalised as in Fig.4.) to travelling triangular pulse deficit in $q(\bar{s})$ of duration $\sim$ 1 sec. and magnitude $\simeq$ 0.15C. [Mean wave parameters: $\lambda_b = 0.40$m; $q_e = 1.15$, $\bar{s}_e = 0.24$].

From an ensemble of 25 realizations, the mean relaxation time ($T_R$) for $\lambda_b = 0.40$m and fairly gentle breaking, was observed to be 1.21 seconds with a standard deviation of 0.10 seconds. The corresponding computed values ranged from 1.1 to 1.25 seconds, as the degree of breaking varied from very light to moderate. This close quantitative correspondence, together with the qualitative agreement apparent in Figures 4 and 5, offers encouraging initial support for the model and an extension to longer and shorter breaking wavelengths is in progress.

## Discussion

This simple model suggests that the effect of the unsteady spilling region may be modelled to lowest order by an equivalent force system consisting of a shearing force $F_T \sim \frac{\rho' g B \sin\theta}{C}[\dot{\bar{s}}+q]$ and a normal force $F_N = \rho' g B \cos\theta$, localized near the centroid of the spilling zone. The rate of dissipation of wave energy D may be estimated from the rate of working of these forces on the fluid in the underlying wave. For quasi-steady breaking,

$$D \sim 2\rho' g BC \sin\theta \sim 2\rho' grL^2 C \sin\theta .$$

Over a wave period, the total energy dissipated as a fraction of the wave energy is then $\frac{DT}{E\lambda_b} \sim 0(1)$ for typical steady breaking geometrical data, indicating the strength of the process. For unsteady breaking the integral of D(t) over the history of the breaking event would be needed: it will be significantly less that for the steady spilling breaker in view of the quadratic dependence of D(t) on L(t). Clearly more detailed information on the underlying velocity $q(\bar{s})$ is needed for transient breaking and this is left to a future study. However, the model suggests the possibility of using the geometry history of ocean whitecaps to estimate the dissipation rate in the wave field.

## References


1. Phillips, O.M.: Hasselmann, K. (eds.) Wave Dynamics and Radio Probing of the Ocean Surface. New York: Plenum Press 1986.
2. Toba, Y.; Mitsayasu, H. (eds.) The Ocean Surface. Dordrecht: D. Reidel 1985.
3. Longuet-Higgins, M.S.; Turner, J.S.: An 'entraining plume' model of a spilling breaker. J. Fluid Mech.,63 (1974) 1-20.
4. Battjes, J.A.; Sakai, T.: Velocity field in a steady breaker. J. Fluid Mech., 111 (1981) 421-437.
5. Duncan, J.H.: An experimental investigation of breaking waves produced by a towed hydrofoil. Proc. Roy. Soc. A, 377 (1981) 331-348.
6. Banner, M.L.: On the mechanics of spilling zones of quasi-steady breaking waves. Proc. NATO A.R.W. on Natural Mechanisms of Surface Generated Noise in the Ocean, La Spezia, June 1987.

# Steady and Unsteady Spilling Breakers : Theory

M.P. TULIN and R. COINTE*

Ocean Engineering Laboratory, University of California at Santa Barbara
*Currently, Institut Français du Pétrole; Rueil Malmaison

## INTRODUCTION

Background. The phenomena of wave breaking is one of the most important basic unsolved problems in hydrodynamics, see Longuet-Higgins 1980a. Therefore we have begun the study of spilling breakers at the OEL (UCSB) in 1985. The goal of this initial theoretical study, forming the dissertation of R. Cointe (1987), is to explain and understand the earlier experimental studies of J. Duncan (1981,83), on steady breaking and to answer a number of fundamental questions. This work has been reported elsewhere while in progress, Tulin & Cointe (1986), which also contained a brief review of previous work, while a more detailed review is given in Cointe (1986). More experimental work on this subject is underway at the OEL (UCSB), particularly on 2D & 3D unsteady spilling breakers, and more theoretical work at IFP, where R. Cointe is currently employed.

Modelling the Spilling Breaker. We model the unsteady spilling breaker as a turbulent density current, following Longuet-Higgins & Turner (1974). This current is driven down the wave face by gravity, growing by turbulent entrainment of water from the underlying potential wave, and by the entrainment of air from above; its motion is determined by the balance between hydrostatic and inertial forces, and turbulent shear due to friction between breaker and wave. The breaker is thought to be thin in comparison to its length so that "shallow water" approximations can be made. The governing equations of hyperbolic type can be readily solved numerically, given initial conditions.

When energy is continuously supplied to the wave (as by Duncan's hydrofoil), the breaker becomes steady (time averaged), although turbulent. The breaker can then be modelled as a closed recirculating region (eddy) of aerated water in contact with the underlying wave from crest at pt(b), downward over the wave face to its leading edge at pt(a), see figure 1. As shown below here, the dynamic pressures within the eddy are negligible so that the eddy may be modelled as an essentially stagnant aerated mass of water supported by the shear on the streamline dividing the turbulent eddy from the underlying potential flow. At and near the leading edge (a) the flow around the dividing streamline is

K. Horikawa, H. Maruo (Eds.)
Nonlinear Water Waves
IUTAM Symposium Tokyo/Japan 1987

modelled as a turbulent mixing layer, figure 1 bottom; in consequence the speed of the flow on the dividing streamline is suddenly reduced at (a) by about a factor of 1/2 and a turbulent shear arises which in the idealized case is constant and has been measured (shear stress/dynamic pressure) to be $0(10^{-2})$.

As discussed below this model seems well supported by existing experimental evidence and allows various analyses outlined below providing very important results, notably the existence of two breaking solutions (one stable) prior to the Stokes limiting wave.

## ANALYSES CONDUCTED AND RESULTS

### Steady Spilling Breakers:

Theory of the Breaker. The vertical and horizontal force equilibria of the breaker lead to

$$\int_a^b p_{dyn}\, dx + \int_a^b \tau\, dy = 0 \qquad \int_a^b \{ (\rho_e\, gh + p_{dyn}) \tan\theta - \tau \}\, dx = 0$$

where $p_{dyn}$ is the dynamic pressure due to the circulatory motion within the eddy, as distinct from the hydrostatic pressure acting on the free streamline, $\rho_e gh$. An estimate of the relative importance of the dynamic pressure follows from assuming that the face of the wave is flat, constant $\theta$ (the angle between the dividing streamline and the horizontal). Then:

$$\int_a^b p_{dyn}\, dx \Big/ \int_a^b \rho_e gh\, dx \approx -\sin^2\theta$$

Duncan (1981) measured values of $\theta$ comprised between 10° and 15° so that we estimate the dynamic pressures within the eddy to be only approximately 5% of the hydrostatic pressures in magnitude. This leads to the conclusion that the circulation within the eddy is weak.

The horizontal force equilibrium also leads to an estimate of the mean friction coefficient along the dividing streamline by assuming the flow before the leading edge (a) to be potential: $c_f = \tau/\rho q^2(a) = (A/4d^2)(\rho_e/\rho)\cos^2\theta$, where d is the horizontal length of the breaker. Duncan (1981) found experimentally $A/d^2 = 0.11$ (a constant) and estimated the density ratio to be equal to 0.6, leading to the estimate that the friction coefficient is of order $10^{-2}$. This value is in excellent agreement with experimental data for the mixing zone behind a bluff body. This result and the preceeding justify our physical picture of the flow.

Neglecting dynamic effects leads to a very simple theory in which the shape of the breaker is determined by the horizontal balance between the hydrostatic pressure and the turbulent shear stresses acting on the side and bottom boundaries of a vertical slice of the breaker. If one assumes the density of the aerated water within the breaker to be constant, this analysis shows that the breaker has a parabolic shape near the leading edge and has to be inscribed in a right triangle, as in figure 1. In this limiting case of a triangular, flat-topped breaker, the integral of the shear stress along the dividing streamline, the drag, is simply related to the total height of the breaker by: $\tau^* = (\rho_e/2)gh^{*2}$.

Duncan's experiments show a linear relationship between the height of the breaker and the height of the following waves, see figure 2: $gh^*/c^2 = [1/2 - ga_f/4c^2]$. This relationship can be justified theoretically by assuming a constant head along the dividing streamline, as in a classical mixing layer (see Tulin & Cointe, 1986). In the limiting case of a flat-topped breaker, it also leads to the conclusion that the total drag scales with the height of the breaker and the velocity just upsteam of the leading edge of the breaker, point (a), i.e., $\tau^* = \rho C_D^* h^* q^2(a)$ with $C_D^* = \rho_e/8\rho$.

The assumption that the breaker is flat-topped is well supported by Duncan's experiments and will be made here. This assumption is convenient, but not crucial, for the following computations. It allows us to express the total drag in terms of only one parameter, the "apparent" density ratio $\rho_e/\rho$ - which, in fact, represents the ratio of the mass of the breaker to that of water within the circumscribing right triangle.

Effect of the Wake: Boundary Layer Analysis. According to the measurements of Duncan (1981) and others, the trailing wake behind the spilling breaker seems to behave very much like the turbulent wake behind a two-dimensional body. Using the method of matched asymptotic expansions and boundary layer arguments, it is possible to show that for a suitably small shear stress applied at the dividing streamline, its effect may be calculated by finding the outer potential flow - the pseudo flow - corresponding to the actual physical wave. This analysis is closely related to that of Longuet-Higgins (1969) who investigated the effect of wind shear on wave propagation. It shows that: the momentum and displacement thicknesses of the wake have to satisfy Von Karman's momentum equation; the free surface of the pseudo flow is displaced downward by a distance corresponding to the displacement thickness of the wake, $\delta_1$; an additional pressure has to be applied at the free surface of the pseudo flow to account for the weight of the displaced layer of fluid and curvature effects.

In the case of a triangular, flat-topped breaker, and assuming that the fluctuations of the pseudo flow speed are small, it is possible to show that the mean momentum thickness, $\Delta$, is related to the total height of the breaker, $h^*$, by $g\Delta/c^2 = (\rho_e/2\rho)(gh^*/c^2)^2$.

It can be further shown that the effect of the modulation of the boundary layer thickness by the orbital velocity of the wave is to induce a modification of the dispersion relation according to: $\lambda_f = (2\pi c^2/g)(1 - 4g\Delta/c^2)$. This theoretical result compares well, see figure 3, with Duncan's measurements of the shortened following wavelength for an apparent density ratio in the range [0.2,0.6].

"Shock" Analysis. Modeling the real physical wave by a potential flow, the pseudo flow, allows us to use potential flow theory to compute the repressive effect of the breaker on the underlying wave. This calculation can be done in an indirect manner, by applying a momentum balance, see figure 4. It leads to the conclusion that the total resistance $R_o$ resulting in the entire wave system is the sum of the breaking resistance, $R_b$, and of the following wave resistance, $R_f$. If one neglects the effect of the wake, the breaking resistance is just equal to the drag, $\tau^*$, while the following wave resistance is equal to the wave resistance associated with a non-breaking wave train of height $a_f$. The effect of the wake can be taken into account, it is small but noticeable.

The results of this analysis, utilizing non-linear numerical results of Longuet-Higgins for large wave steepness, are shown in figure 4. Note that the non-breaking wave resistance (the lower curve) reaches a maximum, $R^* \approx 0.02\ \rho c^4/g$, for $a_f \approx 0.69\ c^2/g$. The upper curves correspond to the total wave resistance for two different values of the parameter: $\rho_e/\rho = 0.22$ and $\rho_e/\rho = 0.5$. Duncan (1983) observed the onset of breaking at $R_o/R^* = 0.83$. Note that the present theory predicts that, near incipient breaking, two breaking solutions can exist: a strong solution (with a larger breaker) and a weak solution. All measured values of $ga_f/c^2$ (Duncan, 1981) are in the strong breaking regime, and the largest one corresponds to onset. It is shown below that, indeed, only strong breakers are stable. Duncan (1983) observed a maximum total resistance equal to 3.1 R* and noticed in this case that the following waves were almost suppressed. The maximum total resistance predicted by the present theory is equal to 6.25 $\rho_e/\rho$, so that we need a value of $\rho_e/\rho$ approximately equal to 0.5 for large breakers to account for this observation. This seems to imply that large breakers are more triangular and less aerated than small breakers, a result which might be explained by the fact that the slope of the dividing streamline, $\theta$, is smaller for large breakers.

Direct Calculation. A direct potential flow calculation of the repressive effect of the steady breaker on the following wave is also possible and has been performed (Tulin & Cointe, 1986). It agrees very well with the preceeding "shock" analysis and gives further information on the geometry of the breaking wave.

Unsteady Spilling Breakers:

Stability of Steady Breakers, Oscillations and Natural Frequencies. The analysis of the stability of steady breakers is based on a linearization of the nonlinear density current equations which have been derived to model the unsteady breaker (not given here). When the steady shape of the breaker is assumed to be triangular, this problem is somehow similar to the sloshing problem in a triangular tank, except that entrainment from the underlying flow is taken into account and that, of course, the boundary condition at the leading edge is different.

Two families of solutions are found. They both have Bessel-like modal shapes in space, but have either an exponential (unstable) or a sinusoidal (stable) behavior in time. The existence of unstable modes is shown to depend critically upon the boundary condition at the leading edge of the breaker. In fact, a detailed analysis (Cointe, 1987) shows that the breaker is unstable if its height at the leading edge falls below the height of the crest when its length increases, see figure 5.

An approximate leading edge boundary condition has been derived assuming that the underlying wave evolves in a quasi-steady way. It leads to the conclusion that weak breakers are unstable and that strong breakers are stable. It also allows us to compute the first natural frequency of the system breaker-breaking wave, see figure 5, leading to a value of period of oscillation approximately equal to 4.5 times the wave period for large breakers, a value remarkably close to that measured experimentally by Duncan (1981), $T_{br}/T_w \approx 4.4$. However, small resonant oscillations around the steady state are expected for impulsively started wave motions even without breaking at a period $T_{cr} = 4T_w$ (Maruo, 1957; Duncan, 1981) and since this period is almost equal to $T_{br}$ resonant phenomena of this different type cannot be excluded.

Establishment of the Underlying Wave. Direct numerical computations of the evolution of the free surface elevation of an initially undisturbed wave on which the pressure distribution corresponding to the fully formed breaker was applied impulsively have been performed at IFP (Nays, 1987) and show that a steady state is reached below the breaker on a time scale $T_D = d/c_g$, where d is the length of the breaker and $c_g$ the group velocity of the wave. Since d is of the order of a wavelength, $T_D/T_{br} \approx 0.2$ and the hypothesis that the underlying wave evolves in a quasi-steady way appears well justified.

Direct Calculation, Establishment of the Unsteady Breaker. Assuming that the underlying wave evolves in a quasi-steady way, the nonlinear equations describing the dynamics of the breaker have been solved numerically. The numerical results confirm the stability analysis carried out analytically and lead to the same estimate of the period of oscillations of strong breakers. These numerical computations show that, if disturbed, a weak breaker will either disappear or further grow to form a strong breaker. It allows us to estimate the time scale for the establishment of the breaker, approximately a few wave periods. These numerical computations allow us to propose a speculative description of inception of breaking and of the breaking cycle, see figure 6.

## SUMMARY AND CONCLUSIONS

### Steady Breakers

* the breaker is modeled as an almost stagnant mass of aerated water which rides the

front of the breaking wave, supported by shear stresses;

* breaking first appears when the total resistance can be balanced by the following wave resistance plus the breaking resistance; this threshold resistance is not a direct consequence of Stokes' limiting wave;

* near incipient breaking, two steady breaking solutions are possible: a weak solution with a smaller breaker and a strong solution with a larger breaker.

Unsteady Breakers

* the breaker is modelled as a thin unsteady density current which grows due to turbulent entrainment from the underlying wave. This theory includes the theory for steady breakers;

* weak breakers are shown to be unstable and strong breakers stable. This explains why a fully formed breaker appears at incipient breaking;

* it is suggested that, above threshold, a local disturbance can lead to the creation of a weak breaker which is unstable and further grows to form a strong breaker within a few wave periods. Numerical computations of the evolution of a weak into a strong breaker support this statement.

Acknowledgements. This work has been partially supported by the Office of Naval Research ocean engineering research program (Dr. Gene Silva), by the California SeaGrant Program of NOAA (Dr. James Sullivan), and by the Institut Francais du Petrole.

References.

1. Cointe, R., 1987. "A Theory of Breakers and Breaking Waves," Ph.D. Dissertation, Department of Mechanical and Environmental Engineering, University of California at Santa Barbara

2. Duncan, J.H., 1981. "An Experimental Investigation of Breaking Waves Produced by a Towed Hydrofoil," Proc. R. Soc. London A, Vol. 377, pp. 331-348.

3. Duncan, J.H., 1983. "The Breaking and Non-Breaking Resistance of a Two-Dimensional Hydrofoil," J. Fluid Mechanics, Vol. 126, pp. 507-520.

4. Longuet-Higgins, M.S., 1969. "Action of a Variable Stress at the Surface of Water Waves," The Physics of Fluids, Vol. 12, pp. 737-740.

5. Longuet-Higgins, M.S., and Turner, J.S., 1974. "An Entraining Plume Model of a Spilling Breaker," J. Fluid Mech., Vol. 63, pp. 1-20.

6. Longuet-Higgins, M.S., 1980. "The Unsolved Problem of Wave Breaking," Proceedings, 17th Int. Conf. Coastal Eng., Sydney, Australia, pp. 1-28.

7. Maruo, H., 1957. "Modern Developments of the Theory of Wave-Making Resistance in Non-Uniform Motion," SNAME 60th Anniversary Series, Vol. 2.

8. Tulin, M.P. and Cointe, R., 1986. "A Theory of Spilling Breakers," Proceedings of the 16th Symposium on Naval Hydrodynamics, Berkeley, CA. National Academy Press, Washington, D.C.

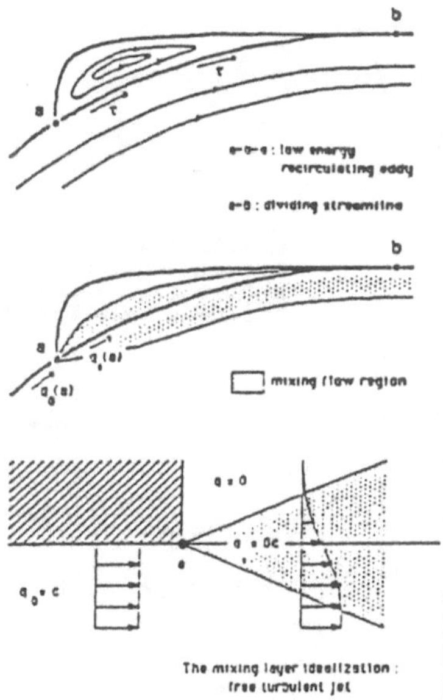


Figure 1. Physical Model of a Spilling Breaker

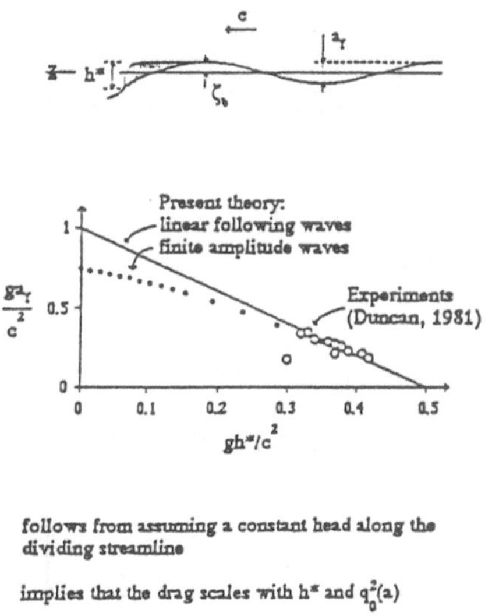


Figure 2. Height of the breaker vs. following wave height

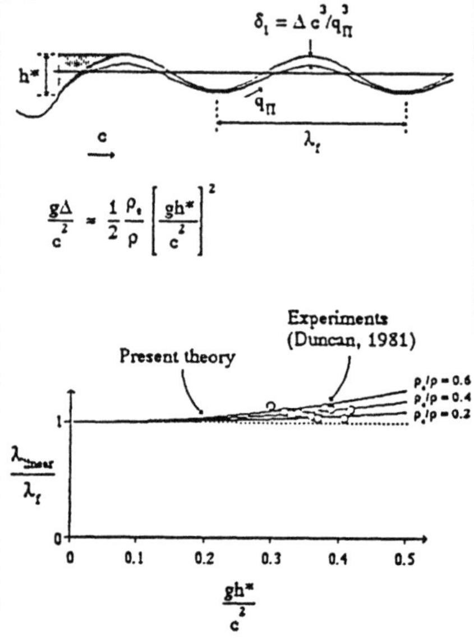


Figure 3. Boundary layer analysis

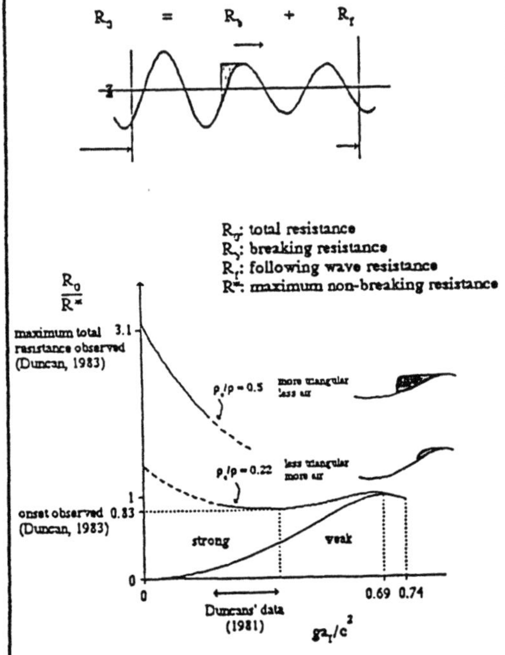


Figure 4. "Shock" analysis

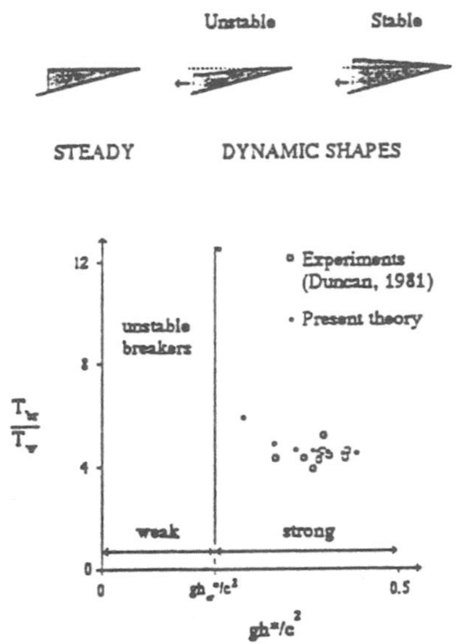


**Figure 5.** **Stability and free oscillations**

**Analysis based on linearization of the non-linear density current equations (analytic solution) and confirmed by nonlinear numerical computations**

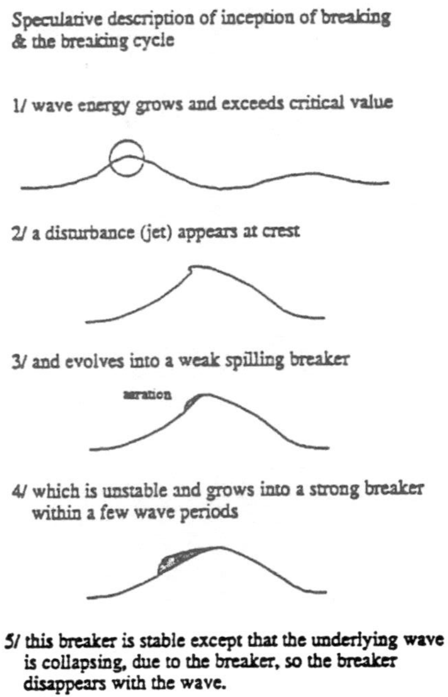


**Figure 6**

# Surf Zone Breakers with Current

J. BUHR HANSEN

Institute of Hydrodynamics and
Hydraulic Engineering
Technical University, Denmark.

IB A. SVENDSEN

Department of Civil Engineering,
University of Delaware,
USA.

## Abstract

A two-dimensional model for waves and steady current in the surf zone is developed. It is based on a depth integrated and time averaged version of the equations for the conservation of mass, momentum, and wave energy. A numerical solution gives the variation of wave height, set-up, and steady current in the surf zone, taking into account the mass flux in the waves and a possible net volume flux. It is assumed that a bottom boundary layer exists with an eddy viscosity $\nu_{tb}$ much smaller than the eddy viscosity $\nu_t$ generated over the rest of the water depth by the breaking.

The paper concentrates on the determination of some of the basic parameters in the model and on the comparison between the model results and experiments.

## 1. Introduction

On a beach with a longshore bar the wave breaking over the bar is usually combined with a net shoreward volume flux, $\overline{Q}$, which, in the three-dimensional flow situation, feeds longshore currents. On a plane beach the problem is in principle also three-dimensional and does also involve longshore components of the current.

In the present paper we will concentrate on the two-dimensional combination of wave and current motion with or without a net volume flux. In choosing this arrangement we implicitly assume that in the prototype situation the magnitude of $\overline{Q}$ will be determined by the distance between the rips and by the position of our two-dimensional flow section relative to the position of the rip currents.

The mathematical numerical model developed for the wave height/set-up variation in the surf zone is based on the depth-integrated time-averaged equations of continuity, momentum, and energy as derived by, e.g. Phillips [7]. In the model, the wave properties, such as radiation stress, mass flux, and energy flux are represented by dimensionless coefficients chosen in a way which makes them depend only weakly on water depth, wave height and wave period. The model is coupled with a model for the Eulerian mean current (i.e. the undertow for $\overline{Q}$ = 0) used to determine the bottom mean current velocity and the bottom shear stress.

K. Horikawa, H. Maruo (Eds.)
Nonlinear Water Waves
IUTAM Symposium Tokyo/Japan 1987

The general form of the model allows the dimensionless coefficients to be determined by any chosen wave theory. In the present application, however, the coefficients are developed using the description of breaking waves suggested by Svendsen [11]. In that paper, and in Svendsen [12],Hansen & Svendsen [1, 2, 3], Svendsen & Hansen [13, 14 ,15], wave height, set up, and net current velocity were analysed for breaking waves with or without a net flow. Finally is presented a comparison with experimental results for wave height, set-up, and net current velocity profiles over a barred bottom profile (see Hansen & Svendsen [2]).

## 2. Description of the model

The mathematical-numerical model is based on the two-dimensional depth-integrated and time-averaged equations for the conservation of mass, momentum and energy. Using the definitions in Fig. 1, these equations read (see Phillips [7]):

Mass:

$$\bar{Q} = \overline{\int_{-h_o}^{b+\eta} u dz} = Q_s + Ud_t \tag{2.1}$$

Where $Q_s$ is the mass flux above trough level in the wave-current motion, and U the trough to bottom depth-average of the mean Eulerian velocity $U(\zeta)$.

Momentum:

$$\frac{d}{dx}\{\rho \frac{\bar{Q}^2}{h} + S'_{xx}\} + \rho g h \frac{db}{dx} + \bar{\tau}_b = 0 \tag{2.2}$$

Here $S'_{xx}$ is the radiation stress and $\bar{\tau}_b$ the mean bottom shear stress. Following Phillips we define the radiation stress by

$$S'_{xx} = \overline{\int_{-h_o}^{b+\eta} \rho \; (u_w^2 - w^2) dz} + \frac{1}{2} \rho g \overline{\eta^2} - \frac{\rho Q_s^2}{h} \tag{2.3}$$

where $u_w$ and w are the horizontal and vertical particle velocities in the wave motion ($\bar{u}_w = 0$), and ' in $S'_{xx}$ indicates the inclusion of the $Q_s$-term, which also in the surf zone turns out to be rather unimportant.

Except for the $\bar{\tau}_b$-term (2.2) is the same as Phillips' expression (3.6.11).

Energy:

$$\frac{d}{dx} \; (E_f + UE - \frac{1}{2} \rho \bar{Q} (\frac{Q_s}{h})^2) + S'_{xx} \frac{dU}{dx} - \bar{\tau}_b U = \mathcal{D} \tag{2.4}$$

where $E_f$ is the energy flux and E the energy density in the oscillatory part of the motion. $\mathcal{D}$ is the mean energy dissipation per unit area of the bottom. Except for the $\bar{\tau}_b U$ term, which is the energy dissipation due to the current, this expression corresponds to Phillips' eq. (3.6.19). Although the $\bar{\tau}_b U$-term is small compared to the total dissipation $\mathcal{D}$ we shall maintain this term because of the cumulative effect and the increasing importance with increasing total mass flux.

The mean Eulerian current profile below trough level represents turbulent shear stresses which at each point balance the difference between the gradients of radiation stress and set-up. The turbulent shear stresses are modelled by means of an eddy viscosity $\nu_t$ created by the breaker turbubulence so that $\bar{\tau} = \rho \nu_t \, \partial u / \partial z$, except near the bottom where a thin (oscillatory) boundary layer is assumed to have an eddy viscosity $\nu_{tb} \ll \nu_t$ (Svendsen et al. [14]).

With an assumed constant $\nu_t$ above the boundary layer this arrangement results in a parabolic variation with z of U with a finite velocity $U_b$ at the top of the bottom boundary layer. The mean bottom shear stress $\bar{\tau}_b$ may be determined from $U_b$ in combination with the oscillatory velocity amplitude $u_{bm}$

$$\bar{\tau}_b = \frac{1}{2} \rho f_w \, u_{bm} \, U_b \tag{2.5}$$

where $f_w$ is a friction factor.

## 3. Determination of wave quantities

In the present application we use the surf zone model suggested by Svendsen [11] - Fig. 2 - to determine the quantities $Q_s$, $S'_{xx}$, $E_f$, E in the basic equations. However, the following three modifications are applied.

1) The wave celerity, $c_r$, relative to U is in Hansen & Svendsen [2] found to agree well with the celerity of a bore with the same H, h, and $d_t$.

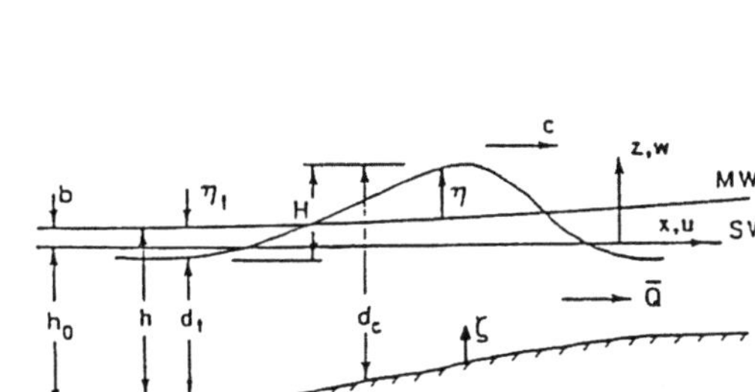


Fig. 1. Definitions used.

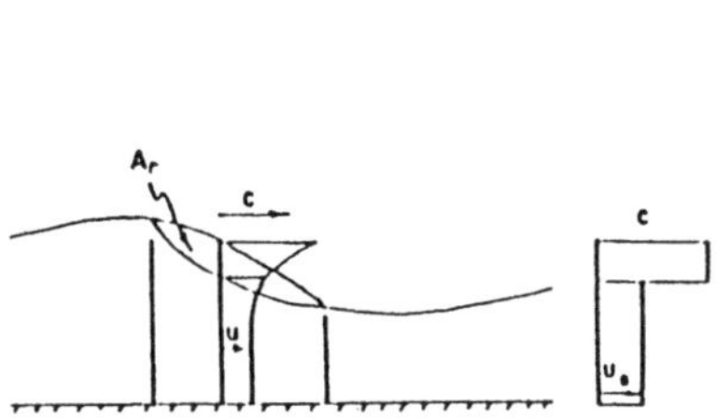


Fig. 2. Actual and assumed oscillatory velocity profiles.

2) The oscillatory particle velocities, $u_w$, are from experiments found to be significantly smaller than predicted by linear theory ($u = c\ \eta/h$). Hansen & Svendsen [3] show that $u_w = \beta\ u$ with $\beta = 0.6 - 0.85$. Analysis of Stive & Wind's tests [8] shows $\beta$ factors within the same range. An average $\beta = 0.7$ is applied in the example below. It should be noticed that the profile shape factors of the type $B_o = \overline{\eta^2}/H^2$ are in general 0 to 10% larger for the velocity recordings than for the surface recordings.

3) The vertical cross-sectional area of the roller, $A_r$, is in Hansen & Svendsen [3] and Okayasu et al. [6] found to be $A_r/HL \simeq 0.06$. When the above $\beta = 0.7$ is applied, however, $A_r/HL = 0.07$ seems to fit the data better.

The surf zone model yields the following dimensionless parameters:

$$\frac{Q_s\ h}{c_r\ H^2} = \left( \beta\ B_o + \frac{A_r}{HL} \cdot \frac{h}{H} \right) \frac{d_t}{h} \tag{3.1}$$

where $B_o = \overline{\eta^2}/H^2$.

$$\frac{S'_{xx}}{\rho g H^2} = B_o\left\{\frac{1}{2} + \frac{c_r^2}{gh}\ (\beta^2 - \beta(\frac{H}{h})^2\right\} + B_r\left\{1-(\frac{H}{h})^2\right\} \tag{3.2}$$

where

$$B_r = \frac{A_r}{HL} \cdot \frac{h}{H} \cdot \frac{c_r^2}{gh} \tag{3.3}$$

$$E_f/\rho g c_r H^2 = \beta\ B_o + \frac{1}{2}\ B_r \tag{3.4}$$

$$E/\rho g H^2 = \frac{1}{2}\ \left\{B_o(1 + \beta^2 \frac{c_r^2}{gh}) + B_r\right\} \tag{3.5}$$

Furthermore, the energy dissipation $\mathcal{D}$ is approximated by the dissipation in a bore of the same height as the wave, see e.g. Svendsen [11]

$$\frac{4\,\mathcal{D}\,h\ T_a}{\rho\ g\ H^2} = -\frac{h^2}{d_t d_c} \cdot \frac{1}{(1 + U/c_r)} \tag{3.6}$$

where $T_a$ is the absolute wave period.

## 4. Determination of bottom shear stress and current profile

If we assume a constant $\nu$ above the bottom boundary layer, and assume the mean current driving force

$$\alpha_1 = \frac{\partial}{\partial x}\ (\overline{u_w^2} - \overline{w^2} + U^2 + gb) \tag{4.1}$$

(see e.g. Stive & Wind (10) and Svendsen et al. [14]) also to be constant over the depth - the set-up pressure gradient being the dominating term - the Eulerian velocity profile may be written as

$$U(\zeta) = \frac{1}{2}\,\frac{\alpha_1}{\nu_t}\,\zeta^2 + \frac{A_1}{\nu_t}\zeta + U_b \tag{4.2}$$

where $\zeta = z + h$ and $A_1$ is a (fictive) shear stress term defined as

$$A_1 = \overline{\tau}_{bU}/\rho = \nu_t (\frac{\partial U}{\partial \zeta})_{\zeta = 0} \tag{4.3}$$

If we neglect the contribution from the undertow forcing term $\alpha_1$ over the boundary layer and assume that the shear stress from the steady streaming at the upper limit of the boundary layer is negligible too, the momentum equation over the boundary layer becomes

$$\overline{\tau}_b = \overline{\tau}_{oU} + \overline{\tau}_{sb} \tag{4.4}$$

where $\overline{\tau}_b$ is the total bottom shear stress and $\overline{\tau}_{sb} = \rho \, \overline{u_w w}$ is the contribution from the organised oscillatory motion responsible for the steady streaming. $\tau_{oU}$ is equivalent to (4.3) but at the upper limit of the boundary layer. If we further neglect the small difference between $\overline{\tau}_{bU}$ and $\overline{\tau}_{oU}$ and introduce $\overline{\tau}_b$ from (2.5), the following expression is obtained for $U_b$

$$\frac{U_b}{c_r} = \left( \frac{U_m}{c_r} - \frac{1}{6} A + \frac{1}{2} \frac{\overline{\tau}_{sb}}{\rho \alpha_1 d_t} A \right) (1 + R)^{-1} \tag{4.5}$$

with

$$A = \frac{\alpha_1 d_t^{\,2}}{\nu_t \, c_r} \tag{4.6}$$

$$R \equiv \frac{f_w}{8} \frac{u_{bm} \, d_t}{\nu_t} \tag{4.7}$$

and $f_w$ determined from the expression

$$\frac{1}{4\sqrt{f_w}} + \log \frac{1}{4\sqrt{f_w}} = -0.08 + \log \frac{a_b}{k_N} \tag{4.8}$$

(Jonsson [4]) where $a_b$ is the oscillatory bottom amplitude and $k_N$ the Nikuradse bottom roughness.

In the literature is reported a wide range of $\nu_t$ values determined in different ways (e.g. Okayasu et al. [6], Stive & Wind [10], and Svendsen et al. [14,15]). If the measured velocity profiles are approximated by the best fitting parabola of the form (4.2), $\frac{1}{2} \alpha_1/\nu_t$ is determined directly as a parameter in the parabolic fit. Published data are analysed, and the results are given in Fig. 3. The variations of $\nu_t$ seems not correlated to any other parameter in the experiments, and a "mean" value of $\nu_t/h\sqrt{gh} = 0.007$ will be adopted here.

## 5. Comparison with experimental results

The breaker/set-up model (chapter 3) may now be combined with the model for the bottom shear stress and velocity profile (chapter 4) as follows:

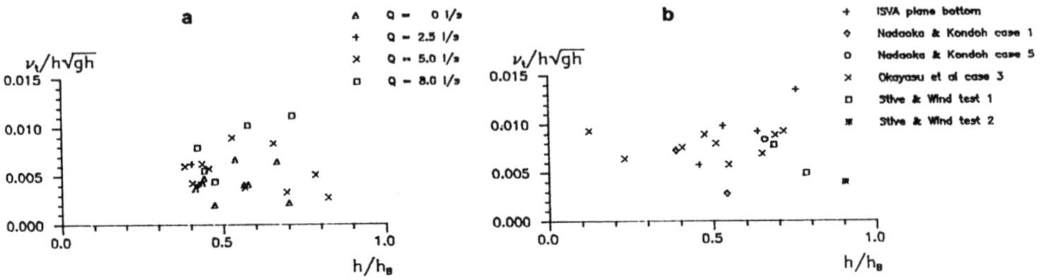


Fig. 3 $\nu_t$ determined from parabolic approximation to measured velocity profiles. a) ISVA tests on bar bottom; b) tests on plane beaches.

From the breaker/set-up model the necessary information for calculating $U_b$ from eq. 4.5 is obtained. Eventually $\bar{\tau}_b$ is determined from (2.5) and included in the subsequent solution of (2.2) and (2.4).

The wave profile shape parameter $B_o$ appearing in most of the terms of (2.2) and (2.4) is for the present comparison taken constant $B_o = 0.09$ throughout the surf zone (Hansen & Svendsen [2]). This is obviously an approximation, but the importance is reduced by the fact that the $\beta B_o$ values are found almost constant. For the present comparison further $K_N = 0.5$ mm and $\eta_c/H = 0.67$ are used.

In Figs. 4 and 5 the numerical results are compared with experiments in cases of a net flow $\bar{Q}/h\sqrt{gh} = 0$ and 0.181 at the bar crest from Hansen & Svendsen [2]. The results of two sets of computations are shown. Comp. 1 starts at the breaking point with the dissipation made artificially to grow gradually from 0 to the value given by (3.6) at the end of the transition region where comp. 2 is started (at $h = 0.8\ h_B$). In the transition region, furthermore, the set-down is artificially held constant.

While comp. 2 follows the measured wave height variations quite closely, comp. 1 gives the best agreement to the measured set-up, but both computations show too small set-up values, may be because the transition region has been extended too far. The agreement between measured and calculated Eulerian mean velocity profiles is generally quite convincing.

## 6. Conclusions

The two-dimensional model for wave-current interaction in the surf zone shows promising results for both wave height/set-up variation and mean velocities. For further improvements some of the parameters in the model need to be investigated further:

1) The size and possible variations of $B_o$ and $\eta_c/H$, and 2) the relation between particle velocities and surface elevations (the $\beta$-factor).

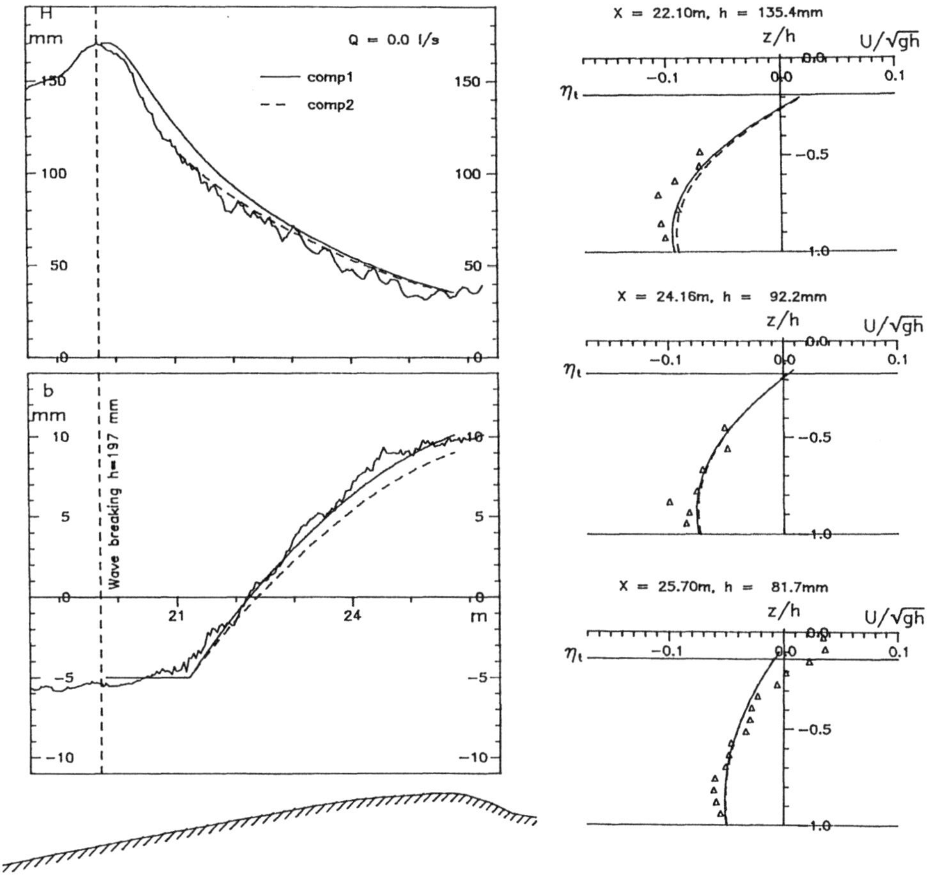


Fig. 4. Comparison of measured and computed wave height, set-up and Eulerian mean velocity profiles without a net flow over a bar bottom profile.

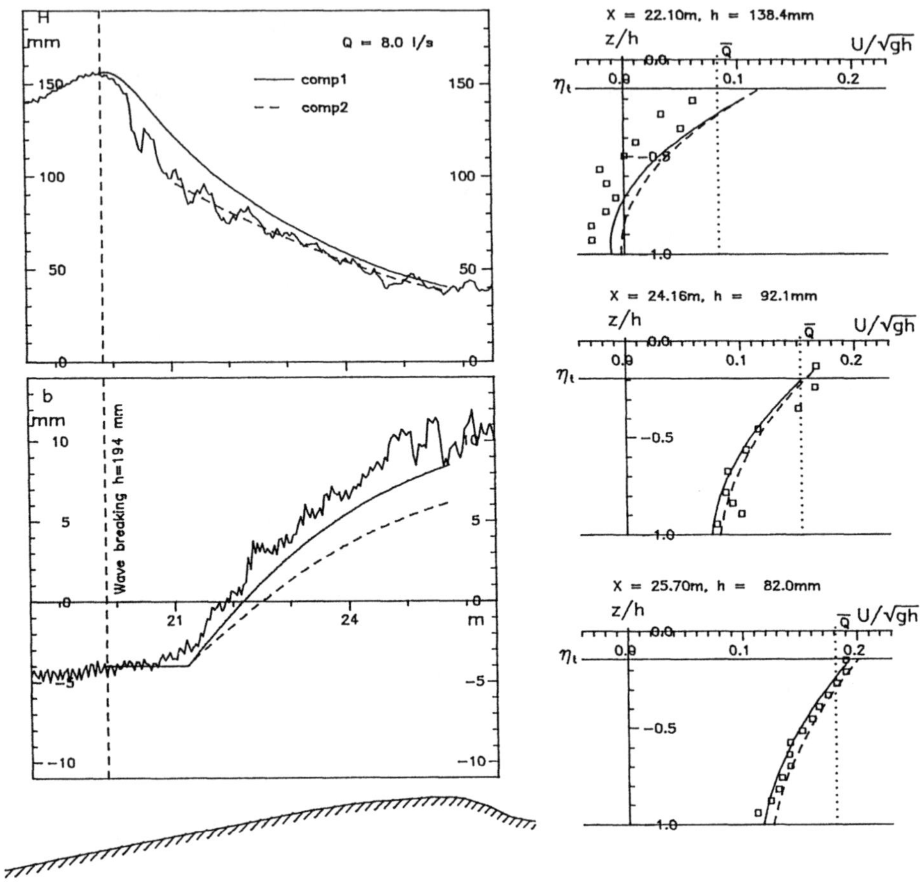


Fig. 5. Comparison of measured and computed wave height, set-up and Eulerian mean velocity profiles with a net flow over a bar bottom profile.

References:

1. Hansen, J. Buhr & I.A. Svendsen. A theoretical and experimental study of undertow, Proc. 19th Int. Conf. Coastal Eng., Houston 1984, ch. 151, 2246-2262.
2. Hansen, J. Buhr & I. A. Svendsen. Experimental investigation of the wave and current motion over a longshore bar, Proc. 20th Int. Conf., Coastal Eng., Taipei 1986.
3. Hansen, J. Buhr & I.A. Svendsen. The cross shore motion over a barred beach profile, Proc. 2nd Int. Conf. Coast. and Port Eng. Developing Countries, Beijing 1987.
4. Jonsson, I.G. Wave boundary layer and friction factors, Proc. 10th Int. Conf. Coastal Eng., Tokyo 1966, ch. 10, 127-148.
5. Nadaoka, K. & T. Kondoh. Laboratory measurements of velocity field structure in the surf zone by LDV, Coastal Eng. Jap. 25 (1982), 125-146.
6. Okayasu, A., T. Shibayama & N. Mimura. Velocity field under plunging waves, Proc. 20th Int. Conf. Coastal Eng., Taipei 1986.
7. Phillips, O.M. The dynamics of the upper ocean, Cambridge University Press, Cambridge 1977.
8. Stive, M.J.F. Two dimensional breaking of waves on a beach, Report M 1585, part I...IV, Delft Hydraulics Laboratory 1983.
9. Stive, M.J.F. & H.G. Wind. A study of radiation stress and set-up in the nearshore region, Coastal Eng. 6 (1982), 1-25.
10. Stive, M.J.F. & H.G. Wind. Cross-shore mean flow in the surf zone, Coastal Eng. 10 (1986), 325-340.
11. Svendsen, I.A. Wave heights and set-up in a surf zone, Coastal Eng. 8 (1984) 4, 303-329.
12. Svendsen, I.A. Mass flux and undertow in a surf zone. Coastal Eng. 8 (1984), 4, 347-365.
13. Svendsen, I.A. & J. Buhr Hansen. The interaction of waves and currents over a longshore bar. Proc. 20th Int. Conf. Coastal Eng., Taipei 1986.
14. Svendsen, I.A., H.A. Schäffer & J. Buhr Hansen. The interaction between the undertow and the bottom boundary layer flow on a beach. To appear in J. Geophys. Res. 1987.
15. Svendsen, I.A., J. Buhr Hansen & H.A. Schäffer. Analysis of cross-shore circulation on a beach. Proc. ASCE Conf. on Coastal Hydrodynamics, Newark, Delaware 1987.

# Session 5

# Nonlinear Water Waves

# Subharmonic Resonance of Nonlinear Cross-waves: Theory and Experiment

SETH LICHTER

Department of Aerospace and Mechanical Engineering
University of Arizona, Tucson

Abstract

Recent analytical and experimental results on the parametric resonance of cross-waves are presented. The results of a matched asymptotic expansion including the viscous effects from the free surface, sidewalls, bottom, and wavemaker boundary layers are presented. Results from a bifurcation analysis reveal a transition to hysteretic behavior which is observed experimentally. Finally, the equations governing cross-wave evolution for two coupled unstable modes are presented.

## 1 Introduction

Analytical and experimental results on the parametric resonance of cross-waves are presented. These results are divided into three overlapping topics: (1) Viscous Effects--The evolution equation governing wavemaker-generated viscous cross-waves near a cut-off frequency in a channel is presented. Viscous effects from the sidewalls, bottom, free surface, bulk, and the wavemaker are included. (2) Bifurcations--The results of a bifurcation analysis reveal a transition from subcritical to supercritical flow at a critical value of the detuning. The predicted hysteretic behavior is observed experimentally. A similarity scaling in the inviscid limit is also predicted and observed experimentally. (3) Coupled Modes--Finally, results are presented for the case when two modes are present and the possibility exists for the exchange of energy between the two modes. The coupling produces a modulation across the span of the tank and also detunes the resonant frequency. These effects are observed experimentally.

## 2 Viscous Effects

Previous work [1, 2] has coped with viscosity in an ad hoc manner, usually incorporating a linear term with an empirically determined damping coefficient. Here, results from a formal analysis in which viscous effects are

K. Horikawa, H. Maruo (Eds.)
Nonlinear Water Waves
IUTAM Symposium Tokyo/Japan 1987

incorporated into a matched asymptotic expansion are presented. The relative contributions of the free-surface, sidewalls, bottom, and wavemaker viscous boundary layers are computed.

Surface tension effects are considered negligible. The channel is considered to be of depth $z = -d$ and width $0 < y < W$, with a wavemaker at $X = 0$ and extending to $X \to \infty$. The number of wavelengths which span the tank is $2N$, where $N$ is the mode number. The effects of viscosity, $\nu$, are parameterized by the Reynolds number, Re,

$$Re = W^2\omega/\nu N^2\pi^2 \ , \tag{1}$$

where $2\omega$ is the frequency of the wavemaker. Cross-waves are most easily excited at the cut-off frequency, $\omega_0 = (gN\pi\tanh(d)/W)^{1/2}$, where $g$ is the acceleration due to gravity. The perturbation parameter, $\varepsilon$, is the forcing amplitude nondimensionalized with respect to the characteristic length of the cross-wave. The complex amplitude, A, of the modulation varies on a slow temporal, T, and spatial, X, scale. Details of the calculation may be found in Bernoff et al. [3]. They found that the nonlinear Schrödinger equation for inviscid cross-waves is modified by viscous damping and detuning. It is illuminating to consider the linearized results:

$$- i\, A_T + A_{\tilde{X}\tilde{X}} + (\tilde{\sigma} - i\mathcal{C})A = 0 \ , \tag{2}$$

where

$$\tilde{\sigma} = \frac{J_a}{2} + \sigma_2 + \lambda S + \frac{1}{\varepsilon^2\sqrt{2Re}}\left[\frac{S}{N\pi} + \frac{1}{\sinh(2d)}\right] , \tag{3}$$

$$\mathcal{C} = \frac{1}{2\varepsilon^2 Re}\left\{4 + \sqrt{2Re}\left[\frac{S}{N\pi} + \frac{1}{\sinh(2d)}\right] + J_c\right\} , \tag{4}$$

and

$$S = 1 - d[\tanh(d) - 1/\tanh(d)] \ . \tag{5}$$

The detuning due to the presence of the progressing wave is represented by $J_a$ [4], frequency detuning is $\sigma_2 = (\sigma - \sigma_0)/\varepsilon^2$, $\lambda$ is the detuning due to changes in tank width [4], and $J_b$ is due to self-interaction of the cross-wave and is the same as that obtained by Larraza and Putterman [5]. The wavemaker boundary condition is

$$A_{\tilde{X}} = \tilde{a}A^* \quad \text{at } \tilde{X} = 0 \ , \tag{6}$$

where the coefficient of $A^*$ depends only on tank geometry and wavemaker amplitude and shape. In equation (4), the first term is due to the viscous effects at the free surface, the second term is due to sidewall damping, the third term is due to the bottom boundary layer, and $J_c$ represents the interaction between boundary layers and is small if the boundary layers are thin.

Curves of marginal stability are given by

$$-\Gamma + (\tilde{a}^4 - \tilde{\sigma}^2)^{1/2} = 0. \qquad (7)$$

The asymptotes of the hyperbole, for which $\tilde{a}^2 = |\tilde{\sigma}|$, represent the stability boundary as $\Gamma \to 0$, as considered by Jones [4]. Equation (4) shows that viscous effects will increase the forcing amplitude necessary to yield instability. The effect of viscosity on the marginal stability curves is shown schematically in Fig. 1. From equation (3), it can be seen that viscous effects will also reduce the cut-off frequency from its inviscid value, cf. Fig. 1. So viscous effects both detune the resonant frequency and increase the onset of instability.

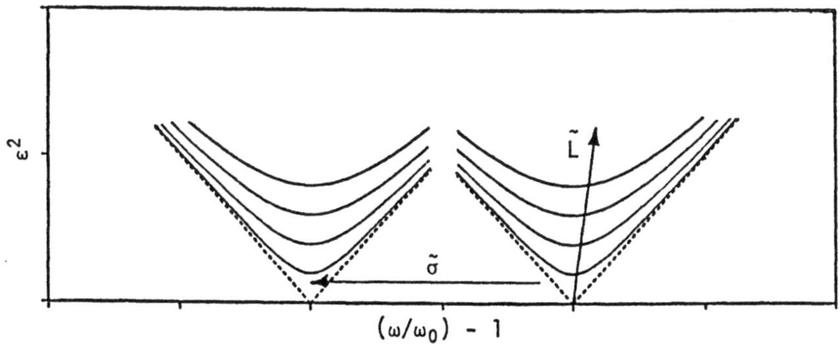


Fig. 1. Neutral stability curves for the linearized equation (5) for various values of the damping, $\Gamma$, and detuning $\tilde{\sigma}$. The dashed lines represent the limit $\Gamma \to 0$.

## 3 Bifurcation Analysis

While contained within the full weakly nonlinear and weakly damped equations given in Section 2, a bifurcation analysis can indicate the type of solution and bifurcation sequence more simply than numerical techniques. Here, the results from a bifurcation analysis reveal a supercritical-to-subcritical bifurcation. The details of the bifurcation analysis can be found in Lichter and Bernoff [6]. Here, we outline the approach and give some results. The complex cross-wave amplitude is written $A = C + iD$, where C and D are real. Then, the suitably rescaled equations are

$$\underline{C}_T = \mathcal{L}\underline{C} + \mathcal{N}(\underline{C}) \tag{8a}$$

and

$$\underline{C}_X = \mathcal{B}\underline{C} \qquad \text{at } X = 0 \text{ ,} \tag{8b}$$

where

$$\underline{C} = \begin{bmatrix} C \\ D \end{bmatrix} . \tag{9}$$

The operators $\mathcal{L}$, $\mathcal{N}$, and $\mathcal{B}$ are defined by

$$\mathcal{L} = \begin{bmatrix} -L & -\lambda - \partial_{XX} \\ \lambda + \partial_{XX} & -L \end{bmatrix} , \tag{10}$$

$$\mathcal{N}(\underline{C}) = \begin{bmatrix} -(C^2 + D^2)D \\ (C^2 + D^2)C \end{bmatrix} , \tag{11}$$

and

$$\mathcal{B} = \begin{bmatrix} 1 & 0 \\ 0 & -1 \end{bmatrix} , \tag{12}$$

Here, the detuning $\lambda$ can represent changes in tank width (as in Section 2), or changes in frequency detuning (i.e., $\sigma_2$ in Section 2), or both. In this problem, near a cut-off frequency, only one eigenvector becomes unstable as the forcing amplitude is increased. For this most unstable eigenvector, $C_0$, the eigenvalue associated with the linearization of equation (8) is

$$S_0 = - L + \sqrt{1 - \lambda^2} \ . \tag{13}$$

The remainder of the spectrum is damped. The development of the most unstable eigenvector on the center manifold can be followed by considering

$$\underline{C} = r(t)\underline{C}_0 \, , \tag{14}$$

where r(t) is a time-dependent amplitude. On substituting equation (14) into equation (8), the projection onto the $\underline{C}_0$ direction on the center manifold is performed by first defining the adjoint eigenvector $\underline{C}_0^\dagger$,

$$\mathcal{L}^\dagger \underline{C}_0^\dagger = S_0 \underline{C}_0^\dagger \, , \tag{15}$$

and then using $\langle \underline{C}_0^\dagger, \cdot \rangle$ applied to equation (8) to project onto the $\underline{C}_0$ subspace. This yields the Landau equation

$$r_t - S_0 r = H r^3 \, , \tag{16}$$

where

$$H = \frac{1 - 3\lambda}{5 - 3\lambda} \, \frac{1}{\sqrt{1 - \lambda^2}} \, . \tag{17}$$

For $H < 0$ (i.e., $\lambda > 1/3$), the bifurcation is a supercritical pitchfork (Fig. 2a). When $H > 0$ (i.e., $\lambda < 1/3$), the bifurcation is a subcritical pitchfork (Fig. 2b). When $s_0 > 0$, the nonlinear term augments the linear growth term and the solution grows without bound. When $s_0 < 0$, the behavior is hysteretic.

For $|H| \ll 1$, the fifth-order correction to equation (16) will become important. In the case when this term is stabilizing, the subcritical branch terminates in a saddle node bifurcation (Fig. 2b).

In the limit $L \ll \lambda$, the dynamics of cross-waves are dominated by the inviscid effects [3]. For fixed Reynolds number, this corresponds to the limit of large forcing, $\varepsilon$. If this limit is regular, bifurcations should only depend upon $\lambda$ at leading order. Assuming that a bifurcation occurs at some $\lambda_c$, then it follows from the definition of the detuning that, considered as a function of $\varepsilon$ and $\omega$, the bifurcation will occur on the curve

$$\varepsilon^2 = \frac{1}{\lambda_c \tilde{a}^2} \left( \frac{\omega}{\omega_0} - 1 \right) . \tag{18}$$

This suggests that $(\omega/\omega_0 - 1, \varepsilon^2)$ may be convenient experimental parameters; with these coordinates, bifurcation curves should asymptote to straight lines in the limit of large $\varepsilon$.

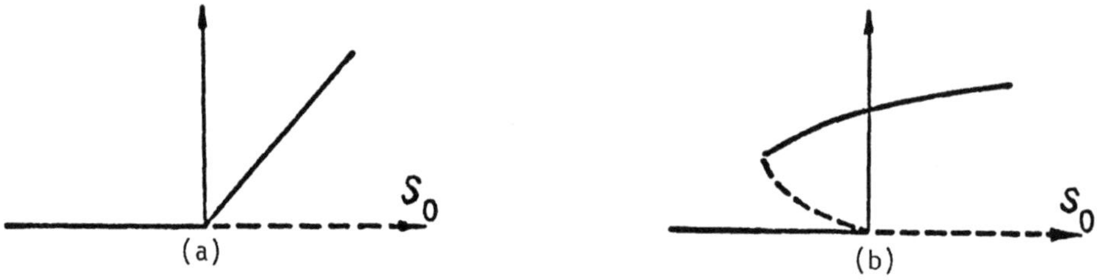


Fig. 2 (Wave height)$^2$ of the most unstable eigenvector as a function of the eigenvalue, $s_0$, for (a) supercritical ($H < 0$) and (b) subcritical ($H > 0$) solutions. Solid lines show stable solutions, and dashed lines are unstable solutions.

The experimental results are described in detail in Lichter and Bernoff [6]. The up-triangles in Fig. 3 indicate the appearance of cross-waves for increasing wavemaker amplitude, whereas the down-triangles mark the disappearance of cross-waves with decreasing wavemaker amplitude. To the left of the line $\lambda = 1/3$, these two curves diverge, indicating the appearance of hysteretic behavior in agreement with the prediction. Figure 4 shows cross-wave amplitude as a function of wavemaker forcing at constant detuning. In Fig. 4a, a detuning to the right of the hysteretic transition is shown; note that a bifurcation to a steady cross-wave is observed, in agreement with Fig. 2a. In Fig. 4b, a detuning to the left of the hysteretic transition is shown; here, a stable zero solution and a stable steady cross-wave exist at the same parameter values, in agreement with Fig. 2b.

For $\lambda \gg L$, the analysis indicates that inviscid effects should dominate and the various bifurcation curves may be approximated by straight lines. Figure 3 shows that the transition curves qualitatively agree with this scaling.

## 4 Coupling Between Modes

The partitioning of energy between the two spatial modes introduces an additional frequency. This solution is observed experimentally as a wave propagating transversely from side to side. This Y-modulation is a precursor to chaotic solutions.

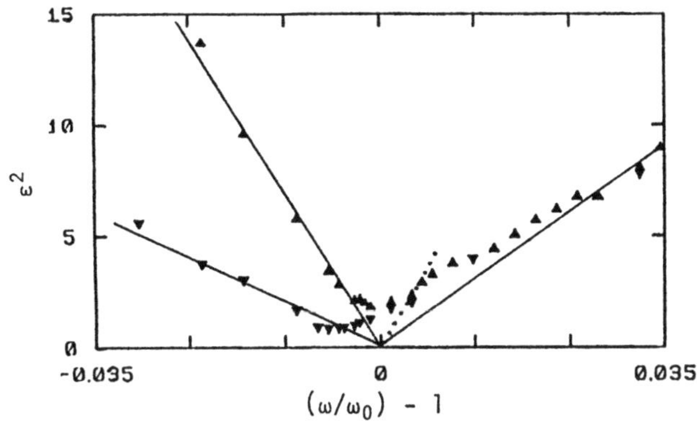


Fig. 3 Stability diagram for the experimental data. The up- (down-) triangles indicate that the forcing amplitude was increased (decreased) at constant forcing frequency. The solid lines converging to the inviscid cut-off frequency reveal the similarity scaling of the inviscid limit. The dotted line is the theoretical prediction $\lambda = 1/3$. The scale is in arbitrary units.

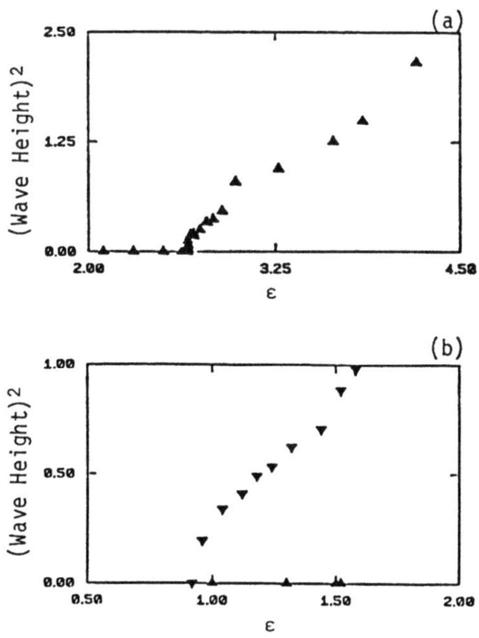


Fig. 4 (Wave height)$^2$ as a function of forcing amplitude for (a) 8.1 Hz showing behavior similar to the supercritical case, Fig. 2a, and (b) for 7.81 Hz showing behavior similar to the subcritical solution, Fig. 2b. The dashed line is hypothesized. The scale is in arbitrary units.

In this case, two modes are introduced. The solution is now allowed to depend on a slow y scale

$$Y = \varepsilon^2 y \tag{19}$$

so that the first-order solution for the velocity potential, $\phi_1$, is of the form

$$\phi_1 = A(X, T)e^{iMY} e^{i(\alpha y+2\sigma t)+\alpha z} + B(X, T)e^{iMY}e^{i(\alpha y-2\sigma t)+\alpha z} + \text{c.c.} \tag{20}$$

where $\alpha$ is the square of the cut-off frequency. The analysis follows that of Jones [4] to yield

$$- iA_T + A_{XX} + iA_Y + \lambda A - iLA - |A|^2A + 2|B|^2A = 0 \tag{21}$$

$$iB_T + B_{XX} + iB_Y + \lambda B + iLB - |B|^2B + 2|A|^2B = 0 \,. \tag{22}$$

The boundary condition of zero normal velocity on the sidewalls means that the transverse slow modulation will appear with a wavelength equal to twice the width of the tank.

From the linearized set of equations, the eigenvalue is found to be

$$S_0 = - L + \sqrt{1 - (\lambda - M)^2} \,. \tag{23}$$

Note that the presence of the Y-modulation will detune the resonant frequency to a higher value at the cut-off.

The transverse modulation can be observed easily in a wave tank. As predicted, the wavelength of the modulation is twice that of the width of the wave tank. Shemer and Lichter [7] observed this modulation and the region of transverse modulation relative to the neutral stability curve is shown. In their Fig. 1, the stability curve for the modulation is shifted to the right of the neutral stability curve, as predicted.

Acknowledgments

The original work presented in this chapter was funded by the National Science Foundation and the Office of Naval Research. This work was also made possible by the diligence of my graduate students Jerry Chen, Loong-Piu Kwok, Hassan Ayanle, and William B. Underhill and by Dr. Andrew J. Bernoff of the Department of Mathematics.

References


1. Barnard, B. J. S.; Mahony, J. J.; Pritchard, W. G.: The excitation of surface waves near a cut-off frequency. Phil. Trans. Roy. Soc. London, Ser. A 286 (1977) 87-123.

2. Lichter, S.; Chen, J.: Subharmonic resonance of nonlinear cross-waves. J. Fluid Mech. 183 (1987) 451-465.

3. Bernoff, A. J.; Kwok, L. P.; Lichter, S.: Viscous cross-waves: an analytical treatment. J. Fluid Mech. (submitted for publication).

4. Jones, A. F.: The generation of cross-waves in a long deep channel by parametric resonance. J. Fluid Mech. 138 (1984) 53-74.

5. Larraza, A.; Putterman, S.: Theory of non-propagating surface-wave solitons. J. Fluid Mech. 148 (1984) 443-449.

6. Lichter, S.; Bernoff, A. J.: Stability of steady cross-waves: theory and experiment. Phys. Rev. A (submitted for publication).

7. Shemer, L.; Lichter, S.: Identification of cross wave regimes in the vicinity of a cut-off frequency. Phys. Fluids (to appear).

# Pulse Interactions and Wave Evolutions in an Unstable Dissipative Dispersive System

Takuji Kawahara, Sadayoshi Toh, and Masanori Takaoka

Department of Physics, Faculty of Science
Kyoto University, Kyoto 606 Japan

Summary

Wave evolutions in a nonlinear equation with instability, dispersion, and dissipation are investigated both numerically and analytically. It is shown that equilibrium states or chaotic evolutions in the initial value problem can be described by interactions of localized soliton-like pulses arranged in a string. Nonlinear lattice equations governing the inter-pulse distances involve chaotic motions in accordance with the numerical results.

## Introduction

The necessity of research on nonlinear evolutions of instability waves is obvious for a wide class of wave problems. Not a few works, particularly directing attention to transition to chaos in unstable wave systems, have been made recently on the Ginzburg-Landau(envelope) equation governing nonlinear evolutions of quasi-monochromatic waves[1,2] or the Kuramoto-Sivashinsky type equations governing long wave evolutions[3]. Here we investigate several properties of solutions to the latter type nonlinear evolution equation of the form

$$u_t + uu_x + \alpha u_{xx} + \beta u_{xxx} + \gamma u_{xxxx} = 0, \qquad (1)$$

where $\alpha$, $\beta$, and $\gamma$ are positive constants characterizing instability(self-excitation), dispersion, and dissipation, respectively. This type of equation describes asymptotic behaviours of unstable long waves in fluid systems, e.g., the long surface waves on a viscous liquid film flowing down an inclined plane [4,5] and the long waves at the interface between two cocurrent superposed viscous fluids[6].

K. Horikawa, H. Maruo (Eds.)
Nonlinear Water Waves
IUTAM Symposium Tokyo/Japan 1987

Several features of this equation have been studied recently[7-12]. Numerical initial value problem(IVP) exhibits that an array of localized soliton-like pulses develops for strongly dispersive case, whereas the wave evolutions become chaotic for weakly dispersive case. In this connection, it is shown that the localized pulses play essential roles in such wave evolutions and the qualitative differences in wave evolutions for strongly or weakly dispersive cases are well understood by means of the perturbation approximation of weakly interacting localized pulses. Several features of wave evolutions are discussed in terms of the nonlinear lattice equations governing the inter-pulse distances.

Result of Initial Value Problem

Equation (1) can be normalized as

$$u_t + uu_x + u_{xx} + \delta u_{xxx} + u_{xxxx} = 0, \qquad (2)$$

where $\delta \equiv \beta/(\alpha\gamma)^{1/2}$ is a parameter representing relative importance of the dispersion. Typical consequences of numerical IVP for various values of $\delta$ with periodic boundary condition are summarized as follows:

i) Generation of pulses: Generally, an array of pulses is generated. For strongly dispersive case(i.e., large $\delta$), the asymptotic stages of time evolutions are equilibria consisting of a string of soliton-like pulses of equal(saturated) amplitudes. According to decrease in $\delta$, the amplitude of pulses decreases and eventually chaotic behaviours are observed.

ii) Multiplicity of solutions: Different number of pulses are generated depending on the initial conditions even for a fixed value of $\delta$. Various equilibrium states with different pulse configurations are observed even when the number of generated pulses is the same.

iii) Configuration of pulses: For large $\delta(\delta \gtrsim 1.2)$, the inter-pulse distances tend asymptotically to an equal value and a periodic equilibrium state is attained. For intermediate $\delta(0.35 \lesssim \delta \lesssim 0.8)$, equilibrium states with definite inter-pulse distances are possible(Fig.1). For smaller values of $\delta(\delta \lesssim 0.35)$, pulse intervals begin to fluctuate while the individual pulses are still nearly of the same shape(Fig.2). For much smaller values of $\delta$($\delta \lesssim 0.15$), wave evolutions become chaotic and incessant generations or annihilations of pulses take place[10,11].

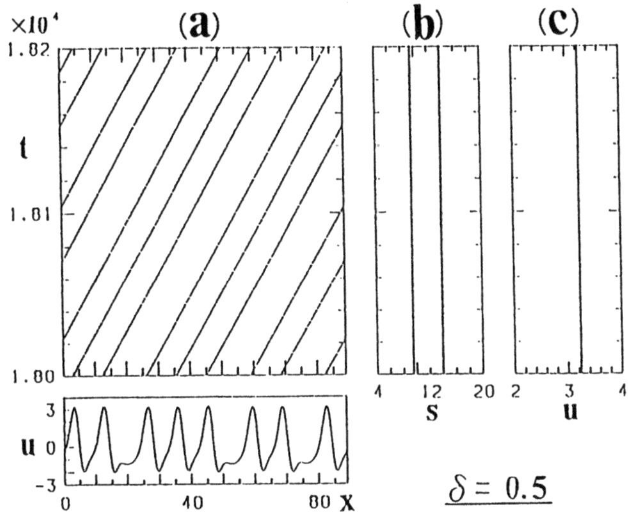


Fig.1. Numerical solution of Eq.(2) for $\delta = 0.5$. (a): Waveform at $t = 1.80\times10^4$ and diagram of pulse peak evolution. (b): Inter-pulse distances. (c): Pulse amplitudes. Eight pulses with the same amplitude propagate.

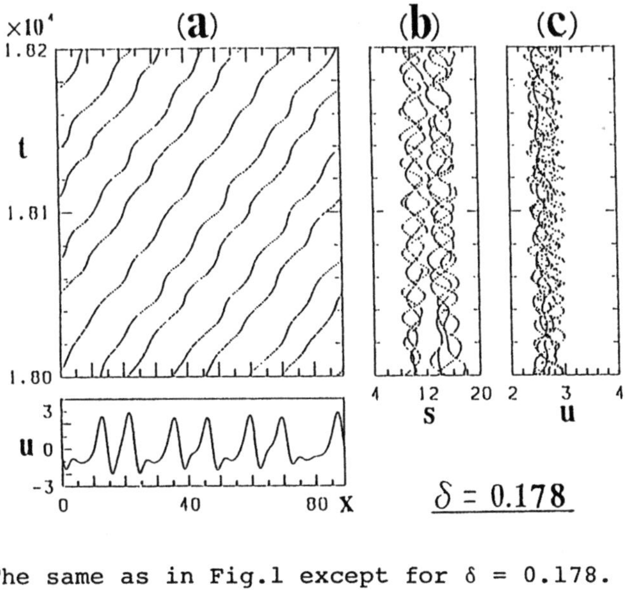


Fig.2. The same as in Fig.1 except for $\delta = 0.178$.

## Steady Pulse Solution

The steady pulse solutions to Eq.(2) can be obtained as shown in Fig.3. For large $\delta$, pulses are of large amplitude and their shapes are close to a symmetric soliton for the Korteweg-de Vries equation. For small $\delta$, asymmetry develops in pulse shape and then an oscillatory structure arises on the front side tail. This oscillatory structure increases until the limiting case $\delta = 0$. We observe that the pulses generated in numerical IVP coincide well with the steady pulse solutions[11]. This result suggests that the steady pulse becomes an elementary entity and the evolutions of the system can be described by an aggregation of such pulses.

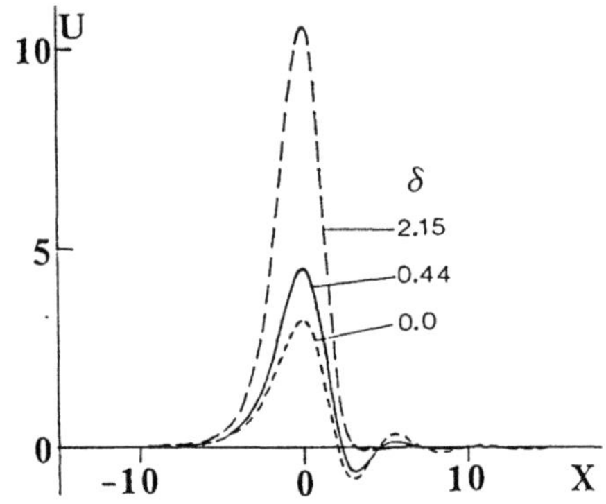


Fig.3. Steady pulse solutions.

## Pulse Interaction

Since the amplitude and the velocity of individual pulses tend to be identical, the pulses propagate as a whole once they have been generated and arranged appropriately. Thus they interact weakly with each other through those tail parts of adjacent pulses which exhibit either monotone or oscillatory decay. Small perturbations due to adjacent tails cause appreciable differences in velocities of neighbouring pulses, which determines the inter-pulse distances.

We consider a row of pulses and look for a solution in the form

$$u(x,t) \cong U_i(\xi_i,\tau) + \sum_{k\neq i} \tilde{U}_k(\xi_k,\tau) \ ,$$
$$\xi_i \equiv x - \int c_i(\tau)dt \ , \quad \tau \equiv \varepsilon t \ , \tag{3}$$

where $U_i$ denotes the i-th pulse and $\tilde{U}_k$, which is assumed to be a small quantity of order $\varepsilon$, denotes a tail part of an adjacent pulse. Substituting (3) into Eq.(2), we obtain

$$\frac{dE_i}{d\tau} + \sum_{k\neq i} \int_{\infty}^{\infty} U_i^2(\xi_i) \frac{\partial U_k(\xi_k)}{\partial \xi_k} d\xi_i = 0 \ ,$$

$$E_i \equiv \frac{1}{2} \int_{-\infty}^{\infty} U_i^2(\xi_i) d\xi_i \ . \tag{4}$$

First of all, we consider mutual interactions of two pulses i and j in an infinite region. Then we can derive an equation which governs the inter-pulse distance $s \equiv s_{ij}$ between the i- and j-pulse as follows

$$a \frac{d^2 s}{d\tau^2} + \frac{1}{2} \int_{\infty}^{\infty} U_i^2(\xi_i) \frac{\partial U_i(\xi_i + s)}{\partial \xi_i} d\xi_i$$

$$- \frac{1}{2} \int_{\infty}^{\infty} U_j^2(\xi_j) \frac{\partial U_j(\xi_j - s)}{\partial \xi_j} d\xi_j = 0, \tag{5}$$

where an approximation that $E_i$ depends on $\tau$ only through the pulse velocity $c_i(\tau)$ and $a \equiv dE_i/dc_i \cong dE_j/dc_j$ has been used.

In this case, the main part of the i-pulse interacts with the oscillatory tail of the j-pulse and that of the j-pulse does with the monotone tail of the i-pulse. Those tail parts can be expressed as

$$\tilde{U}_j = \tilde{u}_0 \exp(-\lambda_1 \xi) \cos(\lambda_2 \xi + \phi_0),$$

$$\tilde{U}_i = \tilde{u}_0' \exp(\lambda' \xi) \ , \tag{6}$$

where $\lambda_1$, $\lambda_2$, $\lambda'$ are eigenvalues of the linear equation and $\tilde{u}_0$, $\phi_0$, $\tilde{u}_0'$ are constants to be determined by fitting to the pulse solution. Inserting (6) into (5) and carrying out the integration with respect to $\xi$, Eq.(5) can be reduced to an oscillation equation for the pulse interval s. Further integration yields an energy integral

$$\frac{1}{2} a \left(\frac{ds}{d\tau}\right)^2 + F(s) = E = \text{const.},$$

$$F(s) \equiv \frac{\tilde{u}_0}{4} \exp[-(\lambda_1 + i\lambda_2)s] \int_{\infty}^{\infty} U^2(\xi) \exp[-(\lambda_1 + i\lambda_2)\xi] d\xi$$

$$+ \text{c.c.} + \frac{\tilde{u}_0}{2} \exp(-\lambda' s) \int_{\infty}^{\infty} U^2(\xi) \exp(\lambda' \xi) d\xi \ . \tag{7}$$

The change of the inter-pulse distance s is determined from (7). When the oscillatory structure is prominent, regions of closed trajectories exist as is depicted in Fig.4, which indicates the existence of bound states of two pulses. Definite or fluctuating inter-pulse distances in Figs.1 and 2 lie within the regions of closed trajectories. When the oscillatory structure becomes small for dispersion dominant case, regions of closed trajectories disappear and pulses cannot be bounded. This tendency is consistent with the observed numerical results where pulses are rearranged into a periodic state in the long time evolution. This fact is due to the repulsive effect of monotone tails[11]. When a tail of the localized pulse has a prominent oscillatory structure, bound states of pulses exist, so that chaotic behaviours are expected for interactions of pulses more than two.

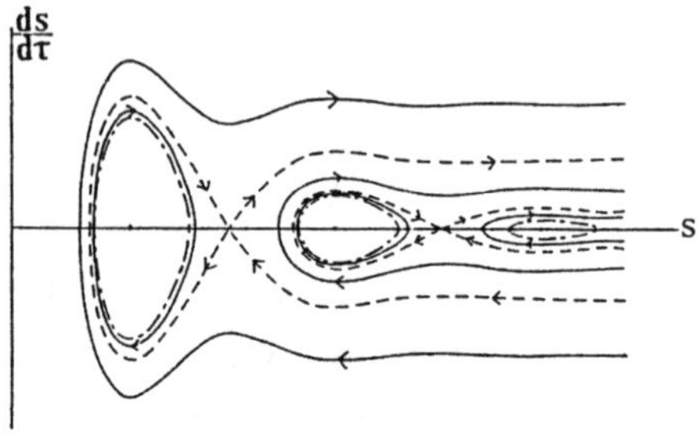


Fig.4. Phase trajectories for mutual interaction of two pulses with prominent oscillatory structure.

Soliton Lattice

For an array of pulses, a similar perturbation approximation is applicable and it leads to the nonlinear lattice equation

$$\frac{d^2 s_n}{dt^2} = f(s_n - s_{n-1}) - g(s_{n+1} - s_n) \ ,$$

$$f(s) \equiv K_1 \exp(-\lambda_1 s)\cos(\lambda_2 s + \phi), \quad g(s) \equiv K_2 \exp(-\lambda' s) \ , \tag{8}$$

where $s_n$ denotes the position of the n-th pulse. This is a nonlinear lattice equation with asymmetric and oscillatory forces. When $\lambda_2 = \phi = 0$, $\lambda_1 = \lambda'$, $K_1 = K_2$, Eq.(8) reduces to the Toda lattice with symmetric exponential forces. Interactions of oscillatory solitons in the fifth-order KdV equation are described by a similar lattice equation with symmetric oscillatory forces[12].

It can be shown that the asymmetry in forces makes nonlinear lattice equations to be non-integrable and chaotic behaviours occur[12]. For a three-pulse periodic system, we can derive, from Eq.(8), coupled equations for the inter-pulse distances x, y, and z with the total length L. They can be reduced to two coupled oscillation equations thanks to the momentum integral(the energy integral is not admitted), so that the degree of freedom is four. These equations possess three types of fixed points. Numerical solutions show that both periodic and chaotic motions are available around a fixed point with center-like singularity. When the deviation of the initial condition from this fixed point is small, the motion is quasi-periodic with two fundamental frequencies $\omega_1$ and $\omega_2$. However, the motion becomes chaotic as the deviation is increased and then a number of frequency components appear in the power spectrum. An example of such chaotic motions is shown in Fig.5, where the inter-pulse distances x, y, z and the power spectrum of x are illustrated.

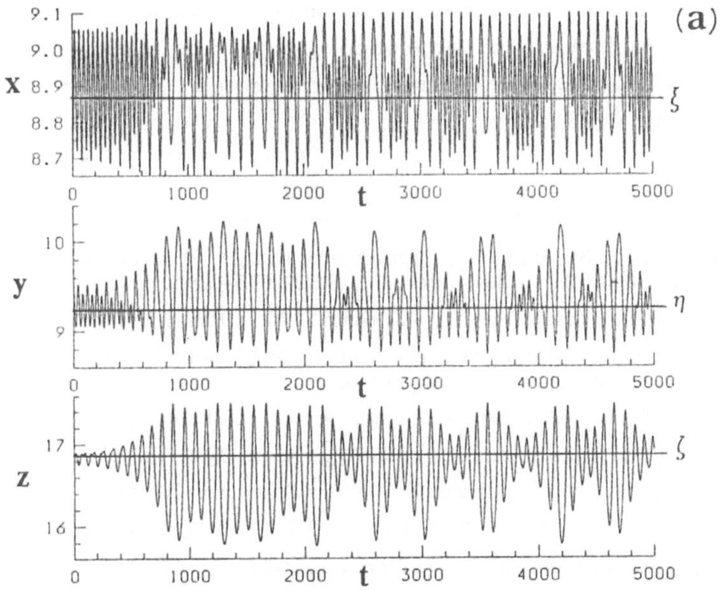


Fig.5a. Chaotic solution in three-pulse periodic system for $\delta = 0.2$ and $L = 35$. Variations of the inter-pulse distances x, y, z. $(\xi,\eta,\zeta)$ show a fixed point with center-like singularity.

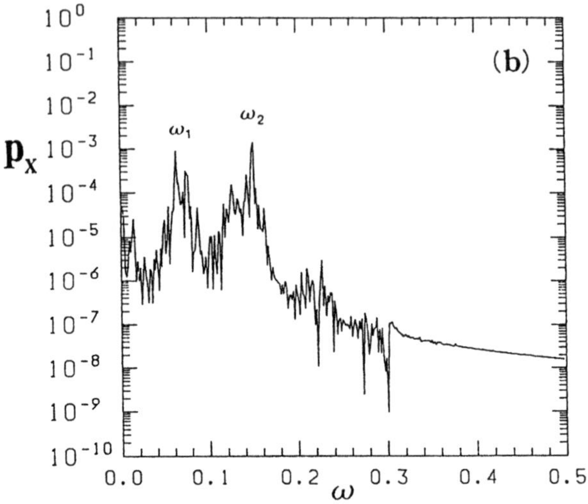


Fig.5b. Power spectrum of x. Eigen frequencies corresponding to the fixed point are $(\omega_1, \omega_2) = (0.08407, 0.15261)$.

Although the present analysis is the lowest approximation, it is expected to be applicable for understanding chaotic behaviours. The direction of approach given here provides a possibility of reducing the degree of freedom of chaos in partial differential equations.

References

1. Moon,H.T.; Huerre,P.; Redekopp,L.G.: Phys.Rev.Lett. 51 (1982)458-460; Physica 7D(1983)135-150.
2. Keefe,L.R.: Stud.in Appl.Math. 73(1985)91-153.
3. Kuramoto,Y.: Chemical oscillations, waves and turbulence. Berlin: Springer-Verlag 1984.
4. Benney,D.J.: J.Math.& Phys. 45(1966)150-155.
5. Topper,J.; Kawahara,T.: J.Phys.Soc.Jpn. 44(1978)663-666.
6. Hooper,A.P.; Grimshaw,R.: Phys.Fluids 28(1985)37-45.
7. Kawahara, T.: Phys.Rev.Lett. 51(1983)381-383.
8. Toh,S.; Kawahara,T.: J.Phys.Soc.Jpn. 54(1985)1257-1269.
9. Kawahara,T.; Toh,S.: Phys.Fluids 28(1985)1636-1638.
10. Toh,S.: J.Phys.Soc.Jpn. 56(1987)949-962.
11. Kawahara,T.; Toh,S.: submitted to Phys.Fluids.
12. Kawahara,T.; Takaoka,M.: submitted to Phys.Fluids.

# Nonlinear Wave Deformation on a Beach with Arbitrary Profile

K. NADAOKA and M. HINO

Dept. of Civil Eng., Tokyo Institute of Technology, Japan

Summary

A new wave equation has been derived for fully nonlinear dispersive waves propagating over an arbitrarily shaped sea bed. The method of the derivation of the equation uses a conformal mapping technique by which the original domain can be transformed onto a uniform depth region to make the basic equation easily integrable vertically. By taking an inverse Fourier transform, the velocity potential obtained by the integration can be expressed in the form which can construct the exact wave equation from the water surface boundary conditions. An example of the numerical solution is presented to demonstrate the availability of the proposed wave equation. Further, for practical use, a WKB approximate solution is also presented.

## 1. INTRODUCTION

Nearshore wave deformation is characterized by the effects of nonlinearity, dispersivity, bottom topography, etc. Among these the effect of bottom topography has not been well incorporated in the existing wave theories so that they are all limited to slowly varying bottom topography (see, e.g., C.C. Mei [1]).

The present paper proposes a new wave equation which can be applied to a wave field on the arbitrarily shaped sea bottom. The method of derivation of the wave equation uses a technique of conformal mapping together with that of Fourier transform. By the mapping, the original domain can be transformed onto a uniform depth region without change of the form of the basic equation, so that it becomes easily integrable vertically with the simplified bottom boundary condition. An example of the numerical solution is presented to demonstrate the availability

K. Horikawa, H. Maruo (Eds.)
Nonlinear Water Waves
IUTAM Symposium Tokyo/Japan 1987

of the proposed wave equation. Further, for practical use, a WKB approximate solution of the basic equations in a mapped space is also presented.

## 2. FORMULATION OF EQUATION

### 2.1 Conformal Mapping of Basic Equation and Boundary Conditions

Original domain defined on the $z$-plane in Fig.1 can be mapped conformally into a constant depth region on the $\zeta$-plane. By this mapping the coordinates $(x,y)$ in the $z$-plane may be written as functions of the coordinates $(\alpha,\beta)$ in the $\zeta$-plane,

$$x = \psi_1(\alpha,\beta)$$
$$y = \psi_2(\alpha,\beta). \tag{1}$$

The boundary conditions are transformed through this mapping as follows,

$$\beta = -\beta_0 \;;\; \frac{\partial\Phi}{\partial\beta} = 0, \tag{2}$$

$$\beta = \tilde{\eta} \;;\; \frac{\partial\tilde{\eta}}{\partial t} + s^2\left[\frac{\partial\Phi}{\partial\alpha}\cdot\frac{\partial\tilde{\eta}}{\partial\alpha} - \frac{\partial\Phi}{\partial\beta}\right] = 0, \tag{3}$$

$$\frac{\partial\Phi}{\partial t} + \frac{s^2}{2}\left[\left(\frac{\partial\Phi}{\partial\alpha}\right)^2+\left(\frac{\partial\Phi}{\partial\beta}\right)^2\right] + g\psi_2(\alpha,\tilde{\eta}) = 0, \tag{4}$$

where $\tilde{\eta}$ is a $\beta$ coordinate of the water surface in the $\zeta$-plane, $\Phi$ is a velocity potential, and $s$ is a scale factor related with the mapping defined as

$$s = \left|\frac{d\zeta}{dz}\right| = \left[\left(\frac{\partial\alpha}{\partial x}\right)^2 + \left(\frac{\partial\alpha}{\partial y}\right)^2\right]^{1/2} = \left[\left(\frac{\partial\beta}{\partial x}\right)^2 + \left(\frac{\partial\beta}{\partial y}\right)^2\right]^{1/2}. \tag{5}$$

The form of the basic equation, on the other hand, remains unchanged through the mapping by the nature of the conformal transform. Therefore, the basic equation can be easily integrated vertically on the $\zeta$-plane with the simplified bottom boundary condition (2) to yield

$$\bar{\Phi}(k,\beta,t) = A(k,t)\frac{\cosh k(\beta+\beta_0)}{\cosh k\beta_0} \tag{6}$$

or

$$\Phi(\alpha,\beta,t) = \int_{-\infty}^{\infty} A(k,t)\frac{\cosh k(\beta+\beta_0)}{\cosh k\beta_0}\cdot e^{ik\alpha}dk, \tag{7}$$

where $\bar{\Phi}$ is the Fourier transform of $\Phi$ with respect to $\alpha$ and $A(k,t)$ is a time-varying wave-number spectrum to be determined by the water surface boundary conditions.

In the water surface boundary conditions (3) and (4) the scale factor $s$ is coupled with the nonlinear terms. Hence it can be said that the scale factor $s$ directly affects the nonlinearity of wave motion. Further, $s$ represents also *nonuniformity of the gravity* in the mapped space, because the linearized water surface boundary condition on $\Phi$ in $\zeta$-plane can be written as

$$\frac{\partial^2\Phi}{\partial t^2} + g\cdot s_0(\alpha)\cdot\frac{\partial\Phi}{\partial\beta} = 0, \qquad (\beta = 0) \tag{8}$$

where $s_0(\alpha)$ is the scale factor at $\beta = 0$. The above equation is the same as the usual linear boundary condition for waves on a horizontal bed except that the gravity $g$ is coupled with the scale factor $s_0$.

## 2.2 Derivation of Wave Equation

Substituting the temporal and spatial derivatives of $\Phi$ into eqs.(3) and (4), we can obtain the following integro-differential equations for $\tilde{\eta}$ and $\chi$,

$$\frac{\partial\tilde{\eta}}{\partial t} + \frac{1}{2\pi}\cdot s^2(\alpha,\tilde{\eta})\cdot\Big[\int_{-\infty}^{\infty} K_1(\xi;\tilde{\eta})\cdot\frac{\partial}{\partial\alpha}\chi(\alpha-\xi,t)d\xi\cdot\frac{\partial\tilde{\eta}}{\partial\alpha} - \int_{-\infty}^{\infty} K_2(\xi;\tilde{\eta})\cdot\frac{\partial}{\partial\alpha}\chi(\alpha-\xi,t)d\xi\Big] = 0, \tag{9}$$

$$\int_{-\infty}^{\infty} K_1(\xi;\tilde{\eta})\cdot\frac{\partial}{\partial t}\chi(\alpha-\xi,t)d\xi + 2\pi g\psi_2(\alpha,\tilde{\eta}) + \frac{1}{4\pi}\cdot s^2(\alpha,\tilde{\eta})\cdot\Big[\Big\{\int_{-\infty}^{\infty} K_1(\xi;\tilde{\eta})\frac{\partial}{\partial\alpha}\chi(\alpha-\xi,t)d\xi\Big\}^2 + \Big\{\int_{-\infty}^{\infty} K_2(\xi;\tilde{\eta})\frac{\partial}{\partial\alpha}\chi(\alpha-\xi,t)d\xi\Big\}^2\Big] = 0, \tag{10}$$

where $\chi(\alpha,t) = F^{-1}\{A(k,t)\}, \qquad (11)$

$$K_1(\alpha;\beta) = F^{-1}\Big\{\frac{\cosh k(\beta+\beta_0)}{\cosh k\beta_0}\Big\}, \tag{12}$$

$$K_2(\alpha;\beta) = -iF^{-1}\{\frac{\sinh k(\beta+\beta_0)}{\cosh k\beta_0}\}, \qquad (13)$$

and $F^{-1}\{\}$ is an operator to take inverse Fourier transform. Since the equations are derived without any assumptions and constraints, they constitute a set of exact wave equations for $\tilde{\eta}$ and $\chi$. Therefore they can express principally arbitrary degree of nonlinearity, dispersivity and the effect of nonuniformity of the water depth.

Fenton & Rienecker [2] proposed a numerical method for accurate solutions of nonlinear equation for water waves over a horizontal bed. The method uses a finite Fourier series approximation on the velocity potential $\Phi$ and hence assumes horizontal periodicity of $\Phi$ and $\eta$. On the contrary, the wave equations (9) and (10) include all components of the continuous wave-number spectrum $A(k,t)$ and they can describe directly the evolution of $\tilde{\eta}$ and $\chi$ in the physical space.

## 3. EXAMPLE OF NUMERICAL COMPUTATION OF THE WAVE EQUATION

Figure 2 shows an example of the numerical solution for waves propagating over a stepped sea bed with a slope of arbitrary inclination as shown in Fig. 3. The wave solution shows that the longer waves propagate faster than the shorter waves. This is a direct manifestation of the wave dispersivity. The figure also shows that the right-going waves divide into reflected and transmitted waves on the slope, revealing steepening their front face as they propagate over the slope and fission into the several solitons on the step.

## 4. APPROXIMATE WAVE SOLUTIONS BASED ON CONFORMAL MAPPING TECHNIQUE

If the bottom variation is gradual, it becomes not so essential to use the conformal mapping technique and hence approximate solutions can be obtained by, e.g., a usual WKB method. For the problems such as sediment transport, however, conformal mapping solution gives more useful estimation of bottom velocity field compared with other methods, because the mapping solution, even its lowest order solution, satisfies exactly the bottom boundary condition. Hence, in the present study, approximate solutions of the basic equation in the mapped space are obtained by using a WKB perturbation technique developed by Hamanaka & Kato [3].

Introducing a small parameter $\delta$ which characterizes the horizontal scale of the bottom variation, we can normalize the variables as

$$(\alpha',\beta') = (\delta\alpha,\beta)\cdot\frac{\omega^2}{g}, \qquad t' = \omega t,$$
$$(\tilde{\eta}',\beta_0') = (\tilde{\eta},\beta_0)\cdot\frac{\omega^2}{g}, \qquad \Phi' = \Phi\cdot\frac{\omega^3}{g^2}. \tag{14}$$

For convenience the primes will be dropped from here on. In the next step, let us introduce a local wave number $k$ and transform the independent variables as

$$(\alpha,\beta,t) \rightarrow (\alpha,\beta,\xi), \qquad \xi = \delta^{-1}\int k d\alpha - t. \tag{15}$$

The unknown variables $\Phi$, $\tilde{\eta}$, and $k$ may be expanded by $\delta$ as

$$\Phi = \Phi_0 + \delta\Phi_1 + \delta^2\Phi_2 + \cdots\cdots\cdots ,$$
$$\tilde{\eta} = \tilde{\eta}_0 + \delta\tilde{\eta}_1 + \delta^2\tilde{\eta}_2 + \cdots\cdots\cdots , \tag{16}$$
$$k = k_0 + \delta k_1 + \delta^2 k_2 + \cdots\cdots\cdots .$$

Substituting Eq.(16) into the basic equation and linearized boundary conditions in the mapped space and separating the orders, we obtain a set of systematically solvable equations. Their solutions can be summarized as follows.

At $O(\delta^0)$:

$$\Phi_0 = a\cdot\cosh k_0(\beta+\beta_0)\cdot\sin\xi,$$

$$\tilde{\eta}_0 = a\cdot s_0(\alpha)\cdot\cosh k_0\beta_0\cdot\cos\xi, \tag{17}$$

$$s_0(\alpha)\cdot k_0\cdot\tanh k_0\beta_0 = 0.$$

At $O(\delta^1)$:

$$\Phi_1 = -\{\frac{1}{2}a(\beta+\beta_0)^2\frac{\partial k_0}{\partial\alpha}\cosh k_0(\beta+\beta_0) + \frac{\partial a}{\partial\alpha}(\beta+\beta_0)\sinh k_0(\beta+\beta_0)\}\cdot\cos\xi,$$

$$\tilde{\eta}_1 = s_0(\alpha)\cdot\{\frac{1}{2}a\beta_0^2\frac{\partial k_0}{\partial\alpha}\cosh k_0\beta_0 + \frac{\partial a}{\partial\alpha}\beta_0\sinh k_0\beta_0\}\cdot\sin\xi, \qquad (18)$$

$$k_1 = 0.$$

$$\vdots$$

The water particle trajectories for the lowest order solution ($O(\delta^0)$), are represented in Fig.4 with, for comparison, the higher order solutions obtained by Hamanaka & Kato[3] for which we need the solution of the order $\delta^2$ or more to obtain satisfactory estimation of the bottom velocity field.

For nonlinear waves, the solution can be obtained by almost the same manner as the above by introducing another small parameter $\varepsilon$ which characterizes nonlinearity of waves and by representing the unknown variables in the form of double perturbation with respect to $\delta$ and $\varepsilon$.

## 5. CONCLUSIONS

The one dimensional wave equation which can represents the arbitrary degree of uneven bottom effect as well as that of nonlinearity and dispersivity has been developed. Availability of the new wave equation is confirmed through several numerical computations. Further, for practical use, WKB approximate solutions are also presented. More detailed description on the present study can be found in Nadaoka & Hino [4] or Nadaoka[5].

## References


1. Mei, C.C.: The Applied Dynamics of Ocean Surface Waves, John Wiley & Sons, Inc., New York 1983.
2. Fenton, J.D. ; Rienecker, M.M.: Accurate numerical solutions for nonlinear waves, Proc. 17th Int. Conf. on Coastal Eng., ASCE, (1980) 50-69.

3. Hamanaka, K. & Kato, K.: Perturbation analysis of finite amplitude waves with slowly varying depth, Proc. 29th Japanese Conf. on Coastal Eng., JSCE, (1982) 65-69 (in Japanese).

4. Nadaoka, K. ; Hino, M.: Conformal mapping solution of a wave field on the arbitrarily shaped sea bed, Proc. 19th Conf. on Coastal Eng., ASCE, (1984) 1192-1208.

5. Nadaoka, K.: A fundamental study on shoaling and velocity field structure of water waves in the nearshore zone, Doctoral Dissertation, Tokyo Inst. of Tech. 1986. ( reproduced in Tech. Rept. of Dept. Civil Eng., Tokyo Inst. of Tech., 36, (1986) 33-125).


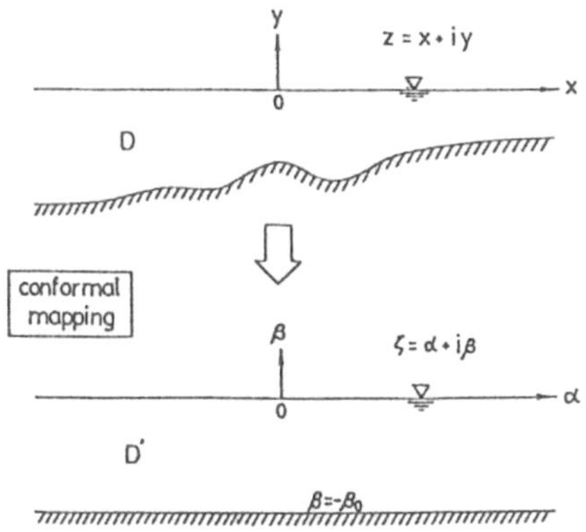


Fig.1 Conformal mapping of a wave field.

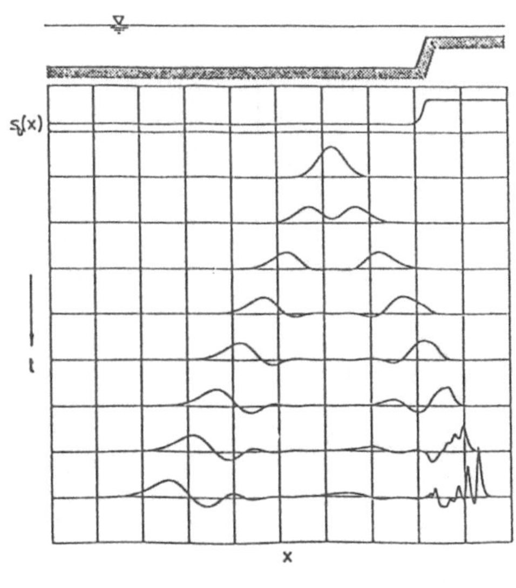


Fig.2 Nonlinear wave evolution over a stepped sea bed ($\theta = \pi/4$, $b/h = 1/4$, see Fig.3).

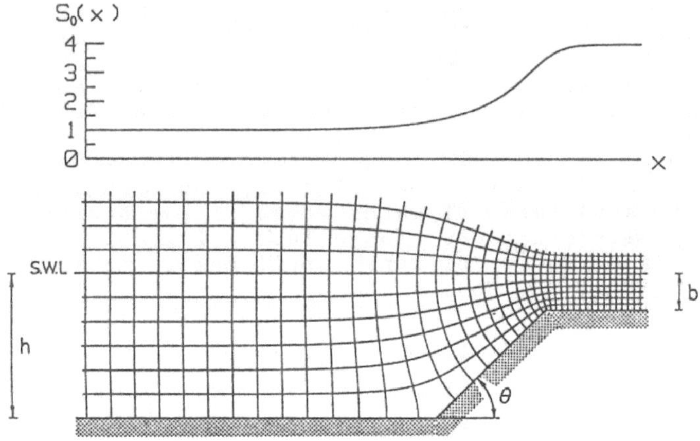


Fig.3 Orthogonal curvilinear grid system and scale factor $s_0$ for a stepped bottom profile.

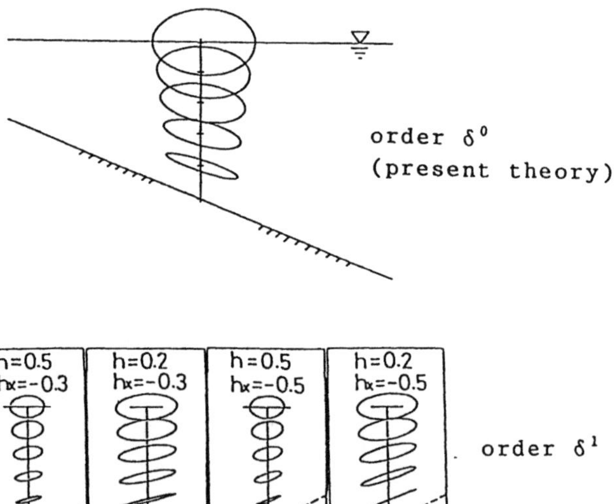


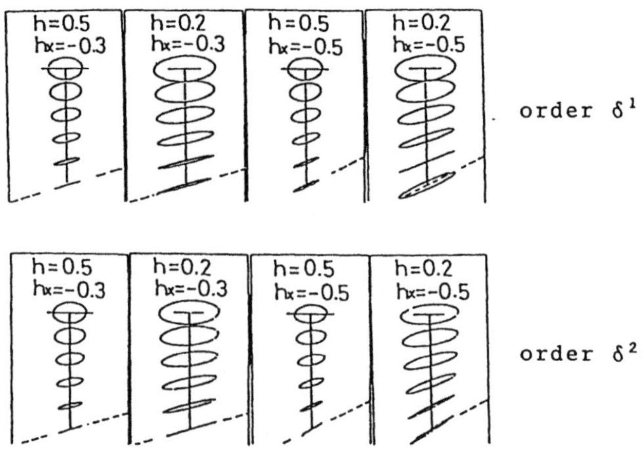


Fig.4 Water particle trajectories for the velocity field solution of the order $\delta^0$ by the present method and of the order $\delta^1$ and $\delta^2$ by Hamanaka & Kato [3].

# Application of Hamiltonian Methods to the Structure and Stability of Water Waves of Permanent Form

P.G. SAFFMAN

California Institute of Technology, Pasadena, California

## Introduction

Extensive computations of the equations describing irrotational surface waves on an inviscid fluid have uncovered a number of interesting and in some cases surprising phenomena. In particular, an unexpected non-uniqueness of gravity waves was found by Chen and Saffman [1]. Calculations of the stability of finite amplitude water waves of permanent form to infinitesimal disturbances (pioneered by Longuet-Higgins [2]) have demonstrated a wide range of sub and super harmonic unstable disturbances, of both two and three dimensional kinds. But there are unexpected differences between gravity and capillary waves, and changes of stability to super harmonic disturbances are found generally to coincide with extrema of the energy of the steady waves as a function of wave height (Tanaka [3], [4]), and not with the extrema of phase speed as predicted by an elementary argument. To some extent, the unexpected results were found serendipiditiously while investigating related problems by large scale computing, and attempts to answer other questions by computing without a clear theoretical idea, such as the existence of nonsymmetric waves, have not succeeded. There is therefore a need for a general theory, both to understand the structure of the results and to provide guidance for the numerical studies attacking open questions.

## Hamiltonian formulation for stability

We report on the application of Hamiltonian methods to these problems. A Hamiltonian formulation for the general case of gravity-capillary waves was given by Zakharov [5]. He showed that the equations of motion for a two-dimensional periodic motion of wavelength $2\pi$ in the x-direction could be written in the compact and elegant Hamiltonian form

$$\frac{da}{dt} = -i\frac{\partial E}{\partial a^*}$$

K. Horikawa, H. Maruo (Eds.)
Nonlinear Water Waves
IUTAM Symposium Tokyo/Japan 1987

where $a = \{a_n\}$, the $a_n$ are linear combinations of the complex Fourier components of the surface elevation $\eta(x,y,t)$ and surface velocity potential $\phi(x,\eta,t)$, and $E(a, a^*)$ is the total energy density (kinetic plus potential plus surface per unit horizontal length and unit span). The nice feature of this approach is that all the complexities are hidden in the calculation of the energy and that using approximations or even models for the dependence of the energy on the shape and speed of the surface will not violate any conservation laws or break any of the symmetries of the flow.

Equilibrium solutions describing waves of permanent form travelling with speed c are given by solutions of the equations

$$\frac{\partial E}{\partial A} - c I_n A = 0, \quad A = A^*$$

where $I_n$ denotes a diagonal matrix with n on the nth row and column. It can then be shown (Saffman [6]) that infinitesimal superharmonic perturbations of the form $a = A + \epsilon b e^{i\sigma t}$ satisfy an eigenvalue problem

$$i\sigma b = M b + N b^*$$

where $M = \partial^2 E / \partial A \partial A^*$ and $N = -\partial^2 E/\partial A^2$. It is now an exercise in linear algebra to demonstrate that the eigenvalue problem has an eigenvector such that $\sigma^2$ goes from positive to negative when the equilibrium energy $E(A, A)$ has an extremum as a function of wave height. It can also be shown that there is no exchange of stability at those wave heights for which the phase speed is an extremum. This argument shows that the superharmonic exchange of stability occurs for any Hamiltonian system, and is not just a property of gravity waves on water of infinite depth. It can also be generalized to solitary waves on water of finite depth (Zufiria and Saffman [7]). Associated with the stability of Hamiltonian systems is the concept of signature. This is an idea which can be traced to studies of Weierstrass in 1858, and has been studied extensively by several workers (e.g. Moser [8]). Briefly, the signature is an invariant of the eigenvector which is like the sign of the energy of the disturbance. Then there is a fundamental result: "If two pure imaginary simple eigenvalues of the same signature collide at a point other than zero, then they cannot leave the imaginary axis." It is known from the work of McLean et al [9] that the instabilities of finite amplitude waterwaves are associated with the collisions of eigenvalues of the disturbances to the wave as the wave height increases. In the case of gravity waves, this collision always leads to the appearance of bands of instability, while on the other hand for capillary waves it is usually the case that the collision does not lead to instability (Chen and Saffman [10]). The signature notion suggests that this phenomenon is due to the fact that collisions are between disturbances of opposite signature for gravity waves and the same signature for capillary waves. It is non-trivial to calculate the signature of disturbances to waves of finite amplitude, but the signature is an invariant (in the absence of collisions of eigenvalues) and hence the signature of a disturbance is the same as its signature for the case when the undisturbed wave is of small amplitude, in which case it is trivial to find the signature. This can be done, and predicts successfully with a few lines of algebra the qualitative results found by McLean et al [9] and Chen and Saffman [10]. Of course, for quantitative results, it is necessary to carry out the extensive computing. Further, the idea predicts a new kind of super-harmonic instability. Longuet-Higgins found that there existed a collision of the eigenvalues of super harmonic disturbances away from zero at a wave height (ka = 0.25) somewhat less than that at which super harmonic instability was found by Tanaka (ka = 0.43). MacKay and Saffman [11] showed that this was a collision of eigenvalues of opposite signature, and should therefore correspond to a band of instability. This was verified by direct computation. The band is extremely narrow, of order $10^{-5}$, and it would have been practically impossible to find without a good theoretical idea of its existence.

SYMMETRY BREAKING

The bifurcations discovered by Chen and Saffman retain the fore-aft symmetry of the periodic Stokes waves or finite amplitude capillary-gravity waves. It is not possible to tell the direction of wave propagation just from the shape of the wave. Chen and Saffman attempted to find bifurcations into non-symmetrical waves, both gravity and gravity-capillary, but were unsuccessful. Recently, Zufiria [12],[13] has used the Hamiltonian formulation to find symmetry breaking for gravity waves and calculate waves which do not have the fore-aft symmetry. The key step was the transformation of the problem into a form with the same Hamiltonian as for a class of area preserving maps, for which a bifurcation tree into symmetry breaking maps had been discovered by Green et al [14]. The procedure to find non-symmetric waves is not simple. First one follows the period tripling branch which bifurcates from regular periodic waves and then follows a branch which results from a period doubling onto a branch which has a period six times that of the original. As one follows this period six branch, it is found that there is a coalescence of Floquet multipliers at unity which gives rise to a symmetry breaking bifurcation into a period six wave without fore-aft symmetry. The actual amount of symmetry breaking is small, and it is expected that these waves are unstable and the results are not expected to be of practical significance, but they are important in that they demonstrate the existence of non-symmetric waves and there are expected to be physical situations where this may be important. Zufiria [15] has shown further that this bifurcation tree exists also for gravity-capillary waves on water of finite depth, and moreover also exists for the solitary waves. He has shown using the Hamiltonian methods that solitary waves are not unique and need not be symmetrical.

Acknowledgement

This work was supported by the office of Naval Research, (N00014-79-C-0412,NR062-639) and by the National Science Foundation (OCE-841598).

References

1. Chen B. and Saffman, P.G. Numerical evidence for the existence of new types of gravity waves of permanent form. Stud. App. Math. 62 (1980) 1-21.

2. Longuet-Higgins, M.S. The instabilities of gravity waves of finite amplitude. I Superharmonics. II Subharmonics. Proc. Roy. Soc. A 360 (1978) 471-505.

3. Tanaka, M. The stability of steep gravity waves. J. Phys. Soc. Japan 52 (1983) 3047-3055.

4. Tanaka, M. The stability of solitary waves. Phys. of Fluids 29 (1986) 650-655.

5. Zakharov, V.E. Stability of periodic waves of finite amplitude on the surface of a deep fluid. J. Appl. Mech. Tech. Phys. 2 (1968) 190-194.

6. Saffman, P.G. The superharmonic instability of finite-amplitude water waves. J. Fluid Mech. 159 (1985) 169-174.

7. Zufiria, J.A. and Saffman, P.G. (1986) The super-harmonic instability of finite-amplitude surface waves. Stud. App. Math. 74 (1986) 259-266.

8. Moser, J.K. New aspects in the theory of stability of Hamiltonian systems. Comm. Pure App. Math. 11 (1958) 81-114.

9. McLean, J.W., M, Y.C., Martin, D.U., Saffman, P.G. and Yuen, H.C. Three-dimensional instability of finite amplitude water waves. Phys. Rev. Lett. 46 (1981) 817-820.

10. Chen, B. and Saffman, P.G. Three-dimensional stability and bifurcation of capillary and gravity waves on deep water. Stud. App. Math. 72 (1985) 125-147.

11. MacKay, R.S. and Saffman, P.G. Stability of water waves. Proc. Roy. Soc. A 406 (1986) 115-125.

12. Zufiria, J.A. Weakly nonlinear nonsymmetric gravity waves on water of finite depth. J. Fluid Mech. (to appear).

13. Zufiria, J.A. Nonsymmetric gravity waves on water of infinite depth. J. Fluid Mech. (to appear).

14. Green, J.M., MacKay, R.S., Vivaldi, F. and Feigenbaum, M.J. Universal behavior in families of area-preserving maps. Physica. 7D (1981) 468-486.

15. Zufiria, J.A. Symmetry breaking in periodic and solitary gravity-capillary waves on water of finite depth. J. Fluid Mech. (to appear).

# Session 6

# Wave-current Interaction and Interfacial Waves

# Effects of Velocity Shear on the Stability of Surface Deep Water Wave Trains

J. C. Li
Chinese Academy of Science, Beijing, China
W. H. Hui
University of Waterloo, Ontario, Canada
M. A. Donelan
National Water Research Institute, Burlington, Canada

## SUMMARY

The side-band instability of a Stokes wave train in uniform velocity shear is studied. In particular, theory predicts that small velocity shear tends to enhance instability whereas large velocity shear (tends to) suppress it. Laboratory experiments have also been conducted on mechanically generated waves with wind blowing over them. The observed side-band growths are in qualitative, but not quantitative, agreement with the theoretical model in that gentle wind (hence smaller velocity shear generated) tends to enhance growth whereas strong wind (hence larger velocity shear) suppress it.

## §1. Introduction

It is well-known that the Stokes wave train on deep water is subject to Benjamin-Feir side-band instability. Recently, however, experiments by Bliven, Huang and Long (Ref 1) at NASA Goddard Space Flight Center, and by Donelan at Canada Centre for Inland Waters (CCIW) both show that the efficiency of the Benjamin-Feir instability mechanism is reduced by wind blowing over waves. This is a very interesting and far-reaching finding, and the purpose of this paper is to propose a simple mathematical model to explain the effects of wind blowing over waves on the stability of finite amplitude wave trains.

Wind blowing over surface water waves is known to produce a phase-shifted pressure force, causing the waves to grow. At the same time it imposes a tangential force, resulting in velocity shear in the water. As a simple model to study the effects of wind on the stability of water waves, the action of the wind is represented solely by a velocity shear distributed in the water. That is to say, that we assume the action of the wind to alter the surface pressure is not important in the growth of side-band instabilities. The problem is then to investigate the effects of velocity shear on the stability of a Stokes wave train. For simplicity in the mathematical analysis, a uniform velocity shear $\Omega$ is assumed.

This model, although obviously over-simplifies the effects of wind, is shown to be capable of predicting the qualitative effects of wind on the stability of a uniform wave train.

K. Horikawa, H. Maruo (Eds.)
Nonlinear Water Waves
IUTAM Symposium Tokyo/Japan 1987

## §2. Stokes Wave Trains in Uniform Velocity Shear

### 2.1 Mathematical Formulation

As mentioned in the Introduction, we consider the motion of wave on water of infinite depth in the presence of uniform velocity shear; the wave motion being otherwise irrotational. Thus the velocity is of the form

$$\vec{V} = \Omega y \vec{i} + \nabla\phi(x,y,t) \tag{2.1}$$

where $\Omega$ is constant, $x$ and $y$ are, respectively, horizontal and upward vertical Cartesian coordinates, with unit vectors $\vec{i}$ and $\vec{j}$. The governing equations are

$$\nabla\cdot\vec{V} = 0 \tag{2.2}$$

$$\frac{\partial \vec{V}}{\partial t} + \vec{V}\cdot\nabla\vec{V} + \frac{\nabla p}{\rho} = -g\vec{j} \tag{2.3}$$

Substitution of Eq. (2.1) into (2.2) yields the usual Laplace equation for the potential $\phi$

$$\nabla^2\phi = 0 \tag{2.4}$$

Although the flow under study is unsteady and rotational, with velocity given by Eq. (2.1), a first integral of Eq. (2.3) still exists. Indeed, substituting (2.1) into (2.3) and integrating with respect to $x$ and $y$ yields the Bernoulli integral as follows

$$\frac{\partial \phi}{\partial t} + \frac{1}{2}\vec{V}^2 + \frac{p}{\rho} + gy - \frac{1}{2}\Omega^2 y^2 - \Omega\psi = \text{const.} \tag{2.5}$$

where $\psi$ is related to $\phi$ by the Cauchy-Riemann conditions

$$\frac{\partial \psi}{\partial y} = \frac{\partial \phi}{\partial x}, \quad \frac{\partial \psi}{\partial x} = -\frac{\partial \phi}{\partial y} \tag{2.6}$$

The mathematical problem for calculating water wave motion of the type (2.1) is thus to solve Eq. (2.4) subject to the following boundary conditions

$$\frac{\partial \phi}{\partial y} = 0 \qquad \text{at } y \to -\infty \tag{2.7}$$

$$\frac{\partial \eta}{\partial t} + \left(\frac{\partial \phi}{\partial x} + \Omega\eta\right)\frac{\partial \eta}{\partial x} - \frac{\partial \phi}{\partial y} = 0 \qquad \text{at } y = \eta(x,t) \tag{2.8}$$

$$\frac{\partial \phi}{\partial t} + \frac{1}{2}\left\{\left(\frac{\partial \phi}{\partial x}\right)^2 + \left(\frac{\partial \phi}{\partial y}\right)^2\right\} + \Omega\eta\frac{\partial \phi}{\partial x} + g\eta - \Omega\psi = \text{const.} \qquad \text{at } y = \eta(x,t) \tag{2.9}$$

where Eq. (2.8) expresses the kinematic boundary condition that a fluid particle once on the free surface remains there for all time, whereas Eq. (2.9) is the dynamic boundary condition that the surface pressure is constant.

## 2.2 Stokes Wave Trains in Uniform Velocity Shear

We now look for wave motion of Stokes type, i.e. periodic waves of permanent form propagating at constant velocity in the $x$-axis direction. Such waves are steady waves when viewed from a suitable frame of reference $Xy$ where

$$X = x - Ct \tag{2.10}$$

$C$ being an apparent phase velocity of the wave. These steady waves are represented by

$$\phi = -CX + \Phi(X,y) \tag{2.11}$$

$$\eta = \eta(X) \tag{2.12}$$

It is noted that for steady motion the free surface is a streamline corresponding to the constancy of the stream function $\left[\frac{1}{2}\Omega y^2 + \psi\right]_{y=\eta}$. Thus substituting Eqs. (2.11) and (2.12) into Eqs. (2.4), (2.7) to (2.9), we get

$$\frac{\partial^2\Phi}{\partial X^2} + \frac{\partial^2\Phi}{\partial y^2} = 0 \tag{2.13}$$

$$\frac{\partial\Phi}{\partial y} = 0 \qquad y \to -\infty \tag{2.14}$$

$$\left(\frac{\partial\Phi}{\partial X} + \Omega\eta - C\right)\frac{\partial\eta}{\partial X} - \frac{\partial\Phi}{\partial y} = 0 \qquad y = \eta(X) \tag{2.15}$$

$$\frac{1}{2}\left\{\left(\frac{\partial\Phi}{\partial X} + \Omega\eta - C\right)^2 + \left(\frac{\partial\Phi}{\partial y}\right)^2\right\} + g\eta = \text{const.} \qquad y = \eta(X) \tag{2.16}$$

Solutions to Eqs. (2.13) to (2.16) are sought in the form of a power series in a small parameter $\epsilon$ characterising the steepness of the waves.

$$\Phi = \epsilon\Phi_1 + \epsilon^2\Phi_2 + \epsilon^3\Phi_3 + \cdots \tag{2.17}$$

$$\eta = \epsilon\eta_1 + \epsilon^2\eta_2 + \epsilon^3\eta_3 + \cdots \tag{2.18}$$

$$C = C_0(1 + \epsilon\delta_1 + \epsilon^2\delta_2 + \cdots) \tag{2.19}$$

By use of standard singular perturbation technique for calculating Stokes wave trains, the terms in the series (2.17) - (2.19) can be calculated successively. The results are

$$\Phi_1 = e^{ky}\sin kX$$

$$\eta_1 = \cos kX$$

$$\Phi_2 = \frac{1}{4}\overline{\Omega}(\overline{\Omega} + 3)e^{2ky}\sin 2kX \tag{2.20}$$

$$\eta_2 = -\frac{1}{4}\overline{\Omega}(\overline{\Omega}+2) + \frac{1}{4}(\overline{\Omega}^2+4\overline{\Omega}+2)\cos 2kX$$

and

$$C = C_0\left[1 + \epsilon^2\frac{1+2\overline{\Omega} + \frac{5}{4}\overline{\Omega}^2-\frac{1}{8}\overline{\Omega}^4}{2+\overline{\Omega}}\right] \tag{2.21}$$

where

$$\overline{\Omega} = \frac{\Omega}{\omega} \tag{2.22}$$

$k$ being the wave number, $\omega = kC_o$ and

$$C_0 = \sqrt{\frac{g}{k(1+\overline{\Omega})}} \tag{2.23}$$

## §3. Evolution of Stokes Wave Trains in Uniform Velocity Shear

The slow modulation of the Stokes wave train obtained in §2 can be studied by the usual method of multiple scales (Refs 2 & 3) applied to Eqs. (2.4), (2.7) - (2.9). In particular, let

$$\eta = \epsilon A(\xi,\tau)e^{i(kx-\omega t)} + \text{complex conjugate} \tag{3.1}$$

together with a compatible expression for $\phi$, where

$$\xi = \epsilon(x-c_g t) \tag{3.2}$$

$$\tau = \epsilon^2 t \tag{3.3}$$

$c_g$ being the linear group velocity. Then the evolution of $A$ is governed by the following equation

$$i\frac{\partial A}{\partial \tau} + \mu\frac{\partial^2 A}{\partial \xi^2} = q|A|^2 A \tag{3.4}$$

where

$$\mu = \frac{\omega}{8k^2}\frac{(1+\overline{\Omega})^2}{(1+\overline{\Omega}/2)^3} \tag{3.5}$$

$$q = -\frac{1+2\overline{\Omega}+\frac{5}{4}\overline{\Omega}^2-\frac{1}{8}\overline{\Omega}^4}{2+\overline{\Omega}}\omega k^2 \tag{3.6}$$

The Stokes wave solution of §2 corresponds to the special solution of Eq. (3.4) with

$$A = a_0 e^{-ia_0^2 q\tau} \tag{3.7}$$

## §4. Stability of Stokes Wave Trains in Uniform Velocity Shear

A standard linear stability analysis (e.g. Ref 4) shows that the stability of the Stokes wave train depends solely on the sign of $\mu q$: it is stable when $\mu q > 0$ and unstable when $\mu q < 0$. The maximum growth rate of the side-band, when normalized by $\frac{1}{2}\omega a_0^2 k^2$, is equal to

$$\frac{1+2\overline{\Omega}+\frac{5}{4}\overline{\Omega}^2-\frac{1}{8}\overline{\Omega}^4}{1+\overline{\Omega}/2} \tag{3.8}$$

This is plotted in Fig 1, which shows that small velocity shear $\Omega/\omega$ tends to enhance side-band instability, whereas large velocity shear tends to suppress instability.

## §5. Comparisons with Experiments

A series of experiments on mechanically generated water waves with wind blowing over them have been conducted in the wind-wave flume of Canada Centre for Inland Waters, Burlington, Ontario. Figure 2 shows the frequency spectra at fetches of 12.0, 24.1 and 34.2 m for wind speeds of 1.4 m/s and 10.5 m/s. It is clear that gentle wind at 1.4 m/s (hence smaller velocity shear generated in the wave field) tends to enhance side-band instability whereas stronger wind at 10.5 m/s (hence larger velocity shear) tends to suppress side-band instability. These, and the similar experimental results of Ref 1, are in qualitative agreement with our theoretical predictions.

Attempts to give a quantitative comparison between experiments and our simplified theoretical model are, however, met with the difficulty of assigning an appropriate linearized velocity shear to the experimental results. The wind-induced velocity shear is highly nonlinear and the best one can do is estimate lower and upper bounds for the equivalent linear shear. An estimate of the upper bound may be deduced from Kinsman's formula (Ref 5) for the surface shear. This yields a range of $\Omega/\omega$ from -0.1 to 18.3 for our experiments. An estimate of the lower bound may be based on a linear gradient from the drift velocity of the surface, approximately 3% of the wind speed, to the point of zero velocity in the tank (about 20% of the depth from the surface). This yields a range of $\Omega/\omega$ from 0.02 to 0.17. Thus, appropriate linear velocity shear may be in the range of 0 to 4.0 in which the theoretical calculations yield first an increase of side-band growth rates and then a decrease. It is encouraging that this is the behavior observed, but quantitatively the prediction and experimental results (Fig 3) differ by a factor of 3. It is clear that a more realistic velocity shear, e.g. an exponential one, will have to be incorporated in the theory before more precise agreement with observations is to be expected.

## References

1. Bliven, L.F., Huang, N.E. and Long, S.R., J. Fluid Mechanics, **162**, 237, 1986.

2. Hasimoto, H. and Ono, H., J. Phys. Soc. Japan, **33**, 805, 1972.

3. Hui, W.H. and Hamilton, J., J. Fluid Mechanics, **93**, 117, 1979.

4. Lonquet-Higgins, M.S., Proc. Roy. Soc. Lond., **A347**, 311, 1976.

5. Kinsman, B., *Wind Waves*, Prentice Hall, 1965.

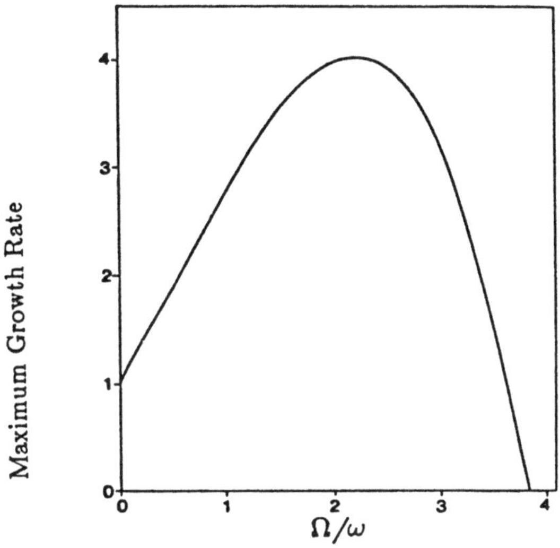


**Fig 1.** Maximum side-band growth rate vs vorticity.

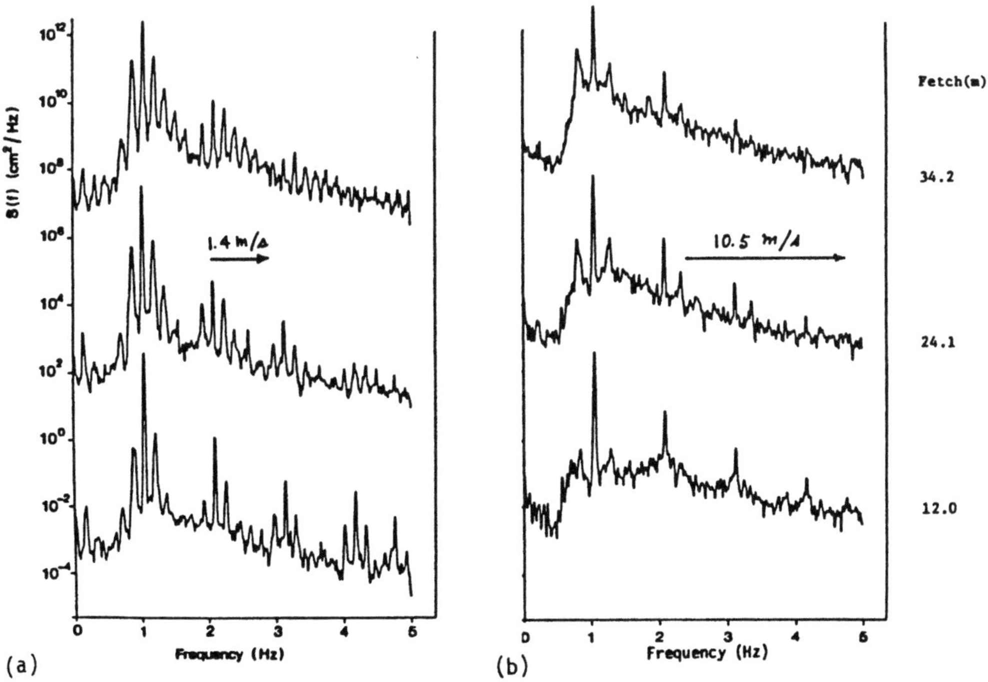


**Fig 2.** Frequency spectra at fetches of 12.0, 24.1, 34.2 m. The logarithmic ordinate corresponds to the bottom spectrum (at 12.0 m fetch); the others are offset in increments of 5 decades. Resolution bandwidth = 0.0186 Hz, Nyquist frequency = 10 Hz, 32 degrees of freedom.

(a) Low wind speed, 1.4 m/s at 35cm height, (b) High wind speed, 10.5 m/s.

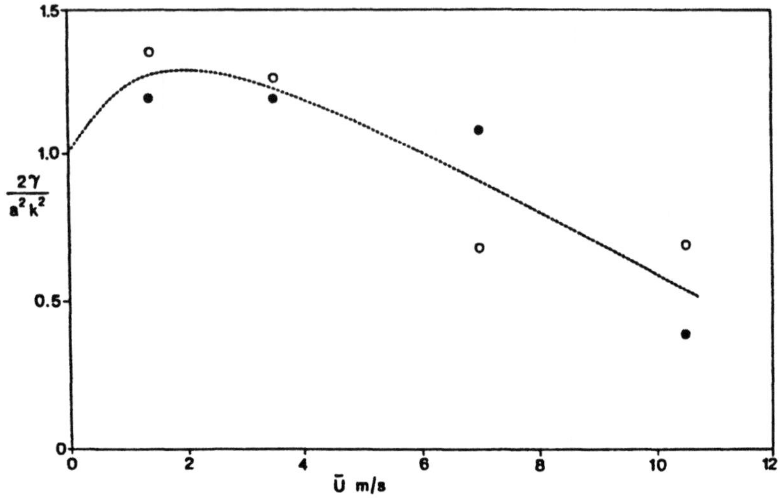


**Fig 3.** The observed growth rate for the fastest growing side-bands for various wind speeds. The open circles correspond to the low frequency side-band; the closed circles to the high frequency side-band.

# Large Amplitude Interfacial Solitary Waves

D. I. PULLIN

Department of Mechanical Engineering
University of Queensland
St. Lucia 4067, Australia

R.H.J. GRIMSHAW

School of Applied Mathematics
University of New South Wales
Kensington 2033, Australia

Summary
We consider finite amplitude interfacial solitary waves in a two-layer fluid. Wave profiles for symmetrically shaped waves are found numerically by solving a system of nonlinear integral equations obtained following a conformal mapping of the half-space below the rigid boundary into the interior of a circle. When the flow is irrotational and the density ratio across the interface lies between zero and one, the results suggest a limiting wave which exhibits a bubble-capped profile. In the Boussinesq limit when the density ratio approaches unity, wave profiles remain single valued and the numerical solutions indicate that the wave amplitude may be unbounded.

## Introduction

It is well known that fluids of shallow depth support large amplitude solitary waves. At small or moderate amplitudes these have a characteristic "sech²" profile, while the limiting wave of maximum amplitude exhibits a $120^0$ corner at the crest, of an amplitude to depth at infinity ratio A = 0.83322 [1]. Benjamin [2] and Davis and Acrivos [3] showed that solitary waves may exist in infinitely deep density stratified fluids and that the tails of these waves exhibit algebraic decay at large horizontal distances from the wave crest.

This paper will describe the results of a study of numerical solutions for large amplitude solitary waves at the interface between two inviscid incompressible fluids of different density in the presence of gravity. Referring to figure 1 the upper fluid has density $\rho_1$ and is bounded above by a rigid boundary at a height d from the horizontal unperturbed

K. Horikawa, H. Maruo (Eds.)
Nonlinear Water Waves
IUTAM Symposium Tokyo/Japan 1987

interface position. The lower fluid is of density $\rho_2$ and is of unbounded depth. We nondimensionalize by choosing a length scale d and a time scale $(d/\alpha g)^{1/2}$ where $\alpha=(\rho_2-\rho_1)/(\rho_2+\rho_1)$ is the Boussinesq parameter and g is the acceleration due to gravity. In solutions to be discussed presently, the gravity vector acts in the y direction and always points into the more dense fluid. Hence $\alpha g>0$ always. We allow for shear in the basic flow by letting the upper (lower) fluid have uniform dimensionless vorticity $\Omega_1$ $(\Omega_2)$.

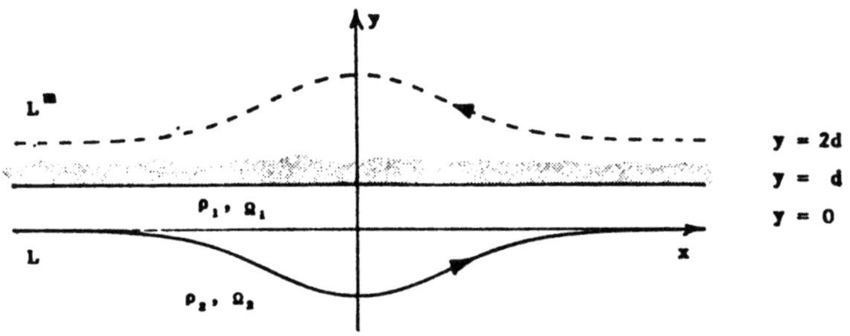


Figure 1. Progressive solitary wave configuration in a reference frame at rest with respect to the wave. The fluid velocity at $y \to -\infty$ is $(-C,0)$. The wave profile is symmetrical in shape about $X = 0$.

## Formulation

We shall view the presence of the solitary wave as a two-dimensional, irrotational shape preserving perturbation of finite amplitude, to the basic flow. In axes (X,Y) the flow in the presence of the wave may be described in dimensionless co-ordinates by

$$(U_j,V_j) = (-\Omega_j Y-C+\hat{U},\hat{V}), \quad j = 1,2, \tag{1a}$$

$$\Psi_j = -\tfrac{1}{2}\Omega_j Y^2-CY+\hat{\Psi}_j, \quad j = 1,2. \tag{1b}$$

In (1), $(U_j,V_j)$ are the (X,Y) components of fluid velocity in layer j, measured in a frame of reference moving with the wave speed C in the X-direction with respect to a laboratory reference frame, and $\Psi_j$ is the corresponding stream function.

The unperturbed velocity field is $(-\Omega_j Y-C, 0)$. Quantities with a caret are those associated with perturbations to the basic flow.

In the reference frame of (1), we assume that the flow corresponding to progressive solitary waves is strictly steady. We must then solve the problem of determining the displacement of the stationary interface from its undisturbed level $Y = 0$ for a given set of parameters comprising $\alpha, \Omega_1, \Omega_2$ together with parameters which define particular solitary wave solutions of the equations of motion. We denote the perturbed interface in the complex $Z=X+iY$ plane by L as shown in figure 1, and specifically describe the shape of L by the complex function $Z_c(a)=X_c(a)+iY_c(a)$, where a is a parameter on L, $-\pi \le a \le \pi$. For solitary wave solutions we require

$$|X_c(a\to \pm\pi)| \to \infty, \quad Y_c(a\to \pm\pi) \to 0. \tag{2}$$

Introducing the dimensionless pressure $P_j(X,Y)$ in each fluid the boundary conditions may now be stated as

$$P_1(X_c, Y_c) = P_2(X_c, Y_c), \tag{3a}$$

$$\Psi_j(X_c, Y_c) = 0, \quad j=1,2, \tag{3b}$$

$$\Psi_1(X,1) = -\tfrac{1}{2}\Omega_1 - C. \tag{3c}$$

Equations (3a) and (3b) are the well known dynamic and kinematic conditions on L respectively, and (3c) requires that the solid boundary $Y = 1$ is a streamline. Using the appropriate Bernoulli equation in each fluid layer [4], (3a) may be written as

$$(1-\alpha)[U_1^2+V_1^2-C^2]-(1+\alpha)[U_2^2+V_2^2-C^2]-4Y_c = 0. \tag{4}$$

on $Z=Z_c$.

In order to incorporate a self consistent treatment of the wave tails which are of infinite extent in the Z-plane and which at the outset are of unknown shape, we make use of the transformation

$$\zeta = -\frac{Z}{Z-2i}, \quad \zeta = \xi + i\eta . \tag{5}$$

Equation (5) maps the region $Y<1$, $-\infty<X<\infty$ into $|\zeta|<1$ and the half plane $Y<0$ into the interior of the circle $|\zeta+\frac{1}{2}| = \frac{1}{2}$. $Z_c(a)$ is mapped into the curve $\zeta = \zeta_c(a)$; the image of L in the $\zeta$-plane is denoted by $\mathcal{L}$. $Z = \infty$ is mapped to $\zeta = -1$.

We proceed by adapting the method used in [5] to the present solitary wave geometry. Briefly, the complex velocity $U_j - iV_j$ and $\Psi_j$ are first expressed as contour integrals on $\mathcal{L}$ in the $\zeta$-plane, using an image principle to satisfy (3c) automatically. When these integrals are used in (3b) and (4), coupled integro-differential equations result containing as unknowns $\zeta_c(a)$, the local jump in tangential velocity across L in the Z-plane which we denote by $\gamma(a)$, and the wave speed C. Closure equations are provided by the specification of the parameter a via arclength $\sigma$ on $\mathcal{L}$, and by the use of pseudo-arc length continuation [6] on a solution branch in $(\zeta_c, \gamma, C)$ space, with fixed $\Omega_1$, $\Omega_2$ and $\alpha$. We consider only symmetrically shaped solitary wave solutions satisfying $\zeta_c(-a) = \zeta_c^*(a)$, $0 \le a \le \pi$.

## Numerical Method

A finite difference numerical method was used. Computations were carried out in the $\zeta$-plane, where $\mathcal{L}$ is defined for numerical purposes by 2N points with equispaced a given by $a_n = n\pi/N$, $n = -N...N$. The boundary conditions were implemented by satisfying (3b), (4) and a chord length approximation to $\sigma$ at the $a_n$, $n = -N...0$. Derivatives were evaluated using a 5-point rule and the integrals were calculated using standard methods [4]. This resulted in 3N+1 equations for the 3N+1 unknowns, which were solved numerically using Newton's

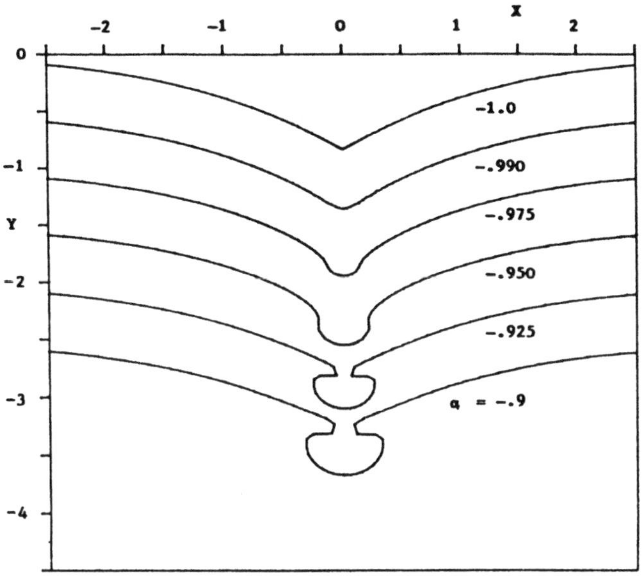


Figure 2. Extreme wave profiles near the wave crest for various α as indicated. The gravity vector points in the positive Y direction. Only the profile with α = -1 is at the correct Y-level and other profiles are displaced successively 0.5 units in the -Y direction.

method, subject to mean residual decreasing below $10^{-10}$. All computations reported presently have N = 40 and $\Omega_1 = \Omega_2 = 0$. The present formulation and numerical method was tested by comparing results obtained for free-surface solitary waves (α = -1) with accurate calculations made using conformal mapping methods [7].

Results for $-1 \leq \alpha \leq 0.2$

Calculations were performed by tracking numerically along solution branches corresponding to fixed α, $\Omega_1 = \Omega_2 = 0$, in the 3N+1-dimensional space of the unknowns. In figure 2 we depict a sequence of extreme wave profiles for α approaching α = -1. By "extreme wave" we mean the numerical solution with the largest value of the pseudo-arc length parameter which could be obtained with the present numerical method (see [5] for further discussion). Only that portion of each profile near the wave crest is shown in figure 2. When

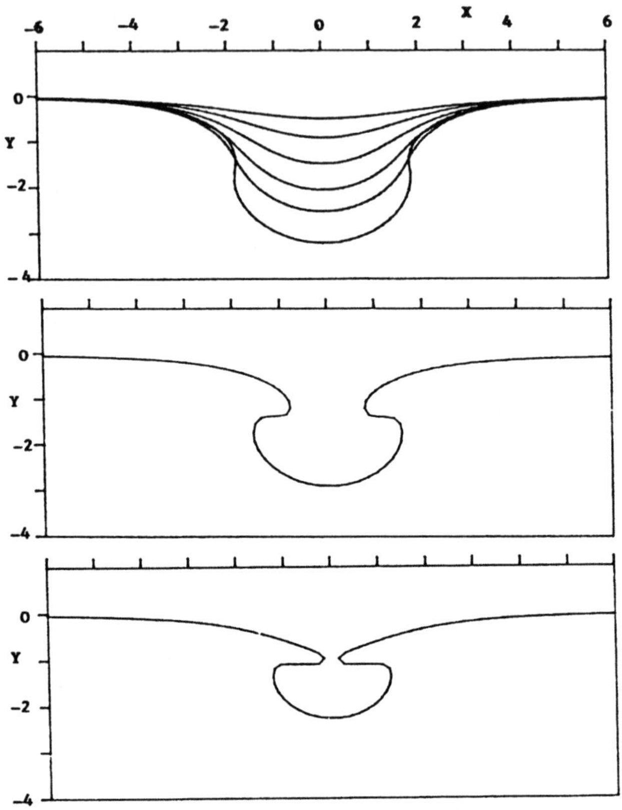


Figure 3. Wave profiles with $\alpha = -0.7$. Gravity points in the positive Y direction.

$\alpha = -0.9$ and $\alpha = -0.925$, the extreme wave profile indicates a limiting wave capped by a bubble of stagnant fluid of density $\rho_1$. Such limiting wave forms were suggested by Meiron and Saffman [8] for periodic interfacial waves. Evidence supporting this hypothesis was given in [5] for the special case $\alpha = 0$. The extreme wave solutions found presently with $\alpha = -0.950$, $-0.975$ and $-0.99$ do not exhibit an overhanging region, and there is no firm indication of a bubble-capped limiting wave configuration. We believe however that this is a numerical resolution problem associated with oscillation of the solution as the pseudo arc-length parameter varies when both the limiting wave configuration and $\alpha = -1$ are approached simultaneously. We speculate that all solution

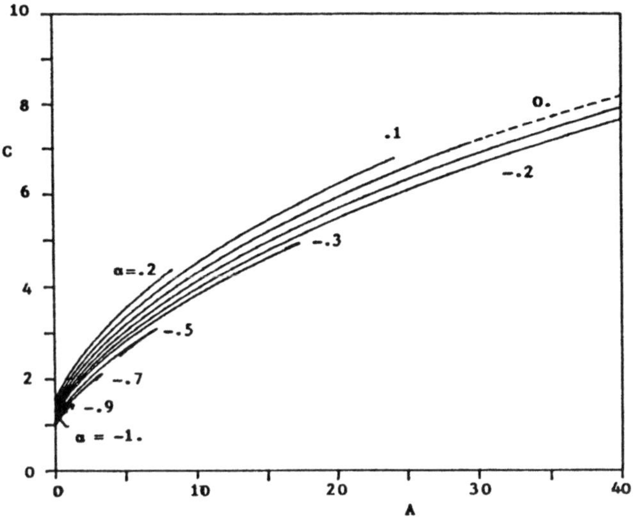


Figure 4. Wave speed C versus amplitude A for values of α as shown. Dashed lines indicate hypothetical extension of present results.

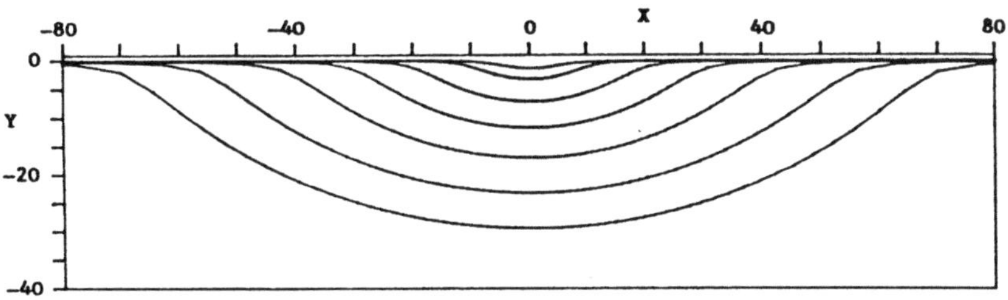


Figure 5. Wave profiles with α = 0. Gravity may act in either the positive or negative Y-direction.

branches with α fixed near α = -1 are terminated by a bubble-capped wave and that further, the bubble dimension vanishes when $\alpha \to -1$.

Figure 3 shows wave profiles with α = -0.7. The onset of overhanging and the progression towards a bubble-capped limiting wave profile is again evident. This appears to be generally associated with the appearance of a limit point in the C-A curves as can be seen (on a small scale) in figure 4 for $\alpha = -0.9, -0.7, -0.5$ and $-0.3$. We note that the first appearance of overhanging, and the occurrence of the limit point are not necessarily coincident. When $\alpha = -0.2, -0.1, 0$ and $0.1$, numerical solutions corresponding to waves of very large amplitude were found. The first indication of overhanging was found with $\alpha = -0.2$ at $A \approx 44.63$, but the further continuation of this solution branch was not feasible owing to resolution problems with $N = 40$. In figure 4 we have indicated those solution branches which we believe may be continued beyond the extreme wave solution obtained presently. Figure 5 shows wave profiles for the Boussinesq limit $\alpha = 0$. We were able to calculate waves of amplitude larger than the maximum $A = 29.02$ in figure 5, but with somewhat degraded accuracy owing to lack of resolution in the wave tail with $N = 40$. This, together with the present results at small $\alpha$ suggests that solitary waves with $\alpha = 0$ may be of unbounded amplitude. This is consistent with numerical results for long periodic interfacial waves : see for example figure 6 of [4].

## References


1. Hunter, J.K., and Vanden-Broeck, J.-M., J.Fluid Mech., 136, (1983), 63 - 71.
2. Benjamin, T.B., J.Fluid Mech., 29, (1967), 559 - 592.
3. Davis, R.E., and Acrivos, A., J.Fluid Mech. 29, (1967), 593 - 607.
4. Pullin, D.I., and Grimshaw, R.H.J., Phys. Fluids, 26, (1983), 1731 - 1739.
5. Grimshaw, R.H.J., and Pullin, D.I., Phys.Fluids 29, (1986), 2802 - 2807.
6. Keller, H.B. in *Applications of Bifurcation Theory*, Academic Press (1977), 359 - 384.
7. Tanaka, M., Phys. Fluids, 29, (1986), 650 - 655.
8. Meiron, D.I., and Saffman, P.G., J.Fluid Mech., 129, (1983), 213 - 218.

# Session 7

# Second-order Wave-body Interaction

# Nonlinear Hydrodynamic Forces Due to Two-dimensional Forced Oscillation

J.H.HWANG*, Y.J.KIM** and S.Y.KIM***

* Seoul National University, Seoul 151, Korea
** National Fisheries University of Pusan, Pusan 608, Korea
*** Korea Institute of Machinery & Metals, Daejeon 300-31, Korea

Summary

Large-amplitude forced-heaving motion of a circular cylinder is treated by using the semi-Lagrangian time-stepping method, with application of a free-surface regriding technique. Numerical computations are performed for various combinations of the motion amplitudes and frequencies. Through the comparisons with the previous experimental results, it is found that the effects of large-amplitude motion on hydrodynamic force can be reasonably predicted by solving a fully-nonlinear potential-flow model. Also, a novel type of jet, which is generated in the body-emerging process, is shown and its generation mechanism is described.

## INTRODUCTION

Two experimental studies for the large-amplitude forced oscillation have been reported by Tasai & Koterayama[1] and Yamashita[2]. In their measurements, it is shown that the effects of large-amplitude motion on the 1st-harmonic force, as well as the other nonlinear forces, are significant. Yamashita[2] and Higo et al.[3] tried theoretical predictions of these experimental results by their own calculation methods. However, in their calculations the free-surface condition was linearized.

In the present work, two-dimensional forced oscillation problem is treated by solving a fully-nonlinear potential-flow model in time-domain. The obtained results of hydrodynamic forces are compared with the previous experimental results, and discussed. The numerical method used here is, in principle, that of Faltinsen[4]. However, to treat a large and rapid body motion a free-surface regriding algorithm is inevitably necessary. In the present computation, this algorithm is applied over a partial region of the free-surface near the body.

K. Horikawa, H. Maruo (Eds.)
Nonlinear Water Waves
IUTAM Symposium Tokyo/Japan 1987

Problem Formulation and Numerical Implementation

The coordinate system is fixed in space with the x-axis along the undisturbed waterline and y-axis positive upward. A velocity potential $\Phi(x,y,t)$ is introduced under the usual assumption of potential-flow. The time-dependent boundary-value problem to be treated can be written as,

$$\nabla^2\Phi = 0 \qquad \text{in fluid region} \qquad (1)$$

$$\frac{D\Phi}{Dt} = \left(\frac{\partial}{\partial t} + \nabla\Phi\cdot\nabla\right)\Phi = -gy + \tfrac{1}{2}\nabla\Phi\cdot\nabla\Phi \qquad \text{at instantaneous free surface} \qquad (2)$$

$$\frac{D^x f}{Dt} = \frac{\partial\Phi}{\partial x}, \quad \frac{D^y f}{Dt} = \frac{\partial\Phi}{\partial y} \qquad \text{at instantaneous free surface} \qquad (3)$$

$$\frac{\partial\Phi}{\partial n} = \bar{V}\cdot\bar{n} \qquad \text{at instantaneous body surface} \qquad (4)$$

with zero initial and infinity conditions. Here, g denotes the gravitational acceleration, $\bar{n}$ unit normal vector into the fluid, $\bar{V}$ velocity vector of the body motion. $(x_f, y_f)$ is the x,y coordinates of a fluid particle on the free surface.

To solve this problem, the semi-Lagrangian time-stepping method using singularity distribution on boundary surface [4] was employed. In this numerical scheme, the position and the potential value of free-surface are updated by time-integration of the complete nonlinear free-surface conditions, Eq.(2) and Eq.(3). And to obtain the potential value on body surface and the normal velocity on free-surface, the following integral equation is solved at each time step.

$$-\pi\Phi(x',y') = \fint_{S_f+S_b} \left\{\Phi(x,y)\frac{\partial}{\partial n(x,y)}\operatorname{Log}\sqrt{(x-x')^2+(y-y')^2} - \operatorname{Log}\sqrt{(x-x')^2+(y-y')^2}\,\frac{\partial\Phi(x,y)}{\partial n(x,y)}\right\} ds(x,y) \qquad (5)$$

$$,(x',y');\ \text{point on boundary surface}$$

Here, $\fint$ means a principal value integral, $S_f$ and $S_b$ denote free surface and body surface respectively. The tangential velocities at the boudary surfaces are evaluated by numerical differentiation of $\Phi$ . More details can be found in the references[4,5].

The particulars in our numerical computation can be summarized as follows. (1) The free-surface segments in a partial region near the body are regenerated at every time step so as to sustain appropriate spacing. This simple but important algorithm prevents the invasion of marked free-surface particle into the body. Similar regriding algorithm was employed by Dommermuth & Yue [6] to avoid a numerical instability. No artificial smoothing of free surface is required in the present computations. (2) The intersection point between the body and the free surface is treated as a part of the free surface. In practice, this point is updated using Eq.(2) and Eq.(3) and corrected by the kinematic condition as a point on the body surface. (3) For the time-integration, the fourth-order Runge-Kutta algorithm was used. And for the interpolation and differentiation of $\Phi$ along the boundary surface, a linear function and a central difference algorithm were used in the computation of large-rapid body motion, but in the computation of moderate motion a cubic spline algorithm was used.

## Extreme Fluid-Body Interaction

The present numerical scheme is really robust and stable, and can be applied for the extreme fluid-body interaction problems. A typical example of large-rapid body motion is shown in Fig.1. In this example, a circular cylinder of radius r starts a heaving motion of cosine-mode at t=0. Motion amplitude A is 0.4r and nondimensional wave number Kr(= $\omega^2$ r/g) is 2.0. T denotes the motion period and $\Delta x_{init}$ denotes the initial length of the free-surface segments near the body. It is remarkable that a jet is generated in the emerging process of (c), not in the submerging process of (a). The computed fluid velocity reveals that, in the process of this jet generation, the fluid particles on the free surface rush to the body and this fluid-sloshing motion results in the jet. The fluid particle at the tip of this jet has an extremely large velocity and at the next time step it goes away off the body surface. And the numerical computation breaks down. In practice, this implies that the jet turns over and breaks into a spray. We note that such an extreme phenomenon in numerical computation depends on not only the motion amplitude and frequency but also the resolution rate of free-surface.

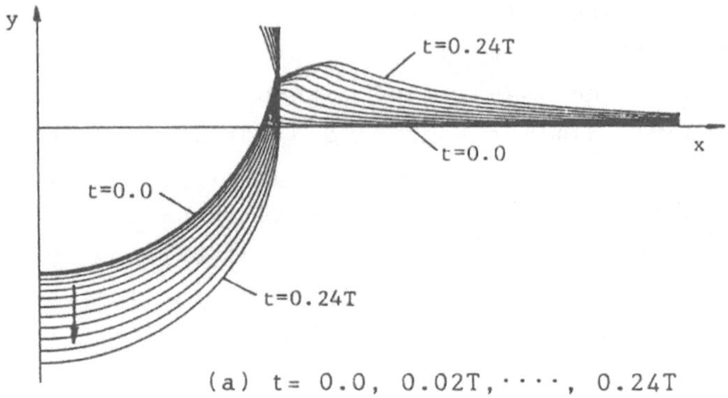


(a) t= 0.0, 0.02T, ····, 0.24T

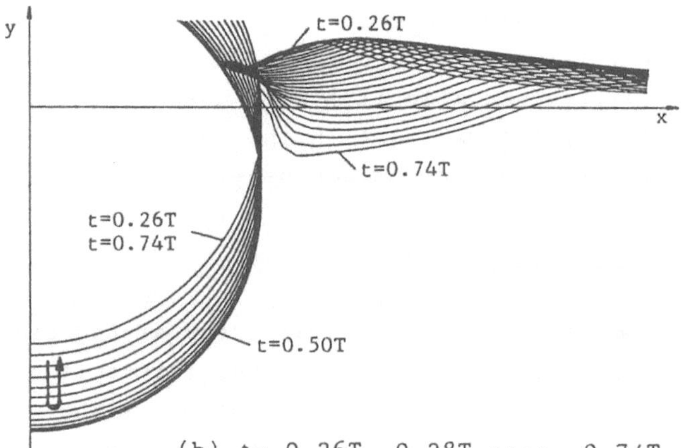


(b) t= 0.26T, 0.28T, ····, 0.74T

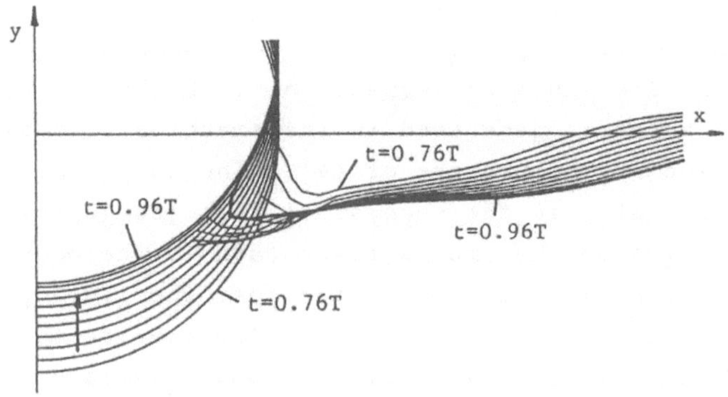


(c) t= 0.76T, 0.78T, ····, 0.96T
( ——— shows a jet due to fluid-sloshing)

Fig.1. Free-surface profiles generated by a circular cylinder heaving with large amplitude and high frequency. (vertical body displacement; $Y_b$=A$\cos\omega t$, Kr=2.0, A=0.4r, $\Delta x_{init}$ =0.03r)

## Quasi-Steady Hydrodynamic Forces

Our primary interest here is the quasi-steady results of hydrodynamic forces. Forced heaving motion of a floating circular cylinder is treated. Numerical computations are performed for two motion amplitudes and eight frequencies. In these computations, the time step size is T/40 and the number of segments per one wave length is 40 except in the vicinity of the body. The length of free-surface segments near the body was reduced in low-frequency for comparableness with body segments and increased in high-frequency to avoid breakdowns due to jet generation.

In the present work, another time-domain model of an intermediate level is also solved, in which only the free-surface condition is linearized. For efficient computation of this model, the method of free-surface spectral representation with simple source distribution on body surface (Chapman[7], Kim & Hwang[8]) was used.

An example of the calculated forces are shown in Fig.2. By the Fourier-transform of these forces, each harmonic force components and time-mean force are obtained. The obtained results of component forces are shown in Fig.3 through Fig.5, and compared with Yamashita's experimental measurements[2] and other computational results[9].

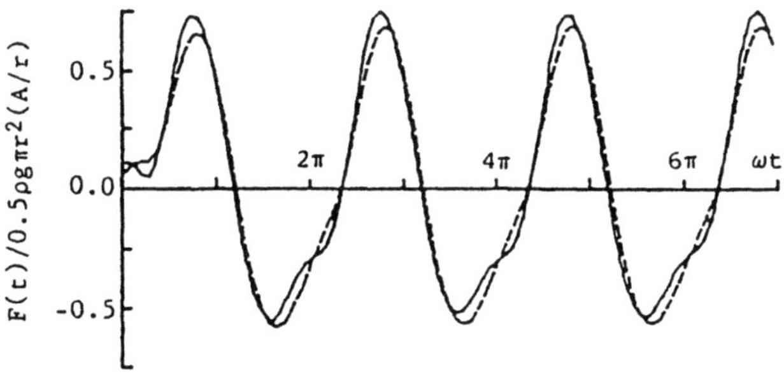


Fig.2. Time history of hydrodynamic forces for a heaving circular cylinder. ($Y_b$=Asinωt,Kr=0.64,A=0.4r, ———— : Fully Nonlinear, - - - - - - - : Free-Surface Linear)

In the figures, the fully-nonlinear results of the present calculation show, in general, good agreements with the experimental measurements of Yamashita except for the time-mean force. Particularly it is remarkable that, in both of the present fully-nonlinear results and Yamashita's measurements, the acceleration-phase 1st-harmonic component (added mass) considerably increases as the motion amplitude increases, and the velocity-phase 1st-harmonic component (damping coefficient) changes with a reverse trend. On the other hand, in the results

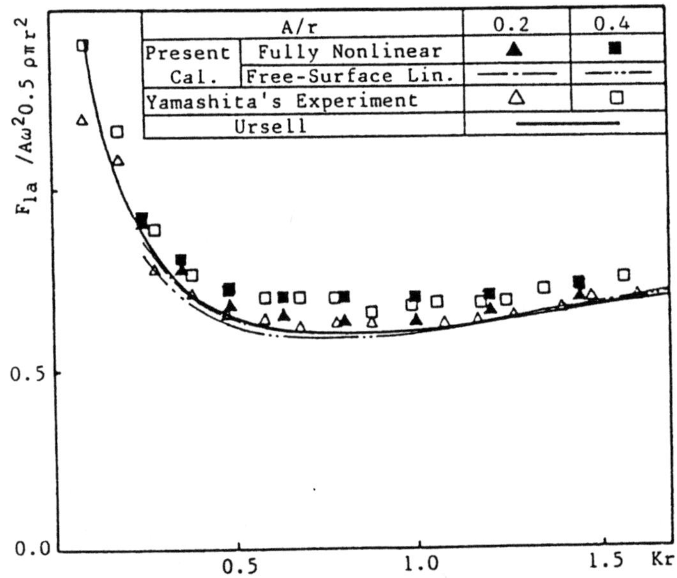


(a) acceleration-phase component (added mass)

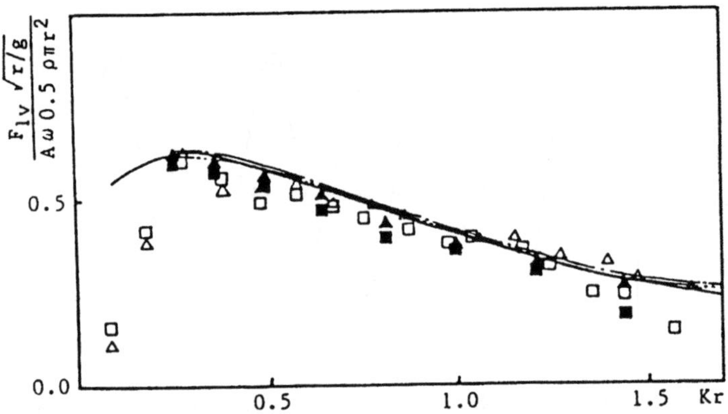


(b) velocity-phase component (damping coefficient)

Fig.3. 1st-harmonic forces of a heaving circular cylinder.

calculated by using the linear free-surface condition the 1st-harmonic force components hardly change as the motion amplitude increases. Therefore, it is supposed that the above variations of the hydrodynamic coefficients according to the increase of motion-amplitudes are mainly attributed to the free-surface nonlinearity. As for the time-mean force, we feel that more experimental data are necessary for a reasonable judging. (The experimental measurements of Tasai et al. [1] and Kyozuka [10] show somewhat larger time-mean force than those of Yamashita.) It is also found that the neglection of the free-surface non-linearity results in under-estimations of the 2nd-harmonic force amplitude in high-frequency range and the 3rd-harmonic force amplitude over the entire frequency range.

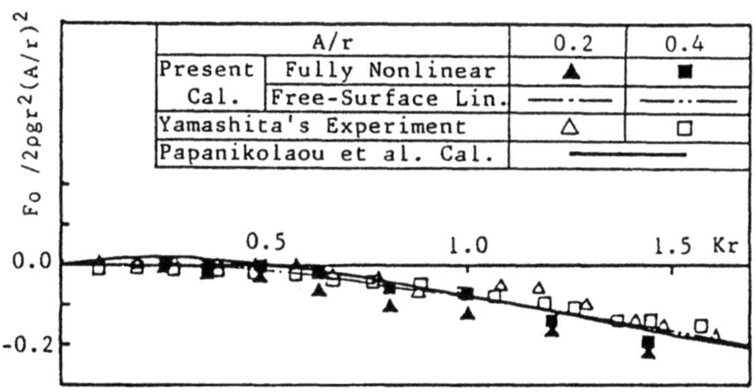


Fig.4. Time-mean force of a heaving circular cylinder. (upward force being positive)

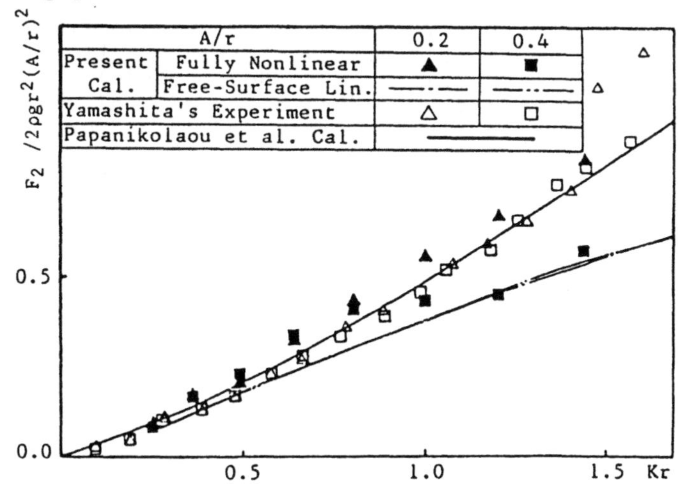


Fig.5. 2nd-harmonic force amplitude of a heaving circular cylinder.

## Concluding Remarks

In the present computation, the free-surface regriding algorithm was adopted. A novel type of jet due to a fluid sloshing motion was calculated realistically. Through the comparisons of quasi-steady hydrodynamic force with the previous experimental measurements, it was found that the variations of the 1st-harmonic forces due to large-amplitude motion, as well as the other nonlinear forces, can be reasonably predicted by solving a fully-nonlinear potential-flow model. As a future research, a fundamental treatment of the singular character of the intersection point between the body and the free-surface is desirable. And more experimental studies for the large-amplitude motion are also required.

## References

1. Tasai, F. and W. Koterayama: Nonlinear Hydrodynamic Forces Acting on Cylinders Heaving on the Surface of a Fluid. Rep. of Res. Inst. for Appl. Mech. Vol.24, No.77, (1976).

2. Yamashita, S.: Calculations of the Hydrodynamic Forces Acting upon Thin Cylinders Oscillating Vertically with Large Amplitude. J. of Soc. Naval Arch. of Japan. Vol. 141, (1977).

3. Higo, Y., Y. Mochioka and I. Nakamura: A Study on the Non-Linear Hydrodynamic Forces for Two-Dimensional Floating Body Heaving with Large-Amplitude. J. of Kansai Soc. Naval Arch., Japan. Vol. 194, (1984).

4. Faltinsen, O.M.: Numerical Solution of Transient Nonlinear Free-Surface Motion Outside or Inside Moving Bodies. Proc. of 2nd Intl. Conf. on Num. Ship Hydrodyn., Berkeley, (1977).

5. Hwang,J.H., Y.J.Kim and S.Y.Kim: On the Nonlinear Hydrodynamic Forces due to Large Amplitude Forced Oscillations. J. of Soc. Naval Arch. of Korea. Vol.23, No.2, (1986).

6. Dommermuth,D.G. and D.K. Yue: Study of Nonlinear Axisymmetric Body-Wave Interactions. Proc. of 16th Symp. on Naval Hydrodyn., Berkeley, (1986).

7. Chapman, R.B.: Large-Amplitude Transient Motion of Two-Dimensional Floating Bodies. J.of Ship Res.,Vol.23,No.1,(1979).

8. Kim, Y.J. and J.H. Hwang: Two-Dimensional Transient Motions with Large Amplitude by Time-Domain Method. Proc. of 16th Symp. on Naval Hydrodyn., Berkeley, (1986).

9. Papanikolaou, A. and H. Nowacki: Second-Order Theory of Oscillating Cylinders in a Regular Steep Wave. Proc. of 13th Symp. on Naval Hydrodyn., Tokyo, (1980).

10. Kyozuka, Y.: Experimental Study on Second-Order Forces Acting on a Cylindrical Body in Waves. Proc. of 14th Symp. on Naval Hydrodyn., Ann Arbor, (1982).

# Nonlinear Diffraction Loads upon Three-dimensional Bodies of Arbitrary Shape

W. KIOKA

Department of Civil Engineering
Nagoya Institute of Technology, Japan

Summary
A numerical method for calculating second-order forces due to the interaction of moderately steep waves, both periodic and modulated, with a large structure of general shape is described using the multiple-scale perturbation approach as well as the straightforward perturbation procedure. The structures considered here are assumed at rest, and only the wave exciting forces are found. Numerical results are presented for vertical cylindrical structures. These steady-state solutions are compared with the fully nonlinear numerical solutions involving the time-stepping procedure developed by Isaacson (1982).

## Second-Order Approximations

The fluid surrounding the body is assumed incompressible and the flow irrotational. For simplicity, the fluid bottom is taken to be horizontal at a finite depth h. We shall use a right-handed Cartesian coordinates OXYZ, with OZ vertically upward, and OXY corresponding to the undisturbed free surface. Expanding in a power series with respect to a perturbation parameter $\varepsilon = k\eta_o$ (k: wave number, $\eta_o$: wave amplitude) , the dynamic force in the j-th direction of interest up to the second order in $\varepsilon$ may be obtained in the form

$$F_j = F_{1j} + F_{dj} + F_{wj} + F_{qj} \tag{1}$$

The classical first-order solution $F_{1j}$ is given by

$$F_{1j} = \mathrm{Re}[\, i\omega\rho e^{-i\omega t} \int_{S_b} \phi_1 n_j \, dS] \tag{2}$$

where $\Phi_1 = \mathrm{Re}[\phi_1 e^{-i\omega t}]$ is the linear potential, $S_b$ is the body surface and $n_j$ is the direction normal in the j-th direction. The second-order forces are given respectively by

$$F_{dj} = \mathrm{Re}[\, -\rho e^{-2i\omega t} \int_{S_b} \frac{(\nabla\phi_1)^2}{4} n_j \, dS] - \rho \int_{S_b} \frac{\phi_1 \, \phi_1^*}{4} n_j \, dS \tag{3}$$

K. Horikawa, H. Maruo (Eds.)
Nonlinear Water Waves
IUTAM Symposium Tokyo/Japan 1987

$$F_{wj} = \mathrm{Re}[-\rho e^{-2i\omega t} \int_W \frac{(\phi_1)^2}{4g} n_j dl] - \rho \int_W \frac{\phi_1 \phi_1^*}{4g} n_j dl \tag{4}$$

$$F_{qj} = \mathrm{Re}[2i\omega\rho e^{-2i\omega t} \int_{S_b} \phi_q n_j dS] \tag{5}$$

where * denotes the complex conjugate, W is the waterline, with the assumption that the body intersects the free surface at right angle, and $\Phi_q = \mathrm{Re}[\phi_q e^{-2i\omega t}]$ the quadratic potential.

The linear force $F_{1j}$ and the second-order forces $F_{dj}$ and $F_{wj}$ can be readily computed once the linear potential $\phi_1$ is known. To calculate the quadratic force $F_{qj}$, we adopt the approach developed by Lighthill (1979), in which only radiation potential at double frequency due to a unit translation oscillation in the j-th direction, $\tilde{\Phi}_j = \mathrm{Re}[\tilde{\phi}_j e^{-2i\omega t}]$, is required to be given along with the evaluation of the inhomogeneous term $\gamma$ appeared in the free surface condition for $\phi_q$, i.e.

$$F_{qj} = \mathrm{Re}[2i\omega\rho e^{-2i\omega t} \int_{S_f} \gamma \tilde{\phi}_j dS] \tag{6}$$

where $S_f$ is the free surface (z = 0) extending from the body surface to a sufficiently large distance, and

$$\gamma = -\frac{i\omega}{2g^2} \phi_1 (-\omega^2 \frac{\partial \phi_1}{\partial z} + g \frac{\partial^2 \phi_1}{\partial z^2}) + \frac{i\omega}{g} (\nabla \phi_1)^2 \tag{7}$$

$\tilde{\phi}_j$ satisfies the radiation condition at infinity and the following boundary conditions:

$$\frac{\partial \tilde{\phi}_j}{\partial n} = n_j \qquad \text{on } S_b \tag{8}$$

$$\frac{\partial \tilde{\phi}_j}{\partial z} - \frac{4\omega^2}{g} \tilde{\phi}_j = 0 \qquad \text{at } z=0 \tag{9}$$

with

$$\frac{\omega^2}{g} = k \tanh(kh) \tag{10}$$

Equation (6) has been already used by Rahman (1984) and by Kioka and Ishida (1984) to calculate the second-order forces on a vertical circular cylinder. We note that the second-order moments $M_j$ can be found by the same procedure if one replaces $n_j$ by the components of $\mathbf{r} \times \mathbf{n}$, with $\mathbf{r}$ being the position vector on $S_b$. Since the body of arbitrary shape are treated here, a source distribution method is incorporated to compute the linear diffraction and radiation potentials.

Unsteady Waves

We consider next two incident wavetrains propagating in the positive x-direction with frequency $\omega_1$ and $\omega_2$ respectively. The first order potential $\Phi_1$ may be written in the form

$$\Phi_1 = \mathrm{Re}[\ \phi_{11} e^{-i\omega_1 t} + \phi_{12} e^{-i\omega_2 t}\ ] \tag{11}$$

In this case, the second-order interactions between two free waves render the progressive waves unsteady. Expression for the second-order incident potential corresponding to such unsteady waves have been obtained by Hamada (1965). The results are rather lengthy and are not reproduced here.

The forces $F_{1j}$, $F_{dj}$ and $F_{wj}$ can be readily found by substituting eq. (11) into eqs. (2) -(4). The quadratic force $F_{qj}$ may be evaluated from

$$\begin{aligned} F_{qj} = \mathrm{Re}[\ & 2i\omega_1 \rho e^{-2i\omega_1 t} \int_{S_f} \gamma_1 \tilde{\phi}_{j1} dS \\ & + 2i\omega_2 \rho e^{-2i\omega_2 t} \int_{S_f} \gamma_2 \tilde{\phi}_{j2} dS \\ & + i(\omega_1 + \omega_2)\rho\ e^{-i(\omega_1+\omega_2)t} \int_{S_f} \gamma_3 \tilde{\phi}_{j3} dS \\ & + i(\omega_1 - \omega_2)\rho\ e^{-i(\omega_1-\omega_2)t} \int_{S_f} \gamma_4 \tilde{\phi}_{j4} dS\ ] \end{aligned} \tag{12}$$

where $\tilde{\phi}_{j1}$, $\tilde{\phi}_{j2}$, $\tilde{\phi}_{j3}$ and $\tilde{\phi}_{j4}$ satisfy eqs. (8) - (10) but with $\omega$ being replaced by $\omega_1$, $\omega_2$, $\omega_1 + \omega_2$, $\omega_1 - \omega_2$ respectively. $\gamma_1$ and $\gamma_2$ are given by eq. (7) with $\omega_1$ and $\omega_2$ instead of $\omega$ respectively. $\gamma_3$ and $\gamma_4$ are

$$\begin{aligned} \gamma_3 = & -\frac{i\omega_1}{2g^2} \phi_{11} \left(-\omega_2^2 \frac{\partial \phi_{12}}{\partial z} + g \frac{\partial^2 \phi_{12}}{\partial z^2}\right) + \frac{i\omega_2}{2g^2} \phi_{12} \left(-\omega_1^2 \frac{\partial \phi_{11}}{\partial z} \right. \\ & \left. + g \frac{\partial^2 \phi_{11}}{\partial z^2}\right) - \frac{i(\omega_1 + \omega_2)}{g} \nabla \phi_{11} \nabla \phi_{12} \end{aligned} \tag{13}$$

$$\begin{aligned} \gamma_4 = & -\frac{i\omega_1}{2g^2} \phi_{11} \left(-\omega_2^2 \frac{\partial \phi_{12}^*}{\partial z} + g \frac{\partial^2 \phi_{12}^*}{\partial z^2}\right) + \frac{i\omega_2}{2g^2} \phi_{12}^* \left(-\omega_1^2 \frac{\partial \phi_{11}}{\partial z} \right. \\ & \left. + g \frac{\partial^2 \phi_{11}}{\partial z^2}\right) - \frac{i(\omega_1 - \omega_2)}{g} \nabla \phi_{11} \nabla \phi_{12}^* \end{aligned} \tag{14}$$

The resultant second-order forces $F_{dj}$, $F_{wj}$ and $F_{qj}$ contain the components corresponding to the frequency difference $\omega_1 - \omega_2$ and sum $\omega_1 + \omega_2$ as well as the double frequency $2\omega_1$ and $2\omega_2$. When eq. (12) is used to calculate the quadratic force $F_{qj}$, the second-order incident potential is not necessarily prescribed.

Slowly Varying Forces

When the envelope of the incident wavetrain modulates slowly in both x-direction and time, one is often interested primarily in the slowly-varying second-order wave forces. In this case, the numerical task for calculating second-order forces can be significantly reduced by using the multiple-scale perturbation method. The procedure employed here has been considered by Agnon and Mei (1985) for a two-dimensional problem and by Zhou and Liu (1987) for the case of a vertical circular cylinder. We introduce the fast (x, y, z, t) and the slow ($x_1 = \varepsilon x$, $y_1 = \varepsilon y$, $t_1 = \varepsilon t$) variables and express the solution in terms of harmonics with respect to the fast time,

$$\Phi^{(n)} = \mathrm{Re}\left[ \sum_{m=0}^{n} \Phi^{(n,m)}(x,y,z,t;x_1,y_1,t_1)\, e^{-im\omega t} \right] \tag{15}$$

It can be shown that the first-harmonic potential $\Phi^{(11)}$ obeys the boundary-value problem for the linear interaction. $\Phi^{(11)}$ can be then readily evaluated if the incident-wave envelope $A(x_1, t_1)$ is prescribed. The evaluation of the forces due to the self-interactions of the first-harmaonic component $\Phi^{(11)}$ is straightforward through eqs. (3) and (4).

To obtain the complete solution for the low-frequency forces, we need to evaluate the forces induced by the set-down waves, i.e.

$$\bar{F}_j = -\rho \int_{S_b} \frac{\partial \Phi^{(10)}}{\partial t_1} n_j dS \tag{16}$$

We note that the zeroth harmonic $\Phi^{(10)}$ is independent of short scales. The long-scale equation for $\Phi^{(10)}$ can be obtained from the third-order zeroth harmonic equations as follows (see Mei, 1983):

$$\frac{\partial^2 \Phi^{(10)}}{\partial x_1^2} + \frac{\partial^2 \Phi^{(10)}}{\partial y_1^2} - \frac{1}{gh}\frac{\partial^2 \Phi^{(10)}}{\partial t_1^2} = -\frac{1}{gh}\left(\frac{\omega^3}{k\tanh^2 kh} + \frac{C_g\omega^2}{2\sinh^2 kh}\right)(|A|^2)_{x_1} \tag{17}$$

where we have used

$$\frac{\partial A}{\partial t_1} + C_g \frac{\partial A}{\partial x_1} = 0 \tag{18}$$

In addition, $\Phi^{(10)}$ is required to satisfy

$$\frac{\partial \Phi^{(10)}}{\partial n} = 0 \qquad \text{on } S_b \tag{19}$$

Since the right-hand side of eq. (17) is regarded as the forcing terms for the set-down waves associated with the incident waves, we assume the total potential to be written in the form

$$\phi^{(10)} = \mathrm{Re}[\, \phi_I^{(10)} e^{-2i\omega_o t_1} + \phi_S^{(10)} e^{-2i\omega_o t_1}] + \bar{\phi}^{(10)} \tag{20}$$

where $\phi_I^{(10)}$ and $\phi_S^{(10)}$ denote the oscillating parts of the incident and scattered waves respectively, with $\omega_o$ being the frequency of the envelope of incident waves, and $\bar{\phi}^{(10)}$ is the potential independent of $t_1$. The scattered low-frequency wave potential $\phi_S^{(10)}$ satisfies

$$\frac{\partial^2 \phi_S^{(10)}}{\partial x_1^2} + \frac{\partial^2 \phi_S^{(10)}}{\partial y_1^2} + \alpha^2 \phi_S^{(10)} = 0 \tag{21}$$

$$\frac{\partial \phi_S^{(10)}}{\partial n} = -\frac{\partial \phi_I^{(10)}}{\partial n} \qquad \text{on } S_b \tag{22}$$

with the notation of

$$\alpha = \frac{2\omega_o}{(gh)^{1/2}} \tag{23}$$

$\phi_S^{(10)}$ may be represented by distributing point wave sources G over $S_b$ according to the source strength function f. The Green function appropriate for the present case may be given in the two-dimensional form

$$G = i\pi H_0^{(1)}(\alpha r_1) \tag{24}$$

where $H_0^{(1)}$ denotes the Hankel function of the first kind and zero order, and

$$r_1 = ((x_1 - \xi_1)^2 + (y_1 - \eta_1)^2)^{1/2} \tag{25}$$

If we further suppose that the body forms vertical cylinders of arbitrary section, the solution can be expressed in the two-dimensional form as the incident wave potential $\phi_I^{(10)}$ is independent of z, i.e.

$$\phi_S^{(10)}(x_1, y_1) = \frac{1}{4\pi} \int_W f(\xi_1, \eta_1) G(x_1, y_1; \xi_1, \eta_1)\, dl \tag{26}$$

The source strength distribution function f may be determined from eq. (22);

$$-\frac{1}{2} f(x_1, y_1) + \frac{1}{4\pi} \int_W f(\xi_1, \eta_1) \frac{\partial G}{\partial n}(x_1, y_1; \xi_1, \eta_1)\, dl = -\frac{\partial \phi_I^{(10)}}{\partial n}(x_1, y_1) \tag{27}$$

## Results and Discussion

The numerical calculations based on the straightforward perturbation method and the multiple-scale perturbation method were performed for the case of surface-piercing circular and rectangular cylinders. The full second-order solutions obtained from the straightforward perturbation are compared with the results from the fully nonlinear time-stepping method without any perturbation. In the latter numerical approach, the wave diffraction is treated as a transient problem with given initial conditions and the flow field is calculated by the integral equation method at each time step (Isaacson 1982).

Figure 1 shows an example of the results for the periodic incident wavetrain. In the fully nonlinear time-stepping calculation, the spatial periodicity is assumed in the x-direction and the impermeable wall in the y-direction. In the case (a), both results become graphically indistinguishable within the duration of two wave periods. In the case (b), where nonlinear effects are seen to be more pronounced, the second-order solution differs from the fully nonlinear one in the profile.

The full second-order solution involving sum and difference frequencies for the modulated incident wave is presented in fig. 2. In the transient analysis, the boundary values on the imaginary (open) boundary surrounding the body were prescribed using the results given by Hamada (1965). For this case, the numerical stability was difficult to achieve for sufficiently long duration.

Finally, fig. 3 indicates the results for the low-frequency wave forces defined as eq. (16). The incident-wave envelope is assumed sinusoidal;

$$A = A_o \sin k_o (x_1 - C_g t_1) \tag{28}$$

where $k_o$ denotes the wavenumber of the envelope. The amplitude of the oscillations of the set-down waves is then given by

$$\phi_I^{(10)} = \frac{i}{2k_o} A_o^2 \frac{1}{C_g^2 - gh} \left( \frac{\omega^3}{k \tanh^2 kh} + \frac{C_g \omega^2}{2 \sinh^2 kh} \right) \tag{29}$$

Limited comparisons indicate that the results based on the multi-scale perturbation method approach the values predicted from the straightforward perturbation when $k_o a$ becomes small.

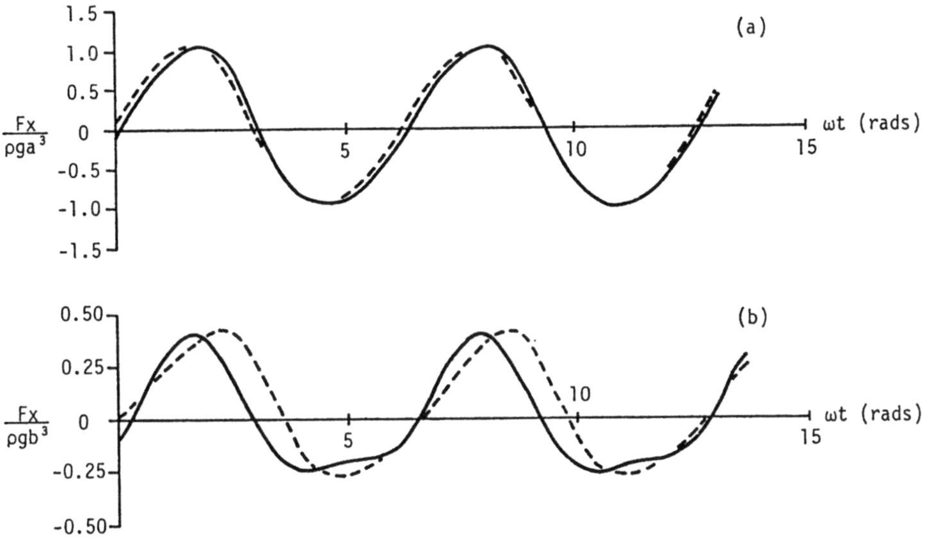


Fig. 1 Temporal variations of horizontal wave force for a circular cylinder of radius a with the amplitude $\eta_o/a=0.2$, and $ka=0.5$ in the depth $kh=1.5$ (a), and for a rectangular cylinder of side length b with the orientation angle $\Theta=45°$, $\eta_o/b=0.22$, $kb=0.3$, $kh=0.72$ (b). The broken line indicates the results from the time-stepping calculation.

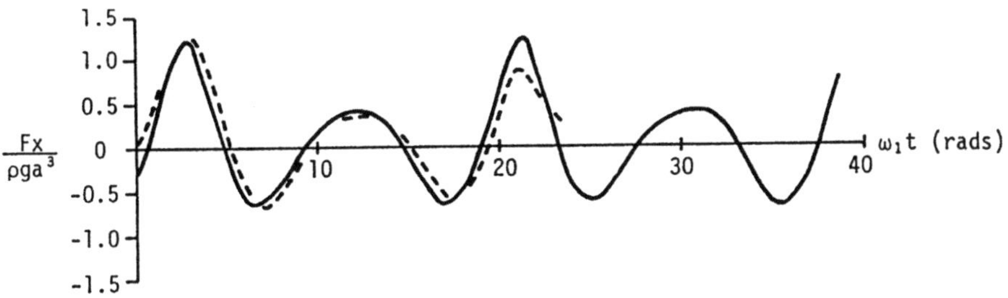


Fig. 2 Temporal variations of horizontal wave force induced by unsteady Stokes waves with $\eta_{o1}/a=0.4$, $k_1a=0.5$, $k_1h=2.0$, $\eta_{o2}/a=0.2$, $k_2a=1.1$ and $k_2h=4.0$. The broken line indicates the results from the time-stepping calculation.

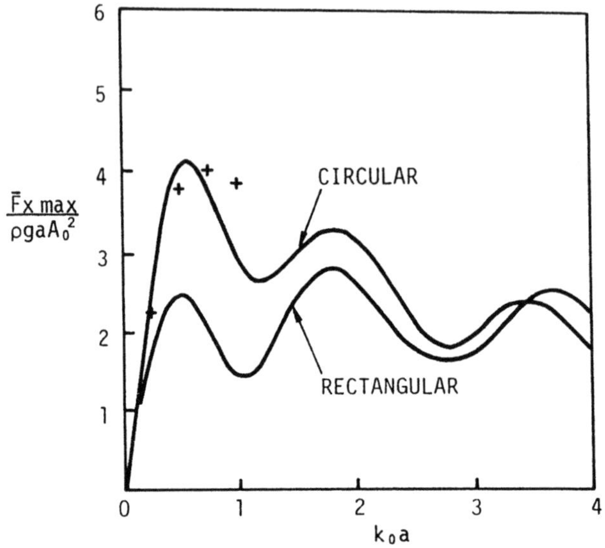


Fig. 3 Maximum horizontal low-frequency force $\bar{F}$ as a function of $k_0a$ for a circular cylinder of radius a and for a rectangular cylinder with the same sectional area; i.e. $b=a(\pi)^{1/2}$, and $\Theta = 0°$. In the case of circular cylinder, the results based on the straightforward perturbation scheme (ka = 4.0) are indicated by + .


References

1. Agnon, Y. and Mei, C. C.: Slow-drift motion of a two-dimensional block in beam sea. J. Fluid Mech. 151 (1985) 279-294.
2. Hamada, T.: The secondary interactions of surface waves. Rept. Port and Harbour Tech. Res. Inst. 10 (1965).
3. Isaacson, M. de St. Q.: Nonlinear-wave effects on fixed and floating bodies. J. Fluid Mech. 120 (1982) 267-281.
4. Kioka, W. and Ishida, A.: Second-order diffraction loads upon vertical circular cylinders. Proc. 31th Japanese Conf. on Coastal Engng. (1984) 620-624 (in Japanese).
5. Lighthill, M. J.: Waves and hydrodynamic loading. Proc. 2nd Int. Conf. on Behaviour of Offshore Structure (1979) 1-40.
6. Mei, C. C.: The Applied Dynamics of Ocean Surface Waves. (1983) Wiley-Interscience.
7. Rahman, M.: Wave diffraction by large offshore structures-an exact second-order theory. Applied Ocean Res. 6 (1984) 90-100.
8. Zhou, C. and Liu, P. L.-F.: Second-order low-frequency wave forces on a vertical circular cylinder. J. Fluid Mech. 175 (1987) 143-155.

# On Calculations of Nonlinear Wave-body Interaction Effects

A. PAPANIKOLAOU

Department of Naval Architecture and Marine Engineering
National Technical University of Athens, Greece

ABSTRACT

The paper deals with the evaluation of the nonlinear hydrodynamic effects arising at oscillations of two-dimensional, arbitrarily shaped cylinders floating in or being fully submerged under the free surface and in response to a regular wave train. The theoretical model, introduced earlier by the author(1980) has been extended herein to the quasi third-order. The employed potential theory is based on a regular perturbation expansion in terms of small parameters to the third-order and Green´s Function Integral-Equation-Method. The resulting set of Fredholm Integral Equations, to be solved for the velocity potentials to the third-order, is formulated as a Combined-Helmholtz-Integral-Equation, which is free of irregular frequencies (author, 1984). Comparisons of numerical results for the hydrodynamic forces with experimental data by Tasai-Koterayama(1976) and Yamashita (1977), especially for the simple frequency forces, added-mass and hydrodynamic(wave) damping of cylinders in forced heaving, show the significant influence of the amplitude of oscillation over the whole frequency range of interest.

THEORETICAL MODEL

The mathematical modelling of the hydrodynamics of a cylinder, oscillating with finite amplitude in response to a regular wave train, Fig. 1, has been described in earlier publications ( Papanikolaou and Nowacki, 1980). Herein the theory will be extended to the quasi third-order without going into details for the sake of brevity.

Perturbation Expansion

The velocity potential describing the fluid motion resulting from the interaction of an incident wave train of frequency $\omega$ with an arbitrarily shaped cylinder can be expressed in power series of the small parameter $\varepsilon$ (wave steepness) in the sense of a regular perturbation expansion:

K. Horikawa, H. Maruo (Eds.)
Nonlinear Water Waves
IUTAM Symposium Tokyo/Japan 1987

$$\Phi(x,y;t;\varepsilon) = \sum_{n=1}^{N} \sum_{k=0}^{K} \varepsilon^n \varphi_k^{(n)}(x,y) \cdot e^{-jk\omega t} \tag{1}$$

In (1), going to the third-order, we can set $N = K = 3$ and consider that potential functions with $n \neq k$ are trivial, except for $k = 0$, $n = 2$ and $k = 1$, $n = 3$. Thus we obtain to the third-order:

$$\Phi(x,y;t;\varepsilon) = \varepsilon \varphi_1^{(1)} e^{-j\omega t} + \varepsilon^2 (\varphi_0^{(2)} + \varphi_2^{(2)} e^{-2j\omega t})$$

$$+ \varepsilon^3 (\varphi_1^{(3)} e^{-j\omega t} + \varphi_3^{(3)} e^{-3j\omega t}) + O(\varepsilon^4) \tag{2}$$

Boundary-Value-Problem

The exact boundary conditions have been presented earlier and can be omitted herein.

The BVP for the quasi third-order velocity potential $\varphi_1^{(3)}$, which results from the perturbation expansion of the free-surface and body boundary conditions about the positions of these boundaries at rest, is given in the following (see Fig. 2):

$$\Delta \varphi_1^{(3)} = 0 \quad , \quad (x,y) \in D \tag{3.1}$$

$$\left\{\frac{\partial}{\partial y} - \nu\right\} \varphi_1^{(3)} = \ell_1^{(3)} \quad , \quad (x,y) \in S_F \tag{3.2}$$

$$\frac{\partial}{\partial n}(\varphi_1^{(3)}) = r_1^{(3)} \quad , \quad (x,y) \in S_o \tag{3.3}$$

$$\frac{\partial}{\partial y}(\varphi_1^{(3)}) = 0 \quad , \quad (x,y) \in S_B \tag{3.4}$$

$$\left\{\frac{\partial}{\partial x} \mp j\nu\right\} \varphi_1^{(3)} = 0 \quad , \quad (x,y) \in S_R, S_L \tag{3.5}$$

where $\nu = \omega^2/g$ means the frequency number.

The right-hand-sides of equs. (3.2) and (3.3) are given in the Appendix.

Integral-Equation-Method

Applying Green's third theorem to the BVP (3.1) to (3.5), stated before, the following Fredholm Integral Equation of second kind for $\varphi_1^{(3)}$ is obtained:

$$-\pi\varphi_1^{(3)}(P)+\int_{S_0}\varphi_1^{(3)}(Q)(\vec{n}\vec{\nabla})_Q G^{(1)}(P,Q)dS_Q$$

$$=\int_{S_0} G^{(1)} r_1^{(3)} dS_Q - \int_{S_F} G^{(1)} \ell_1^{(3)} dS_Q \quad , \quad P(x,y)\in S_0 \, , \tag{4}$$

where $G^{(1)}(P,Q)$ stands for the Green´s function satisfying the adjoint, homogeneous BVP, corresponding to (3.1) to (3.5), and which is identical to the Green´s function of the first-order velocity potential.

For avoiding the well known irregular frequencies phenomenon, arising at the solution of free-surface problems by Green´s function method, the aforementioned Green´s function has been modified by adding pulsating sources and dipoles at the coordinate system´s origin, as proposed by Ogilvie and Shin(1978) and modified by the author later on (1984).

Line-Integral

The most critical part in the solution of any of the higher-order free-surface BVP´s, formulated in the course to the third-order, seems to be the proper evaluation of the line integral

$$I_{F_1}^{(3)} = \int_{S_F} \ell_1^{(3)}(\xi,0)\, G^{(1)}(x,y;\xi,0)\, d\xi \tag{5}$$

which has to be evaluated along the free-surface $S_F$ from its intersection with the body side and up to infinity.

Two fundamental problems arise in evaluating the inhomogeneity $\ell_1^{(3)}$:

- At the intersection of the body with the free-surface there exists a nonintegrable singularity due to the singular behaviour of the first and second derivatives of the velocity potentials in the horizontal direction. Besides an analytical treatment, which remains to be developed, a practical way out of this problem has been to assign the relevant quantities there finite values according to the body boundary condition near the intersection point.

- The truncation of the line integral at infinity, i.e. at a sufficient distance away from the body, causes usually no serious problems, since of the harmonic behaviour of all members of the integrand a few half-beams away from the body. For certain BVP´s, like the second-order forced motion problem, the integrand goes in fact to zero, whereas for the second-order diffraction and the third-order forced motion problem the integrand is oscillatory harmonic.

Body-Inhomogeneity

The evaluation of the body-contour inhomogeneity, $r_1^{(3)}$ (A.2), and of the corresponding RHS to equ.(4)

$$I_{R1}^{(3)} = \int_{S_0} r_1^{(3)}(\xi,\eta)\, G^{(1)}(x,y;\xi,\eta)\, d\xi\, d\eta \tag{6}$$

is less problematic. The required derivatives of the lower order velocity potentials can be implicitly calculated from the corresponding lower order kinematic boundary conditions along the body contour and numerical differentiation in terms of $s$.

Pressures

The hydrodynamic pressure $P(x,y,t)$ on the moving body contour $S(t)$ is obtained through Bernoulli´s equation and Taylor series expansions of the contributing velocity potentials about the equilibrium position of the body contour $S_o \equiv S(t=0)$. According to the perturbation expansion for $\Phi$ we obtain for the pressure to the third-order:

$$P(x,y;t;\varepsilon) = p^{(0)} + \varepsilon p_1^{(1)} e^{-j\omega t} + \varepsilon^2 (p_0^{(2)} + p_2^{(2)} e^{-2j\omega t}) + $$
$$+ \varepsilon^3 (p_1^{(3)} e^{-j\omega t} + p_3^{(3)} e^{-3j\omega t}) + O(\varepsilon^4) \tag{7}$$

The individual pressure terms $p_k^{(n)}$ to the second-order have been given earlier(author, 1980). The new terms to the third-order, which are complicated functions of lower order potentials and their derivatives, can be found in /7/, and will be omitted herein since of the limited space available.

Forces

Integration of the pressure over the wetted body contour S(t) leads to the hydrodynamic forces acting on the body. Thereby the influence of the above mean-water-level wetted part of the body contour has been evaluated by means of Leibnitz´s integration rule, as shown earlier for the second-order(author, 1980).

According to the perturbation expansion employed for the velocity potential(2) we obtain for the amplitude of the quasi third -order force oscillating with simple frequency:

$$\vec{f}_1^{(3)} = -\int_{-L}^{+L} (p^{(0)}\vec{n}_1^{(3)} + p_1^{(1)}\vec{n}_0^{(2)} + \tfrac{1}{2}\,\vec{n}_2^{(2)}\bar{p}_1^{(1)} + \tfrac{1}{2}p_2^{(2)}\overline{\vec{n}_1^{(1)}}$$

$$+\, p_0^{(2)}\vec{n}_1^{(1)} + p_1^{(3)}\vec{n}^{(0)})\,ds + \vec{f}_{1R}^{(3)} \tag{8}$$

where the individual pressures, normales and the rest-force term $\vec{f}_{1R}^{(3)}$ are given in /7/. In evaluating those parts of $\vec{f}_1^{(3)}$, associated with $p_1^{(3)}\vec{n}^{(0)}$, and especially those terms including the quasi third-order potential $\varphi_1^{(3)}$ we can exploit Green´s theorems over the fluid domain D(Fig. 2) and the boundary conditions for $\varphi_1^{(1)}$ and $\varphi_1^{(3)}$ and obtain, as shown before for the second-order by Söding(1976):

$$\int_{S_O} \varphi_1^{(3)} r_1^{(1)} dS = \int_{S_O} \varphi_1^{(1)} r_1^{(3)} dS - \int_{S_F} \varphi_1^{(1)} \ell_1^{(3)} dS \tag{9}$$

DISCUSSION OF RESULTS

The presented theoretical model has been studied systematically by extensive numerical evaluations of various nonlinear effects, as they appear when certain sectional(beam to draft ratio, area coefficient, flare angle) and physical parameters(frequency of oscillation, amplitude, depth of submergence) are varied systematically. These results, to the second-order, have been shown earlier by the author(1980, 1984, 1986).

Going with the present study to the quasi third-order some preliminary results of third-order effects on simple frequency quantities, like added-mass and damping of "first-order", are presented.

In Fig. 3, resp. 4, the dimensionless amplitude, resp. phase, of the simple frequency hydrodynamic force acting on a circular cylinder in forced heaving are presented. The depicted experimental data for various amplitudes of oscillation stem from Tasai and Koterayama(1976). Besides the third-order hydrodynamic effects, as expressed by equ. (8), the third-order restoring force term has been included in both the numerical results and the experimental data analysis. The agreement between theory and experiment is remarkably good, both in the tendency as well as in the absolute value level, even for higher $\varepsilon$.

In Fig. 5, resp. 6, the dimensionless added-mass and damping of the circular cylinder in heaving, corresponding to the analysis of the simple frequency hydrodynamic force in real and imaginary parts, but with the restoring term excluded, are depicted. The experimental values are due to Yamashita(1977). It is clearly shown, at least in this particular case, that an increase of the amplitude of oscillation increases the added-mass and lowers the hydrodynamic damping over the entire frequency range of practical interest. Further results for other cross sections are similar , but not so clear, as to the dependence on the frequency. This will be given in future publications.

The present two-dimensional, quasi third-order theory and the related computer program, can form the basis for the development of an improved strip-theory, as known from applications to ship motion problems. Solving at first for the first-order motions of the vessel by a common linear strip-theory the motion of each strip along the vessel´s length will be known. Therefrom the quasi third-order problem for each strip can be solved, or will have been solved in advance for a unit motion, and the results of additional simple frequency, third-order forces will contribute to "corrected" added-mass and damping coefficients. For these new coefficients the ship motions problem will be recalculated leading to an iteration scheme. It is expected, for strips at the ends and for frequencies near the first order resonances the changes to be most significant. However, it remains to be clarified, how 3D and forward speed effects influence the suggested theoretical model.

REFERENCES

1. Tasai, F.; Koterayama, W.: Nonlinear hydrodynamic forces acting on cylinders heaving on the surface of a fluid. Rep. No. 77, Res. Inst. of Appl. Mech., Kyushu Univ. 1976.

2. Yamashita, S.: Calculations of the hydrodynamic forces acting upon thin cylinders oscillating vertically with large amplitude. Journ. Soc. Naval Arch. of Japan 141 (1977) 67-76.

3. Papanikolaou, A. ; Nowacki, H.: Second-order theory of oscillating cylinders in a regular steep wave. Proc. 13th Symp. on Naval Hydrod., Tokyo, (1980) 303-333.

4. Ogilvie, T.F.; Shin, Y.S.: Integral equation solutions for time-dependent free surface problems. Journ. Soc. Naval Arch. of Japan 142 (1978) 43-53.

5. Papanikolaou, A.: On the nonlinear wave forces acting on partially or fully submerged cylinders in waves. Proc. 1st Int. Workshop on Water Waves and Floating Bodies, Boston, (1986) 129-135.

6. Papanikolaou, A.: On calculations of nonlinear hydrodynamic effects in ship motions. Journ. Schiffstechnik 31 (1984) 89-129.

7. Papanikolaou, A.: Third-order theory of oscillating cylinders in a regular steep wave. Rep. Dep. of Naval Arch. and Marine Eng., NTUA, 1987.

8. Söding, H.: Second-order forces on oscillating cylinders in waves. Journ. Schiffstechnik 23 (1976) 205-209.


CONVENTIONS

a) Bars over complex values denote the conjugate complex values:

$$(\bar{a}) = (\overline{a_c + ja_s}) = a_c - ja_s$$

b) Triple products of complex values in { } (braces) mean:

$$\{a \cdot b \cdot c\} = (a \cdot \bar{b})c + (\bar{a} \cdot b)c + (a \cdot b)\bar{c}$$

c) Double products of complex values in { } (braces) mean:

$$\{a \cdot b\} = (a \cdot \bar{b}) + (\bar{a} \cdot b) + (a \cdot b)$$

d) Indexing by $x$, $y$, $\xi$, $\eta$ or $n$ and $s$ means partial derivative with respect to $x$, $y$, $\xi$, $\eta$ or the normal and arclength of the body contour.

e) The complex motion amplitudes $x_i^{(n)}$ appearing in the formulas are defined through:

$$x_i(t) = x_i^{(1)} e^{-j\omega t} + (x_{0i}^{(2)} + x_i^{(2)} e^{-2j\omega t})$$

where i=2 for sway, i=3 for heave and i=4 for roll.

f) In (A.2) $y'$ means the directional sine and $x'$ the cosine.

APPENDIX

The RHS of the free surface boundary condition, equ. (3.2), is given by:

$$\ell_1^{(3)} = j\frac{\omega}{g}\varphi_2^{(2)}(\nu^2\bar{\varphi}_1^{(1)} + \bar{\varphi}_{1xx}^{(1)})$$

$$- j\frac{\omega}{2g}\bar{\varphi}_1^{(1)}(4\nu\varphi_{2y}^{(2)} + \varphi_{2xx}^{(2)})$$

$$+ j\frac{\omega}{g}\varphi_{2x}^{(2)}\bar{\varphi}_{1x}^{(1)} + j\frac{\omega}{g}\nu\varphi_{2y}^{(2)}\bar{\varphi}_1^{(1)}$$

$$+ 2j\frac{\omega}{g}\varphi_{0x}^{(2)}\varphi_{1x}^{(1)} + 2j\frac{\omega}{g}\nu\varphi_{0y}^{(2)}\varphi_1^{(1)} + j\frac{\omega}{g}\varphi_1^{(1)}\varphi_{0xx}^{(2)}$$

$$+ \frac{\nu}{8g}\{(j\varphi_1^{(1)})(j\varphi_1^{(1)})(2\nu^3\varphi_1^{(1)} - 3\nu\varphi_{1xx}^{(1)} - \varphi_{1xxy}^{(1)})\}$$

$$- \frac{1}{8g}\{(\varphi_{1x}^{(1)})(\varphi_{1x}^{(1)})(5\nu^2\varphi_1^{(1)} + 3\varphi_{1xx}^{(1)})\}$$

$$- \frac{\nu^2}{8g}\{(\varphi_1^{(1)})(\varphi_1^{(1)})(\nu^2\varphi_1^{(1)} - \varphi_{1xx}^{(1)})\}$$

$$+ \frac{\nu^2}{g}\{(j\varphi_1^{(1)})(j\varphi_{1x}^{(1)})(\varphi_{1x}^{(1)})\}$$

$$- \frac{\nu^2}{2g}\{(\varphi_1^{(1)})(j\varphi_1^{(1)})(j\varphi_{1xx}^{(1)})\} \qquad (A.1)$$

Accordingly, the RHS of the body boundary condition, equ. (3.3), is given by:

$$r_1^{(3)} = -j\omega[y'x_{12}^{(3)} - x'x_{13}^{(3)} - (xx'+yy')x_{14}^{(3)}]$$

$$+\frac{1}{8}\omega(xx'+yy')\{(-jx_4^{(1)})(x_4^{(1)})(\bar{x}_4^{(1)})\}$$

$$-\frac{1}{4}j\omega x'(2x_2^{(2)}\bar{x}_4^{(1)} - x_4^{(2)}\bar{x}_2^{(1)} + 2x_2^{(1)}x_{04}^{(2)})$$

$$-\frac{1}{8}\omega y'\{(-jx_2^{(1)})(x_4^{(1)})\}(x_4^{(1)}+\bar{x}_4^{(1)}) \qquad - \ldots$$

$$-\frac{1}{4}j\omega y'(2x_3^{(2)}\bar{x}_4^{(1)}-x_4^{(2)}\bar{x}_3^{(1)}+2x_3^{(1)}x_{04}^{(2)})$$

$$+\frac{1}{8}\omega x'\{(-jx_3^{(1)})(x_4^{(1)})\}(x_4^{(1)}+\bar{x}_4^{(1)})$$

$$-\frac{1}{2}(x_2^{(2)}-yx_4^{(2)})\bar{\varphi}_{1nx}^{(1)}+\frac{1}{8}x\{(x_4^{(1)})(x_4^{(1)})(\varphi_{1nx}^{(1)})\}$$

$$-\frac{1}{2}(x_3^{(2)}+xx_4^{(2)})\bar{\varphi}_{1ny}^{(1)}+\frac{1}{8}x\{(x_4^{(1)})(x_4^{(1)})(\varphi_{1ny}^{(1)})\}$$

$$-\frac{1}{2}(\bar{x}_2^{(1)}-y\bar{x}_4^{(1)})\varphi_{2nx}^{(2)}-\frac{1}{2}(\bar{x}_3^{(1)}+x\bar{x}_4^{(1)})\varphi_{2ny}^{(2)}$$

$$-(x_2^{(1)}-yx_4^{(1)})\varphi_{0nx}^{(2)}-(x_3^{(1)}+xx_4^{(1)})\varphi_{0ny}^{(2)}$$

$$-\frac{1}{8}\{[(x_2^{(1)})(x_2^{(1)})-2y(x_2^{(1)})(x_4^{(1)})+$$

$$+y^2(x_4^{(1)})(x_4^{(1)})]\varphi_{1nxx}^{(1)}\}$$

$$-\frac{1}{8}\{[(x_3^{(1)})(x_3^{(1)})+2x(x_3^{(1)})(x_4^{(1)})+$$

$$+x^2(x_4^{(1)})(x_4^{(1)})]\varphi_{1nyy}^{(1)}\}$$

$$-\frac{1}{4}\{[(x_2^{(1)})(x_3^{(1)})+x(x_2^{(1)})(x_4^{(1)})-y(x_3^{(1)})(x_4^{(1)})$$

$$-xy(x_4^{(1)})(x_4^{(1)})]\varphi_{1nxy}^{(1)}\} \qquad (A.2)$$

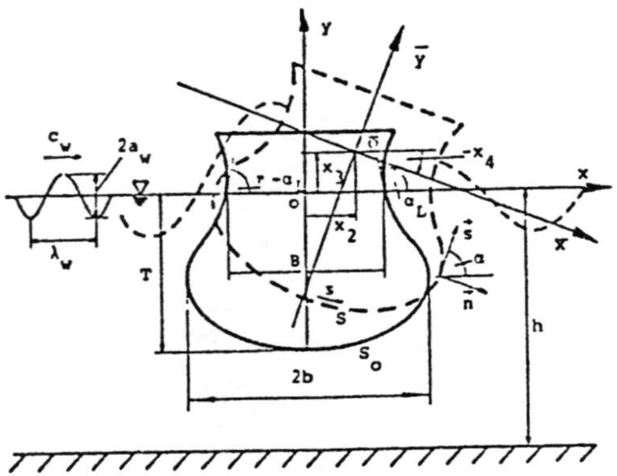


Fig.1 : Coordinate Systems

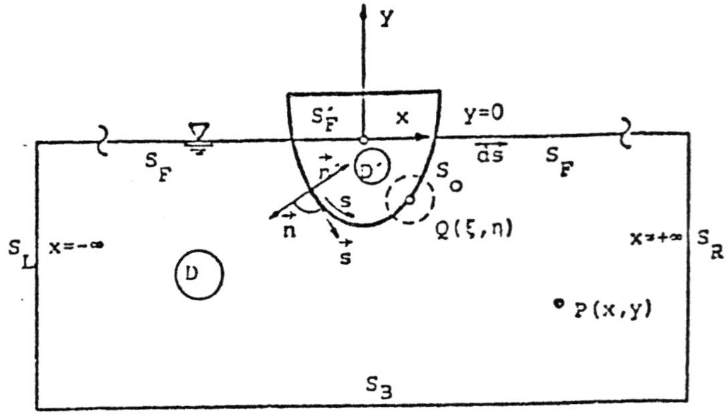


Fig.2 : Geometry of the BVP Problems

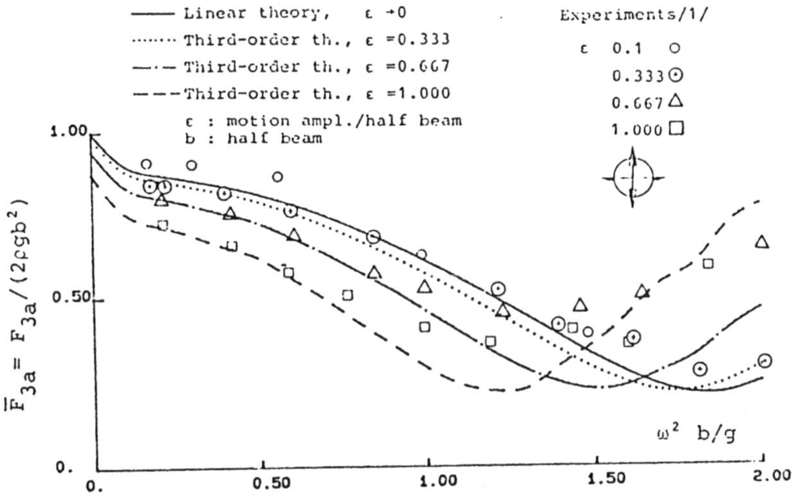


Fig. 3. Amplitude of the simple frequency force acting on a circular cylinder in forced heaving, numerical results by present theory, experiments by Tasai and Koterayama(1976)

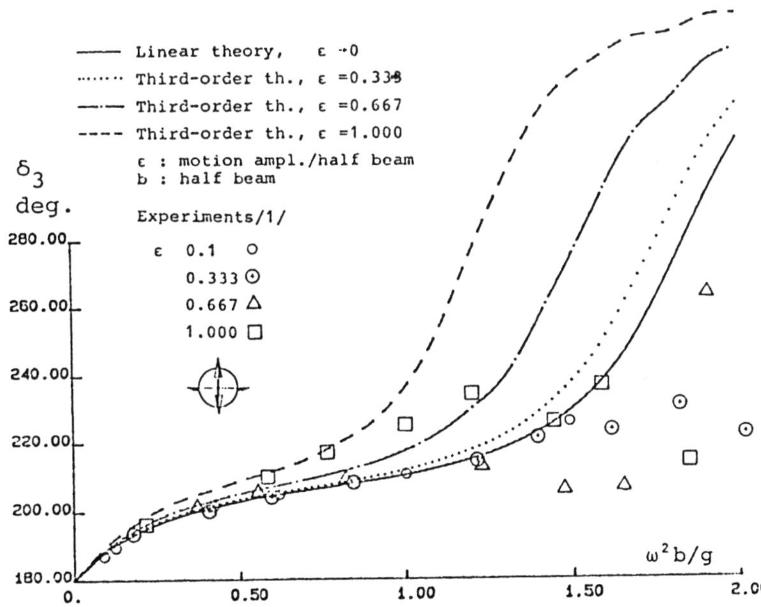


Fig. 4. Phase difference between heaving motion and simple frequency force acting on a circular cylinder in forced heaving, experiments by Tasai and Koterayama(1976)

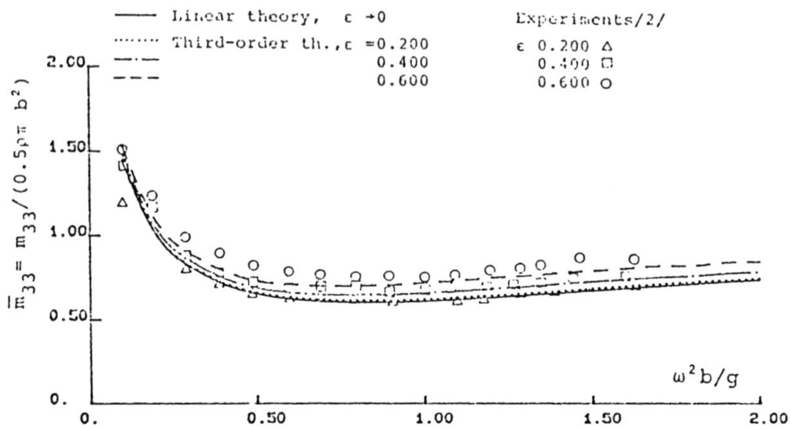


Fig. 5. Added-mass coefficient of circular cylinder in heaving motion after linear and present third-order theory, experiments by Yamashita(1977)

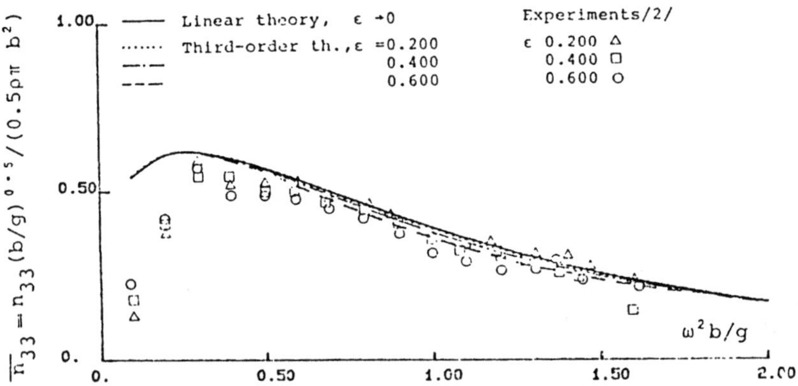


Fig. 6. Damping coefficient of circular cylinder in heaving motion after linear and present third-order theory, experiments by Yamashita(1977)

# Session 8

# Slow-drift Wave Force

# Second-order Wave Forces Acting on a Horizontal Circular Cylinder in Irregular Waves

Yusaku KYOZUKA

Research Institute for Applied Mechanics,
Kyushu University, Kasuga, Japan

Summary

A consistent theory of the second-order wave forces acting on a horizontal cylinder in irregular waves is presented. Second-order boundary conditions imposed on body and free surface are exactly taken into account. One emphasis in this study is placed on the treatment of the second-order free surface condition which oscillates with the difference of wave numbers of two regular waves and never decay even at infinity. Free surface integral method is applied for obtaining second-order forces, in which asymptotic solutions of the first- and second-order quantities are effectively used. The results of calculations are discussed in comparison with experiments.

## Introduction

It is well known that the complete second-order forces acting on a body in waves appear at the sum and different frequencies of the incident waves. The steady drift forces and slowly drifting forces are due to the low frequency part of them and they have been studied extensively by many researchers. Faltinsen and Løken[1] found that second-order forces due to the second-order potential would be small in two dimensional problem and that Pinkster's or Newman's approximation has a good reliability to estimate the slowly drifting oscillation of a body in waves. On the contrary, the high frequency part has been usually omitted in the problems.
Recently, Herfjord and Nielsen[2] studied on the high frequency part relating to fatigue problems of a cylindrical member, but the second-order force due to the second-order diffraction potentials is not taken into account. Lee[3] showed the complete solutions in heaving oscillation of a cylinder at bichromatic frequencies. It would be true that the second-order forces are very small compared with the first-order forces, but they become important when they concern with a resonant phenomenon in dynamic responses.
In this paper, theoretical calculations for the second-order forces acting on a circular cylinder in regular waves and bichromatic waves are compared with experiments. The second-order forces due to the second-order diffraction potentials are obtained by the free-surface integral which enable us to calculate the second-order forces without solving the boundary value problems.

K. Horikawa, H. Maruo (Eds.)
Nonlinear Water Waves
IUTAM Symposium Tokyo/Japan 1987

## Second-order Theory in Irregular Waves

Let us suppose wave forces are acting on a semi-submerged cylinder as shown in Fig.1. The fluid is assumed to be incompressible, irrotational and infinitely deep. Hence, there exists a velocity potential $\phi(x,y,t)$ satisfying Laplace's equation. We assume the velocity potential can be expanded in a perturbation series in terms of $\varepsilon$ :

$$\Phi(x, y, t) = \varepsilon\Phi^{(1)}(x, y, t) + \varepsilon^2\Phi^{(2)} + O(\varepsilon^3) \tag{1}$$

where $\varepsilon$ denotes a small perturbation parameter.
By using the regular perturbation theory, we obtain the first- and second-order free surface conditions as

$$\left.\begin{aligned}
&\varepsilon \text{ ----- } \Phi^{(1)}_{tt} - g\Phi^{(1)}_{y} = 0 && \text{on } y=0\\
&\varepsilon^2 \text{ ----- } \Phi^{(2)}_{tt} - g\Phi^{(2)}_{y} = Q^{(2)}(x, t) && \text{on } y=0\\
&\qquad = -2(\Phi^{(1)}_{x}\Phi^{(1)}_{xt} + \Phi^{(1)}_{y}\Phi^{(1)}_{yt}) + \Phi^{(1)}_{t}(\Phi^{(1)}_{yy} - \frac{1}{g}\Phi^{(1)}_{tty}) &&
\end{aligned}\right\} \tag{2}$$

We assume the velocity potential of irregular waves to be described as a sum of harmonic components, i.e.

$$\varepsilon\Phi^{(1)}_{0}(x, y, t) = Re\{\sum_{j=1}^{N} \frac{iga_j}{\omega_j} e^{-K_j y + i\theta_j}\} \tag{3}$$

where $\omega_j$ : circular frequency, $K_j = \omega_j^2/g$
$a_j$ : amplitude of wave , $\theta_j = K_j x + \omega_j t + \delta_j$
$g$ : gravity constant , $\delta_j$ : random phase

First-order potential is expressed by superposing the diffraction potential on that of the incident waves as

$$\varepsilon\Phi^{(1)}(x, y, t) = Re\{\varepsilon\sum_{j=1}^{N} \varphi_j(x, y)e^{i\omega_j t}\} \tag{4}$$

By substituting the eq.(4) into eq.(2), the second-order free surface condition in eq.(2) yields as

$$\Phi^{(2)}_{tt} - g\Phi^{(2)}_{y} = Re\{\sum_{j=1}^{N}\sum_{k=1}^{j}(Q^{-}_{jk}e^{i(\omega_j-\omega_k)t} + Q^{+}_{jk}e^{i(\omega_j+\omega_k)t}\} \tag{5}$$

where superscripts (+) and (-) denote the sum and difference of frequencies.
Referring to eq.(5), second-order potential can be written as

$$\Phi^{(2)}(x, y, t) = Re\{\sum_{j=1}^{N}\sum_{k=1}^{j}(\varphi^{-}_{jk}e^{i(\omega_j-\omega_k)t} + \varphi^{+}_{jk}e^{i(\omega_j+\omega_k)t})\} \tag{6}$$

From above considerations, we conclude that the second order problem in irregular waves could be obtained by the sum of the solutions such as $\varphi^{-}_{jk}$ and $\varphi^{+}_{jk}$. For simplicity, we describe the problem in bichromatic waves hereafter.

In bichromatic waves, the second-order potential is written as

$$\Phi^{(2)} = Re\{\varphi_0^{(2)} + \varphi_1^{(2)} e^{i2\omega_1 t} + \varphi_2^{(2)} e^{i2\omega_2 t} + \varphi_3^{(2)} e^{i(\omega_1-\omega_2)t} + \varphi_4^{(2)} e^{i(\omega_1+\omega_2)t}\} \tag{7}$$

where $\varphi_0^{(2)}$ is not concerned with the pressure or forces, $\varphi_1^{(2)}$ and $\varphi_2^{(2)}$ are the second-order potentials which appear in regular wave problems of the frequencies $\omega_j$ and $\omega_k$.

The second-order incident wave potential in bichromatic waves can be expressed as follows

$$\varepsilon^2\Phi_0^{(2)} = Re\{-i\omega_1 a_1 a_2 e^{-(K_1-K_2)y+i(\theta_1-\theta_2)}\} \qquad (\omega_1 > \omega_2) \tag{8}$$

It should be noted that the second-order potential of the incident waves appears only in the different frequency problem.

Next, the second-order pressure is also expressed by

$$P^{(2)}(x, y, t) = Re\{P_0^{(2)} + P_1^{(2)} e^{i2\omega_1 t} + P_2^{(2)} e^{i2\omega_2 t} + P_3^{(2)} e^{i(\omega_1-\omega_2)t} + P_4^{(2)} e^{i(\omega_1+\omega_2)t}\} \tag{9}$$

The second-order hydrodynamic force in direction k is formally written as

$$F_{jk}^{(2)} = -\int_{C+\Delta C} P_j^{(2)} n_k ds \tag{10}$$

where $n_k$ is the component of unit normal into fluid

$\Delta C$ denotes wetted-surface change in time

## Free Surface Integral and Numerical Results

Second-order forces given in eq.(10) are decomposed into three terms in the diffraction problem, that is, force due to quadratic term in Bernoulli equation, wetted-surface effect term and second-order potential term. First two terms are calculated by the first-order potential and its derivatives. Third term is the force due to the second-order potential, but it can be calculated without solving the boundary value problem. This method was derived by Söding[4], and Molin[5], and is applied by Kyozuka[6,7] in two-dimensional diffraction problems. Recently, Eatock Taylor and Hung[8] present detailed numerical results by this method for a vertical cylinder in regular waves.

Let us denote the second-order force due to the second-order potential $\Phi^{(2)}(F)$, attributed to the second-order free surface condition as

$$f_{jk}^{(2)}(F) = \frac{F_{jk}^{(2)}(F)}{\rho g a_1 a_2} = -\int_C \Phi_j^{(2)}(F)\ n_k\ ds \tag{11}$$

and we introduce a new potential $\phi_k^R$, which satisfies Laplace equation, homogeneous free-surface condition, radiation condition and

$$\frac{\partial}{\partial n}\phi_k^R = n_k \qquad on\ C \tag{12}$$

Substituting eq.(12) into (11) and applying Green's theorem for $\phi_j^{(2)}(F)$ and $\phi_k^R$, we obtain

$$f_{jk}^{(2)}(F) = -\int_C \phi_j^{(2)}\frac{\partial}{\partial n}\phi_k^R\, ds = \int_F q_j(x)\,\phi_k^R\, dx \tag{13}$$

where

$$\left.\begin{aligned} q_3(x) &= q_{3_c}(x)+iq_{3_s}(x) = \frac{i(\omega_1-\omega_2)}{g^2 a_1 a_2}Q_3(x) \\ q_4(x) &= q_{4_c}(x)+iq_{4_s}(x) = \frac{i(\omega_1+\omega_2)}{g^2 a_1 a_2}Q_4(x) \end{aligned}\right\} \tag{14}$$

Therefore, this force could be calculated by integrating the product of pressure distribution $q_j(x)$ and the radiation potential $\phi_k^R$ on free-surface. Asymptotic solutions for potential and pressure distribution are very useful to carry out this integral, and they are written by using Kochin functions as follows:

$$\phi(P) \to e^{-Ky+iKx} + iH^{\pm}(K)e^{-Ky\mp iKx}, \quad \text{as } x \to \pm\infty$$

$$\left.\begin{aligned} q_3(x) &\to K_{12}^{-}\{e^{i(K_1-K_2)x} + H^{+}(K_1)H^{+*}(K_2)e^{-i(K_1-K_2)x}\}, && \text{as } x \to +\infty \\ &\to K_{12}^{-}\{1 + iH^{-}(K_1)\}\{1 - iH^{-*}(K_2)\}e^{i(K_1-K_2)x}, && \text{as } x \to -\infty \\ q_4(x) &\to -iK_{12}^{+}\{H^{+}(K_1)e^{-i(K_1-K_2)x} + H^{+}(K_2)e^{i(K_1-K_2)x}\}, && \text{as } x \to +\infty \\ &\to 0, && \text{as } x \to -\infty \end{aligned}\right\} \tag{15}$$

where

$$K_{12}^{\pm} = 2\sqrt{K_1K_2}(K_1 + K_2 + \sqrt{K_1K_2})$$

$$H^{\pm}(K) = -\int_C \phi(Q)\frac{\partial}{\partial n_Q}e^{-Ky\pm iKx}\, ds_Q$$

Figures 2 and 3 show the distribution of $q_3(x)$ and $q_4(x)$ in bichromatic waves of $K_1b=1.0$ and $K_2b=0.75$, where b denotes radius of cylinder. Horizontal axis denotes the distance from the origin normalized by the radius of the cylinder. In Fig.2, both of $q_3(x)$ and $q_3(-x)$ oscillate with $e^{i(K_1-K_2)x}$ and they do not decay even at infinity. In the same figure, asymptotic values given by eq.(15) are also plotted and they are found to be good approximations where x takes a larger value than 6 or 7.

In Fig.3, the vertical scale for $q_4(x)$ is a hundred times larger than that for $q_3(x)$. $q_4(x)$ for negative x tends to zero at some distance away from the cylinder, but, $q_4(x)$ for positive x oscillates with the difference of two wave numbers of bichromatic waves in the same manner as $q_3(x)$. From these results, the second-order free-surface condition for the sum frequency problem would be more severe than that for the different frequency problem. In Figs.4(a) and 4(b), total second-order force of the different frequencies in horizontal direction is shown in a map of equi-contours, where the horizontal axis denotes $K_1b$ and the vertical axis $K_2b$. So that, the diagonal line denotes regular waves of the amplitude of $(a_1+a_2)$, and the steady drifting force is obtained along the line. In Fig.4(a), we have simple symmetry relations for the second-order forces, so that, the real and imaginary part of the force is given for a half combinations of $K_1b$ and $K_2b$. Fig.4(b) shows contours of the absolute values. In Fig.5, a sectional plan for the constant value of $K_1b=1.0$ is shown, which is taken from Fig.4(a). Three curves denote the total force, the force due to the second-order free-surface condition and the force due to the second-order incident wave potential. From these results, the second-order force due to the free-surface condition would be small and negligible for irregular waves of a narrow band width.

In the same manner, the contour maps and sectional plans are presented in Figs.6(a),(b) and 7 for the vertical force of different frequencies, Figs.8(a),(b) and 9 for the horizontal force of the sum frequencies, and Figs.10(a),(b) and 11 for the vertical force of the sum frequencies.

In the sum frequency problems, the incident wave potential takes no contribution to the second-order forces but, the forces due to the free-surface condition are very significant as shown in Figs.9 and 11.

## Experiments and Discussion

Experiments were carried out in narrow wave tank(LxBxd=60mx1.5mx1.2m) of Tsuyazaki Laboratory of our Institute as shown in Fig.12. The radius and length of a circular cylinder are 0.13 m and 0.5 m respectively.

In Figs. 13 and 14, second-order wave forces of double frequency in regular waves are shown, where dashed lines are taken from Miao and Liu[9] and + marks are experiments taken from Kyozuka[6].

Miao and Liu solve the second-order boundary value problems in which inhomogeneous free-surface condition is considered by applying Delta function. Although they claim that the free-surface integral method is inconsistent, their last results with respect to the radiation condition for the second-order potential is same as the one used in the free-surface integral method. Therefore, both calculations should coincide with each other, but some small discrepancies are observed in horizontal force(Fig.13). + marks are experiments carried out in 9 m wave tank, but they are found to be in good agreement with present experiments. Figure 15 shows one of measured records in bichromatic waves and Fig.16 shows computed time series by the first- and second-order theories assuming the same incident waves as Fig.15. By comparing those time series, we could confirm the reliability of our theory. However, if we compare them in Fourier coefficients as shown in Figs.17 and 18, some discrepancies are observed in high frequency responses in horizontal force and low frequency response in vertical force. In those figures, approximate second-order results obtained by ignoring the second-order potentials are also shown and we would conclude that they could be applied to low frequency responses but fail in high frequency responses.

## Conclusions

Throughout this study, results obtained are summarized as follows;

(1) The second-order forces in irregular waves are generally classified into two parts, i.e., low frequency force and high frequency force. The second-order free-surface condition in both problems oscillates with the difference of wave-numbers of the component regular waves.

(2) The second-order forces could be effectively obtained by the free-surface integral method by making use of asymptotic solutions of the first- and second-order quantities.

(3) The second-order forces due to second-order potential are very small in low frequency part, but they are significant in high frequency part.

(4) Numerical results on a semi-submerged circular cylinder in regular and bichromatic waves show good agreements with experiments.

## References


1. Faltinsen, O.M.; Løken, A.E.: Slow drift oscillations of a ship in irregular waves, Appl. Ocean Res., Vol. 1, No. 1, (1979) 3-13
2. Herfjord, K.; Nielsen F.G.: Nonlinear wave forces on a fixed vertical cylinder due to the sum frequency of waves in irregular seas, Appl. Ocean Res., Vol. 8, No.1, (1986) 8-21
3. Lee, C.M.: The second-order theory for nonsinusoidal oscillations of a cylinder in a free surface, 8th Symp. Naval Hydrodyn., (1970), 905-950
4. Söding, H.: Second-order forces on oscillating cylinders in waves, Schiffstechnik, Vol.23, (1976) 205-209
5. Molin, B.: Second-order diffraction loads upon three-dimensional bodies, Appl. Ocean Res., Vol.1, No.4, (1979) 197-202
6. Kyozuka,Y.: Non-linear hydrodynamic forces acting on two-dimensional

bodies (first report), J. Soc. Naval Arch. of Japan, Vol.148,(1980) 49-57

7. Kyozuka, Y.: Experimental study on second-order forces acting on a cylindrical body in waves, 14th Symp. Naval Hydrodyn. (1982) 319-379

8. Eatock Taylor,R.; Hung, S.M.: Second-order diffraction forces on a vertical cylinder in irregular waves, Appl. Ocean Res., Vol.9, No.1, (1987) 19-30

9. Miao, G.P.;Liu Y.Z.: A theoretical study on the second-order wave forces for two-dimensional bodies, 5th Int. OMAE Symp., Vol.1, (1986) 330-336


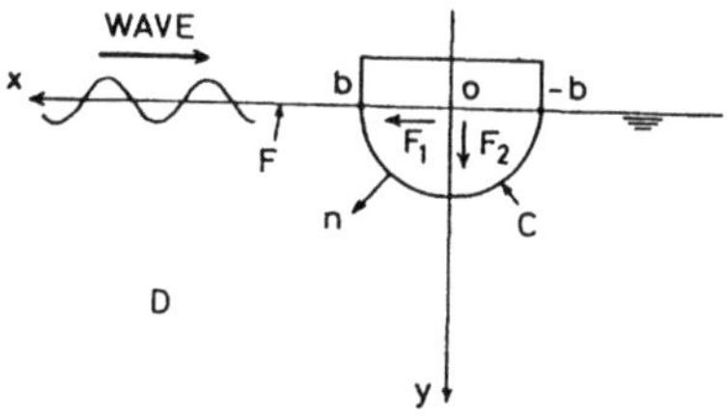


Fig.1 Coordinate system

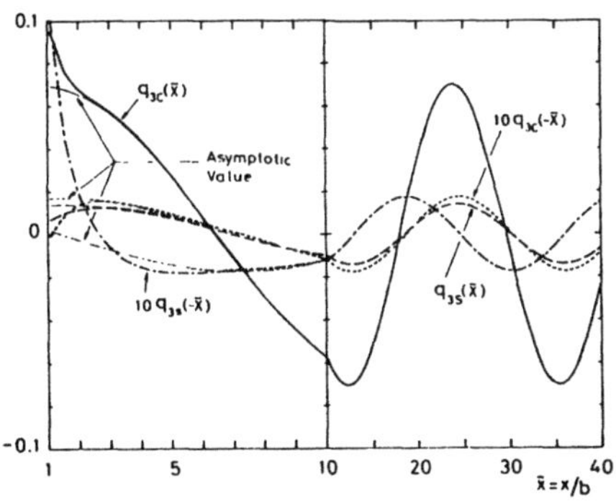


Fig.2 Second-order free-surface condition for $\phi_3^{(2)}$ in bichromatic waves, $K_1b=1.0$, $K_2b=0.75$

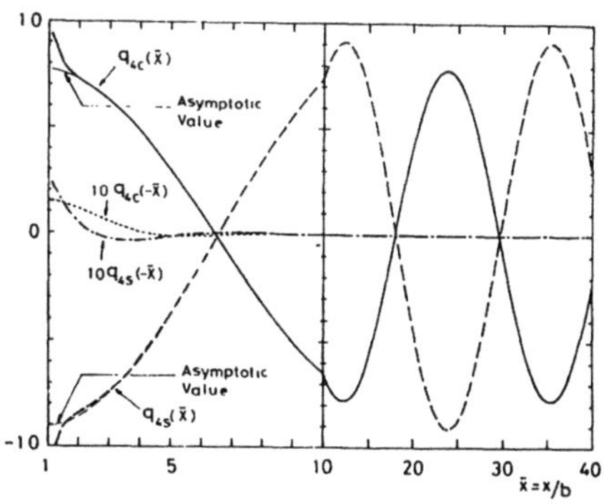


Fig.3 Second-order free-surface condition for $\phi_4^{(2)}$ in bichromatic waves, $K_1b=1.0$, $K_2b=0.75$

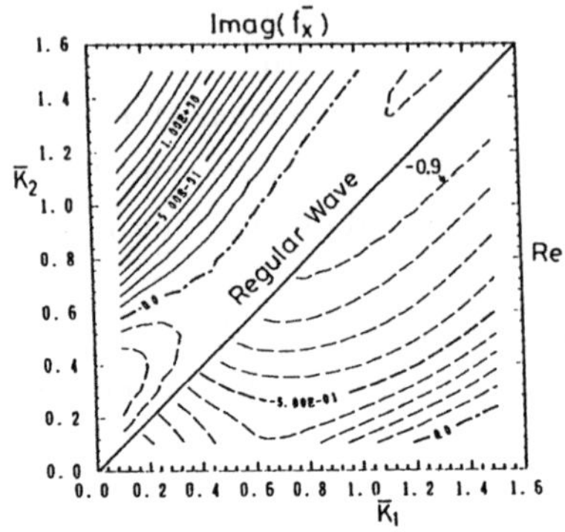


Fig.4(a) Second-order horizontal force of the different frequency in bichromatic waves, the real and imaginary part

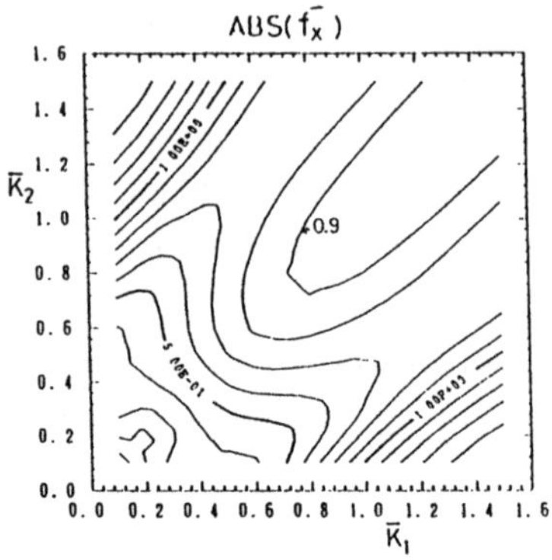


Fig.4(b) Second-order horizontal force of the different frequency in bichromatic waves, the amplitude

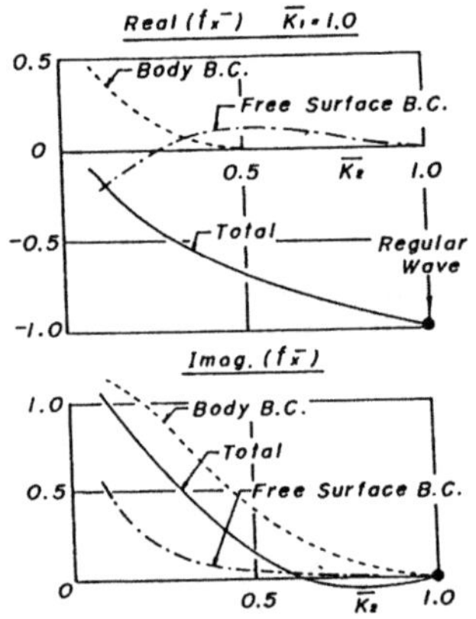


Fig.5 Second-order horizontal force of the different frequency in bichromatic waves, $K_1b=1.0$ and $K_2b$ is varied

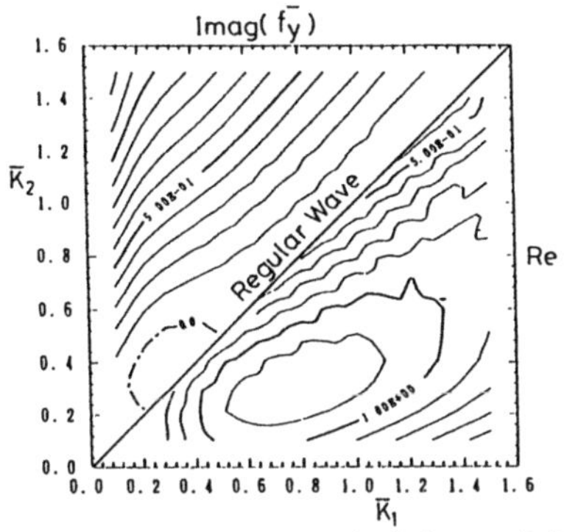


Fig.6(a) Second-order vertical force of the different frequency in bichromatic waves, the real and imaginary part

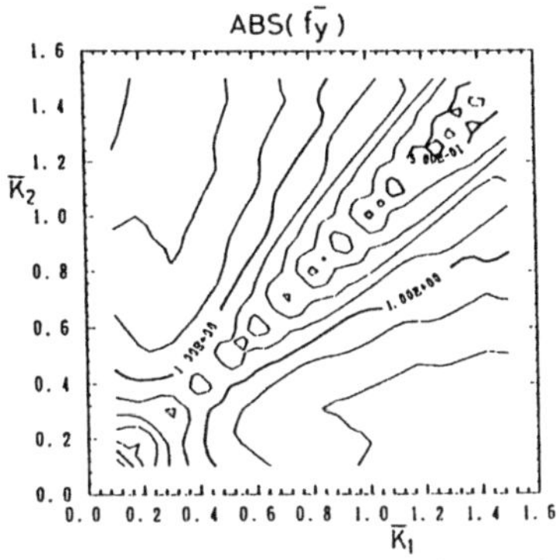


Fig.6(b) Second-order vertical force of the different frequency in bichromatic waves, the amplitude

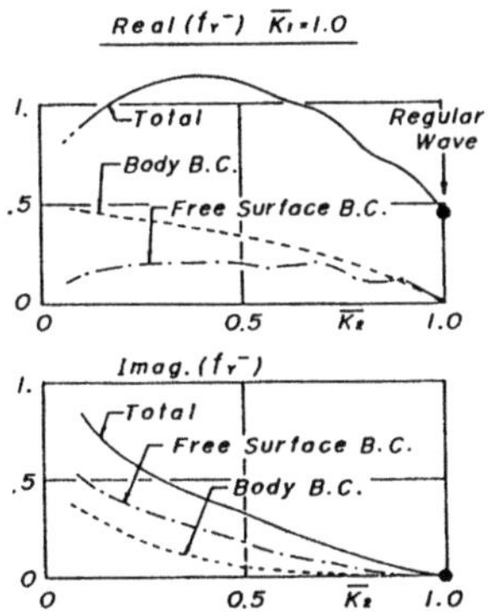


Fig.7 Second-order vertical force of the different frequency in bichromatic waves, $K_1b=1.0$ and $K_2$ is varied

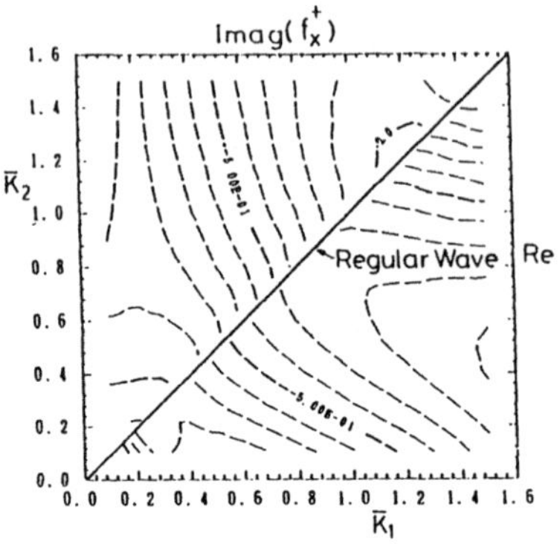


Fig.8(a) Second-order horizontal force of the sum frequency in bichromatic waves, the real and imaginary part

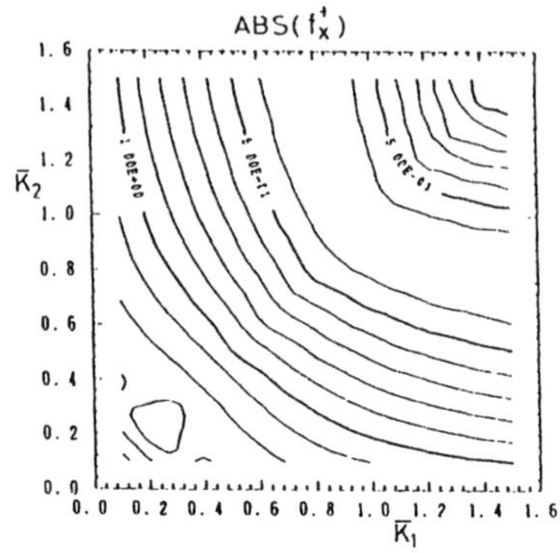


Fig.8(b) Second-order horizontal force of the sum frequency in bichromatic waves, the amplitude

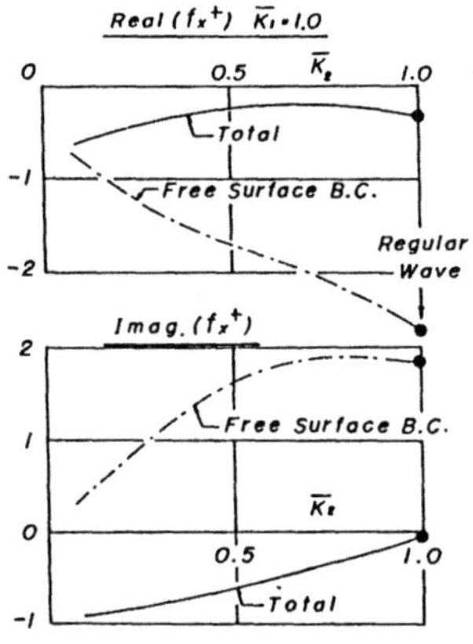


Fig.9 Second-order horizontal force of the sum frequency in bichromatic waves, $K_1b$=1.0 and $K_2b$ is varied

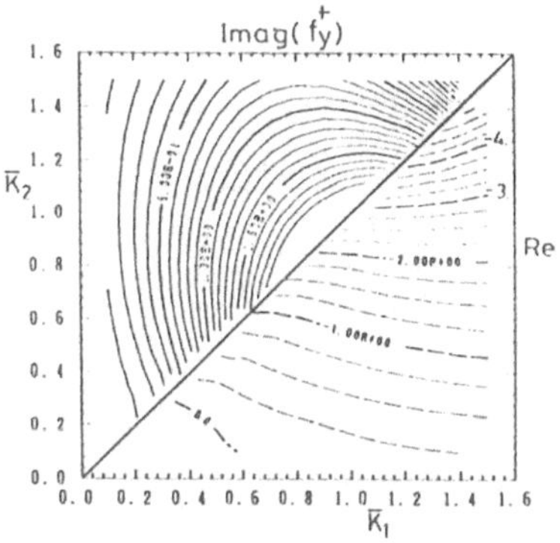


Fig.10(a) Second-order vertical force of the sum frequency in bichromatic waves, the real and imaginary part

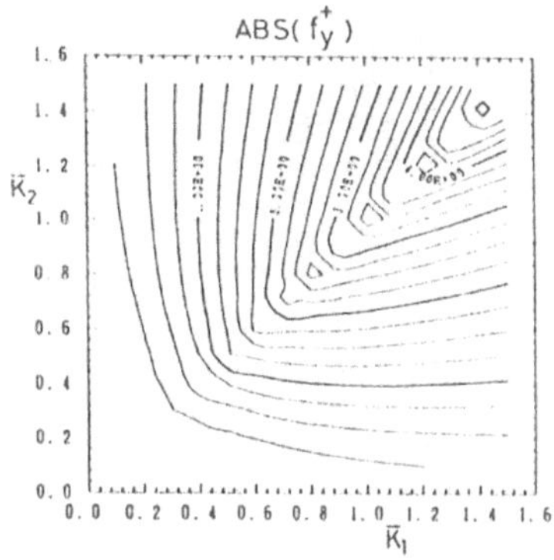


Fig.10(b) Second-order vertical force of the sum frequency in bichromatic waves, the amplitude

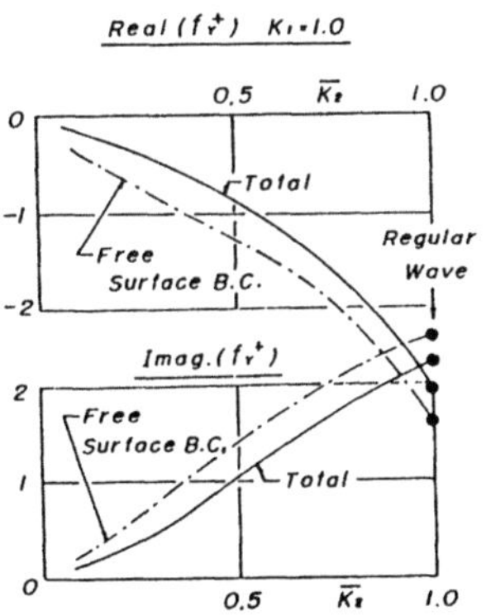


Fig.11 Second-order vertical force of the sum frequency in bichromatic waves, $K_1b$=1.0 and $K_2b$ is varied

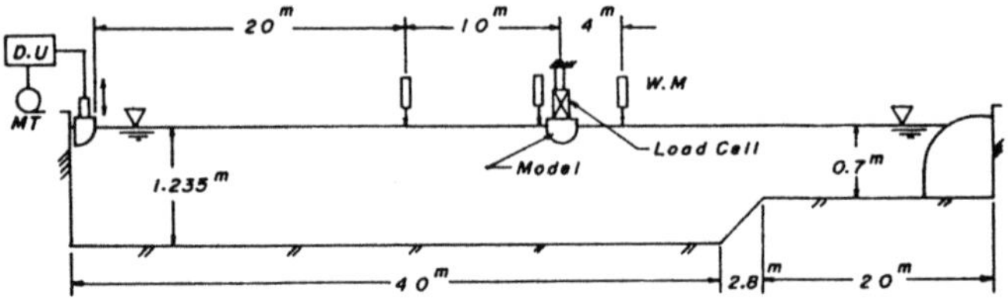


Fig.12 Experimental setup

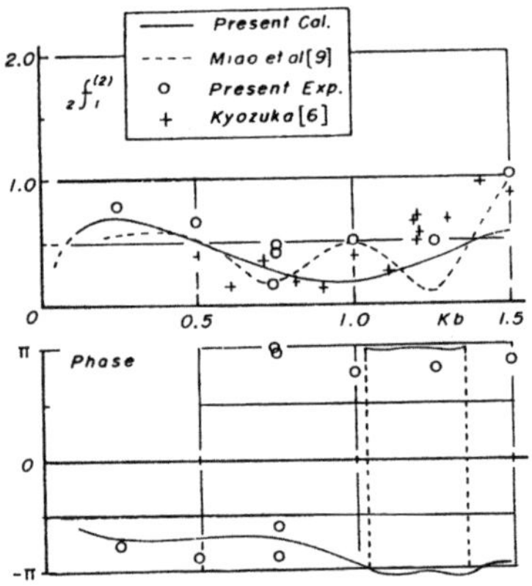


Fig.13 Second-order horizontal force in regular waves

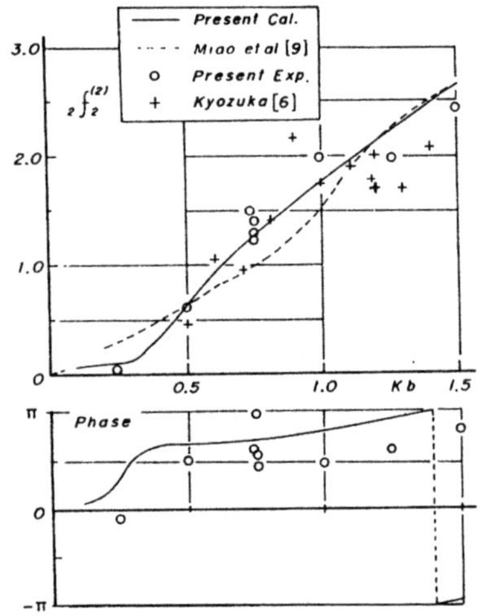


Fig.14 Second-order vertical force in regular waves

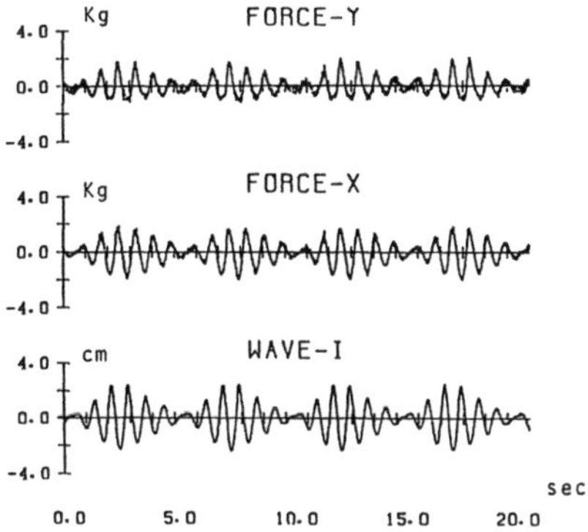


Fig.15 Measured time series in bichromatic waves, $T_1$=0.716(s), $a_1$=1.0(cm), $T_2$=0.835(s), $a_2$=1.04(cm)

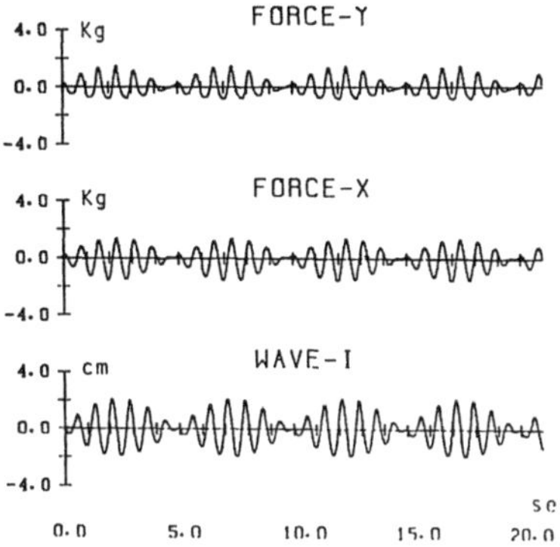


Fig.16 Calculated time series, incident waves are assumed to be same as Fig.15

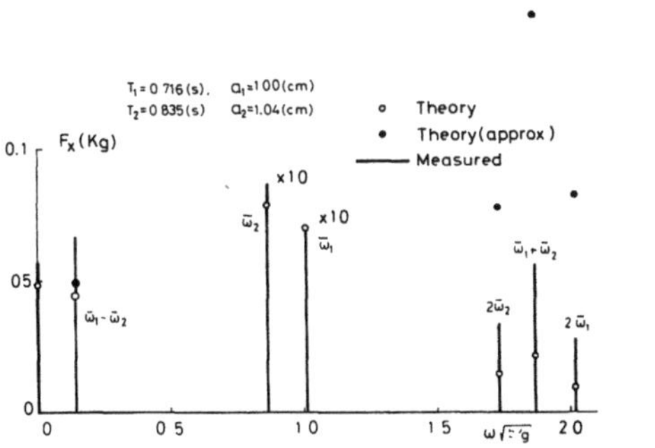


Fig.17 Fourier coefficients of horizontal force in bichromatic waves

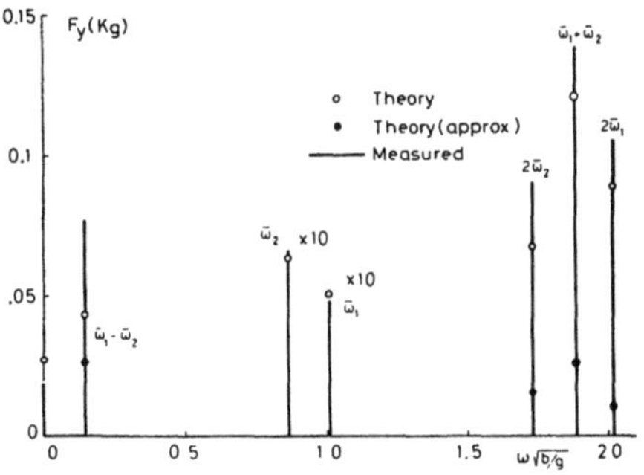


Fig.18 Fourier coefficients of vertical force in bichromatic waves

# Slow Drift and Trapping of Waves on Submerged Bodies

J.A.P.ARANHA
C.P.PESCE
I.P.T.; S.Paulo, Brazil

SUMMARY

Low frequency nonlinear wave forces on ocean structures are usually divided into two distinct components. One, $F_1(t)$, associated with quadratic interaction of the first order (linear) potential; the other, $F_2(t)$, associated with the second order potential. The purpose of the present paper is to indicate that this second parcel must always be computed for a submerged body. On one hand, when the depth of submergence is large, $F_1(t)/F_2(t) \sim O(1)$ and the effect of $F_2(t)$ cannot be neglected; on the other hand, when the depth of submergence is small, the ratio $F_1(t)/F_2(t)$ increases but the likelihood of trapped mode excitation also does. If this happens the second order problem is resonant and then $F_2(t)$ is, by far, the dominating parcel. The analysis in this work is restricted to two dimensional infinite water depth case and only diffraction problem is considered (body fixed in waves).

## 1. EXCITING FORCES IN BEAM SEA

Consider first a regular wave train and let $F_1 = \rho g A B \delta . Q_1(\omega)$ be the steady force in the mode being analysed, where:- $\rho$ = density; $g$ = acceleration of gravity; $A$ = wave amplitude; $B$ = beam; $\delta = A/B$ = small amplitude parameter; $Q_1(\omega)$ = steady force coefficient for a harmonic wave with frequency $\omega$. If $K_o = \omega^2/g$ is the wave number, the wave steepness is given by $K_oA = K_oB.\ \delta \ll 1$ when $K_oB \leq O(1)$. In sway mode a simple expression for $Q_1(\omega)$ has been obtained by Maruo (1960) and it is given by,

$$Q_1(\omega) = \frac{1}{2} |R|^2 (1+O(\delta)),$$

where $R$ = reflection coefficient. This regular wave train excites, at second order, the steady potential $\phi_{2o}(y,z)$ where:

K. Horikawa, H. Maruo (Eds.)
Nonlinear Water Waves
IUTAM Symposium Tokyo/Japan 1987

$$\nabla^2\phi_{2o} = 0,$$
$$\partial\phi_{2o}/\partial z = L_{2o}(y) \text{ on the free surface } z = 0,$$
$$\partial\phi_{2o}/\partial n = 0 \qquad \text{on the body surface } \partial B, \qquad (1\text{-}2)$$
$$\nabla\phi_{2o} \to 0 \quad \text{when } r = \sqrt{y^2 + z^2} \to \infty.$$

In the above expression the exciting term $L_{2o}(y)$ is given by,

$$L_{2o}(y) = \frac{\omega}{4} \cdot \frac{\partial}{\partial y}\left(i.\ \phi_L(y,0) . \frac{\partial \phi_L^*}{\partial y}(y,0)\right) + (*), \qquad (1\text{-}3)$$

$\phi_L(y,z;\omega)$ being the linear diffraction potential at frequency $\omega$. If $U_K$ is the mode boundary condition ($U_K = n_y$ for sway, etc.) the following coefficient will be used later on:

$$Q_2(\omega) = -\omega \int_{\partial B} \phi_{2o} .\ U_K .\ d\partial B. \qquad (1\text{-}4)$$

In reality the potential $\phi_{2o}$ need not be determined to obtain $Q_2(\omega)$. In an actual computation the radiation potentials $\phi_K^{(o)}(y,z)$, at zero frequency, must be determined to compute the added mass. They satisfy the homogeneous boundary condition at the free surface and the mode boundary condition, $\partial\phi_K^{(o)}/\partial n = U_K$, at the body surface. If now,

$$G(\phi;\Psi) = \iint_A \nabla\phi . \nabla\Psi . dA; \qquad A = \text{fluid region},$$
$$V_K(\Psi) = \int_{\partial B} \Psi . U_K .\ d\partial B, \qquad (1\text{-}5)$$
$$V_{2o}(\psi) = \int_{-\infty}^{\infty} L_{2o}(y) . \psi(y,0) . dy,$$

then,

$$G(\phi_K^{(o)};\Psi) = V_K(\Psi), \qquad (1\text{-}6)$$
$$G(\phi_{2o};\Psi) = V_{2o}(\Psi).$$

Since $Q_2(\omega) = -\omega . V_K(\phi_{2o})$ from (1-6) it follows that,

$$Q_2(\omega) = -\omega .\ V_{2o}(\phi_K^{(o)}). \qquad (1\text{-}7)$$

This latter expression is just Haskind's relation applied to the present problem and it shows that the knowledge of $\phi_K^{(o)}(y,z)$ is enough to compute $Q_{20}(\omega)$.

Consider now an irregular incident wavetrain consisting of a slow modulation of the harmonic wave at the frequency $\omega$. The spatial and temporal modulation is described by a function $a(y,z,t) \sim 0(1)$ with $(\partial/\partial_y;\ \partial/\partial_z;\ \partial/\partial_t)\ a \sim 0\ (\Delta\omega) << 1$. The diffraction potential, correct to second order in the small amplitude parameter $\delta$, is given by $\phi(y,z) = \delta.\phi_1(y,z)+$ $+\ \delta^2.\ \phi_2(y,z)$. where, obviously, $\phi_2(y,z)$ has energy in two distinct range of frequencies:- one around the frequency $2\omega$ and another around the zero frequency. Only this latter parcel will be considered in the following. From Bernouilli's equation the low frequency exciting force is given by $F(t) =$ $= F_1(t) + F_2(t)$ where

$$F_1(t) = -0.5.\ \rho.\ \int_{\partial B} (\nabla\phi_1)^2.\ U_K.d\partial B,$$

$$F_2(t) = -\rho.\frac{d}{dt} \int_{\partial B} \phi_2.\ U_K\ .d\partial B. \tag{1-8}$$

Due to the slow modulation, $\phi_1 \sim a.\phi_L(1+0(\Delta\omega))$, where $\phi_L$ is the diffraction potential at frequency $\omega$. Placing this expression into (1-8) one obtains Newman's (1974) approximation,

$$F_1(t) = \rho gAB\delta\ .Q_1(\omega).\ |a(t)|^2\ .(1+0(\Delta\omega)), \tag{1-9}$$

where $Q_1(\omega)$ is the steady force coefficient for harmonic waves. With the same relative error the free surface exciting term for $\phi_2$ is given by $|a|^2.\ L_{20}(y)$, what suggests to write $\phi_2 = |a|^2\ .\phi_{20}.$ $(1+0(\Delta\omega))$. After a detailed analysis of the far field radiation condition Aranha & Pesce (1986) have shown that this approximation is indeed consistent. Using it into (1-8) one obtains,

$$F_2(t) = \rho gAB\delta\ .Q_2(\omega).\ \frac{1}{\omega}\ \frac{d}{dt}(|a|^2).\ (1+(\Delta\omega)), \tag{1-10}$$

with $Q_2(\omega)$ defined in (1-4). Since the derivative of $|a(t)|^2$ is of order $\Delta\omega$ the ratio between $F_1(t)$ and $F_2(t)$ can be gauged by,

$$r_1(\omega) = \frac{Q_1(\omega)}{\Delta\omega . Q_2(\omega)} = \frac{|R|^2}{2\Delta\omega . Q_2(\omega)} , \qquad (1\text{-}11)$$

where the expression on the right is valid only in sway, see (1-1). For a surface-piercing body in deep water, $|R|$ and $Q_2(\omega)$ are both of order 1 and $r_1(\omega) \sim 0(1/\Delta\omega) \gg 1$ or in short:- the effect of $F_2(t)$ can be neglected in comparison to $F_1(t)$, a conclusion consistent with the one numerically obtained by Faltinsen & Løckens. (1980). For a circular submerged body $|R| \equiv 0$, see Ogilvie (1963), and in general $|R| \ll 1$ when the depth of submergence is relatively large. In this case $r_1(\omega)$ can be of order 1 and the effect of the second order potential cannot be neglected. When the body approaches the free surface, $|R|$ increases in general and so $r_1(\omega)$ also does. Simultaneously, however, increases the likelihood of a resonant phenomenon to be explained next.

## 2. TRAPPED MODE EXCITATION

A penetrating physical discussion about trapped waves over submerged bodies in shallow water can be found in Longuet-Higgins (1967), and Aranha (1987)a extended some of these results to the arbitrary water depth case. Mathematically a trapped wave is a function of the form,

$$\hat{T}(x,y,z,t) = T(y,z) . e^{i(K_T x - \omega t)}$$

$$T(y,z) \sim A^{\pm} . e^{-\lambda_o |y|} . e^{K_o z} \quad ; K_o = \omega^2/g, \qquad (2\text{-}1)$$

$$\lambda_o = \sqrt{K_T^2 - K_o^2},$$

that satisfies the linear homogeneous water wave equation. The mode $\{K_T; T(y,z)\}$ can be determined from the following simple eigenvalue problem:- to find a non-trivial $T(y,z)$ for which the Lagrangian,

$$\mathcal{L}(T) = \frac{1}{2} \iint_A \left[ (\nabla T)^2 + K_T^2 T^2 \right] dA - \frac{\omega^2}{2} \int_{-\infty}^{\infty} T^2(y,0)\, dy, \qquad (2\text{-}2)$$

is stationary. Since $K_T > K_o$ these modes can be excited only by non-linear interaction of two incoming waves but, if this happens, the second order potential $\phi_2$ will be of order $\delta^{2/3}$ ,see Aranha (1987)b, and so $Q_2(\omega) \sim 0(\delta^{-4/3}) >> 1$. As it is usual in non-linear resonant phenomena a nonlinear cubic wave equation can be obtained by multiple scales, although the pertinent equation is non-dispersive in the present case, see Aranha (1987)b for details. If $\bar{\sigma}$ is the normalized detuning it can be shown that $\bar{\sigma} \sim 0(\delta^{4/3}/(\lambda_o/K_o))$ or in short:- the more diffuse is the mode (the smaller is $\lambda_o/K_o$) the larger will be $\bar{\sigma}$ and the more unlikely the mode will be excited. Since $\lambda_o/K_o$ increases when the depth of submergence S decreases (see figure (5-1)) then the likelihood of excitation also increases when the body approaches the free-surface.

## 3. VARIATIONAL APPROXIMATION FOR $Q_1(\omega)$; $Q_2(\omega)$; $\lambda_o/K_o$

The present analysis depends on three coefficients:- $Q_1(\omega)$; $Q_2(\omega)$ and $\lambda_o/K_o$, where this latter one gauges the likelihood of trapped mode excitation. It is important to be observed that all of them can be expressed as stationary values of well defined functionals and so an order $\varepsilon$ error in the potential leads to an order $\varepsilon^2$ error in these coefficients. In fact, $\lambda_o/K_o$ can be obtained by standard Rayleigh quotient, see (2-2), and Aranha & Pesce (1987) have shown how to express $Q_1(\omega)$, in its form (1-1), in a similar way. A short demonstration that this is also valid for $Q_2(\omega)$ will be given next. Let $\phi = \{\phi_{2o};\phi_K^{(o)}\}$ be the solution of the weak equation $G(\phi,\Psi) = V(\Psi)$, see (1-6), and $\bar{\phi}$ its projection into the finite dimensional space $\bar{W}$. Then $\phi = \bar{\phi} + \Delta\phi$ where $G(\Delta\phi,\bar{\Psi}) = 0$ for all $\bar{\Psi} \in \bar{W}$. The related approximation for $Q_2(\omega)$ is given by (see (1-7), (1-6))

$$\bar{Q}_2(\omega) = -\omega\, V_{20}(\bar{\phi}_K^{(o)}) = -\omega.G(\bar{\phi}_K^{(o)};\, \bar{\phi}_{20}) = -\omega.G(\phi_K^{(o)};\phi_{20}) - \omega.G(\Delta\phi_K^{(o)};\Delta\phi_{20}) =$$

$$= -\omega.\, V_{20}(\phi_K^{(o)}) - \omega.G(\Delta\phi_K^{(o)};\, \Delta\phi_{20}) = Q_2(\omega) - \omega.G(\Delta\phi_K^{(o)};\, \Delta\phi_{20}).$$

From this latter expression and Schwarz inequality it follows that the error in $Q_2(\omega)$ is smaller than $\varepsilon_{20}\,.\varepsilon_K^{(o)}$ where $\varepsilon = \{\varepsilon_{20};\ \varepsilon_K^{(o)}\} = (G(\Delta\phi;\Delta\phi))^{1/2}$ is the error in the potential.

## 4. NUMERICAL RESULTS

A variational approximation will be designated by the size (NxN) of the real and symmetric matrix that should be constructed and inverted in this formulation. Figure (5-1)b compares the phase of the transmission coefficient for a circular cylinder obtained from a (4x4) variational approximation with the ones obtained by Ogilvie (1963); Figure (5-1)d compares the values of $Q_1(\omega)$ in sway mode (rectangular section) obtained from a (4x4) variational approximation with the ones obtained by Inoue & Kyozuka (1985), where in this last work a (30x30) complex non-symmetric matrix was constructed and inverted. Both results agree quite well and this gives an indication of the power of the present variational approximation.

To compute $Q_2(\omega)$ a (1x1) variational approximation was used to represent $\phi_K^{(o)}$ in sway mode and the integral $V_{20}\ (\phi_K^{(o)})$ has been computed only above the body. This coefficient is not supposed then to be determined with great precision although its order of magnitude - and this is enough in the present analysis - is thought to be correct. Since it can be shown that $T(y,z) \to e^{K_o z}$ when $S \to \infty$ this trial function has been used for a circle and compared with the values obtained by McIver & Evans (1985); see Figure (5-1)e to be noticed the drastic improvement when S changes from 1,1 to 1,5 and that, from minimum principle, the variational approximations are always lower bounds here. For the rectangle the same (1x1) approximation was used and compared with a (2x2) approximation, where the second trial function immitates shallow water trapped modes in the region above the rectangle. Again the discrepancy between both increases drastically when S decreases, see Figure (5-1)f.

As a whole two things are to be noticed:- $Q_2(\omega)$ is the dominating term when S is "large", see Figure (5-1)d where $\Delta\omega = 0.20$ has been used to compute $\bar{Q}_2(\omega) = \Delta\omega . Q_2(\omega)$; $\lambda_o/K_o$ increases when the body approaches the free-surface. So the effect of the second order potential should always be computed for a submerged body. Furthermore, variational approximation seems to be a powerfull tool to determine macroscopic quantities, like the exciting forces on a body. As it was shown here this approach can also be used to compute nonlinear exciting forces and this can affect significantly the required ammount of computing time. In the present case the computation of $Q_1(\omega)$, $Q_2(\omega)$ and $\lambda_o/K_o(\omega)$, in the range of frequencies shown in Figure (5-1), did not take more than 15 minutes in a IBM personal micro-computer.

REFERENCES


1. FALTINSEN, O.M. & LØCKENS, A.F. (1980):-"Slow drift oscillations of a ship in irregular waves", Modeling Identification and Control, Vol. 1, nº 4, pp. 195-213.

2. NEWMAN, J.N. (1974):- "Second-order, slowly varying forces in vessels in irregular waves", Proc. Int. Symp. on Dyna. mics of Marine Vehicles, London, 1974.

3. MARUO, H. (1960):- "The drift of a body floating in waves", Journal of Ship Research, Vol. 4, nº 3, pp. 1-10.

4. ARANHA, J.A.P. & PESCE, C.P. (1980):- "Effect of the second order potential in the slow - drift oscillation of a floating structure in irregular waves", Journal of Ship Research, vol. 30, nº 2, pp. 103-122.

5. NEWMAN, J.N. (1977):- "Marine Hydrodynamics", M.I.T. Press

6. OGILVIE,T.F. (1963):- "First-and second-order forces on a cylinder submerged under a free surface", Jornal of Fluid Mechanics, pp. 451-472.

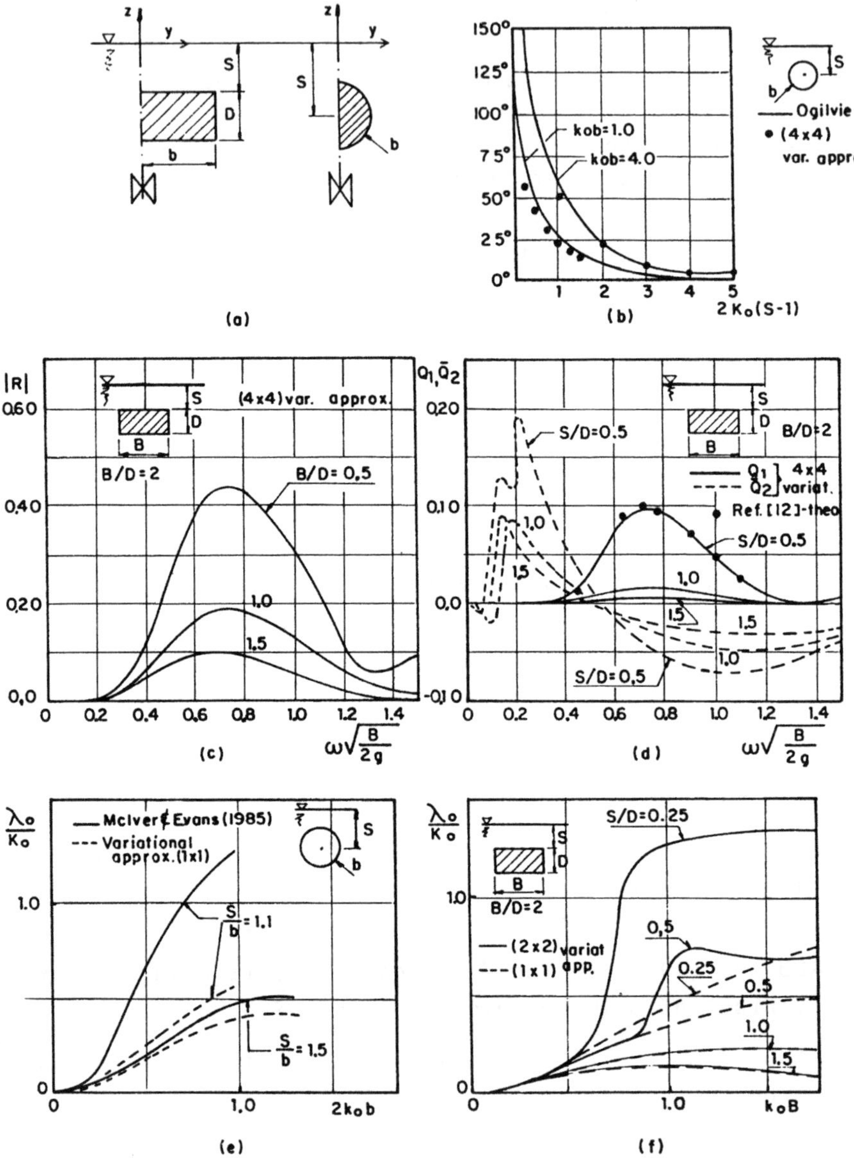

Fig. (5-1):- a) Analysed geometries; b) Phase of T for submerged circular cylinder compared with Ogilvie (1963); c) $|R(\omega)|$ for a rectangle with different submergence depth; d) $Q_1$ & $Q_2$ Drift Force Coefficients for a rectangle; e) $\lambda_o/K_o$ for a circle compared with Mc Iver & Evans (1985); f) $\lambda_o/K_o$ for a rectangle.

# Slowly Varying Wave Drift Forces in Current

R.H.M. HUIJSMANS

Maritime Research Institute Netherlands (MARIN)
P.O. Box 28
6700 AA Wageningen
The Netherlands

Summary

This paper is concerned with the development and application of hydrodynamic theory required for the prediction of wave drift damping phenomena as introduced by Wichers in 1979 and 1982, ref. [1] and [2]. The concept of wave drift damping necessitates the development of forward speed dependent expressions for the mean wave drift forces and a new 3-D diffraction potential algorithm for forward speed (Wichers and Huijsmans [3] and Hearn and Koon [4]). In this paper computed mean wave drift forces are validated with results from model tests as reported by Wichers [5]. Results will be discussed of model tests and computations for this tanker in irregular waves with and without current.

## 1. Introduction

In 1971 Remery and Hermans [6] reported results of the excitation and motion of a barge moored in wave groups. They considered the surge motion only and showed that excitation of the large amplitude low frequency motion was due to the low frequency drift force. In 1980, Pinkster [7] published techniques resulting in excellent agreement of the calculated drift forces at zero forward speed with experiments. An unsolved problem, however, is the estimation of the motion of a moored ship, especially when the low frequency motions have large amplitudes. Wichers and Huijsmans [8] showed that the damping at the natural frequencies of the mooring system have to be considered carefully. Results of model test experiments showed that a large part of the damping at these natural frequencies could be attributed to the velocity dependency of the wave drift forces. Hence, the effect of moderate speed should be accounted for if one wants to evaluate this wave damping phenomenon.

K. Horikawa, H. Maruo (Eds.)
Nonlinear Water Waves
IUTAM Symposium Tokyo/Japan 1987

## 2. Mathematical Formulation

The total potential function will be split in a steady and a non-steady part in a well-known way:

$$\Phi(\underline{x},t) = Ux + \overline{\Phi}(\underline{x};U) + \tilde{\phi}(\underline{x},t;U) \qquad (1)$$

in this formulation U is the incoming unperturbed velocity field, obtained by considering a coordinate system fixed to a ship moving under a drift angle $\beta$. In our approach this angle needs not be small. The time dependent part of the potential consists of an incoming wave at frequency $\omega$, a diffracted and/or a radiated wave contribution. To compute the drift force all these components will be taken into account. In this paper we restrict ourselves to a general theory concerning the wave components. Some results of drift force computations are shown as well.

The equations for the total potential $\Phi$ can be written as:

$$\Delta\Phi = 0 \text{ in the fluid domain } D_e \qquad (2)$$

At the free surface we have the well-known dynamic and kinematic boundary condition.

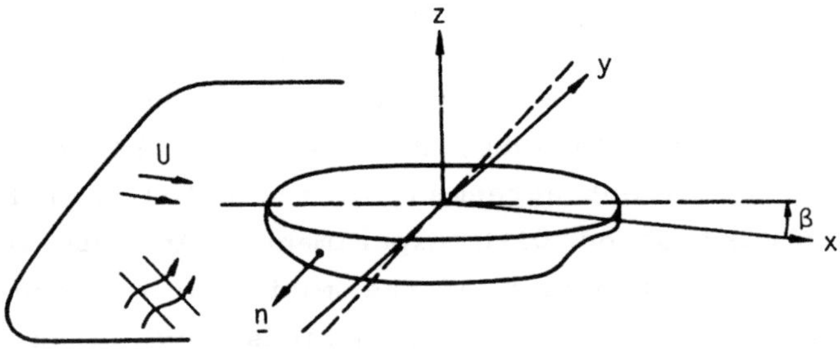


Fig.1. System of axis

We assume that the waves are high compared with the Kelvin wave pattern and that they both are small, hence the free surface condition can be expanded at z = 0. This leads to the following non-linear condition:

$$\frac{\partial^2\Phi}{\partial t^2} + g\frac{\partial\Phi}{\partial z} + \frac{\partial}{\partial t}(\nabla\Phi.\nabla\Phi) + \nabla\Phi.\nabla\left(\frac{\nabla\Phi.\nabla\Phi}{2}\right) = 0 \qquad (3)$$

at z = 0

To compute the wave resistance at low speed the free surface must be treated more carefully, because the wave height is of asymptotically smaller order. This problem is studied extensively by Eggers [9], Baba [10], Brandsma and Hermans [11]. The velocity field is well described by the double body potential with a small wave pattern. Therefore we take the double body potential into account and we neglect the stationary wave pattern.

For the wave potential $\tilde{\phi}(\underline{x},t;U)$ the free surface condition now becomes:

$$\begin{aligned}
&\tilde{\phi}_{tt} + g\tilde{\phi}_z + 2U\tilde{\phi}_{xt} + 2\nabla\overline{\Phi}.\nabla\tilde{\phi}_t + \\
&+ (U^2 + 2U\overline{\Phi}_x + \overline{\Phi}_x{}^2)\tilde{\phi}_{xx} + 2(U + \overline{\Phi}_x)\overline{\Phi}_y\tilde{\Phi}_{xy} + \\
&+ \overline{\Phi}_y{}^2\tilde{\phi}_{yy} + (3U\overline{\Phi}_{xx} + \overline{\Phi}_x\overline{\Phi}_{xx} + \overline{\Phi}_y\overline{\Phi}_{xy})\tilde{\Phi}_x + \\
&+ (2U\overline{\Phi}_{xy} + \overline{\Phi}_x\overline{\Phi}_{xy} + \overline{\Phi}_y\overline{\Phi}_{yy})\tilde{\Phi}_y + L^{(2)}\{\tilde{\phi}\} = 0 \qquad \text{at } z = 0
\end{aligned} \tag{4}$$

The boundary condition on the hull can be written for all the diffracted modes. We have the condition:

$$(n.\nabla\tilde{\phi}) = [\dot{\alpha} + \nabla x(\alpha x\overline{W})].n \tag{5}$$

at the mean position of the hull, in which $\alpha$ is the oscillation motion and $\overline{W} = (u,v,o)^T$, the uniform current vector. The nonlinear operator on $\tilde{\phi}$ will be neglected as well. First we discuss the construction of the regular part of the perturbation problem, with the complete linear free surface condition.

We assume $\tilde{\phi}(\underline{x},t;U)$ to be oscillatory:

$$\tilde{\phi}(\underline{x},t;U) = \phi(\underline{x},U)e^{-i\omega t} \tag{6}$$

The free surface condition is written as:

$$-\omega^2\phi - 2i\omega U\phi_x + U^2\phi_{xx} + g\phi_z = D(U;\overline{\Phi})\{\phi\} \text{ at } z = 0 \tag{7}$$

where $D(U;\overline{\Phi})$ is a linear differential operator acting on $\phi$ as defined in (4). We apply Green's theorem to a problem in $D_i$ inside S and to the problem in $D_e$ outside S where S is the ship's hull. Combining the formulation inside and outside the ship we obtain a description of the potential function defined outside S by means of a source and vortex distribution of the following form:

$$- \iint_S \gamma(\underline{\xi}) \frac{\partial}{\partial n} G(\underline{x},\underline{\xi})dS_\xi - \iint_S \sigma(\underline{\xi})G(\underline{x},\underline{\xi})dS_\xi +$$

$$- \frac{2i\omega U}{g} \int_{WL} \gamma(\underline{\xi})G(\underline{x},\underline{\xi})d\eta + \frac{U^2}{g} \int_{WL} [\gamma(\underline{\xi}) \frac{\partial}{\partial \xi} G(\underline{x},\underline{\xi}) +$$

$$- \{\alpha_t \gamma_t(\underline{\xi}) + \alpha_T \gamma_T(\underline{\xi})\}G(\underline{x},\underline{\xi})]d\eta + \frac{U^2}{g} \int_{WL} \alpha_n \sigma(\underline{\xi})G(\underline{x},\underline{\xi})d\eta +$$

$$+ \frac{i\omega}{g} \iint_{FS} G(\underline{x},\underline{\xi})D\{\phi\}dS_\xi = 4\pi\phi(\underline{x}) \tag{8}$$

$\alpha_t = \cos(0x,\underline{t})$, $\alpha_T = \cos(0x,\underline{T})$ and $\alpha_n = \cos(0x,\underline{n})$ where n is the normal and t the tangent to the waterline and $T = \underline{t}\times\underline{n}$ the bi-normal. With $\gamma(\underline{\xi}) = 0$ we finally arrive at the formulation:

$$- 2\pi\sigma(\underline{x}) - \iint_S \sigma(\underline{\xi}) \frac{\partial}{\partial n_x} G(\underline{x},\underline{\xi})dS_\xi +$$

$$+ \frac{U^2}{g} \int_{WL} \alpha_n \sigma(\underline{\xi}) \frac{\partial}{\partial n} G(\underline{x},\underline{\xi})d\eta +$$

$$+ \frac{i\omega}{g} \iint_{FS} \frac{\partial}{\partial n_x} G(\underline{x},\underline{\xi})D\{\phi\}dS_\xi = 4\pi V(\underline{x}), \qquad x \in S \tag{9}$$

and:

$$4\pi\phi(\underline{x}) = - \iint_S \sigma(\underline{\xi})G(\underline{x},\underline{\xi})dS_\xi + \frac{U^2}{g} \int_{WL} \alpha_n \sigma(\underline{\xi})G(\underline{x},\underline{\xi})d\eta +$$

$$+ \frac{i\omega}{g} \iint_{FS} G(\underline{x},\underline{\xi})D\{\phi\}dS_\zeta, \qquad x \in D_e \tag{10}$$

We now consider small values of U, keeping in mind that there are two dimensionless parameters that play a role in the limit. We consider $\tau = \frac{U\omega}{g} << 1$ and $\nu = \frac{gL}{U^2} >> 1$. It turns out that the source strength and the potential function can be expanded as follows:

$$\sigma(\underline{x};U) = \sigma_0(\underline{x}) + \tau\sigma_1(\underline{x}) + \tilde{\sigma}(\underline{x};U)$$
$$\phi(\underline{x};U) = \phi_0(\underline{x}) + \tau\phi_1(\underline{x}) + \tilde{\phi}(\underline{x};U) \tag{11}$$

where $\tilde{\sigma}$ is $O(\tau^2)$ and $\tilde{\phi}$ is $O(\tau^3)$ as $\tau \to 0$, while the expansion of the Green's function is less trivial, which has been derived by Hermans and Huijsmans [12]:

$$\psi(\underline{x},\underline{\xi};U) = \psi_0(\underline{x},\underline{\xi}) + \tau\psi_1(\underline{x},\underline{\xi}) + \tilde{\psi}(\underline{x},\underline{\xi};U) \tag{12}$$

## 3. Expansion of the Source Strength

In this section an approximation solution of (10) will be derived. Inserting (11) and (12) in (9) one obtains for like powers of $\tau$ the following set of equations:

$$-2\pi\sigma_0(\underline{x}) - \iint_S \sigma_0(\underline{\xi}) \frac{\partial G_0}{\partial n_x}(\underline{x},\underline{\xi})dS_\xi = 4\pi V_0(\underline{x}), \qquad x \in S$$

and:

$$-2\pi\sigma_1(\underline{x}) - \iint_S \sigma_1(\underline{\xi}) \frac{\partial G_0}{\partial n_x}(\underline{x},\underline{\xi})dS_\xi =$$
$$= -\iint_S \sigma_0(\underline{\xi}) \frac{\partial}{\partial n_x} \psi_1(\underline{x},\underline{\xi})dS_\xi + \frac{2i}{g} \iint_{FS} \frac{\partial G_0}{\partial n_x}(\underline{x},\underline{\xi})\nabla\overline{\Phi}.\nabla\phi_0 dS_\xi +$$
$$+ 4\pi V_1(\underline{x}) \tag{13}$$

where $G_0(\underline{x},\underline{\xi}) = -\frac{1}{r} + \frac{1}{r_1} - \psi_0(\underline{x},\underline{\xi})$ is the zero speed pulsating wave source and $V(\underline{x}) = V_0(\underline{x}) + \tau V_1(\underline{x}) + O(\tau^2)$. This perturbation leads to a fast algorithm that takes into account speed effects once a fast method is available for the zero speed diffraction problem. At MARIN the diffraction program has been extended with the Fingreen subroutines of Newman.

The potential functions (11) now become:

$$\phi_0(\underline{x}) = -\frac{1}{4\pi} \iint_S \sigma_0(\underline{\xi})G_0(\underline{x},\underline{\xi})dS_\xi$$

and:

$$\phi_1(\underline{x}) = \frac{1}{4\pi} \iint_S \sigma_0(\underline{\xi})\psi_1(\underline{x},\underline{\xi})dS_\xi - \frac{1}{4\pi} \iint_S \sigma_1(\underline{\xi})G_0(\underline{x},\underline{\xi})dS_\zeta +$$
$$+ \frac{i}{2\pi g} \iint_{FS} G_0(\underline{x},\underline{\xi})\nabla\overline{\Phi}.\nabla\phi_0 dS_\xi \tag{14}$$

## 4. Wave Drift Forces in Current

In order to calculate the wave drift forces a pressure integration technique is used. The validity of this approach has been elucidated by Pinkster [7] for the zero speed case. The resulting expression is:

$$F^{(2)} = -g\int_{WL} \tfrac{1}{2}\rho\,|\zeta_r^{(1)}|^2\,\underline{n}\,dl + \alpha^{(1)} \times (M.\ddot{\underline{X}}_g^{(1)}) +$$
$$+ \iint_{S_0} \tfrac{1}{2}\rho\,|\nabla\tilde{\phi}^{(1)}|^2\,\underline{n}\,dS + \iint_{S_0} \rho(\underline{x}^{(1)}.\nabla\tilde{\phi}_t^{(1)})\underline{n}\,dS +$$
$$+ \iint_{S_0} \rho\,\tilde{\phi}_t^{(2)}\,\underline{n}\,dS \tag{15}$$

In this expression the velocity dependent potentials as derived in the previous section have to be substituted, except for the term depending on $\tilde{\phi}^{(2)}$. For the mean wave drift forces we have shown the validity of [13] and [17] for the non-zero speed case. In the 3-D case it is not a trivial task to calculate the exact contribution of $\tilde{\phi}^{(2)}$ to the wave drift forces (see Benschop et al. [15]). For 2-D cases approaches of Faltinsen and Zhao [16] can be followed. In the following we assume that the slowly varying part of the wave drift forces can be derived from the mean wave drift forces by extrapolation based on Newman's approach [17]. It will be clear that we cannot make a perturbation series with respect to $\tau$. The first order motion $\underline{x}^{(1)}$ depends on U in a complicated way, phase and amplitude are influenced by the small parameter $\tau$. Therefore it is not possible to express (15) in a power series in $\tau$ uniformly. Similar arguments hold for some of the other terms. So, finally, the effect of the speed on the drift forces is computed by evaluating (15) numerically.

## 6. Results of Computations and Model Test Experiments

At MARIN a number of tests have been conducted. The tests comprise in short regular wave tests on a stationary moored 200 kDWT tanker in 206 m water depth with waves and current parallel to each other. Particulars of the tanker are taken from Huijsmans [13]. The results of model tests and computations for head waves and current have been extensively described by

Wichers [5] and Huijsmans [13]. The results of the mean wave drift forces are shown in Fig. 2.

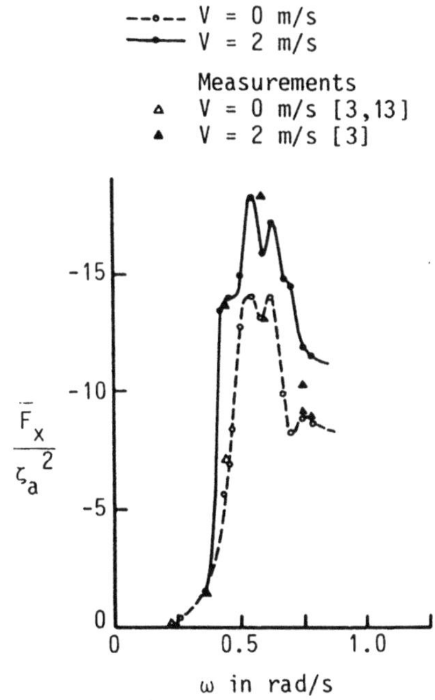


Fig.2. Mean surge drift force in head waves for a 200 kDWT tanker

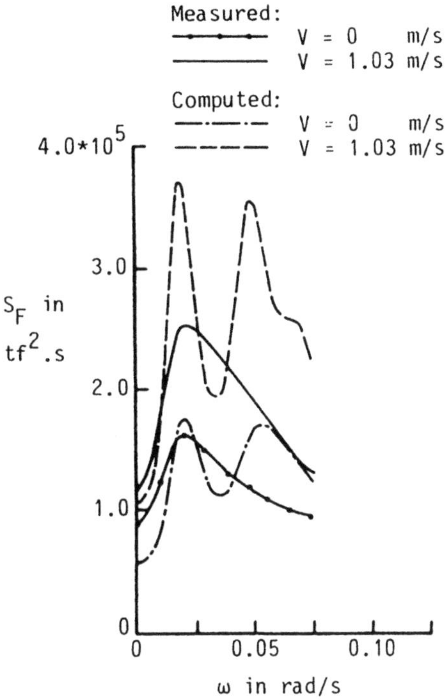


Fig.3. Spectral density of wave drift force

The spectra of the computed and measured surge wave drift forces are presented in Fig. 3. The wave spectra used, were Pierson-Moskowitz spectra with significant wave height of 11.9 m and a peak period of 14.0 s.

For the time domain computation of the low frequency motions of a moored tanker in irregular head waves with and without current the equation of motion has to be solved in the time domain. The computed low frequency motions in the time domain can be compared with the motions as measured during model tests. The tanker was linearly moored in 206 m water depth. The spring constant in surge direction amounts to 19.3 tf/m. The computed and measured results are presented as time domain plots in Fig. 4.

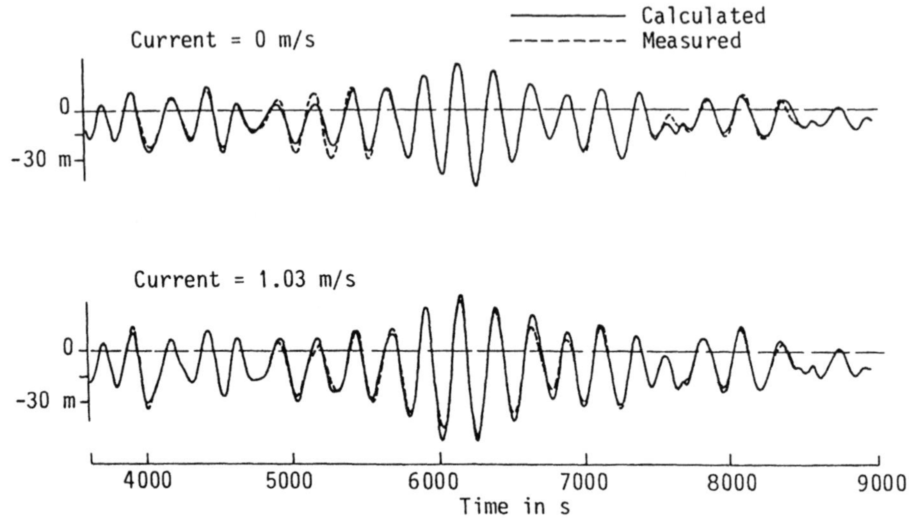


Fig.4. Low frequency surge motion of tanker in irregular waves with and without current

## 7. Discussion

For the case of zero drift angle the contribution of the free surface integrals will be negligible. This is due to the fact that in case of zero drift angle the double body stationary potential may be neglected. For the non-zero drift angle this is no longer true as has been shown by Huijsmans and Wichers [14]. The results of computations and measurements of the slowly varying wave drift forces in currents indicate that Newman's approach to compute the slowly varying wave drift forces is valid for the case of a loaded 200 kDWT tanker in 206 m water. The time domain computations of the surge motions compare favourably with the results of model test experiment.

## References

1. Wichers, J.E.W.: On the low frequency surge motions of vessels mooring in high seas. Offshore Technology Conference, Paper 4437, Houston, 1982.
2. Wichers, J.E.W.; Van Sluijs, M.F.: The influence of waves on the low frequency hydrodynamic coefficients of moored vessels. Offshore Technology Conference, Paper 2625, Houston, 1979.

3. Wichers, J.E.W.; Huijsmans, R.H.M.: On the low frequency hydrodynamic damping forces acting on offshore moored vessels. Offshore Technology Conference, Paper 4813, Houston, 1984.
4. Hearn, G.; Koon, T.: Second order fluid damping. Progress Report, University of Newcastle, July 1985.
5. Wichers, J.E.W.: Progress in computer simulations of SPM moored vessels. Offshore Technology Conference, Paper 5175, Houston, 1986.
6. Remery, G.F.M; Hermans, A.J.: The slow drift oscillations of a moored object in random seas. Offshore Technology Conference, Paper 1500, Houston, 1971.
7. Pinkster, J.A.: Low frequency second order wave exciting forces on floating structures. Ph.D Thesis, Report 650, MARIN (NSMB), 1980.
8. Wichers, J.E.W.; Huijsmans, R.H.M.: On the low frequency hydrodynamic damping forces acting on offshore moored vessels. Offshore Technology Conference, Paper 4813, Houston, 1984.
9. Eggers, K.: Non-Kelvin dispersive waves around non-slender ships. Schiffstechnik, Bd. 8, 1981.
10. Baba, E.: Wave resistance of ships in low speed. Mitsubishi Technical Bulletin, No. 109, 1976.
11. Brandsma, F.J.; Hermans, A.J.: A quasi-linear free surface condition in slow ship theory. Schiffstechnik, April 1985.
12. Hermans, A.J.; Huijsmans, R.H.M.: The effect of moderate speed on the motions of floating bodies. Presented at the occasion of the retirement of Prof. K. Eggers, March 1987.
13. Huijsmans, R.H.M.: Wave drift forces in current. 16th Conference on Naval Hydrodynamics, Berkeley, 1986.
14. Huijsmans, R.H.M.; Wichers, J.E.W.: Considerations on wave drift damping of a moored tanker for zero and non-zero drift angles. Proceedings of PRADS'87 Conference, Trondheim, Norway.
15. Benschop, A.; Hermans, A.J.; Huijsmans, R.H.M.: Second order diffraction forces on a ship in irregular waves. Journal of Applied Ocean Research, Vol. 9, No. 2, 1987.
16. Faltinsen, O.; Zhao, R.: Slow drift motions of a moored 2-D body in irregular waves. To be published.
17. Newman, J.N.: The drift force and moment on ships in waves. Journal of Ship Research, Vol. 11, No. 1, March 1967.

# Second-order Hydrodynamic Forces on Moored Vessels in Random Waves

T. MATSUI

Department of Architecture
Nagoya University
Nagoya, Japan

## Summary

An exact second-order theory is formulated for predicting the slowly-varying second-order hydrodynamic forces on moored vessels in random waves. The near field approach based on direct integration of the fluid pressure on the submerged body surface is employed in conjunction with the Stokes expansion to second order. Green's second identity is exploited to derive the expression for the second-order hydrodynamic forces due to the second-order velocity potential involving first-order quantities and the second-order undisturbed potential. It is shown that the contribution of the second-order potential to the slowly-varying forces may be significant in irregular seas with longer mean wave period.

## Introduction

The wave loads acting on moored vessels in random seas include the second-order, slowly-varying force components at small difference frequencies of the wave group, which are known as the slowly-varying wave drift forces. The magnitude of these forces may not be large compared with the first-order excitation at wave frequency, but if the frequencies are close to any of the natural frequencies of the system, large amplitude motions may occur due to small damping in the system. The forces are of particular importance for compliant systems possessing low restoring forces in horizontal modes and large-volume vessels with small water plane area.

The prediction of the second-order hydrodynamic forces on vessels moored in irregular waves requires not only the knowledge of the first-order velocity potential and the first-order body motions but also of the second-order velocity potential [1]. While the former contribution can be computed using conventional methods for first-order diffraction analysis, the latter requires the solution of the diffraction problem to second order. Due to the complexity of solving the second-order diffraction problem, a number of approximate methods have been employed for evaluating the contribution from the second-order velocity potential [1, 2].

K. Horikawa, H. Maruo (Eds.)
Nonlinear Water Waves
IUTAM Symposium Tokyo/Japan 1987

In the earlier paper [3], the author formulated an exact second-order theory for predicting the slowly-varying second-order hydrodynamic forces on moored vessels in random waves. Green's second identity was exploited to derive the complete expression for the second-order hydrodynamic forces due to the second-order velocity potential involving first-order quantities alone. The theory was developed to give an explicit analytical solution for the slowly-varying wave drift moment on a cylindrical articulated column [4]. The numerical results showed that the contribution of the second-order velocity potential to the drift moment can be significant in irregular seas with longer mean wave period. In the following, a brief description of the formulation is presented together with additional results for a moored semi-submersible, obtained from numerical first-order solutions based on the hybrid finite element technique.

## Hydrodynamic Theory

With the assumption of ideal and irrotational flow, the wave field may be defined by a velocity potential $\Phi$, which may be expanded in a convergent power series with respect to a small parameter $\varepsilon$

$$\Phi = \varepsilon\Phi^{(1)} + \varepsilon^2\Phi^{(2)} + O(\varepsilon^3) \tag{1}$$

The pressure at a point P on the body surface is determined from the Bernoulli equation

$$p = -\rho\left(gX_3 + \Phi_t + \frac{1}{2}|\vec{\nabla}\Phi|^2\right) \tag{2}$$

where $\rho$ is the fluid density, $g$ is the acceleration due to gravity, $X_3$ is the altitude of P above the mean free surface, and the subscript $t$ on $\Phi$ designates the time derivative of the potential. The wave excited force on the body is then obtained by integrating $p$ over the instantaneous submerged body surface S

$$\vec{F} = -\iint_S p\vec{N}dS \tag{3}$$

where $\vec{N}$ is the normal vector directed outwardly from S. Pinkster [1] has shown how equation (3) may be evaluated up to second order by means of a Taylor series expansion of the pressure about the mean submerged body surface $S_0$. The procedure leads to the expansion for the force

$$\vec{F} = \vec{F}^{(0)} + \varepsilon\vec{F}^{(1)} + \varepsilon^2\vec{F}^{(2)} + O(\varepsilon^3) \tag{4}$$

where $\vec{F}^{(0)}$ is the hydrostatic component of the force, and

$$\vec{F}^{(1)} = \rho \iint_{S_0} \Phi_t^{(1)} \vec{n} dS \tag{5}$$

$$\vec{F}^{(2)} = \rho \iint_{S_0} \left[\Phi_t^{(2)} + \frac{1}{2}\left|\vec{\nabla}\Phi^{(1)}\right|^2 + \vec{X}^{(1)} \cdot \vec{\nabla}\Phi_t^{(1)}\right] \vec{n} dS$$
$$- \frac{1}{2}\rho g \int_{C_W} \zeta_R^{(1)2} \vec{n} dC_W + \vec{X}_R^{(1)} \times (m\ddot{\vec{X}}_G^{(1)}) \tag{6}$$

In the above expressions, $\vec{n}$ is the normal vector directed outwardly from $S_0$, $\zeta_R^{(1)}$ is the first-order wave elevation along the mean water line $C_W$ relative to the body in its displaced position, $\vec{X}^{(1)}$ is the first-order translation vector of a point on $S_0$, $\ddot{\vec{X}}_G^{(1)}$ is the first-order translation acceleration vector of the centre of gravity, $\vec{X}_R^{(1)}$ is the first-order rotation vector of the body, and m is the mass of body.

The terms in equation (6) arising from the products of first-order effects can be computed straightforwardly by means of conventional methods for first-order diffraction analysis. In the following, attention is focused on the evaluation of the second-order hydrodynamic forces caused by the second-order velocity potential, namely

$$\vec{F}_q^{(2)} = \rho \iint_{S_0} \Phi_t^{(2)} \vec{n} dS \tag{7}$$

Second-Order Hydrodynamic Forces Due to the Second-Order Potential

It is assumed that the irregular wave system can be approximated by a long-crested regular wave group consisting of regular waves with frequencies $\omega_i$ ($i = 1, 2, \ldots, N$). Hence the instantaneous free surface elevation at a prescribed point in space at time t may be expressed as follows

$$\zeta(t) = \mathrm{Re}\left\{\sum_{i=1}^{N} \zeta_i e^{-i\omega_i t}\right\} \tag{8}$$

where $\zeta_i$ is the complex wave amplitude with random phase. The first-order body motions and the first-order velocity potential may then be defined by

$$\vec{X}^{(1)}(t) = \mathrm{Re}\left\{\sum_{i=1}^{N} \vec{x}_i^{(1)} e^{-i\omega_i t}\right\}, \quad \vec{X}_R^{(1)}(t) = \mathrm{Re}\left\{\sum_{i=1}^{N} \vec{x}_{Ri}^{(1)} e^{-i\omega_i t}\right\} \tag{9}$$

$$\Phi^{(1)}(t) = \mathrm{Re}\left\{\sum_{i=1}^{N} \phi_i^{(1)} e^{-i\omega_i t}\right\} = \mathrm{Re}\left\{\sum_{i=1}^{N} (\phi_{Ii}^{(1)} + \phi_{Di}^{(1)}) e^{-i\omega_i t}\right\} \tag{10}$$

where the first-order complex potential $\phi_i^{(1)}$ may be decomposed into the undisturbed incident wave potential $\phi_{Ii}^{(1)}$ and the potential due to disturbance $\phi_{Di}^{(1)}$, consisting of the scattered wave and waves induced by the first-order body motions. The second-order velocity potential consists of frequency components $\omega_i + \omega_j$ and $\omega_i - \omega_j$. Attention is confined here to the slowly-varying part associated with the difference frequency terms. Hence

$$\phi^{(2)}(t) = \mathrm{Re}\left\{\sum_{i=1}^{N}\sum_{j=1}^{N}(\phi_{Iij}^{(2)} + \phi_{Dij}^{(2)})e^{-i(\omega_i-\omega_j)t}\right\} + \delta^{(2)}t \qquad (11)$$

where $\delta^{(2)}$ is determined from the requirement that the mean water level be $X_3 = 0$. It may easily be shown that $\delta^{(2)}$ contributes to the mean vertical force alone. The second-order incident wave potential $\phi_{Iij}^{(2)}$ may be determined explicitly for water of constant depth [2]. The associated diffraction potential $\phi_{Dij}^{(2)}$ must be determined, such that the following inhomogeneous boundary conditions are satisfied [1]:

$$g\frac{\partial\phi_{Dij}^{(2)}}{\partial X_3} - (\omega_i - \omega_j)^2\phi_{Dij}^{(2)} = \alpha_{ij}^{(2)} \qquad \text{on free surface } S_F \qquad (12)$$

$$\vec{\nabla}\phi_{Dij}^{(2)}\cdot\vec{n} = -\vec{\nabla}\phi_{Iij}^{(2)}\cdot\vec{n} + \beta_{ij}^{(2)} \qquad \text{on body surface } S_0 \qquad (13)$$

where

$$\alpha_{ij}^{(2)} = \frac{i(\omega_i - \omega_j)}{2}\left(\vec{\nabla}\phi_i^{(1)}\cdot\vec{\nabla}\phi_j^{(1)*} - \vec{\nabla}\phi_{Ii}^{(1)}\cdot\vec{\nabla}\phi_{Ij}^{(1)*}\right)$$
$$- \frac{i\omega_i}{2g}\left[\phi_i^{(1)}\frac{\partial}{\partial X_3}\left(g\frac{\partial\phi_j^{(1)*}}{\partial X_3} - \omega_j^2\phi_j^{(1)*}\right)\right.$$
$$\left. - \phi_{Ii}^{(1)}\frac{\partial}{\partial X_3}\left(g\frac{\partial\phi_{Ij}^{(1)*}}{\partial X_3} - \omega_j^2\phi_{Ij}^{(1)*}\right)\right] \qquad (14)$$

$$\beta_{ij}^{(2)} = -\frac{1}{2}\left[(\vec{x}_i^{(1)}\cdot\vec{\nabla})\cdot\vec{\nabla}\phi_j^{(1)*}\right]\cdot\vec{n} - \frac{1}{2}(i\omega_i\vec{x}_i^{(1)} + \vec{\nabla}\phi_i^{(1)})\cdot(\vec{x}_{Rj}^{(1)*} \times \vec{n}) \qquad (15)$$

* denoting the complex conjugate. In addition, $\phi_{Dij}^{(2)}$ satisfies the condition of no flow through the sea bed and the radiation boundary condition associated with outgoing waves at infinity.

The second-order hydrodynamic forces caused by the second-order potential can be evaluated by means of Green's second identity. To do so, use is made of the linearised radiation potential $\Psi_k$ associated with a unit oscillation of the body in the direction $n_k$ at frequency $\omega_i - \omega_j$

$$\Psi_k = \mathrm{Re}\left\{\psi_k e^{-i(\omega_i-\omega_j)t}\right\} \qquad (16)$$

which satisfies $\vec{\nabla}\psi_k\cdot\vec{n} = n_k$ on $S_0$ (17)

$\Psi_k$ satisfies also the linearised free surface condition, the sea bed condition and the Sommerfeld radiation boundary condition. Substitution of equation (17) into equation (7) and application of Green's second identity on the surfaces enclosing the fluid leads to the expression for the slowly-varying part of the second-order hydrodynmaic force caused by the second-order potential

$$\vec{F}_q^{(2)} = \mathrm{Re}\Bigg\{\sum_{i=1}^{N}\sum_{j=1}^{N}\Bigg[-i\rho(\omega_i - \omega_j)\iint_{S_0}\{\phi_{Iij}^{(2)} n_k - \psi_k(\vec{\nabla}\phi_{Iij}^{(2)}\cdot\vec{n}) + \psi_k\beta_{ij}^{(2)}\}\,dS + i\rho(\omega_i - \omega_j)\iint_{S_F}\psi_k\frac{\alpha_{ij}^{(2)}}{g}\,dS\Bigg]e^{-i(\omega_i-\omega_j)t}\Bigg\}\vec{e}_k \quad (18)$$

where $\vec{e}_k$ is the unit vector in the direction of $n_k$. The details of the derivation of equation (18) are given by Matsui [3]. The calculation of $\vec{F}_q^{(2)}$ is thus possible with the knowledge of first-order solutions and the second-order undisturbed potential alone.

## Approximate Evaluation of Contribution Due to the Second-Order Potential

As indicated by the results obtained previously for the articulated column [3, 4], it appears that the force component due to the second-order potential dominates the total drift force only in the low frequency range where diffraction effects due to the first-order waves are small. In the high frequency range where diffraction effects are large, the total drift force is dominated by the components arising from the products of first-order quantities, and the contribution due to the second-order potential decreases in relative importance. This implies that, in computing the contribution of the second-order potential to the slowly-varying drift force, the terms containing the first-order diffraction potential may be neglected compared with the undisturbed incident wave potential. With this simplification, the free surface integral involved in equation (18) vanishes, and as a consequence the computational effort to obtain results is considerably reduced. This simplified method is distinguished from that suggested by Pinkster [1], who ignored also the last term on the body surface integral in equation (18), representing the effect of the first-order body motions on the second-order body surface condition. Bowers [2] adopted a more simplified approach by integrating the second-order wave pressure over the mean submerged body surface in the undisturbed incident wave field alone.

## Results and Discussion

The theory outlined in the foregoing was applied to compute the slowly-varying wave drift force on a moored semi-submersible, of which the principal particulars are given in Fig.1 and Table 1. The expressions derived for the second-order hydrodynamic forces were evaluated from numerical first-order solutions based on the hybrid finite element technique. The calculation of the free surface integral involved in equation (18) was neglected for the reason described in the previous section. Fig.2 shows the computed results for the slowly-varying sway drift force in irregular beam waves. The results are presented in the form of time-independent quadratic transfer function for the case of difference frequency $\omega_i - \omega_j$ = 0.376 rad/sec and plotted for range of mean frequency $(\omega_i + \omega_j)/2$. The results based on the present theory are compared with those from different simplified approaches [1, 2, 5]. It is observed that the contribution from the second-order potential dominates the total drift force only in the low frequency range where diffraction effects due to the first-order waves are expected to be small. In the high frequency range where diffraction effects become large, the total drift force is dominated by the components arising from the products of first-order effects. In this frequency range the slowly-varying drift force can be predicted fairly well with the knowledge of the mean drift force in regular waves, as suggested by Newman [5].

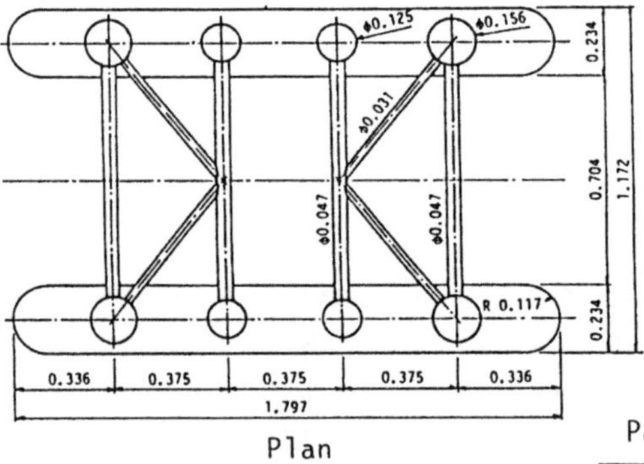


Plan

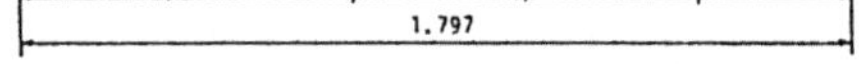


Starboard elevation

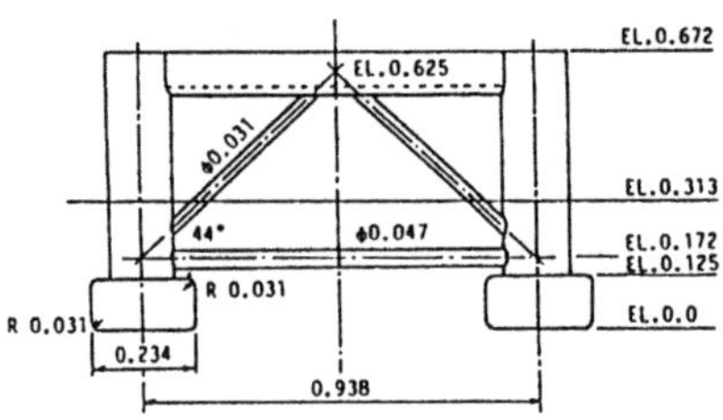


Forward elevation

Fig.1
Layout of the semi-submersible (in metre)

Table 1
Particulars of the semi-submersible

| Description | Value | Unit |
|---|---|---|
| Length of lower hull | 1.797 | m |
| Breadth | 1.172 | m |
| Draft | 0.313 | m |
| Displacement | 130.3 | kg |
| Centre of gravity above keel | 0.273 | m |
| Transverse metacentric height | 0.045 | m |
| Longitudinal metacentric height | 0.037 | m |
| Transverse gyradius in air | 0.536 | m |
| Longitudinal gyradius in air | 0.556 | m |
| Vertical gyradius in air | 0.634 | m |
| Mooring stiffness coeff. in sway | 3.700 | kg/m |
| Viscous damping coeff. in sway | 22.03 | $kg.s^2/m^2$ |
| Natural frequency in sway | 0.376 | rad/s |
| Water depth | 3.000 | m |

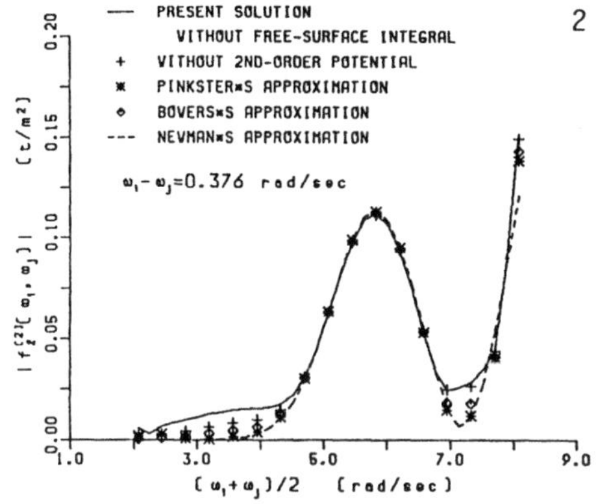


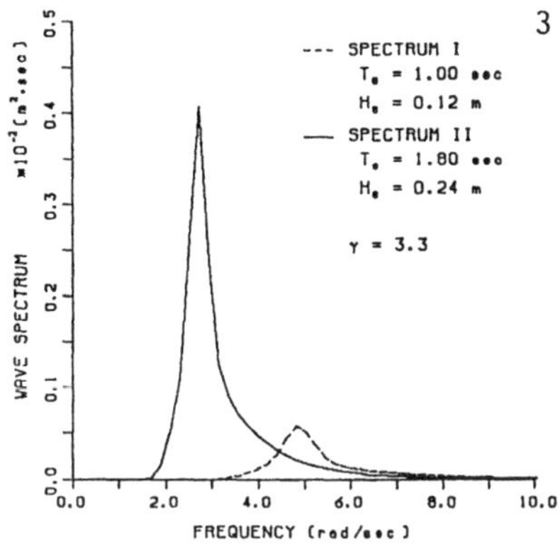


Figs.2 and 3. 2. Quadratic transfer function of sway drift force on the semi-submersible in beam waves; 3. Irregular wave spectra ($T_s$: mean wave period; $H_s$: significant wave height; γ: shape parameter)

Fig.4 illustrates the results of the time domain simulation of the slow drift motion of the semi-submersible under typical sea conditions. Two JONSWAP spectra shown in Fig.3 were used to simulate time histories of the drift forces in irregular waves by means of the impulse response function technique. The drift motions were simulated by direct integration of the

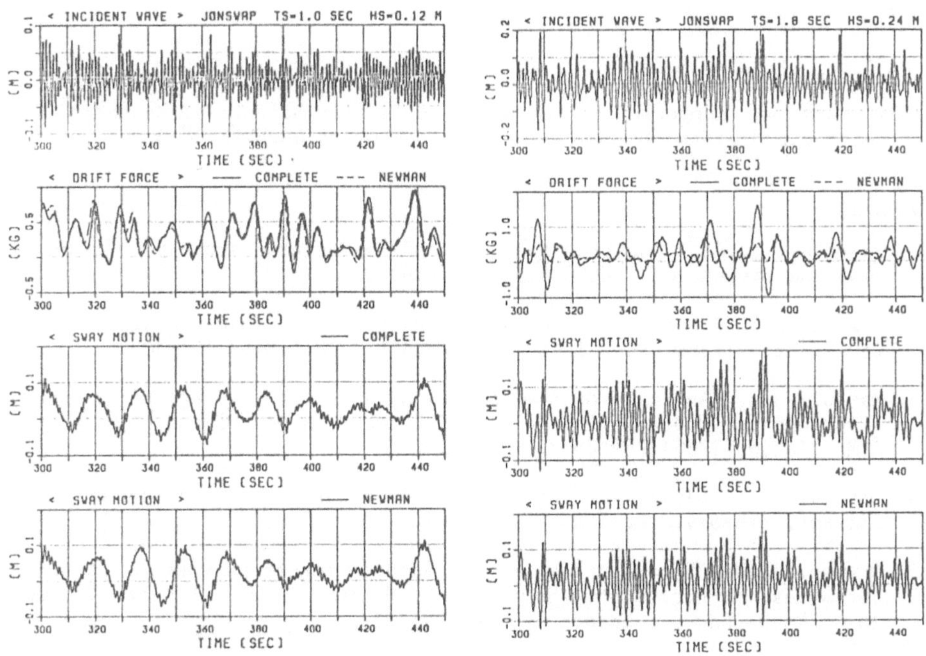


Fig.4 Simulated time histories of sway drift forces and motions for the semi-submersible in irregular beam waves

equation of motion, with the added mass and potential damping terms associated with the sway natural frequency and the viscous damping term based on the drag force coefficients estimated from the Dnv rules [6]. Simulated time histories of the drift forces and motions are compared with those based on Newman's approximation. For Spectrum I, of which the peak period lies in the frequency range where the components due to the products of first-order effects dominates the total drift force, complete agreement is observed between the present and Newman's approaches. On the other hand, for Spectrum II, of which the peak period lies in the frequency range where the second-order potential makes a significant contribution, it can be observed that Newman's method tends to underestimate both the drift forces and motions.

## Conclusion

In the foregoing an exact second-order theory was formulated to compute the slowly-varying second-order hydrodynamic forces on moored vessels in random waves. Numerical results based on the hybrid finite element technique were presented for a semi-submersible platform. It was shown that the contribution of the second-order potential to the slowly-varying forces may be significant in irregular seas with longer mean wave period.

## Acknowledgement

This work was supported by the Grant-in-Aid for Scientific Research from the Japanese Ministry of Education (through Grant No.61460177).

## References

1. Pinkster, J.A.: Low-frequency second order wave exciting forces on floating structures. NSMB Pub. 650 (1980).

2. Bowers, E.C.: Long period oscillations of moored ships subject to short wave seas. Trans. Roy. Inst. Nav. Archit. 118 (1976) 181-191.

3. Matsui, T.: Analysis of slowly varying wave drift forces on compliant structures. Proc. Fifth Int. Offsh. Mech. & Arc. Engng. Symp. Vol.1, ASME, Tokyo (1986) 289-296.

4. Matsui, T.: Slowly varying wave drift forces on an articulated column. 1st Rpt.: Theory. J. Soc. Nav. Archit. Jpn. 159 (1986) 149-164.

5. Newman, J.N.: Second order, slowly varying forces on vessels in irregular waves. Proc. Int. Symp. Dyn. Marine Vehic. & Struct. in Waves, Inst. Mech. Engrs., London (1974) 193-197.

6. Dnv: Rules for classification of mobile offshore units (1981).

# Sessioin 9

# Body-induced Nonlinear Water Waves

# Nonlinear Ship Waves at Low Froude Number

A.J. Hermans, F.J. Brandsma*

Delft University of Technology, Department of Mathematics,
Delft, The Netherlands.

## 1. INTRODUCTION

It is well-known in slow ship theory that the usual linearized free surface condition cannot be applied when obstacles are considered of which the beam to length ratio is not sufficiently small. Several authors (Baba, Eggers, Hermans, Maruo and Ogilvie) have studied the effect of nonlinear terms in the free surface condition. In principle the local velocity determines a local Green's function for the calculation of the wave pattern. A different approach is the application of the ray method (Keller).

In this paper we study the wave pattern of a thick ship-like object with finite bow and stern angles $0 < \beta < \pi/2$. The completely blunt form $\beta = \pi/2$ is excluded. It turns out that the wave pattern is strongly influenced by the nonlinear terms at the free surface. The wave pattern is determined by means of the ray method. The rays are mainly generated at the bow and the stern. A crucial step is the determination of the so-called excitation coefficients. They are constructed by means of an asymptotic evaluation of a distribution of 'local' sources at the free surface. It is shown that for small angles $\beta << 1$ the excitation coefficients are the same as the ones obtained by means of an asymptotic expansion for small values of the Froude number of the results of Michell's thin ship theory. For increasing values of beta the excitation coefficients change asymptotically. Our theory shows a continuous dependence, nevertheless. Similar changes are observed in the far field wave pattern.

The numerical computation of the wave pattern requires as a starting point a very accurate computation of the double body potential. A slight error in this potential gives a magnified error in the final wave pattern. For instance an error near the bow leads to a nonrealistic set of waves bouncing along the ship hull. To obtain a clear view of the method and the results we have chosen a ship form for which the double body potential can be obtained analytically. For

*) Present address: National Aerospace Lab, NLR, The Netherlands

K. Horikawa, H. Maruo (Eds.)
Nonlinear Water Waves
IUTAM Symposium Tokyo/Japan 1987

presentation at this meeting we have chosen lens-shaped cylindrical ship forms. For several values of beta the wave pattern is computed. These results are used to compute the wave resistance.

## 2. NEAR-FIELD FORMULATION

Throughout this paper we consider a fluid, which is non-viscous, incompressible with a constant density, while at the free surface we have the atmospheric pressure. The coordinate system is fixed to the ship and the incoming velocity field is (U,0,0). We introduce the velocity potential $\Phi$

$$\underline{u} = \text{grad}\ \Phi \tag{1}$$

which obeys the law of continuity, hence

$$\Delta\Phi = 0 \qquad \forall \underline{x} \text{ outside the ship and } y < 0. \tag{2}$$

On the free surface we have both the kinematic and the dynamic condition

$$\left.\begin{aligned} 0 &= \Phi_x H_x + \Phi_z H_z - \Phi_y \\ \frac{1}{2} U^2 &= gH + \frac{1}{2}(\Phi_x^2 + \Phi_y^2 + \Phi_z^2) \end{aligned}\right\} \quad \text{on } y = H(x,z). \tag{3}$$

On the hull of the ship, the kinematic condition holds

$$\nabla\Phi.n = 0 \qquad \text{on } z = f(x,y). \tag{4}$$

In our example the bow and stern are wedge-shaped with finite angles and the Froude number $Fn = U/(gL)^{1/2}$ is small.

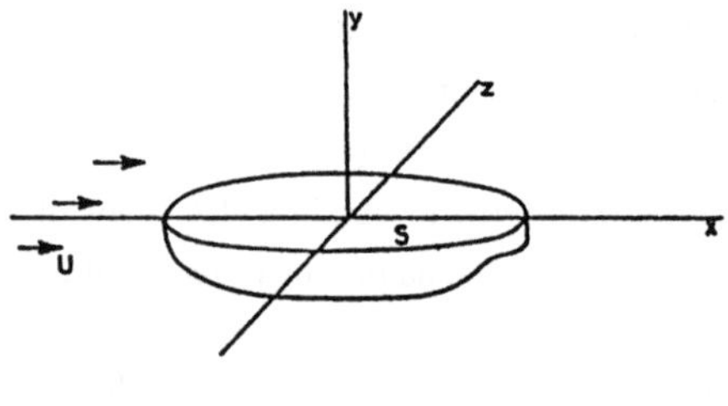


*Figure 1.*

A formal low Froude number expansion can be made. However, it can be shown that no solutions with a wavy character downstream can be found. The lowest order term leads to a good approximation of the fluid velocity near the hull. This so-called double body solution $\Phi_r$ is used as a first step to construct the wave solution.
In the layer near the free surface we assume that the potential has the form

$$\Phi = \Phi_r + \phi. \tag{5}$$

The wave potential $\phi$ is a solution of the Laplace equation for $y < H(x,z) = h_r(x,z) + h(x,z)$, where $h_r(x,z)$ is the surface elevation due to the double body potential

$$h_r(x,z) = \frac{1}{2g}\left[U^2 - \Phi_{rx}^2(x,0,z) - \Phi_{rz}^2(x,0,z)\right]. \tag{6}$$

We introduce a new coordinate system $(x,y,z) := (x,y-h_r(x,z),z)$ and the first approximation to the potential function $\phi(x,y,z)$ is for small values of Fn a solution of the Laplace equation. The free surface condition becomes

$$\phi_y + \frac{1}{g}\left[\Phi_{rx}^2\phi_{xx} + 2\Phi_{rx}\Phi_{rz}\phi_{xz} + \Phi_{rz}^2\phi_{zz}\right] + (3\Phi_{rx}\Phi_{rxx} + 2\Phi_{rz}\Phi_{rxz} + \Phi_{rx}\Phi_{rzz})\phi_x +$$

$$(3\Phi_{rz}\Phi_{rzz} + 2\Phi_{rx}\Phi_{rxz} + \Phi_{rz}\Phi_{rxx})\phi_z = D(x,z) \qquad \text{at } y = 0 \tag{7}$$

where the function $D(x,z)$ only depends on $\Phi_r$

$$D(x,z) = \frac{\partial}{\partial x}\left[\Phi_{rx}(x,0,z)h_r\right] + \frac{\partial}{\partial z}\left[\Phi_{rz}(x,0,z)h_r\right]. \tag{8}$$

The wave potential does not fulfil the condition on the hull. A second order outer solution is needed to correct this error of lower order.
The function $D(x,z)$ can be seen as a pressure distribution over the plane $y = 0$. Consequently the wave potential can be expressed as a source distribution over the free surface

$$c\phi(x,y,z) = \iint_{S_f} D(\xi,\zeta)G(x,y,z;\xi,0,\zeta)d\xi d\zeta. \tag{9}$$

with $c = \lim_{\epsilon\to 0} \int_{\partial B_\epsilon \cap D} \frac{\partial G}{\partial n_\epsilon} ds_\epsilon$

where $\partial B_\epsilon$ denotes the boundary of a ball of radius $\epsilon$ and having the centre at $\underline{x}$ and D denotes the fluid domain.

The source potential is a solution of the homogeneous problem and behaves like

$$G(x,y,z;\xi,\eta,\zeta) \approx \frac{1}{\left[(x-\xi)^2 + (y-\eta)^2 + (z-\zeta)^2\right]^{1/2}} . \tag{10}$$

It will be clear that the source function cannot be constructed exactly. By means of a multiple scale approach a solution can be obtained, which is valid near the ship hull. We consider the case where the wave number $k = g/U^2$ is large. The length scale for the variations of the waves is $1/k$. Therefore we introduce a new set of independent variables $\underline{X} = k\underline{x}$ and consider the potential to be a function of $\underline{x}$ and $\underline{X}$ independently. The solution obtained in this way is only valid in the vicinity of the ship. At finite distance it becomes invalid due to secular terms in the expansion. This is in accordance with the PLK method. In the local stream-line coordinates S

$$\begin{aligned} S &= \frac{\Phi_{rx}(x,0,z)}{Uq(x,z)} X + \frac{\Phi_{rz}(x,0,z)}{Uq(x,z)} Z \\ Y &= Y \\ T &= \frac{\Phi_{rz}(x,0,z)}{Uq(x,z)} X - \frac{\Phi_{rx}(x,0,z)}{Uq(x,z)} Z \end{aligned} \tag{11}$$

where q is the local double body velocity we construct the potential function and the evaluation of the wave part becomes

$$h(x,z) = \mathrm{Re}\, \frac{-k}{2\pi U} \iint_{S_f} D(\xi,\zeta) \int_{\beta-\pi/2}^{\beta+\pi/2} \frac{\exp(ikk_0\omega)}{q^3\cos^3(\theta-\beta)}\, d\xi\, d\zeta\, d\theta \tag{12}$$

where $k_0$ and $\omega$ are given by

$$k_0(x,z;\theta) = \left[q(x,z)\cos(\theta-\beta(x,z))\right]^{-2}$$

and

$$\omega(x,z;\xi,\zeta;\theta) = (x-\xi)\cos\theta + (z-\zeta)\sin\theta \tag{13}$$

while $\beta$ is the angle of the double body velocity vector with respect to the x-axis.

For large values of k (12) can be evaluated asymptotically. The leading term becomes an integral along the waterline C

$$h(x,z) \approx \mathrm{Re}\,\frac{1}{2\pi i U}\int_C\int_{\beta-\pi/2}^{\beta+\pi/2} A\,\exp(ikS)\,d\theta\,ds \tag{14}$$

with

$$A(x,z;\theta,s) = \frac{(n_x\cos\theta + n_z\sin\theta)D(x_0(s),z_0(s))}{q\cos(\theta-\beta)} \tag{15}$$

and

$$S(x,z;\theta,s) = \frac{(x-x_0(s))\cos\theta + (z-z_0(s))\sin\theta}{q^2\cos^2(\theta-\beta)}. \tag{16}$$

It is worthwhile to pay some attention to the evaluation of (14) in the far field. To do so we denote the double body horizontal velocity components at the waterline $(x_0(s),0,z_0(s))$ by $u(s)$ and $w(s)$. We then obtain for the amplitude function

$$A(x,z;\theta,s) = \frac{(w\cos\theta - u\sin\theta)\left(\frac{\partial h_r}{\partial s} + \frac{\mathrm{sgn}(z_0)v_y h_r}{(u^2+w^2)^{1/2}}\right)}{q\cos(\theta-\beta)}. \tag{17}$$

For points far away from the hull the values of q and $\beta$ will approach 1 and 0 respectively. In the limit for $|\underline{x}| \to \infty$ the expressions for A and S in the wave height reduce to

$$\begin{aligned} A^*(\theta,s) &= (w - u\tan\theta)\left(\frac{\partial h_r}{\partial s} + \frac{\mathrm{sgn}(z_0)v_y h_r}{(u^2+w^2)^{1/2}}\right) \\ S^*(x,z;\theta,s) &= \sec^2\theta((x-x_0(s))\cos\theta + (z-z_0(s))\sin\theta). \end{aligned} \tag{18}$$

This result is in agreement with the asymptotic result of Baba, except for the factor $w-u\tan\theta$. This is due to the partial integration that leads from (12) to (14). For details see Bleistein and Handelsman.

The result just obtained is a good approximation in the case of thin ships, where the approximations of q and $\beta$ used here, hold in the whole field. Application to a thick ship leads to an unrealistic ray pattern near the ship. Application of the method of stationary phase in (14) results in a main contribution at the stationary points of the double integral (14). In these points we have $\partial S^*/\partial s$ and $\partial S^*/\partial\theta$ equal to zero. This leads to

$$\frac{dz_0/ds}{dx_0/ds} = \frac{-1}{\tan\theta} \tag{19}$$

and

$$\frac{z-z_0(s)}{x-x_0(s)} = \frac{-\tan\theta}{1+2\tan^2\theta}. \tag{20}$$

In order to look at the solution near the object we introduce the direction angle of the rays as

$$\tan\delta(\underline{x},s) = \frac{z-z_0(s)}{x-x_0(s)}.$$

Due to (20) the angle $\delta$ is confined to the region where $|\tan\delta| \leq 8^{-1/2}$, hence $\delta$ is smaller than the well-known angle of the Kelvin wave pattern, as in the linearized theory.
Introducing $\beta_c(s)$ as the direction angle of the double body velocity at the waterline by

$$\tan\beta_c(s) = \frac{dz_0/ds}{dx_0/ds} \tag{21}$$

we obtain for the stationary contribution

$$\tan\delta(\underline{x},s) = \frac{\tan\beta_c(s)}{2+\tan^2\beta_c(s)}. \tag{22}$$

Hence at the forward part of the ship we have $|\delta| \leq |\beta_c|$, which means that the rays are pointing into the ship. This is not very realistic. These results have no physical meaning at all. Our conclusion at this point is that it is not allowed to simplify the double body velocity components in (14). The ray pattern depends on the local velocities.
Because the waterline is not smooth, further asymptotic evaluations lead to wave excitation coefficients needed for the application of the ray method.

## 3. FAR FIELD FORMULATION

To compute the wave pattern and the wave resistance the continuation of the local solution into the far field must be considered. A powerful tool is the ray method. This method is well-developed in geometrical optics. For large wave numbers k, the ray method suggests the expansion

$$\phi(x,y,z) = \text{Re}\left[\left(\frac{U}{ik}\, a_0(x,y,z) + \frac{U}{(ik)^2 L}\, a_1(x,y,z) + \ldots\right) \times \exp[ikS(x,y,z)]\right] \tag{23}$$

where S is the phase function and the a's are dimensionless amplitude functions. (The length scale L can be chosen to be the length of the ship.)
We substitute this form in the Laplace equation and in (7) and take equal powers of k equal to zero. We obtain the following equations for S and a.

$$|\nabla S|^2 = 0 \qquad \text{for } y < 0$$
$$\frac{\partial S}{\partial y} + i(MS)^2 = 0 \qquad \text{at } y = 0 \tag{24}$$

and

$$2\nabla S.\nabla a_0 + a_0 \Delta S = 0 \qquad \text{for } y < 0$$
$$-i\frac{\partial a_0}{\partial y} + 2(MS)(Ma_0) + \left[M^2 S - \nabla\eta.\nabla S - \frac{\Phi_{ryy}}{U} MS\right] a_0 = 0 \qquad \text{at } y = 0 \tag{25}$$

where M is the differential operator

$$M = \frac{1}{U}\left[\Phi_{rx}(x,0,z)\frac{\partial}{\partial x} + \Phi_{rz}(x,0,z)\frac{\partial}{\partial z}\right] \tag{26}$$

and

$$\eta(x,z) = \frac{g}{U^2} h_r(x,z) = \frac{1}{2}\left[1 - (\phi_{rx}/U)^2 - (\Phi_{rz}/U^2\right]. \tag{27}$$

If the y-derivative in (24) is eliminated we obtain a nonlinear equation for S in the plane y = 0.

$$(MS)^4 - \left(\frac{\partial S}{\partial x}\right)^2 - \left(\frac{\partial S}{\partial y}\right)^2 = 0 \qquad \text{at } y = 0 \tag{28}$$

which is called the eiconal equation. From the asymptotic evaluation of (12) it follows that rays are generated at the bow and the stern. By means of a RK4 integration the rays are calculated starting at the bow and the stern. The wavefronts are also calculated. As an example the wavefronts generated at the stern are shown, they are compared with linear results.

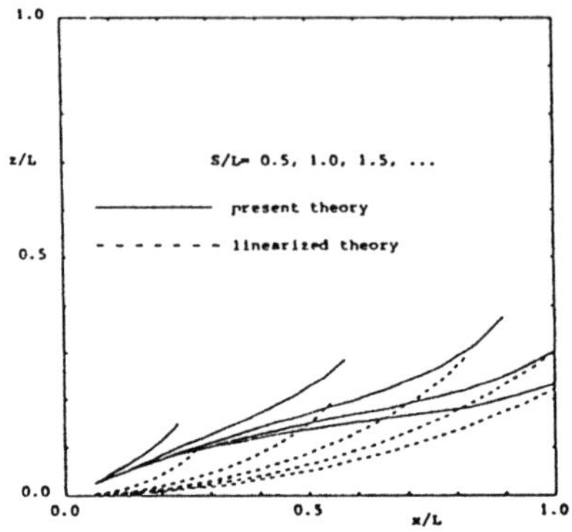


*Figure 2. Some wave fronts for the diverging bow-wave system $\beta_b = \pi/8$.*

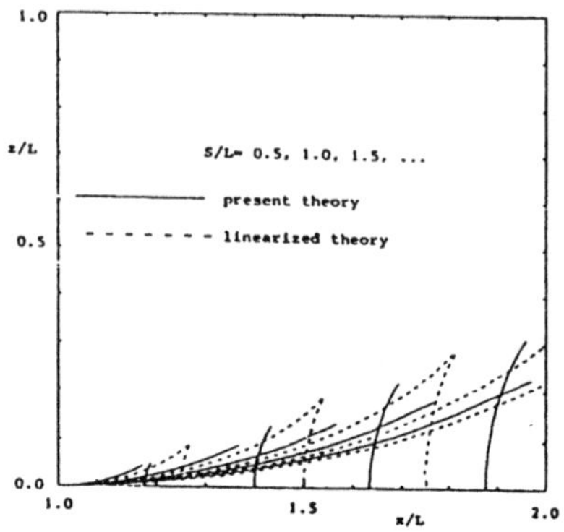


*Figure 3. Some wave fronts for both the diverging- and the transverse stern-wave system $\beta_s = \pi/8$.*

The excitation coefficients obtained from (14) make it possible to compute the waveheight just as well. Finally the wave resistance is calculated. Results are shown for two values of the bow angle $\beta_b$.

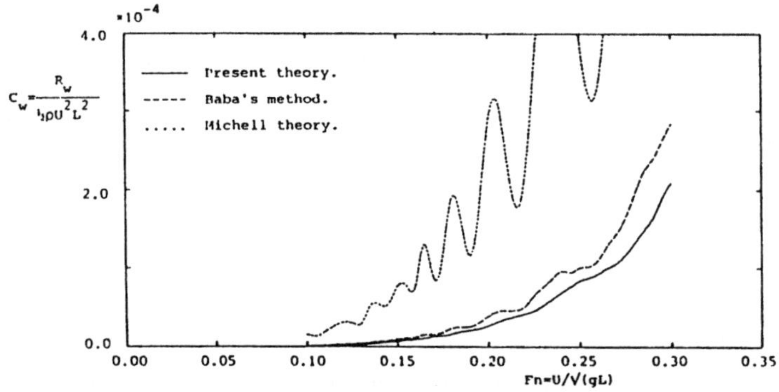


*Figure 4. Comparison of $C_w$ values. $\beta_b = 0.1974$.*

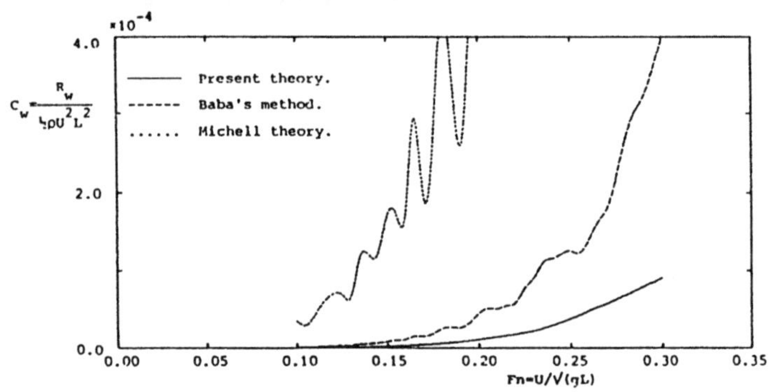


*Figure 5. Comparison of $C_w$ values. $\beta_b = 0.2915$.*

REFERENCES


1. Baba, E. (1976). Wave resistance of ships in low speed, Mitsubishi Technical Bulletin 109.
2. Baba, E. (1978). Numerical evaluation of a wave resistance theory for slow ships, Mitsubishi Technical Report 125.
3. Bleistein, N., Handelsman, R.A. (1975). Asymptotic expansions of integrals, Holt, Rinehart & Winston, New York.
4. Brandsma, F.J., Hermans, A.J. (1985). A quasi linear free surface condition in slow ship theory, *Shiffstechnik* 32, Heft 2, pp. 25-41.
5. Brandsma, F.J. (1987). Low Froude number expansions for the wave pattern and the wave resistance of general ship forms, Ph.D. Thesis, TU Delft.
6. Brard, R. (1972). The representation of a given ship form by singularity distributions when the boundary condition on the free surface is linearized, *J. of Ship Res.* 16, pp. 72-91.
7. Dagan, G. (1975). Waves and wave resistance of thin bodies moving at low speed: the nonlinear free-surface effects, *J. of Fluid Mech.* 69, pp. 405-416.
8. Eggers, K. (1980). Contributions to the wave resistance workshops in Washington (1979) and IZU (1980), Inst. für Schiffbau der Univ. Hamburg, Bericht nr. 396.
9. Eggers, K. (1981). Non-Kelvin dispersive waves around non-slender ships, *Shiffstechnik*, Band 8, pp. 223-251.
10. Erdelyi, A. (1956). Asymptotic expansions, Dover Publ. Inc., New York.
11. Hermans, A.J. (1974). A matching principle in non-linear ship wave theory at low Froude numbers, *Delft Progress Rep.* 1.
12. Hermans, A.J. (1980). The wave pattern of a ship sailing at low speed, Appl. Math. Inst. Univ. of Delaware, Techn. Rep. 84A.
13. Kayo, Y., Takekuma, K., Eggers, K., Sharma, S.D. (1982). Observation of free surface shear flow and its relation to bow wave-breaking of full forms, Inst. für Schiffbau der Univ. Hamburg.
14. Keller, J.B. (1979). The ray theory of ship waves and the class of streamlined ships, *J. of Fluid Mech.* 91, pp. 465-488.
15. Keller, J.B., Ahluwalia, D.S. (1976). Wave resistance and wave patterns of thin ships, *J. of Ship Res.* 20, No. 1, pp. 1-6.
16. Kochin, N.E., Kibel, I.A., Roze, N.V. (1964). Theoretical hydromechanics, Interscience Publ., New York.
17. Korving, C., Hermans, A.J., The wave resistance for flow problems with a free surface, Second Int. Conf. on Num. Ship Hydr., Univ. of California.
18. Lewis, R.M., Keller, J.B. (1964). Asymptotic methods for partial differential equations, N.Y.U. Courant Inst., Res. Report EM-194.

19. Maruo, H. (1979). On the free surface flow around a two-dimensional body fixed in a uniform stream, Proc. of the 29th Japan National Congress for Appl. Math.
20. Maruo, H. (1980). On the free surface flow around full hull forms at low Froude numbers, Bull. of the Fac. of Engineering, 29, Yokohama Nat. Univ.
21. Michell, J.H. (1898). The wave resistance of a ship, *Phil. Mag.* 45, pp. 106-123.
22. O'Malley, R.E. (1970). On non-linear singular perturbation problems with interior non-uniformities, *J. Mat. Mech.* 19, pp. 1103-1112.
23. Ogilvie, T.F. (1968). Wave resistance: the low speed limit, Rep. 002, Dept. of Nav. Arch. and Mar. Engin., Univ. of Michigan.
24. Ogilvie, T.F. (1976). On non-linear wave resistance theory, Int. Seminar on Wave Resist., Tokyo.
25. Salvesen, N. (1969). On higher order wave theory for submerged two-dimensional bodies, *J. of Fluid Mech.* 38, pp. 415-432.
26. Salvesen, N., Von Kerczek, G. (1976). Comparison of numerical and perturbation solutions of two-dimensional non-linear water-wave problems, *J. of Ship Res.* 20, No. 3, pp. 160-170.
27. Stoker, J.J. (1957). Water waves, Interscience Publ., New York.
28. Tuck, E.O. (1965). The effect of non-linearity at the free surface on flow past a submerged cylinder, *J. of Fluid Mech.* 22, pp. 401-414.
29. Van Dyke, M. (1964). Perturbation methods in fluid mechanics, Academic Press, New York.
30. Wehausen, J.V. (1973). The wave resistance of ships, *Adv. in Appl. Mech.* 13, pp. 93-245.
31. Wehausen, J.V., Laitone, E. (1960). Surface waves, *Encyclopedia of Physics* 9, pp. 446-778, Springer Verlag, Berlin.
32. Yim, B. (1981). A ray theory for non-linear ship waves and wave resistance, Third Int. Conf. on Num. Ship Hydr., Paris.

# On Stationary Waves Superposed to the Flow Around a Body in Uniform Stream

KLAUS EGGERS

Institut für Schiffbau der Universität Hamburg, F.R.G.

Over the last decade, certain interest has been focussed on the rate of increase of the stationary ship wave resistance with the speed U when $U \sim 0$. The term "slow ship theory" stands for some attempts to derive flow models and resistance therefrom through more or less rational perturbation analyses governed by an order scheme regarding powers of the Froude number Fn, where the unperturbed state is the "double body flow", corresponding to the limiting case of infinite gravity, where a rigid wall boundary condition can be imposed over the undisturbed free surface.

Such approaches are motivated through the fact that over wide areas of the wetted ship surface this flow provides a far better first approximation than the parallel flow underlying the perturbation analysis in powers of the beam/length ratio (= thin ship theory); it is implicitly hoped that expansions in powers of Fn should have better convergence, even up to $Fn \sim 0.7$ according to Keller |1|.

It has turned out, however, that such analyses require certain bold assumptions and can lead to some debatable ambiguities, which are not reflected in experimental evidence. Thus from the fact that for a simple oscillatory wave potential the spatial derivatives certainly are of lower power in Fn, a postulate is drawn that accordingly certain terms of "small order" have to be disregarded. It is displayed in the following note that such terms may represent essential features of the physical flow model underlying (if not stabilizers against nonuniformity as claimed by Brandsma |2|). In particular, such postulate is incompatible with conventional methods to deal with surface tension, and a systematic treatment of the slow ship problem cannot by-pass surface tension effects. The "ray-tracing" approach from geometrical optics in particular is sensitive as far as rays are assumed to emanate from bow and stern stagnation points, where the lower limit to wave length due to surface tension significantly modifies the purely mathematical abstraction. -

Let us consider a uniform flow $\nabla Ux$, gradient to a potential $Ux$, in a domain

K. Horikawa, H. Maruo (Eds.)
Nonlinear Water Waves
IUTAM Symposium Tokyo/Japan 1987

$z<0$, with $z$ increasing upwards, $y$ sidewise. If we preserve the condition of no vertical flow at the undisturbed free surface $z=0$, but insert a fixed body, we have to consider a flow $\nabla U\phi_r$ of the "double body potential" $U\phi_r$. For finite Froude number, we may decompose the total potential $U\phi$ as $U\phi_r+U\hat{\phi}$; we have the surface deformation $z = \zeta(x,y) = U^2\{1 - (\nabla\Phi\nabla\Phi)\}/2g$, where the condition of zero convective pressure derivative $U\Phi_z = U^2\{\nabla\phi\nabla\{(\nabla\phi\nabla\phi)\}\}$ holds.

Seeking for a linear problem for an approximation $\overset{\dagger}{\phi}$ to the disturbance potential $\hat{\phi}$, adopting the rule of order change for $\overset{\dagger}{\phi}$, it can be concluded that $\overset{\dagger}{\phi} = O(U^4)$ hence $\nabla\overset{\dagger}{\phi} = O(U^2)$, and that $\overset{\dagger}{\phi}$ has to satisfy

$$g\overset{\dagger}{\phi}_z+U^2\{\overset{\dagger}{\phi}{}^2{}_{rx}\phi_{xx}+2\phi_{rx}\phi_{rx}\overset{\dagger}{\phi}_{sy}+\phi^2{}_{ry}\overset{\dagger}{\phi}_{yy}\}=(\zeta_r\phi_{rx})_x+(\zeta_r\phi_{ry})_y$$

on the surface $z = U^2\{1 - (\nabla\phi_r\nabla\phi_r)\}/2g$ in order to keep the error $o(U^3)$. Neglecting terms $o(U^3)$, introducing: $u \doteq U\phi_{rx}(x,y,0)$, $v \doteq U\phi_{ry}(x,y,0)$, $\zeta_r(x,y) \doteq (U^2-u^2-v^2)/2g$, $D_r(x,y) \doteq (u\zeta_r)_x + (v\zeta_r)_y$ this equation may be replaced through

$$g\Phi_z + u^2\Phi_{xx} + 2uv\Phi_{xy} + v^2\Phi_{yy} = D_r(x,y) \tag{A}$$

to be satisfied on the isobaric surface $z = \zeta_r(x,y)$.

Let us now a s s u m e that without significant error we may as well satisfy (A) on the plane $z = 0$. Considering $u$ and $v$ as slowly varying functions, near a point $\{x_p,y_p,0\}$ a wavy solution of (A) must have the form $\sim \exp\{k(z+i(x-x_p)\cos\theta + i(y-y_p)\sin\theta))\}$, with $k = \hat{k}(\gamma) \doteq g/((u^2+v^2)\cos^2\gamma)$, $\gamma \doteq \theta + \beta$, $\beta \doteq \beta(x_p,y_p) \doteq \arctan(u/v)$, so that $\gamma$ is the angle of the wave front normal against the flow tangent. This is the conventional dispersion relation for ship waves in a flow of intensity $Uq$, $q\doteq\sqrt{u^2+v^2)/U^2}$. The requirement of stationarity yields for phase- and group velocity

$$c = Uq\cos\gamma = \sqrt{g/k} \quad ; \qquad c_g \doteq c + k\,\frac{dc}{dk} \doteq \frac{c}{2}$$

But near the stagnation points, with $q \to 0$, we find $k \to \infty$ , i.e. zero wave length ! Adding a term $U\kappa\phi_{zzz}$ for surface tension in (A), a quadratic equation for $k$ is found with roots

$$_1k(\gamma) = \hat{k}\;\frac{2}{1 + \sqrt{1 - c_m{}^2(Uq\cos\gamma)^2}} \quad ; \qquad _2k(\gamma) = g/(\kappa\cdot{}_1k)$$

Here $c_m \doteq \sqrt[4]{4g\kappa}$ is the minimum of $c = \sqrt{g/k+\kappa k}$ . Due to the restriction $\cos\gamma > c_m/Uq$ for the square root, both $_1k$ and $_2k$ remain finite ! For $q<c_m/U$, no waves can exist, as observed by Maruo and Ikehata |3|. However, the inclusion of the capillary term in (A) implies a violation of the order change rule, which would assign $O(U^{-2})$ to this term. Thus this rule, conceived for the case $U \to 0$ , must be dropped for a proper handling of the significant

capillary effects in this limit !

Let us consider an alternative approach under d i s r e g a r d of this rule, based on invariance requirement for the ship wave resistance R |4|. Let S be a control surface enclosing the body from below up to the free surface $z = \zeta(x,y)$, let C stand for the closed line where it intersects z=0, let the normal vector $\vec{n}$ point away from the ship. An exact expression for $R^+ \doteq R/(\rho U^2)$ is

$$R^+ = -1/2\{\iint\{((\nabla\Phi\nabla\Phi) -1)\ n_x - \Phi_n\Phi_x\}dS + \int_C \zeta^2 n_x dC/U^2\}$$

Neglecting terms $o(U^2)$, this may be replaced by

$$R^+ = -1/2\{\iint\{(\nabla\Phi\nabla\Phi)n_x - \Phi_n\Phi_x\}dS_o - \int_C \xi_o^2\ n_x dC/U^2\}$$
$$+\int_C \zeta_o\Phi_x\{\Phi_x n_x + \Phi_y n_y\}\,dC$$

with $\zeta_o \doteq U^2(1-\Phi_x^2(x,y,0)-\Phi_y^2-\Phi_z^2)/2g$ . Here $S_o$ is the part of S with z < 0. The requirement that the above expression should be invariant under change of S (save terms $O(U^4)$) is met if

$$\Phi_x\Delta\Phi = o(U^4)\ ;\quad \Phi_x(U\Phi_z - D^+) = o(U^5)\ \text{at}\ z=0\ ;\quad D^+(x,y) \doteq (\zeta_r\Phi_x)_x + (\zeta_r\Phi_y)_y\ .$$

Up to $O(U^4)$, we obtain

$$u^2\phi_{xx} + 2uv\phi_{xy} + v^2\phi_{yy} + g\phi_z - D_r = g\zeta_r(\phi_{xx}+\phi_{yy}) + (2g\zeta_{rx}+uw_z)\phi_x + (2g\zeta_{ry}+uw_z)\phi_y \quad (A^+)$$

to be satisfied on the horizontal plane z = 0, with $w_z \doteq U\phi_{rzz}(x,y,0)$.

The first term on the r.h.s. of $(A^+)$ could have been obtained from a linear vertical Taylor expansion of $g\phi_z$ in (A). However, the rule of order change makes such expansion divergent, however small $\zeta_r$. The terms containing $\phi_x$ and $\phi_y$ cause some amplitude modulation for periodic wave solutions of $(A^+)$ in a certain analogy with results of Longuet-Higgins and Stewart |5|.

For $(A^+)$ replacing (A), the wave number $k^+$, phase- and group velocities $c^+$ and $c_g^+$ are now related through $k^+(\gamma) = \hat{k}(\gamma)\cdot G$ ,

$$c^+ = Uq\cos\gamma = g/k^+\sqrt{1+\zeta_r k^+}\ ;\quad c_g^+ = c^+ + k^+\frac{dc^+}{dk^+} = c^+\ (1\ -\ 1/2\,G)\ ,$$

where the factor $G \doteq 2q^2\cos^2\gamma/((1+2\cos^2\gamma)q^2-1)$ equals unity for $q^2 = 1$.

We may observe that now over the range $q^2<1/3$ , where G is negative, no waves can exist. At the boundary of the corresponding domains near the stagnation points, the wave number $k^+$ tends to infinity, as for $q^2 = 0$ under (A). Adding a term $U\kappa\phi_{zzz}$ for the influence of surface tension in $(A^+)$ as in (A) before, introducing a factor $W \doteq \sqrt{1-(c_m/c)^2G^2}$ , we find the two roots $k = {}_\nu k(\gamma)$ as

$$ {}_1k(\gamma) = k^+(\gamma)(1+W)/2 \; ; \quad {}_2k(\gamma) = g/\kappa_1 k \; ; \quad {}_\nu c_g = c(1+(-1)^\nu W/2G) \quad \nu = 1,2 $$

with $c = c(\gamma) = Uq\cos\gamma$ as before. $c_g$ exceeds c for the genuine capillary waves where $\nu = 2$. The boundaries of the regions without waves are now given through

$$ q^2 = (1 + 2\,(c_m/U)^2\,)/3 $$

Here we have $\cos_\nu\gamma = (-1)^{\nu+1}$ ; capillary waves emanate where the double body flow is entrant, gravity type waves start sidewise, where this flow is outgoing. Certain rays related to propagation of such waves are tangent to the double body flow, as will be shown later.

The occurrence of wave free zones even ahead of larger blunt ship models, for which capillarity is insignificant, may find a certain explanation from above analysis, though the dominant phenomena of wave breaking certainly result from different effects such as free surface boundary layers and vortices generated therefrom |6|

Let us concentrate on implications from replacing (A) through ($A^+$), let us for simplicity disregard all terms not connected with second derivatives of $\phi$. The 2-D partial differential equation

$$ (u^2-g\zeta_r)\phi_{xx} + 2uv\phi_{xy} + (v^2-g\zeta_r)\phi_{yy} = 0 $$

may be called elliptic, parabolic and hyperbolic in the relevant domains where the discriminant $(u^2-g\zeta_r)(v^2-g\zeta_r)-u^2v^2 = 3(q^2-1)(q^2-1/3)\cdot U^4/4$ is positive, zero or negative; the corresponding term from (A) is identically zero. However, the transition with increasing $q^2$ is continuous, so that no "shocks" should be expected across the lines $q^2 = 1$. This is evident from the dependence of a dimensionless wavelength from the wave angle as displayed in Fig. 2. We may observe that the wavelength remains positive for $q^2<1$ the wave angle is bounded by $\pi/2-\alpha_M$, $\alpha_m \doteq \arcsin c/Uq$ with $c \doteq U\sqrt{(1-q^2)/2}$ ; $c$ and $\alpha_M$ could be interpreted as "critical speed" and "Mach angle" in the hyperbolic domain |4| .

## The geometry of rays for wave propagation

With $q^2$, $\beta$ and $\gamma$ given for a point on the undisturbed free surface, a direction angle $\alpha$ may be defined through the condition of stationary phase

$$ \frac{d}{d\gamma}\,(k\cos(\alpha-\gamma)) = k\,\left(\tan(\alpha-\gamma) - \frac{1}{k}\frac{dk}{d\gamma}\right) = 0 $$

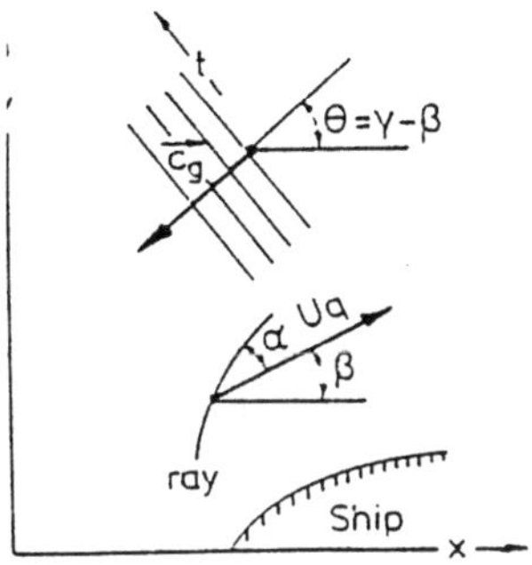


Fig.1 Coordinates and notation

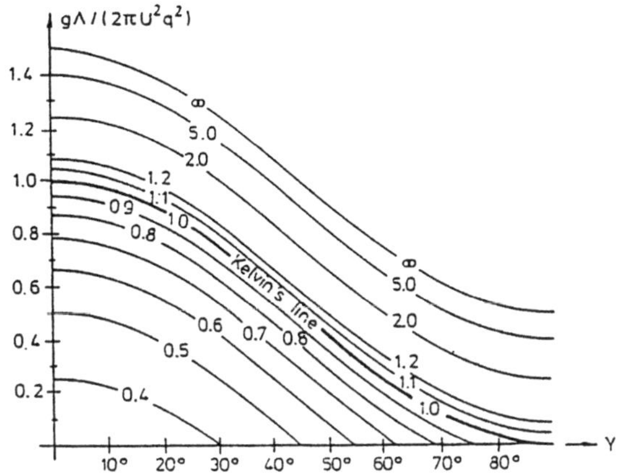


Fig.2 Dimensionless wavelength in dependence from wave angle and local flow intensity q

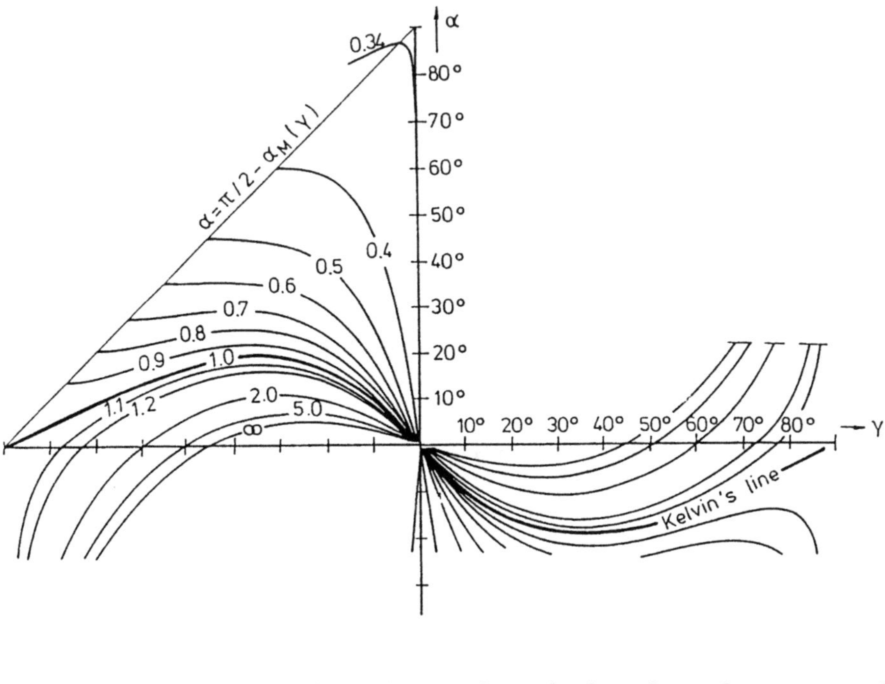


Fig.3 Direction α of stationary phase in dependence from wave angle γ (both measured against double body flow) for various $q^2$

In Fig. 3 we find $\alpha$ plotted against $\gamma$ for different values of $q^2$. Note that for $q^2<1$, the range of $\alpha$ is limited by $\pi/2-\alpha_M$ (otherwise, with $q^2<1/3$, we had negative $k$, no decay downward). Up to $q \sim 0.7$ there is a monotonic dependence of $\alpha$ on $\gamma$ and hence on $k(\gamma)$, we find "transverse waves" only. For larger $q$ however, $d\alpha/d\gamma$ changes sign, for the range of "divergent waves" $k$ decreases with $|\alpha|$ increasing. For $q^2>1$, we find a change of sign for $\alpha$ with larger $|\gamma|$; from Fig. 4, where the ratio of $c_g$ to $c$ is plotted for various $q^2$ against $\gamma$ under $(A^+)$, we find a corresponding change in sign for $c_g$. One may wonder if thus in the domain $q^2>1$ certain short waves cannot propagate further and vanish under some dissipation of energy with $c_g$ approaching zero.

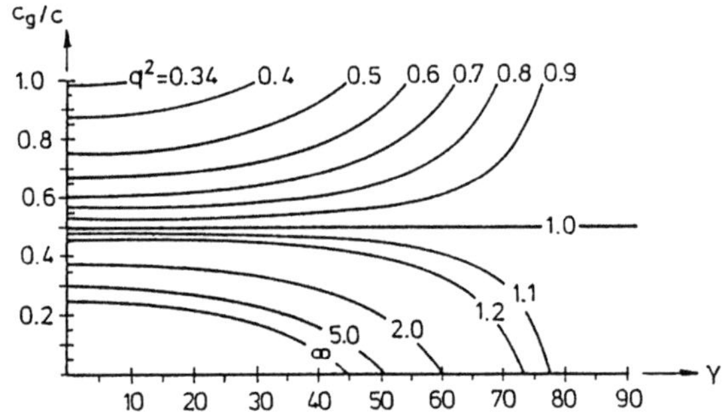


Fig. 4 Ratio of group velocity $c_g$ to phase velocity $c$ for different $q^2$ and $\gamma$

The concept of group velocity is helpful for a qualitative understanding of wave propagation. In particular, the direction $\alpha$ of stationary phase may equivalently be defined as that of the resultant of the double body flow and a group velocity vector $\vec{c}_g \doteq \{-c_g\cos\theta; -c_g\sin\theta\}$, as we shall see; one should be aware, however, that in general this vector is not suitable for representing transport of wave energy. - Observing $c=Uq\cos\gamma$, thus $c_g = Uq(\cos\gamma - k\,d\gamma/dk\,\sin\gamma)$

$$\tan(\gamma-\alpha) = \frac{\sin\theta\cos(\alpha+\beta)-\cos\theta\sin(\alpha+\beta)}{\cos\theta\cos(\alpha+\beta)+\sin\theta\sin(\alpha+\beta)} = D \quad \text{with}$$

$$D \doteq \frac{1}{k}\,\frac{dk}{d\gamma} = \frac{c-c_g}{Uq\,\sin\gamma} \qquad \text{hence}$$

$$\tan(\alpha+\beta) = \frac{\sin\theta - D\cos\theta}{\cos\theta + D\sin\theta} = \frac{(Uq\cos\gamma - c_g)\sin\theta - Uq\cos\gamma\cos\theta}{(Uq\cos\gamma - c_g)\cos\theta + Uq\cos\gamma\cos\theta} = \frac{c_g\sin\theta - Uq\sin\beta}{c_g\cos\theta - Uq\cos\beta}$$

Solving for $\alpha$, we find

$$\cot\alpha = -\cot\gamma\,(1 + \frac{c}{c_g}\tan^2\gamma)$$

This is an extension of Kelvin's case where $c/c_g$ equals two. Note that in order that a ray is "outgoing", i.e. for a positive angle $\alpha$ against the double body flow, the wave angle $\gamma$ must be negative; hence for rays emanating from the ship's water line, we need $\beta < \theta$ at their origin.

Such "rays", i.e. lines with $dy/dx = \tan(\alpha+\beta)$, define a system of characteristics to a first order partial d.e. $F(x,y,S_x,S_y) = 0$, where the function S, the "eikonal", is some generalized phase function. Introducing a 2-D wave number vector $\{k_1;k_2\}$ with $k_1 \doteq k\cos\theta = S_x$, $k_2 \doteq k\sin\theta = S_y$, $F(x,y,k_1,k_2) = 0$ is equivalent to our dispersion relation connecting k and $\gamma$ (and hence $\theta=\gamma-\beta$) for $q^2$ and $\beta$ given. The direction of characteristics may as well be defined as that parallel to the gradient of F (more physically: the gradient of encounter frequency) in the wave number plane. Defining a function K expressing the wave number k in terms of $x,y,k_1$ and $k_2$ through

$$K^2(x,y,k_1,k_2) \doteq F(x,y,k_1,k_2) + k_1^2 + k_2^2, \quad \text{with } \frac{dk}{d\theta} = \frac{dk}{d\gamma},$$

$$F_{k_1} = -2k\cos\theta + 2k\frac{dk}{d\theta}\theta_{k_1};\quad F_{k_2} = -2k\sin\theta + 2k\frac{dk}{d\theta}\theta_{k_2};\quad \theta_{k_1} = -\frac{\sin\theta}{k};\ \theta_{k_2} = \frac{\cos\theta}{k},$$

k depends on $\theta$ only with x and y held constant; hence

$$\frac{dy}{dx} = \frac{F_{k_2}}{F_{k_1}} = \frac{\sin\theta - D\cos\theta}{\cos\theta + D\sin\theta} \qquad \text{as before .}$$

<u>Variation of the wave front angle along a ray</u>

The ray theory of geometrical optics is conceived for the case that the eikonal and hence $k_1$ and $k_2$ are one-valued functions of x and y, though a superposition of transverse and diverging wave systems may be permitted. $k_1$ and $k_2$, though interrelated through the dispersion relation, will vary along characteristics. If we introduce a curve parameter $\tau$ through $d\tau \doteq dx/F_{k_1} = dy/F_{k_2}$, we find

$$\frac{d\theta}{d\tau} = \frac{d}{d\tau}\arctan(k_2/k_1) = (k_1\frac{dk_2}{d\tau} - k_2\frac{dk_1}{d\tau})/k^2$$

$$= \frac{1}{k}\left((\cos\theta\,(k_{2_x}\frac{dx}{d\tau} + k_{2_y}\frac{dy}{d\tau}) - \sin\theta\,(k_{1_x}\frac{dx}{d\tau} + k_{1_y}\frac{dy}{d\tau})\right)$$

$$= \frac{1}{k}(\cos\theta\,(F_{k_1}k_{1_y} + F_{k_2}k_{2_y}) - \sin\theta(F_{k_1}k_{1_x} + F_{k_2}k_{2_x})) = \frac{1}{k}\frac{\partial F}{\partial t} = 2\,\frac{\partial K}{\partial t}$$

where the local variable t is defined through its gradient vector $\{-\sin\theta;\cos\theta\}$ parallel to the wave crest. We have made use above of the irrotationality of the wave number vector as gradient of S; more directly, we could have used $dk_1/d\tau = -F_x$ and $dk_2/d\tau = F_y$ instead. For transverse waves and $y>0$, we have $\partial k/\partial t > 0$ in general and hence $d\theta/d\tau > 0$. With $d\beta/d\tau < 0$ there as well, we must expect that along a ray $\gamma$ will increase continuously finally attaining positive values and hence inward inclination. This explains why rays from transverse waves at the bow become re-entrant |7|,|8| and that the waterline cannot coincide with a ray path.

Thus for transverse waves, one should expect that the local Kelvin angle will finally be decreased, whereas for the divergent bow waves this angle, measured against double body flow, should be widened. Experiments are needed to clarify such predictions and the validity of the ray approach.-

## References


1. Keller, J.B.: The ray theory of ship waves and the class of stream lined ships. J. Fluid Mech. 91 (1979) 465-487

2. Brandsma, F.J. and Hermans, A.J.: A quasi-linear free surface condition in slow ship theory. Schiffstechnik 32 (1985) 25-46

3. Maruo, H. and Ikehata, M.: Some discussion on the free surface flow around the bow. Proc. 16th ONR Symp. on Naval Hydrodynamics (1986)

4. Eggers, K.: Non-Kelvin dispersive waves around non-slender ships. Schiffstechnik 28 (1981) 223-252

5. Longuet Higgins, M.S. and Stewart, R.W.: Changes in the form of short gravity waves on long waves and tidal currents. J. Fluid Mech. 8 (1960) 566-583

6. Takekuma, K. and Eggers, K.: Effect on bow shape on free surface shear flow. Proc. 15th ONR Symp. on Naval Hydrodynamics (1984) 387-405

7. Inui, T. and Kajitani, H.: A study on local non-linear free surface effects in ship waves and wave resistance. Schiffstechnik 24 (1977) 178-213

8. Yim, B.: A ray theory for non-linear ship waves and wave resistance. Proc. Third Int. Conf. on Num. Ship Hydrodynamics Paris (1981) 55-70

# Session 10

# Second-order Nonlinear Water Waves

# Second-order Waves – Importance in Experiment and Nature

V. BARTHEL* AND E.P.D. MANSARD**
*Dr.-Ing., Senior Research Officer
**Dr.-Ing., Associate Research Officer
National Research Council of Canada
Ottawa, Ontario, K1A 0R6, Canada

Summary

Waves of finite amplitude are known to contain harmonics, in the low and high frequency domains, which propagate bound to the first order waves. However, use of first order wave generation technique in laboratory flumes and basins fails to correctly reproduce these second-order components and therefore leads to incorrect test results. The paper demonstrates successful application of second-order wave generation technique for the correct reproduction of these harmonics under different modes of operation of the wave paddle. Knowledge and application of this technique have been proven to be important in model studies and prototype wave data analysis.

## 1. Introduction

The effects of second-order harmonics have long been neglected in the assessment of wave conditions in nature and in the determination of structural response either through numerical or physical models. As early as 1847, Stokes indicated the existence of a second-order harmonic bound to a regular sinusoidal wave. This (high) harmonic, termed here as super-harmonic, has the tendency to sharpen the crest and flatten the trough of the waves. The concept of sub-harmonics in the low frequency domain was first illustrated by Longuet-Higgins and Stewart (4). Using the radiation stress balance, they were able to show that a MWL set-down occurs under the groups of high waves, with a set-up in between groups. Both, the super- and sub-harmonics are bound to their primary components. A simple illustration of these two components is given in Figure 1. The figure shows a single wave with and without its super-harmonics ($2 \times f_1$) as well as a bichromatic wave (composed of two sinusoids) with and without its sub- ($f_2 - f_1$) and super-harmonics ($2f_1$, $2f_2$, $f_1 + f_2$). The correct reproduction of these natural phenomena under laboratory conditions is, therefore, of importance for tests with structures whose response frequency lies within the range of either sub- or super-harmonics.

K. Horikawa, H. Maruo (Eds.)
Nonlinear Water Waves
IUTAM Symposium Tokyo/Japan 1987

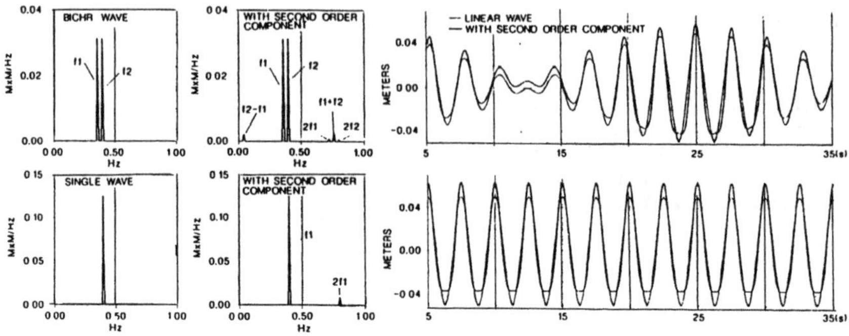

FIGURE 1 WAVES WITHOUT/WITH SECOND ORDER COMPONENTS

The horizontal motions of moored vessels are excited by slow drift oscillations whereas the steepness of waves and the crest heights are affected by the super-harmonics and may lead to higher forces and overtopping on fixed structures.

Second-Order Wave Generation

If the surface elevation of the first order wave train is presented as a Fourier series of sinusoidal components

$$\eta^{(1)}(t) = \sum_{1}^{n} a_n \cos\omega_n t + b_n \sin\omega_n t \tag{1}$$

where $a_n$, $b_n$ = Fourier coefficients; $\omega_n$ = angular frequency

the inclusion of second-order boundary conditions in the solution of the Laplace equation yields the expression for the second order sub- and super-harmonics given here for the simplest wave train composed of two sinusoids:

$$\eta_{nm}^{(2)}(t) = G_{nm}^{(\pm)} \left[(a_n a_m \mp b_n b_m)\cos(\omega_n \pm \omega_m)t + (a_m b_n \pm a_n b_m)\sin(\omega_n \pm \omega_m)t\right] \tag{2}$$

upper signs are super-harmonics (short waves)
lower signs are sub-harmonics (long waves)

$G_{nm}^{(\pm)}$ is the transfer function between first and second-order waves and was first derived by Ottesen-Hansen (7) for the subharmonics. Sand and Mansard (9) computed the transfer function for super-harmonics and verified their existence in flume experiments. Although the magnitudes of these harmonics could be found by equation (2), reproduction in the flumes is difficult due to their bounded nature. If just the first order (classical) wave generation is used, the boundary conditions at the paddle are not satisfied resulting, therefore, in the following spurious components:

- parasitic long wave generated due to the vertical boundary not moving according to second order requirements,
- displacement wave generated due to the paddle being out of its mean position $x = 0$, and
- local disturbance wave generated due to the paddle motion not matching the vertical velocity profile.

All these spurious waves travel at free wave celerity and - dependent on the location in the flume or basin - suppress or augment the natural bounded harmonics. Hence, the concept of correct wave generation has to include second-order terms for the paddle displacement which suppress these free, spurious waves and allow only the bounded (super- and sub-harmonics) waves to propagate with the first order surface elevations.The solution of the Laplace equation with correct boundary conditions and the effectiveness of the suppression of spurious long waves, in a laboratory flume, were shown by Barthel et al. (1). Sand and Mansard (9) illustrated the same for spurious short waves. However, these experiments were related only to piston mode wave generation. Since many laboratories use other generation modes such as flap or combined mode, the validity and the effectiveness of the method had to be proven for various modes as well. Results of investigations carried out to this effect are given below.

## 3. Experiments

Detailed equations for the "general case" of wave paddle motions had already been given in (1) for long waves and were reviewed in (10). The expression given for $\beta_i$ in Appendix A of (1) did not include the general location of the hinged wave maker $(h+\ell)$ and should actually read

$$\beta_i = \frac{k_i(h+\ell)\ [\cosh(k_ih)\ \sinh(k_ih) + k_ih]}{2\ \sinh(k_ih)\ [k_i(h+\ell)\ \sinh(k_ih) - \cosh(k_ih) + 1]}$$

General equations for short waves could be found in Sand and Mansard (9). The verification of the theory was undertaken in a flume whose dimensions are 1.2 x1.2 x67 m and which is equipped with a hydraulically driven wave maker for irregular waves, capable of operating in piston, flap and combined mode. Twin wire capacitance wave probes were used to measure water surface elevations. It has to be emphasized that special attention had to be given to the design of the beach. Since the long wave components cannot easily be dissipated, the porous beach at the end of the 67 m long flume has to be sloped as mildly as possible with a considerable layer of gravel between the toe of the beach and the vertical wall at the extremity of the flume. In order to demonstrate the effectiveness of spurious wave suppression, tests with a combination of water depths, wave board modes, peak frequencies and wave heights were run with and without compensation for spurious waves to demonstrate the overall efficiency of the second-order generation.

## 4. Investigation Results

### 4.1. Second-Order Long Waves (Sub-harmonics)

Data were sampled for a period of 400 s after the generation had been activated. The procedure for isolating the measured long waves and comparing them with their expected counterparts is described in detail in (1). It consists mainly of two operations: separation of measured long waves by filtering process, and computation of the expected long waves by applying the FFT technique and the equation (2) on the primary wave train. Comparison is then done by computing the RMS value of the measured and expected long waves and by superposing their time and spectral domain characteristics for visual observation as illustrated in Fig. 2. It shows, as one example out of a long test series, the variance spectral densities of long waves, the entire wave train and portions of the measured and calculated time series of the first and second-order waves. Although the long waves have amplitudes in the order of only a few millimeters, the match throughout the record is very good as opposed to the lower part of Fig. 2 where wave suppression was not carried out.Special attention was given to a procedure designed to evaluate the phase match between theoretical and measured low frequency oscillations of the mean water level. Phase matching of the bounded long waves with the wave groups is considered to be even more important than accurate amplitude matching since many floating structures respond to the combined impact of groups of primary waves and the set-down connected with their groups. Both, measured and calculated, long waves were cross-correlated resulting in the time shift for maximum correlation. Results demonstrate that,

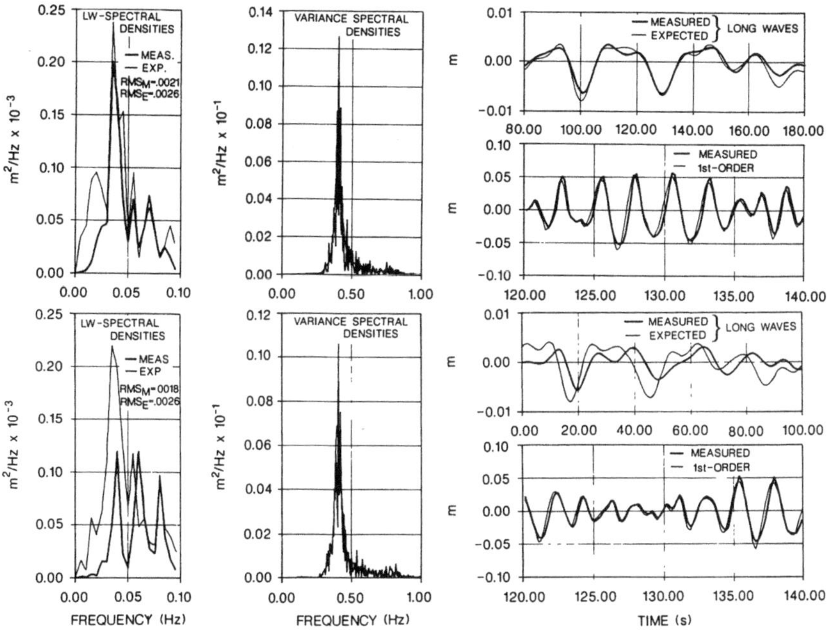


FIGURE 2 IRREGULAR WAVES ($f_p$ = 0.42 Hz) WITH/WITHOUT SUPPRESSION

although inclusion of the suppression yields up to 80% smaller values for the time shift $\tau$, the values for the same wave conditions vary with water depth and with the waveboard mode and are generally higher for bichromatic waves. Among the possible explanations that can be found, is the fact that the bounded long waves with periods of typically around 15-20 s are not easily dissipated by the porous beach at the end of the flume.

## 4.2 Second-Order Short Waves

Unlike the long second-order waves, the super-harmonics are not distinctly separated in the frequency domain from the first order waves,but have overlapping frequency components. Simple filtering techniques could, therefore, not be used to isolate them from the first-order spectrum. Computation of the high harmonics is therefore carried out by calculating these components using a separation technique advanced by Sand and Mansard (9). This technique consists of starting at a low frequency component which could be considered as pure first-order waves and then computing its related second-order wave component. This component is then subtracted from the total wave train. By progressing in this fashion from lower to higher frequencies, the total wave train can be decomposed into first and second order components. The effectiveness of the

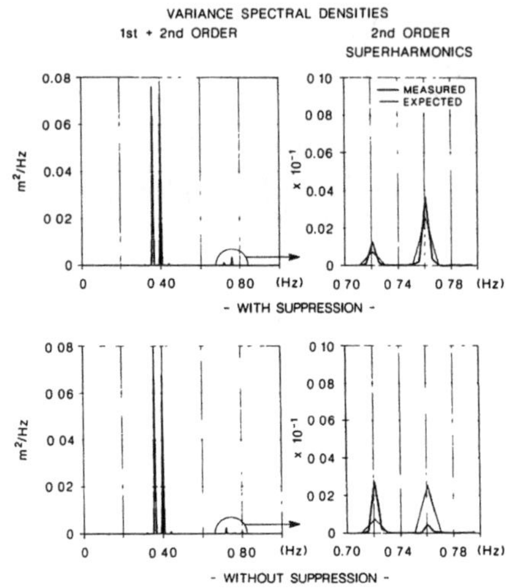


FIG. 3 SPURIOUS SHORT WAVE SUPPRESSION

suppression can be best demonstrated by a test with bichromatic waves. Figure 3 shows the entire spectrum with sub- and super-harmonics clearly indicated. The frequency range between .7 and .8 Hz has been enlarged in the right hand part of the plot in order to illustrate the effect of suppression.

### 4.3 Waveboard Type

Non-linearities are predominant in shallow water waves which are best reproduced with a generator operating in the piston mode since orbital velocities do not vary much over the water depth. Trying to reproduce these by a flapper mode of operation requires a larger stroke which often exceeds the maximum excursion limit of the machine. However, there are some cases in which the piston mode of generation may not be applied. For this type, the effectiveness of suppression tests for sub- and super-harmonics had to be verified. In order to keep the wave conditions similar under all modes of operation, only a significant wave height of less than 10 cm could be generated. Figure 4 shows comparable results for flap, piston and combined mode and clearly demonstrates that the method can be successfully applied for all types of machines, although matching varies with the type of waveboard used. Matching, particularly for flap mode, could be improved if a larger stroke was used. Matching of super-harmonics for all modes could not be included here due to lack of space.

## 5. Application in Model Investigations

Several studies have been performed where suppression of spurious waves was applied and results showed distinctly that the correct reproduction of long waves was necessary in order to have a realistic response of the structures. While investigating the motions and mooring loads of a 227,000 dwt tanker, Mansard and Pratte (5) found that increased horizontal motions and mooring loads were obtained with the suppression of

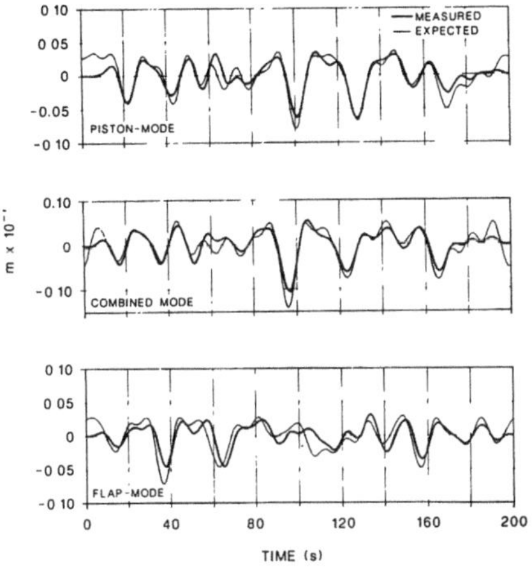


FIG. 4 VARIOUS MODES OF OPERATION

spurious long waves. In a more recent physical model investigation of a container vessel moored behind an offshore terminal (8), spurious long wave suppression had to be included in the wave generation process. Since the response of the vessel, protected by the breakwater against high first-order waves, showed sensitivity to the sub-harmonic energy, neglecting the suppression technique for the worst wave conditions, resulted in an overprediction of mooring and fender forces by up to 40%. Another area of application is the run-up of waves on structures. Tests in a laboratory flume measuring wave run-up on a gently sloping beach (2) showed that omission of spurious long wave suppression led to an exaggeration of statistical run-up parameters.

6. Prototype Measurements

Indications have been given for sub-harmonics (3) and for both (6) that the inclusion of second-order techniques might be useful and important in assessing prototype waves. The implications of the knowledge of the magnitude of the second-order components are within the realm of mathematical modelling and physical modelling. Further investigations were performed on 56 wave records of 45 minutes duration each, which have been obtained at a water depth of 31 m. Analysis was concentrated on significant wave height ($H_{m0}$) and crest elevations which are most affected by second-order components. In addition, the horizontal asymmetry and average crest-front steepness for the measured and the theoretical (first-order) wavetrains were also compared. On the average, the horizontal asymmetry and the crest-front steepness are 13% and 48% higher respectively in the measured wave train. Figure 5 gives another example of the magnitudes of the second-order components.

References

1. Barthel, V., E.P.D. Mansard, S.E. Sand, F.C. Vis, 1983, "Group Bounded Long Waves in Physical Models", Ocean Eng., Vol. 10, No. 4.

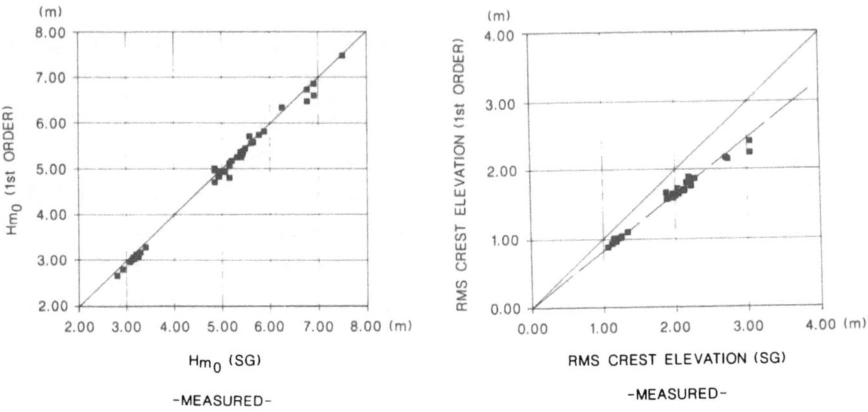


FIGURE 5 MEASURED AND FIRST ORDER WAVES-$H_{m0}$ AND CREST ELEVATION (RMS)

References

2. Barthel, V., E.P.D. Mansard, E.R. Funke, 1983, "Effect of Group-Induced Long Waves on Wave Runup", Proc. ASCE Coastal Structures, Arlington, VA, U.S.A.

3. Barthel, V. and E.R. Funke, 1984, "Estimates of Long Waves in the Weser Estuary", Proc. 19th ICCE, Houston, TX, U.S.A.

4. Longuet-Higgins, M.S. and R.W. Stewart, 1964, "Radiation Stresses in Water Waves, a Physical Discussion with Applications", Deep Sea Research, No. 11.

5. Mansard, E.P.D., B.D. Pratte, 1982, "Moored Ship Response in Irregular Waves", Proc. 18th ICCE, Cape Town, South Africa.

6. Mansard, E.P.D., S.E. Sand and P. Klinting, 1987, "Sub-and Super-harmonics in Natural Waves", Proc. 6th OMAE Symposium, Houston, TX, U.S.A.

7. Ottesen-Hansen, N.E., 1978, "Long Period Waves in Natural Wave Trains", Progress Report No. 46, Inst. Hydrodyn. and Hydraulic Eng., Techn. Univ. of Denmark, 13-24.

8. Pedersen, A.E., V. Barthel, C.S. Birt, 1987, "Port Development for St-Louis, Senegal, to be presented at 2nd COPEDEC, Beijing, China.

9. Sand, S.E. and E.P.D. Mansard, 1986, "Reproduction of Higher Harmonics in Irregular Waves", J. Ocean Eng., Vol. 13, No. 1, 57-83.

10. Sand, S.E., B. Donslund, 1985, "Influence of Wave Board Type on Bounded Long Waves, JHR, Vol. 23, No. 2.

# Deformation of the Two-dimensional Low-frequency Wave Spectrum in the Shallow Water Region

A. KIMURA

Associate Prof., Dept. of Ocean Civil Engineering
Faculty of Engineering, Tottori University
Tottori, Japan

ABSTRACT

This paper deals with the properties of the "averaged" second-order low-frequency directional wave spectrum $E^{(2)}(f_0, \theta_0)$ induced by the non-linear interactions between the component waves in the first-order directional wave spectrum $E^{(1)}(f_{i1}, \theta_{i2})$, and its transformation on a gentle uniform slope due to the wave refraction when the Mitsuyasu type directional wave spectrum is applied for $E^{(1)}(f_{i1}, \theta_{i2})$.

## 1. INTRODUCTION

The importance of the second order low-frequency waves has been recognized for a variety of phenomena such as edge waves, slow drift oscillation of moored vessels and harbor resonances.

Ordinary studies have mainly devoted to explain that the mentioned phenomena were possibly induced by the second order low-frequency waves. Since then, several authors have tried to give the formula for the second order low-frequency wave spectrum. Bowers(1975), Ottesen-Hansen(1978) and Sand(1982a) have proposed the convolution method to give the energy spectrum of the second order low-frequency wave spectrum for the uni-directional irregular waves. Sand(1982b) extended the theory and gave the convolution to give the second order directional low-frequency wave spectrum. This convolution, however, needs the information on phase of the first order component waves. Since the power spectrum does not include the phase information, it is not possible, therefore, to use the ordinary established wave spectrum as an input of the convolution. Some authors have proposed to introduce the

K. Horikawa, H. Maruo (Eds.)
Nonlinear Water Waves
IUTAM Symposium Tokyo/Japan 1987

statistical properties of the phase of the first order component wave (e.g. Sharma and Dean:1979).
The author (Kimura,1984), on the other hand, proposed to average the convolution with respect to phase instead of introducing its statistical property. The averaged second order low-frequency directional wave spectrum for the ordinary spectrum such as Mitsuyasu spectrum can be calculated only by introducing the first order spectrum in the convolution. The change of this spectrum due to the wave shoaling, refraction etc. can be calculated by introducing the first order directional wave spectrum into the convolution, which is changed in the above processes.
In this paper the deformation of the second order low-frequency directional wave spectrum in the shallow water region due to wave refraction is investigated.

## 2. AVERAGED SECOND ORDER LOW-FREQUENCY DIRECTIONAL WAVE SPECTRUM

The averaged second order low frequency directional wave spectrum $E^{(2)}(f_0, \theta_0)$ is given as follows (Kimura,1984),

$$E^{(2)}(f_0,\theta_0)=\int_{f_{i1}}\int_{\theta_{i2}} E^{(1)}(f_{i1},\theta_{i2})E^{(1)}(f_{j1},\theta_{j2}) \cdot\alpha^2(f_{i1},f_{j1},\theta_{i2},\theta_{j2})df_{i1}d\theta_{i2} \tag{1}$$

in which $E^{(1)}(f_0,\theta_0)$ represents the first order directional wave spectrum, $f_i$ and $\theta_j$ (i=i1,j1, j=i2,j2) are frequency and direction of its component wave and $\alpha$ is the transfer function which is given as,

$$\alpha\,(f_{i1},f_{j1},\theta_{i2},\theta_{j2}) =\frac{1}{2}\left[\frac{\bar{D}_{i1i2j1j2}-\{k_{i1}k_{j1}\cos(\theta_{i2}-\theta_{j2})+R_{i1}R_{j1}\}}{\sqrt{R_{i1}R_{j1}}}+R_{i1}+R_{j1}\right] \tag{2}$$

where

$$\bar{D}_{i1i2j1j2}=[(\sqrt{R_{i1}}-\sqrt{R_{j1}})\{\sqrt{R_{j1}}\,(k_{i1}^2-R_{i1}^2)-\sqrt{R_{i1}}\,(k_{j1}^2-R_{j1}^2)\} +2(\sqrt{R_{i1}}-\sqrt{R_{j1}})^2\{k_{i1}k_{j1}\cos(\theta_{i2}-\theta_{j2})+R_{i1}R_{j1}\}]$$

$$/\{(\sqrt{R_{i1}}-\sqrt{R_{j1}})^2-\bar{k}_{i1i2j1j2}\tanh \bar{k}_{i1i2j1j2}h\} \quad (3)$$

$$R_i = k_i \tanh k_i h \quad (4)$$

$$k_i = 4\pi^2 f_i^2/(g \tanh k_i h) \quad (5)$$

$$(i=i1, j1)$$

$$\bar{k}_{i1i2j1j2} = \sqrt{k_{i1}^2+k_{j1}^2-2k_{i1}k_{j1}\cos(\theta_{i2}-\theta_{j2})} \quad (6)$$

in which h is a water depth, $k_i$ $(i=i1,j1)$ is the wave number of the first order component wave. Since the second order low frequency directional wave spectrum ($E^{(2)}$) only consists of the second order difference frequency waves of the first order directional wave spectrum ($E^{(1)}$), frequency $f_0$ and direction $\theta_0$ in $E^{(2)}$ are given as follows.

$$f_{j1} = f_{i1} - f_0$$

$$\theta_{j2} = \theta_0 - \sin^{-1}\{\frac{k_{i1}}{k_{j1}}\sin(\theta_0-\theta_{i2})\} \quad (7)$$

Figure 1 shows schematically the relation between wave numbers of the two first order waves and induced second order low frequency waves. Figure 2 shows schematically another possible combination of the first order waves which induce the second order waves with the same frequency and direction as shown in Fig.1. The radii of the circles are equal to the wave numbers $k_{i1}$ and $k_{j1}$ in both figure. The direction of $k_{j1}$ in Fig.2 is different from $k_{j1}$ in Fig.1. The induced second order wave in Fig.2 has a larger wave number than that shown in Fig.1.
There is an infinitely large number of wave couples of first order component waves which induce the second order low frequency waves of the same frequency and direction. The energies of all of those waves are integrated in the convolution together with the transfer function in eq.(1).

## 3. TRANSFER FUNCTION

There are two kinds of low frequency waves, frequency and direction of which are the same as mentioned in section 2.

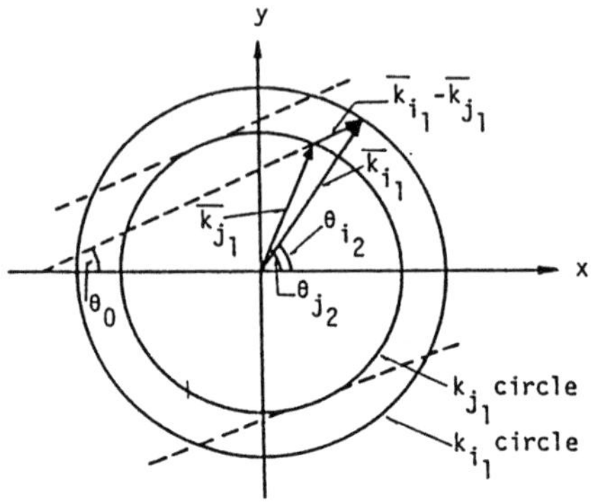


Fig.1 Relation of wave numbers

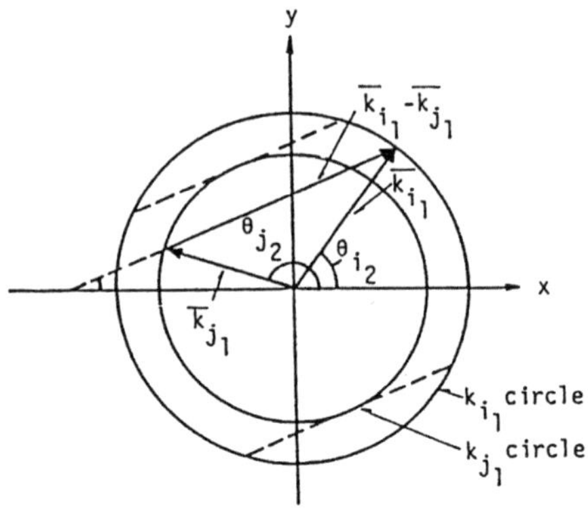


Fig.2 Relation of wave numbers

They are low frequency waves with a small wave number (Fig.1) and a large wave number (Fig.2). To obtain a general properties of those waves, comparisons are made of the normalized transfer function $\alpha_{i1i2j1j2}h$ for these waves in Fig.3 (a), (b) and (c). Solid lines and dotted lines in these figures show the normalized transfer function ($\alpha h$) for the low frequency waves with small and large wave numbers respectively at (a) $k_{i1}h=10\pi$ ,(b) $k_{i1}h=\pi$ and (c) $k_{i1}h=0.1\pi$. Since eq.(2) does not depend on $\theta_0$ , putting $\theta_0=0$, h can be plotted in terms of the normalized angle $\theta_{i2}/\sin^{-1}(k_{j1}/k_{i1})$ and wave number ratio $k_{j1}/k_{i1}$. $\alpha h$ for the low frequency wave with the large wave number decreases with decrease in water depth. However $\alpha h$ for the low frequency with the small wave number, on the other hand, increases with decrease in water depth in a shallow water region. In a very shallow water region, $\alpha h$ for low frequency waves with large wave number reduces to negligibly small. The contribution of the low frequency wave with large wave number to $E^{(2)}$ is negligible in a shallow water region.

## 4. THE CHANGE IN THE LOW FREQUENCY DIRECTIONAL WAVE SPECTRUM DUE TO REFRACTION

Since $E^{(2)}$ is given in terms of $E^{(1)}$ and $\alpha$, the change of $E^{(2)}$ due to the refraction on a slope can be calculated by substituting $E^{(1)}$ in eq.(1), which is changed due to refraction on a slope. The change of $E^{(2)}$ on a slope is

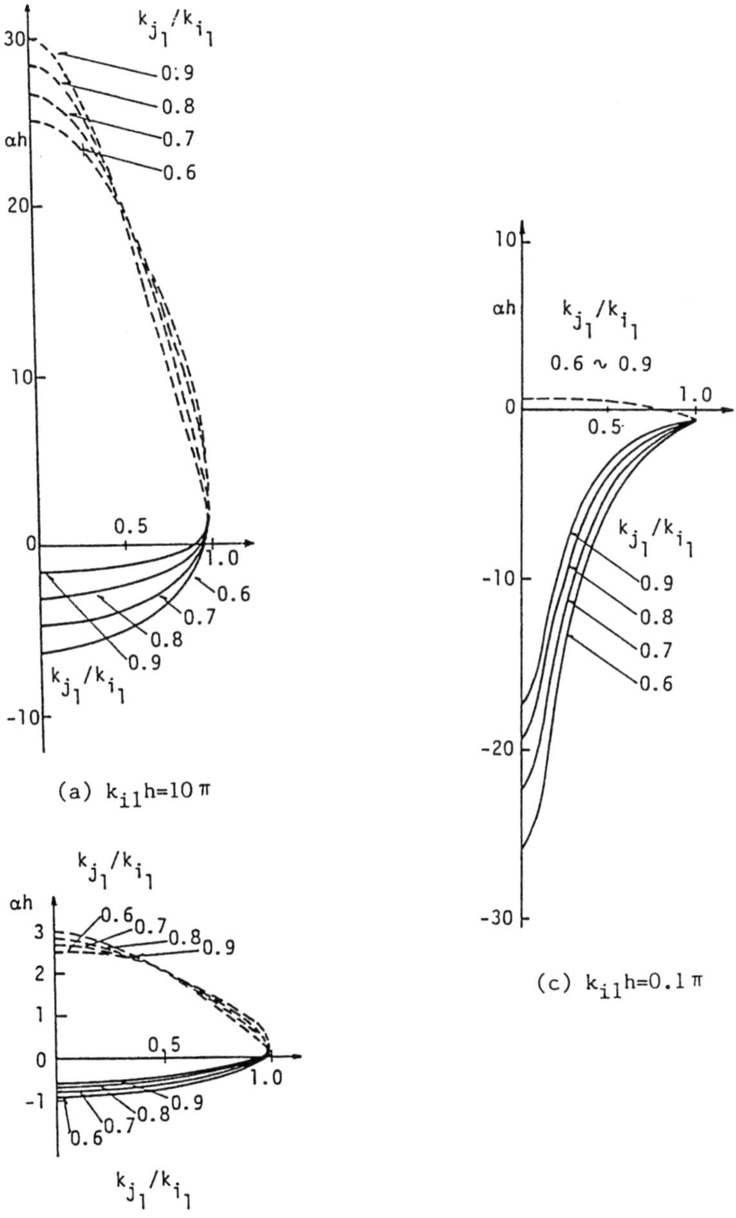


Fig.3 The normalized transfer functions

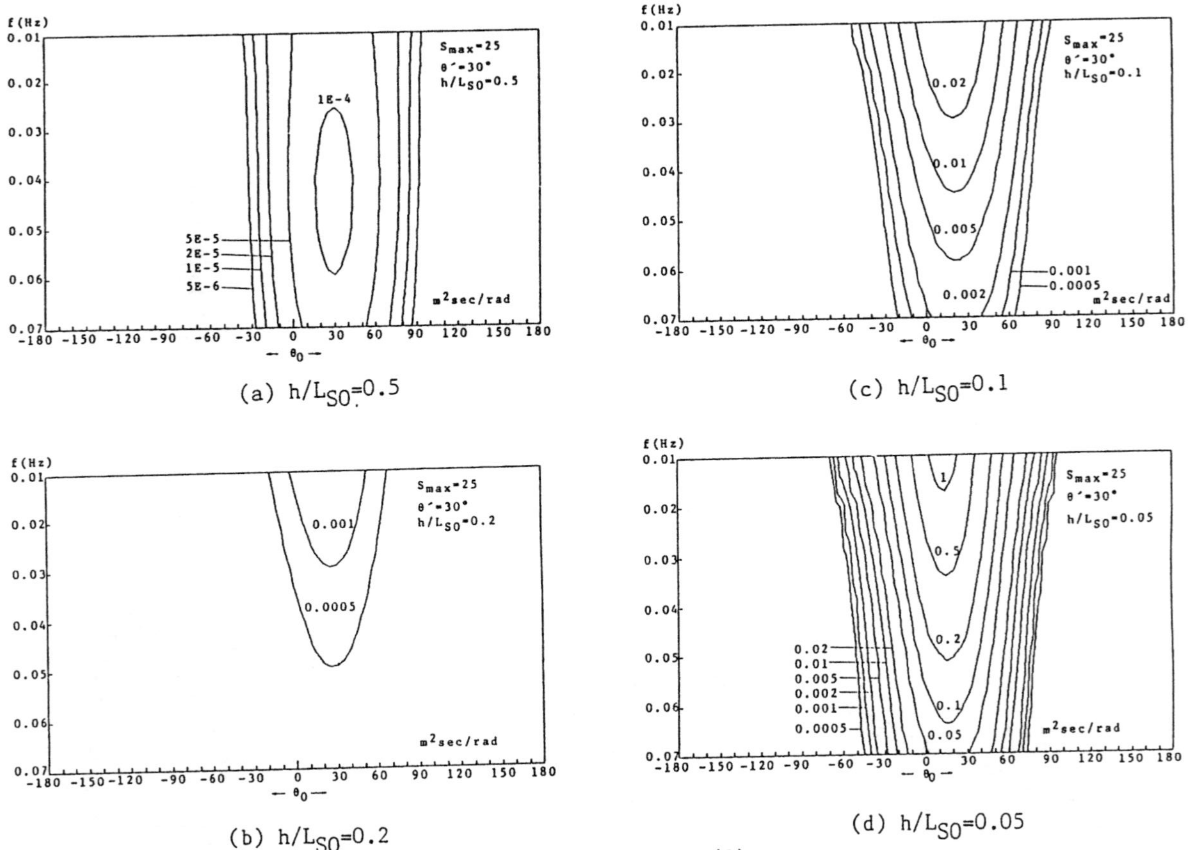


(a) $h/L_{SO}$=0.5

(c) $h/L_{SO}$=0.1

(b) $h/L_{SO}$=0.2

(d) $h/L_{SO}$=0.05

Fig.4 Change of $E^{(2)}$

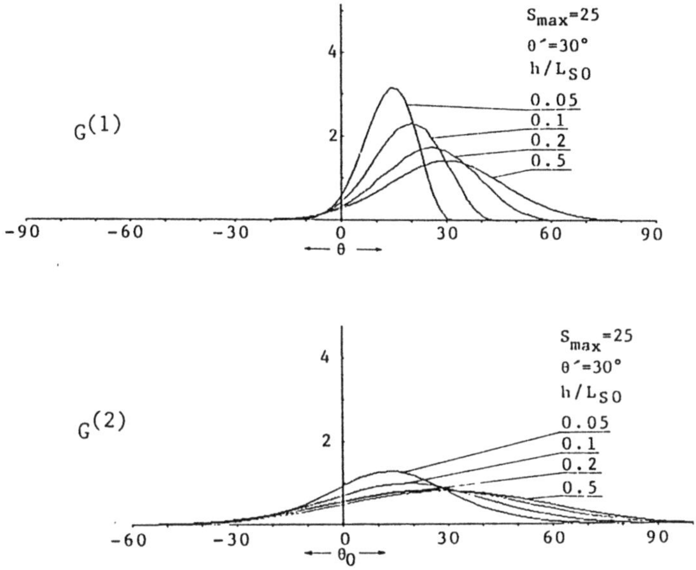


Fig.5 Change of the normalized cross sections $G^{(1)}$ (above), $G^{(2)}$ (below)

calculated in this study by assuming:
(1) the refraction of the component wave follows Snell's theorem; (2) the period of component wave does not change; (3) the wave energy is supplied only from the offshore boundary and dissipated at the shore; and (4) the beach slope is gentle and uniform.

Mitsuyasu spectrum for the swell from a short distance is used for $E^{(1)}$ (Goda et al.,1975). The significant wave height and period are 3.3m and 8s, respectively. Figure 4 shows the change of $E^{(2)}$ at several water depths with an incident angle of 30deg. Figure 5 shows the normalized cross sections of $E^{(1)}$ (above) and $E^{(2)}$ (below) at their peak frequencies, which are given by

$$G^{(i)}=E^{(i)}/S^{(i)}(f \text{ or } f_0) \quad (i=1,2) \tag{8}$$

in which $S^{(i)}$ (i=1,2) is the integrated spectrum of $E^{(i)}$ (i=1,2) with respect to angle. Wave conditions are the same as used in Fig.4.

## 5. CONCLUDING REMARKS

The followings are the major remarks obtained in this study, (1) the wave directions of $E^{(1)}$ and $E^{(2)}$ at their peak frequency coincide with each other; (2) the normalized cross section $G^{(2)}$ becomes narrower with decrease in water depth, however it is always lower in height and wider in width than those of $G^{(1)}$ and its growth rate is rather gradual as compared with that of $G^{(1)}$; and (3) the development of $E^{(2)}$ is prominent in the shallow water region and its growth rate with decrease in water depth is much larger than that of $E^{(1)}$.

## REFERENCES


Bowers, E.C. (1975): Trans. Roy. Inst. Naval Archit., Vol.118.

Goda, Y. and Y. Suzuki (1975): Tech. Note of PHRI, No.230.

Kimura, A. (1984): Comm. Hydraulics, Dept. Civil Engg., Delft Univ., Rept.No.84-3.

Ottesen-Hansen, N.E. (1978): Prog. Rept.46, Inst. Hydrodyn. and Hydraulic Engg. Tech. Univ. Denmark.

Sand, S.E. (1982a): Jour. Waterway, Port, Coastal, Ocean div., ASCE, Vol.108, WW4.

Sand, S.E. (1982b): Coastal Engg., 6.

Sharma, J.N. and R.G. Dean (1979): Ocean Engg. Rept. No.20, Dept. Civil Engg., Univ. of Delaware.

# Poster Session 1

# Nonlinear Water Waves

# An Experimental Investigation into Nonlinear Resonant Wave Interactions in the Ship Model Basin

H. TOMITA

Ship Research Institute, Japan

H. SAWADA

Ship Research Institute, Japan

Summary

The experiment was executed for studying the properties of tertiary wave interactions in a ship model basin. Long-time evolution of resonant wave was observed by linear array of wave gauges located along the direction of propagation of the wave. The behavior of tertiary resonance was found to depend on the amplitudes of two principal waves as well as their wavelengths. The comparison was made in fair agreement with the theoretical prediction after Zakharov equation.

## Introduction

Theoretical studies of the non-linearities in the sea waves have been developed considerably last 25 years since Phillip's investigation on the resonant wave interaction of the finite amplitude surface gravity waves in a seaway. In the recent years, one can treat them as non-linear random process and can estimate various quantities which characterize the properties of wind-generated sea waves. On the other hand, it is said that the experimental and/or observational data to make a verification of the theories are insufficient. Especially, in the case of mutual interaction among waves that intersect each other, phenomena are thought to be essentially two-dimensional, therefore, neither experimental study in the wave flume nor numerical investigation cannot be easily performed.

In this paper, we explain the method of experiment on the resonant wave interaction between two waves(one of which is counted twice so that three trains of waves would be regarded to exist originally) and discuss about the growth of the third wave, for the case of regular waves. The resonance condition among three waves is expressed as follows.

K. Horikawa, H. Maruo (Eds.)
Nonlinear Water Waves
IUTAM Symposium Tokyo/Japan 1987

$$2\underline{K}_1 - \underline{K}_2 = \underline{K}_3, \ 2\omega_1 - \omega_2 = \omega_3, \qquad (1)$$

where $\omega_j^2 = gK_j$ (j = 1, 3) are the dispersion relation of deep water gravity wave. The situations of this experiment are as same as those of Longuet-Higgins et al.[1] and McGoldrick et al.[2] whose data were obtained in the small basin (3m × 3m × 1m) to investigate the initial growth of the third wave. However, the temporal and spatial scale of our experiment is much longer than theirs so that the long time evolution of the third wave could be acquired. From the theoretical point of view, the classical method seems not to be valid in this case. Accordingly, we adopted Zakharov equation (which is considered to be valid to the long time evolution of the surface gravity waves) in the simplest case of three waves mutual interaction and have solved it to compare with the experimental data.

## Experiments

The experiments were carried out in a large basin of the Ship Research Institute (80m × 80m × 4.5m). The work was done selecting the calm weather because the facilities are in the open air. The background noise existing on the surface was measured to be about 0.08 cm throughout the frequencies considered in this experiment. The plan of the basin and the distribution of wave gauges (of capacitance type) for measurement of the surface elevation are presented in Fig. 1. Wave gauges are hung on the wire rope suspended over the basin. Data processing system is schematically drawn in Fig. 2. Measured data are digitized with the sampling time 0.1 second. Waves are generated by two wave-makers in adjacent sides of the basin. They intersect at the right angle in the basin and are absorbed on the beaches installed at the opposite sides. The situation of the relative direction of three waves are presented in Fig. 3, together with the theoretical curve for the strict resonance. Measurements were done for 110 seconds each time at six locations drawn in Fig. 1. Their distances from the wave-maker 1 are about 26 m, 35 m, 40 m, 45 m, 50 m and 60 m. Figure 4 shows the wave profiles of the single wave train detected at ch1(26 m) and ch6(60 m), no appreciable deformation takes place in this case(ak = 0.12). Four cases of combination of the wavelength and wave height($\lambda 2$, H2), which are (5 m, 5 cm), (5 m, 10 cm), (5 m, 20 cm) and (7 m, 10 cm), were chosen for the wavemaker 2. For each case, the second wave by the other wavemaker was generated changing the ratio $R = \lambda 2/\lambda 1$ and the wave height H1. When the ratio R is equal to 3.01, the resonance condition among these waves is strictly satisfied if the linear dispersion relation is to be assumed. For examining the amplitude effect to the dispersion relation, many cases with changing the stroke of the wavemaker 1 were executed precisely. The stroke of the wavemaker was also detected at the same time. The obtained data were analysed by means of FFT technique (usual cosine type data window was adopted and the averaging process was not executed) in order to evaluate the amplitudes of two waves generated by the wavemakers and the third wave which simultaneously grows by the effect of non-linear resonant wave interaction. The precision of the calculation was verified by treating the known data prepared by the electric oscillator. Some exact numerical data were also examined and the relative error for the estimate of amplitudes were known to be less than 0.5%. On the other hand, we solved numerically Zakharov equation up to the third order of the wave steepness ak and compared it with the experimental data.

Results

In Fig. 5, we show an example of the energy spectrum of the water surface of the basin obtained by our experiment. In this figure, one can clearly recognize three lines, two of which in the lower side of the frequency represent the waves generated by the wavemakers. The last line in the upper side corresponds to the tertiary wave which is predicted to grow by the theory of resonant wave interaction. In this example, this tertiary wave was observed by eye on the surface of the basin after two principal waves had passed by. The spatial distribution of the amplitude of the tertiary wave at six points along the direction of propagation of this wave is shown in Fig. 6. The most crucial phenomena of the long time evolution of the tertiary wave is the recurrence of states of energy sharing among three systems of wave train (the energy flow is from A1 to the others and vice versa). The period of recurrence seems to depend upon the amplitudes of each wave. Qualitatively speaking, at nearly on resonance, the maximum value and the period of recurrence of the resonant wave A3 increase with the amplitude A2, on the other hand, A1 shortens the period. In the case of R > 3.2, A1 and A2 exchange their role. This is interpreted by the gradual phase shifting between waves caused by non-linear interaction to the phase velocity. We also show the theoretical prediction by means of the numerical solution of Zakharov equation. Some of them are presented in Fig. 7. The complete recurrence of the states is evident in all cases and the period of recurrence depends upon the amplitudes of each wave (the detailed computational results are presented in Tomita[3 ]). The agreement between theory and experiment is fairly well when the wave steepness ak is less than 0.15. However, when the amplitudes become larger, discrepancies occur between them. The large amplitude resonant waves exceeding this value were not observed in spite of the theoretical prediction. It is presumably caused by the modulation, deformation and reflection of principal waves. Moreover, effects of higher order terms such as wave breaking or quintet interaction between wave components would play some important role to the growth of the tertiary wave.

Considering the most general situation of the oblique interaction in which the transfer of energy would be more intense, one could surmise that the non-linear transfer of energy among gravity waves would have certain influence on the spectrum in a seaway.

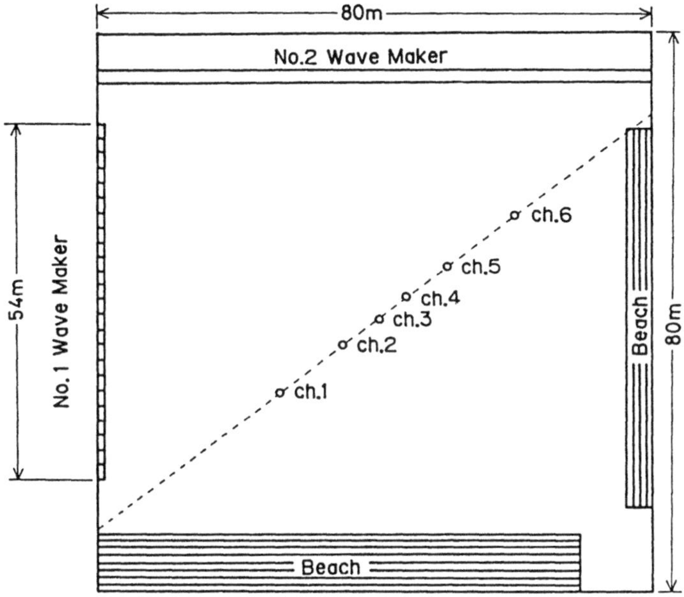


Figure 1. Plan of the wave basin. The surface displacement was recorded at ch1-ch6.

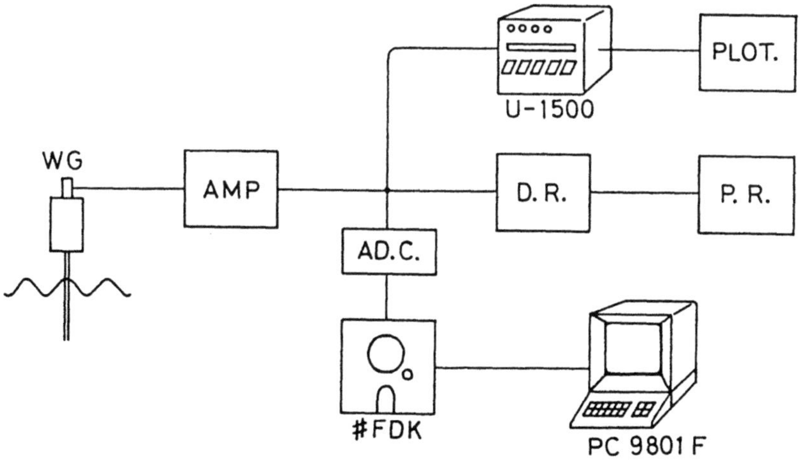


Figure 2. Data acquisition system.

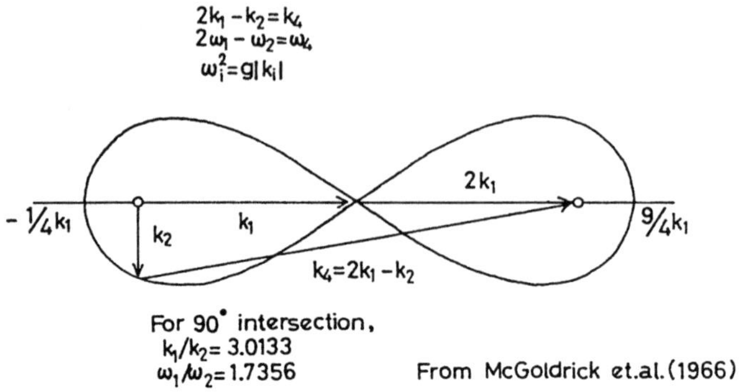


Figure 3. Relative direction of three waves. Curve(solution of resonance condition).

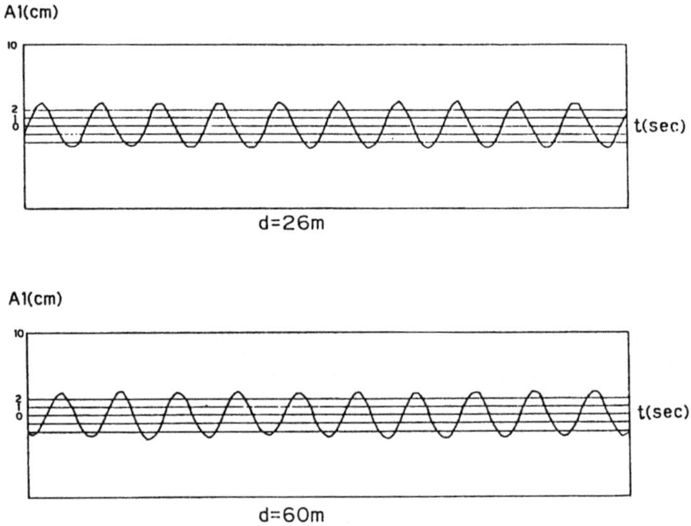


Figure 4. Detected profiles of the single wave train at locations d=26m(upper)and 60m(bottom). T=0.983 second,ak=0.12.

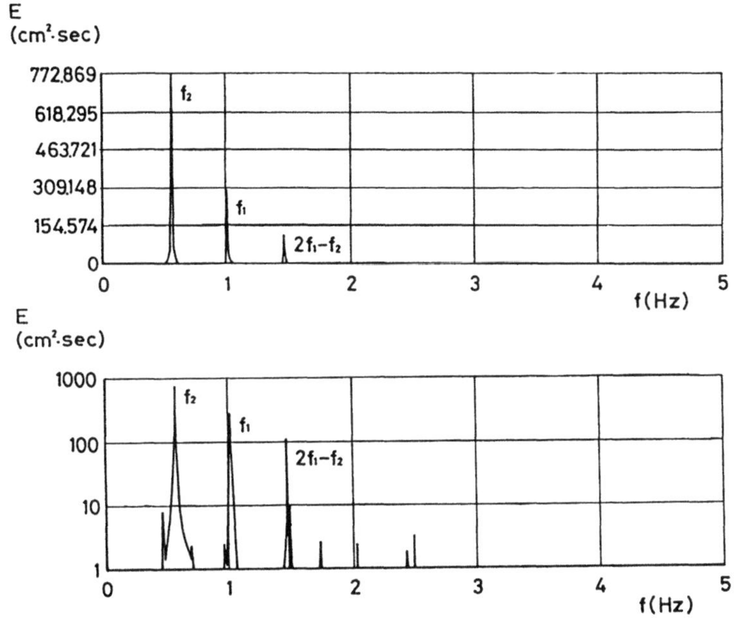


Figure 5. An example of the energy spectrum.
Upper: linear scale
Lower: logarithmic scale
R=3.21(off resonance),d=45m.

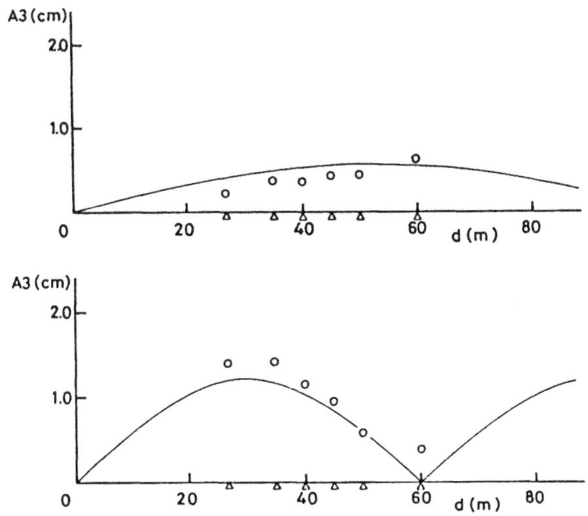


Figure 6. Examples of the evolution of third wave.
Upper: R=2.97(nearly on resonance)
Lower: R=3.21(off resonance)
Solid curves represent the theoretical solutions.

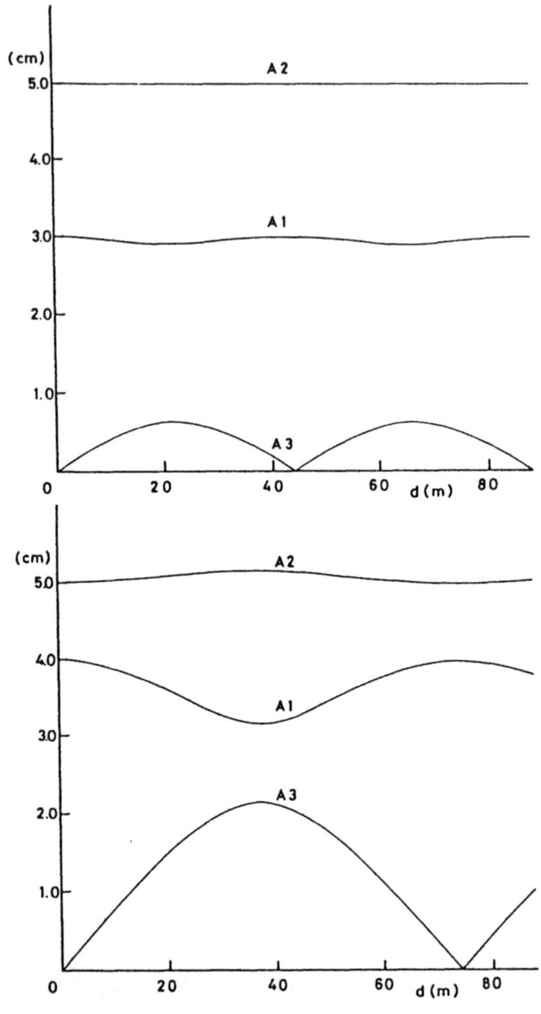


Figure 7. Solutions of Zakharov equation(R=3.24).

References

1. Longuet-Higgins,M.S. and Smith,N.D. (1966). An experiment on third order resonant interactions. J.Fluid Mech. Vol.25, pp.417-435.

2. McGoldrick,L.F.,Phillips,O.M.,Huang,N.E. and Hodgson,T.H. (1966). Measurements of third-order resonant wave interactions. J.Fluid Mech. Vol.25, pp.437-456.

3. Tomita,H. (1987). Etude numérique sur l'interaction resonante entre des vagues d'amplitude finie. (to be published)

# Simulation of Wave-spectra with Boussinesq-type Wave Equations

H.-H. PRÜSER / W. ZIELKE

Institute of Fluid Mechanics

University of Hannover, F.R.G.

Abstract: A numerical model, based on the solution of the Boussinesq equation is applied to simulate the nonlinear transformation of irregular wave trains in shallow water. The main topic of this study is to determine the generation and behaviour of long waves in a flume, which is often a disturbing side effect, especially when hydraulic models are running for a relatively long period of time.

## 1. INTRODUCTION

Linear numerical models are often not sufficient to describe the wave transformation in the coastal zone. This is especially true for the generation of long waves, which are not included in the incoming deep water spectrum. This effect is of particular importance for the planning of harbours and shore protection: Long waves, associated with wave groups and generally referred to as the wave set-down of water level, are one of the critical parameters for the determination of mooring forces and the risk of resonance with the longer natural vessel periods. Also, sediment transport in the coastal zone seems to be strongly dependent on long waves.

Nonlinear time domain numerical models for shallow water wave behaviour were presented by ABBOTT et al. [1], HAUGUEL [2] and SCHAPER / ZIELKE [3]. These models are based on the solution of the Boussinesq or Serre equations and are helpful for simulating the interaction of currents with refraction, diffraction, reflection and shoaling in the coastal zone.

The behaviour of nonlinear, irregular wave trains in a one-dimensional wave flume was investigated by PRÜSER / SCHAPER /

K. Horikawa, H. Maruo (Eds.)
Nonlinear Water Waves
IUTAM Symposium Tokyo/Japan 1987

ZIELKE [4]. They verified their model by means of hydraulic measurements. The paper presented here is a further extension. Its aim is to describe long-wave generation and long-wave resonance in the operation of wave channels. This is frequently a quite disturbing side effect during operation of the facilities with irregular waves. Therefore, it is necessary to implement in the numerical model a boundary condition, which allows to introduce waves into the domain, as well as to simulate the wave re-reflection at the wave-paddle.

## 2. NUMERICAL MODEL

The computations were carried out with a Boussinesq wave model, formulated for the vertically-integrated flux and the surface elevation. The basic equations for the one dimensional case read:

$$\frac{\delta p}{\delta t} + \frac{\delta\zeta}{\delta x}\left(\frac{p^2}{h}\right) + gh\,\frac{\delta}{\delta x} = \frac{Dh}{2}\,\frac{\delta^3}{\delta x^2\,\delta t}\left(\frac{Dp}{h}\right) - \frac{D^2 h}{6}\,\frac{\delta^3}{\delta x^2\,\delta t}\left(\frac{p}{h}\right)$$

$$\frac{\delta\zeta}{\delta t} + \frac{\delta p}{\delta x} = 0$$

| | | |
|---|---|---|
| $h = D + \zeta$ | : | total water depth |
| $D$ | : | mean water level |
| $\zeta$ | : | water elevation |
| $p$ | : | vertically-integrated flux |
| $g$ | : | acceleration due to gravity |

In short shallow water waves, the nonlinear amplitude-dispersive terms on the left hand side and the nonlinear frequency-dispersive terms on the right hand side are of the same order. It is, therefore, necessary to include all these terms into the computation. In the numerical model, the Boussinesq equations are solved with a third-order corrected implicit finite difference method, using two time levels and central differences.

It has been shown, that long wave equations are sufficient to formulate the boundary conditions. Neglecting the third-order terms of the Boussinesq equations, boundary conditions can be formulated which are based on the method of the characteristics. One can, for example, describe waves, which are generated at the wavemaker and propagate into the wave flume and at the same time waves, which are reflected from the wave-paddle. This particular boundary conditon reads:

$$(\zeta^{P} - \zeta^{R})\cdot c_r + q^{P} - q^{R} = 0.$$

$$q^{P} = q(t)$$

$$[-q/h - \sqrt{(g\cdot h)}\,]^{R} = c_r$$

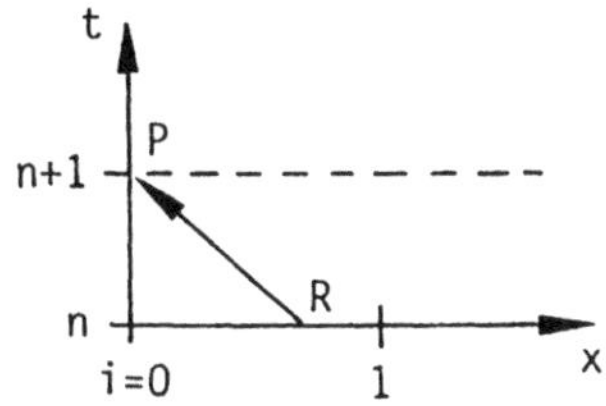


The vertically-integrated flux as an arbitrary function of time is used to simulate waves. At the boundary, the assumed velocity profile is constant over the water depth, i.e. comparable to a wave-paddle, working in piston mode. BARTHEL et al. [5] investigated the wave generation by those wavemakers. He pointed out, that spurious long waves will appear in the flume as the result of a second-order effect; the wave equations are not substituted exactly at the wave-paddle.

## 3. TRANSFORMATION OF PIERSON-MOSKOWITZ SPECTRA

### 3.1. The input signal

The calculations were done with reference to measurements in a hydraulic wave flume, which were performed by DAEMRICH/GÖTSCHENBERG, Franzius Institute of the University of Hannover. Input spectra S were composed of 256 individual frequency components. Their energy distribution in the frequency domain was chosen equal to a Pierson-Moskowitz spectrum, and the amplitudes of the water elevation in the time domain were reduced by a constant factor in order to avoid wave-breaking, which would occur, if these deep water spectra were used in shallow water. The following results have been obtained from a spectrum with a peak-period of 4.5sec and a significant wave height of 0.08m. It has been used repeatedly four times in one test to operate the numerical and the hydraulic wavemaker. The recurrence interval is approximately 140.sec. Figure 2 shows the spectrum just behind the wavemaker in time and in frequency domain. It can be seen that long wave components, which are not included in the original input spectrum, are generated at the boundary.

### 3.2. Experimental set-up

The transformation of the above input signal in shallow water was investigated in a wave flume of 0.5m water depth. Figure 1 displays the experimental set-up with three different boundaries, which were applied at the right hand side.

- System a) : Open boundary condition

In this numerical wave flume the transformation of a spectrum, generated at x = 0.m can be calculated without any disturbances being caused by the right boundary.

- System b) : Slope condition

In a flume with horizontal bottom and a slope on one end, irregular wave trains were generated by the wavemaker. The water elevation was measured and calculated at several gauges, which were positioned in the flume where the bottom is horizontal, as well as on the slope. The Boussinesq wave model does not contain the possibility of simulating breaking waves, which occur in the upper part. A fully dissipating boundary was applied at that point of the slope instead, where the water depth was 0.2 m.

- System c) : Total reflection

Finally a hydraulic and numerical wave flume with a vertical wall was applied. The waves were reflected at the wavemaker as well as at a vertical wall for about ten times, while the test was running. Therefore, it can be expected, that oscillations with longer wave periods will appear.

### 3.3. Results and conclusions

The accuracy of the numerical model depends on the quantity of grid points per wave length. The calculations were carried out with a step in space of 0.15m and a time step of 0.05sec. With these values, the numerical model allows to simulate waves, which are longer than 2.5m (better 3.0m i.e. 20 grid points per wave length); or in other words, which have a frequency smaller than

0.86Hz (0.7Hz) in a 0.5m deep wave flume.

As indicated in chapter 3.1., the wave input is periodic. It contains four repetitions of the spectrum S with a period of 140.sec. Therefore, it is possible to obtain the transformation of one individual spectrum with respect to time and space in system a), with no disturbing influence of the right boundary. Figure 3 shows the results of the water elevation in the frequency domain. It can be seen, that long wave components are generated at the boundary and additionally as an effect of wave grouping in the domain. The strong change of the spectrum in the region of the higher frequencies indicates, that this Pierson-Moskowitz spectrum, being relatively stable in deep water conditions, is unstable in shallow water.

A comparison of measured and calculated time series at three different gauges in system b) displays, that the numerical model reproduces the measured time series with a high degree of agreement in case of constant water depth, as well as on the slope, where shoaling influence leeds to higher, steep crested waves. The generated waves are reaching the slope in 30.sec. They are partially reflected on the slope and superposed on the incoming waves. The reflection coefficient depends on the frequencies: The energy of the short wave components of the spectrum dissappears nearly completely by the effect of wave breaking. Long waves will be reflected instead. Therefore, comparing the input spectrum S of the third repetition (after 280sec) a significant difference in the frequency domain between system a) and system b) can only be observed in the long wave components. The upper part of the slope is not simulated in the numerical model. SYMONDS et al. [6] developed a theory explaining that a time-varying breaking point of short waves on a slope can also generate a long wave. This might be a reason, why the increase of the long waves is determined in the numerical model qualitatively correct, but too small (see figure 3).

In system c) the used input signal was about 10 times longer than the time of beginning reflection at the vertical wall. As expected, seiches motion occurs. The individual modes can be found by placing gauges at the wall and in the middle of the flume. Figure 5 displays the first three modes, which are generated in the numerical model. In a 56.m long flume with a constant water depth of 0.5m they have the frequencies: f1 = 0.02Hz / f2 = 0.04Hz / f3 = 0.06Hz.

## 4. REFERENCES


[1] ABBOTT, M.B., PETERSON, H.M. and SKOOVGAARD, O. (1978): Computation of short waves in shallow water. Proc. 16th ICCE, Hamburg.

[2] HAUGUEL, A. (1980): Adaption of tidal numerical models to shallow water waves problems. Proc. 17th ICCE, Sydney.

[3] SCHAPER, H., ZIELKE, W. (1984): A numerical solution of Boussinesq type wave equations. Proc. 19th ICCE, Houston.

[4] PRÜSER, H.-H., SCHAPER, H., ZIELKE, W. (1986): Irregular wave transformation in a Boussinesq wave model. Proc. 20th ICCE, Taipei.

[5] BARTHEL, V., MANSARD, E.P.D., SAND, S.E. and VIS, F. (1983): Group bounded long waves in physical models. Ocean Eng., Vol. 10.

[6] SYMONDS, HUNTLEY and BOWEN (1982): Two-dimensional surf beat: Long wave Generation by a time-varying breakpoint. J. Geoph. Res., Vol. 87.


APPENDIX: Figures 1-4

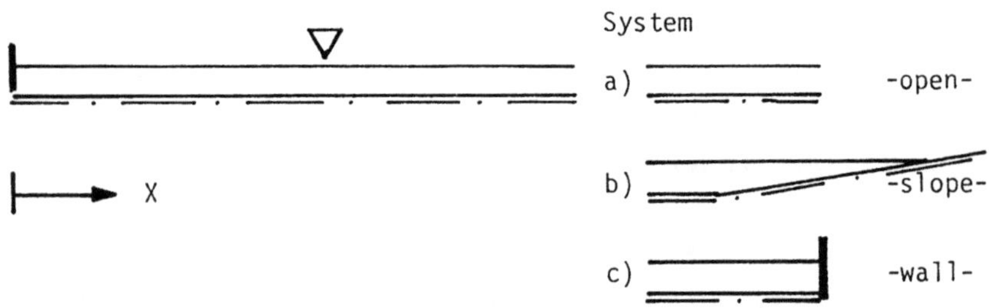


Figure 1: Experimental set-up

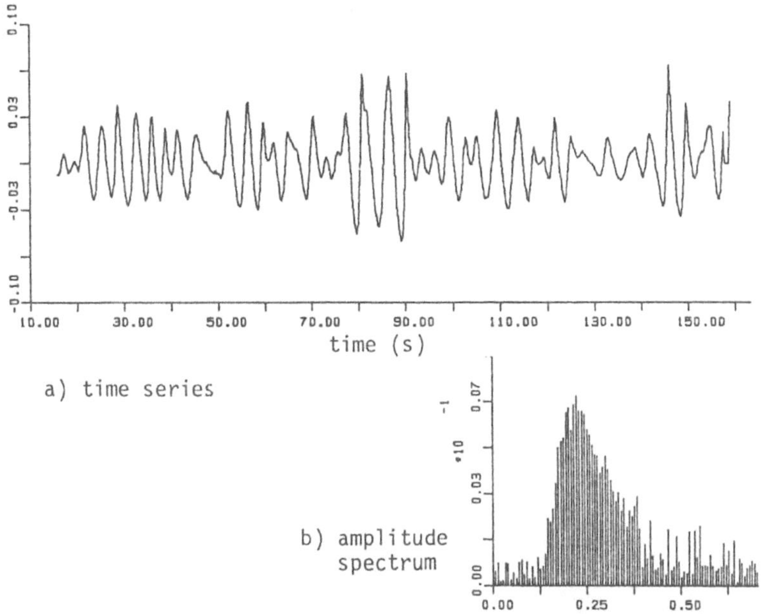


Figure 2: Input signal

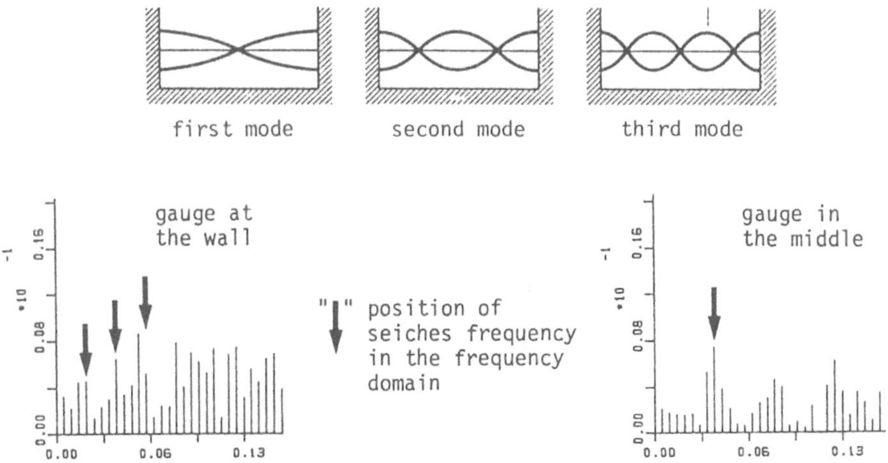


Figure 4: Generation of seiches motion in the flume

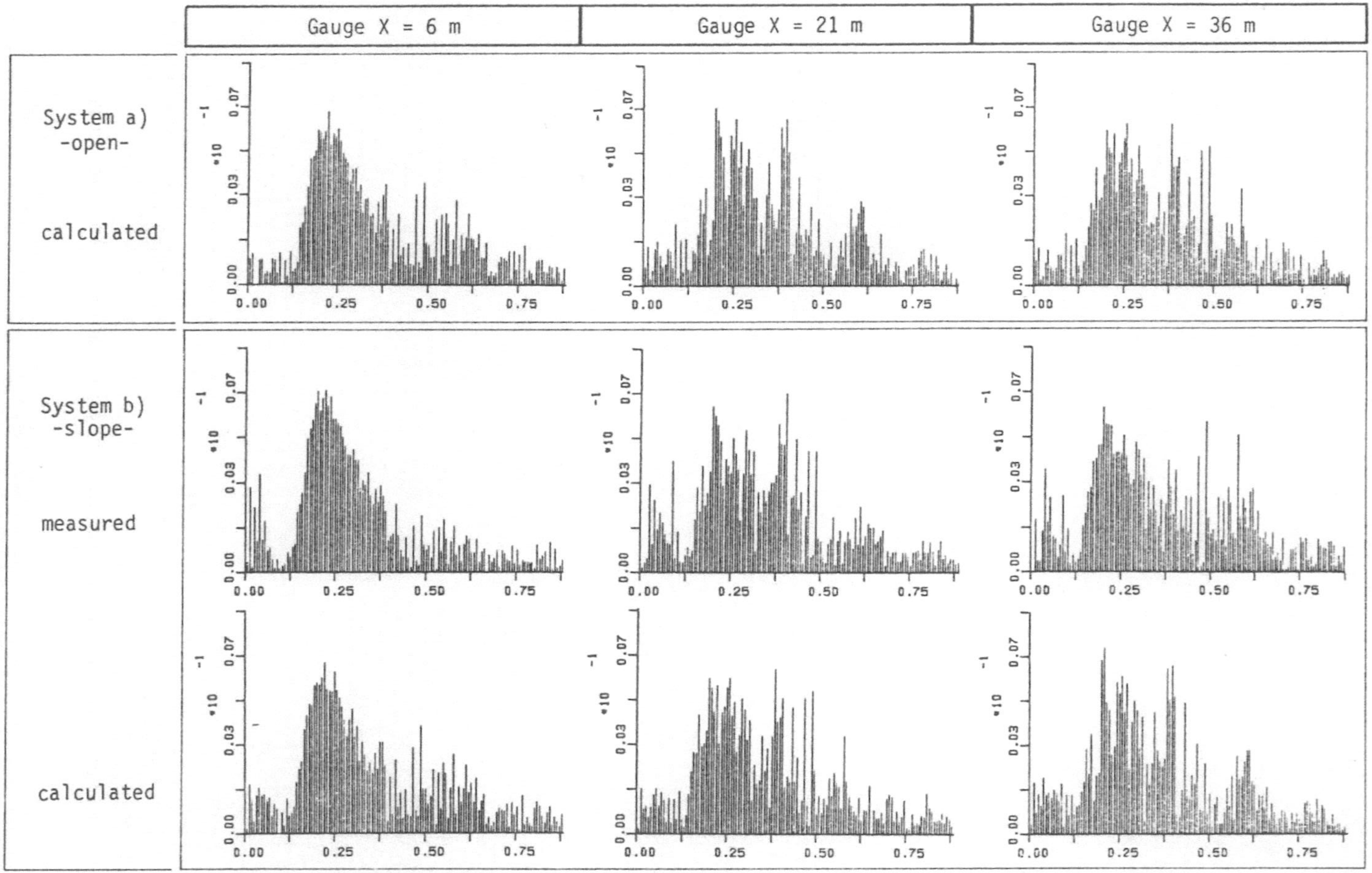


Figure 3: Calculated and measured amplitude spectra at different gauges

# One-dimensional and Weakly Two-dimensional Waves in Varying Channels: Numerical Examples

J. T. KIRBY, R. PHILIP[1] and P. VENGAYIL[2]

[1]Coastal and Oceanographic Engineering Department, University of Florida, Gainesville
[2]presently at Dept. of Civil Engineering, Massachusetts Institute of Technology, Cambridge

Summary

Various examples pertaining to the propagation of long nonlinear waves in channels of varying topography and cross section are presented. Numerical schemes are based on equations of KdV-type for one-dimensional cases and KP-type for weakly two-dimensional cases. Provisions for reflection are made through linear coupling of equations for forward and backward propagation. Examples for surface waves and waves in a two-layer stratification are provided.

## Weakly Two-Dimensional Model for Linear Waves

The introduction of the Kadomtsev-Petviashvili (KP) equation as a weakly two-dimensional extension of the Korteweg-de Vries (KdV) equation has led recently to a number of studies of the evolution of initial or forced pulses in long channels where a predominant propagation direction is apparent [1,2]. The extension of this modelling scheme to the case of channels with varying topography is of obvious importance in modeling waves in the natural environment.

As a starting point, we consider any physical system which may be modelled at leading order by a second order wave equation

$$\eta_{tt} - \nabla_h \cdot (c^2 \nabla_h \eta) = 0 \tag{1}$$

where c denotes the wave phase speed and $\nabla_h$ is the horizontal gradient oeprator in coordinates (x,y). For example, we have

$$c^2 = gh(x,y) \tag{2}$$

for surface waves and

$$c^2 = g\delta \frac{h_1 h_2}{h_1 + h_2} + 0(\delta^2) \quad ; \quad \delta = \frac{\rho_2 - \rho_1}{\rho_2} \tag{3}$$

K. Horikawa, H. Maruo (Eds.)
Nonlinear Water Waves
IUTAM Symposium Tokyo/Japan 1987

for internal waves in a two-layer system with upper (lower) layer depth $h_1$ ($h_2(x,y)$). We denote x as the predominant direction oriented along the channel axis. Using an operator correspondence $\partial/\partial t \sim -i\omega$ in time reduces (1) to a pseudo-elliptic model which may then be split into coupled parabolic equations using standard methods. Letting $\eta^{+(-)}$ denote the forward (backward) propagating mode, and inverting the operator correspondence leads to the coupled equations

$$\frac{1}{c}\,\eta_t^{\pm} \pm \eta_x^{\pm} + \frac{c_x}{2c}\,(\eta^{+} - \eta^{-}) \mp \frac{1}{2}\int_x^{\infty} \frac{1}{c^2}\,(c^2\eta_y^{\pm})_y dx = 0 \qquad (4)$$

where it is assumed that $\eta^{+}, \eta^{-} \to 0$ as $x \to \infty$. Neglecting $\eta^{-}$ gives the linear-nondispersive core of a variable-depth form of the KP equation; retaining both $\eta^{+}$ and $\eta^{-}$ allows for the calculation of reflection by variations in channel depth. For cases of waves in a channel of variable width b(x) and with suitably weak cross-channel dependence in the wave motion, a similar splitting of the cross-channel integrated second order wave equation gives

$$b\eta_t^{\pm} \pm bc\eta_x^{\pm} + \frac{(bc)_x}{2}\,(\eta^{+} - \eta^{-}) = 0 \qquad (5)$$

which is the core of Shuto's [3] model for variable-width channels extended to include a reflected component. The coupled KdV form follows from b = constant in (5) (or no y-dependence in (4)).

Weakly Dispersive Nonlinear Waves

We restrict attention here to the case where variations in the domain (through $c_x$ or $(bc)_x$) and the transverse curvature of the wave surface are of $O(\varepsilon)$ or smaller, where $\varepsilon$ characterizes wave height normalized by depth. Then, nonlinear and dispersive terms of equal magnitude may be added directly to (4) and (5) to obtain the desired model equations. The resulting models are given by

$$\eta_t^{\pm} \pm c\eta_x^{\pm} + \frac{c_x}{2}\,(\eta^{+} - \eta^{-}) \pm \frac{3}{2}\left(\left[\frac{c\alpha(\eta^{\pm})^2}{2}\right] - \left[\frac{c\beta(\eta^{\pm})^3}{3}\right]\right)_x$$

$$- (\gamma\eta_{xt}^{\pm})_x \mp \frac{c}{2}\int_x^{\infty} \frac{1}{c^2}\,(c^2\eta_y^{\pm})_y dx = 0 \qquad (6)$$

for weakly two-dimensional motion, and

$$b\eta_t^{\pm} \pm bc\eta_x^{\pm} + \frac{(bc)_x}{2}(\eta^+ - \eta^-) \pm \frac{3}{2}\left(\frac{bc\alpha(\eta^{\pm})^2}{2} - \frac{bc\beta(\eta^{\pm})^3}{3}\right)_x - (b\gamma\eta_{xt}^{\pm})_x = 0 \qquad (7)$$

for one-dimensional motion. The coefficients $\alpha$, $\beta$, $\gamma$ are given in Table 1 for the two model systems considered. The added terms have been put in mass-conserving form by adding negligibly small terms and in RLW form for computational convenience. The cubic term in the internal wave model is important in transition regions of equal depth $h_1 = h_2$ [4].

The model equations here are solved using a three level implicit scheme taken from Eilbeck and McGuire [5] for the one-dimensional case, and a two-level implicit scheme from the same authors for the two-dimensional case. Due to space limitations, we concentrate here on model verification followed by a speculative example on internal wave evolution in a trapezoidal channel.

One-Dimensional Waves in Diverging or Converging Channels

The first example considers a solitary surface wave evolving in a converging or diverging channel of constant depth. This case was studied experimentally by Chang _et al_. [6]. For the present computations, we use the model equation (7) with channel geometry specified to match the physical layout of the experiment, and include the reflected wave component. The results of this test indicated that the mass-conserving adjustment to the nonlinear and dispersive terms in (7) were important. Results for wave height evolution showed close to a $b^{-2/3}$ dependence in both the converging and diverging channel, which corresponds to the adiabatic solution for a slowly-varying solitary wave in the model of Shuto [3]. The mass-conserving inviscid results are shown in dashed lines in Fig. 1 for both types of channel. Adjustments for frictional damping were made by adding a dissipative term of

Table 1: Coefficients of Model Equation

| | Surface Waves | Two-Layer Internal Waves |
|---|---|---|
| $\alpha$ | $h^{-1}$ | $(h_1 - h_2)/h_1h_2$ |
| $\beta$ | 0 | $2(h_1^3 + h_2^3)/h_1^2h_2^2(h_1 + h_2)$ |
| $\gamma$ | $h^2/6$ | $h_1h_2/6$ |

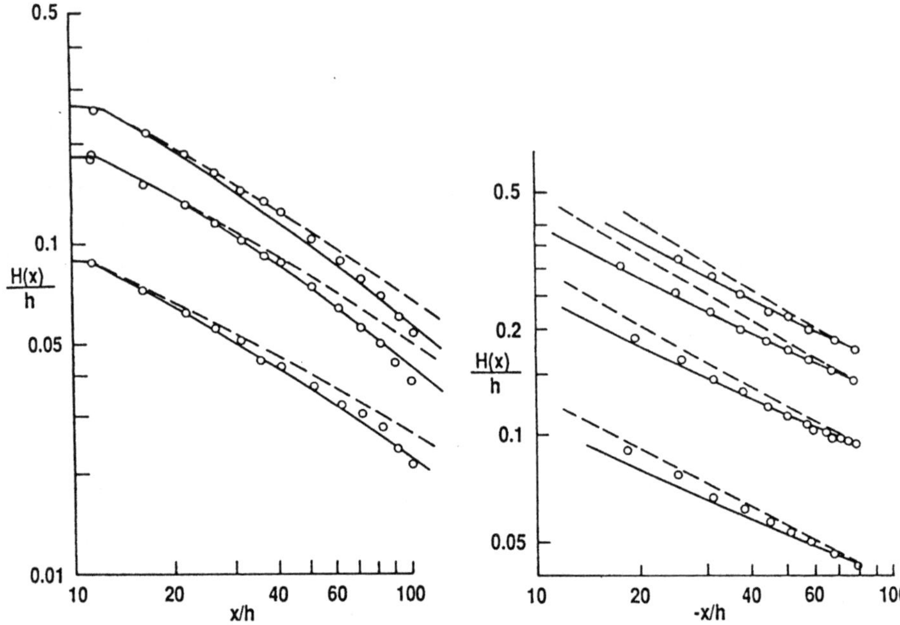


a) Diverging channel b) Converging channel

Fig. 1. Wave height of solitary waves evolving in converging and diverging channels. —— results with calibrated damping. --- inviscid results. o data from Chang et al. [6].

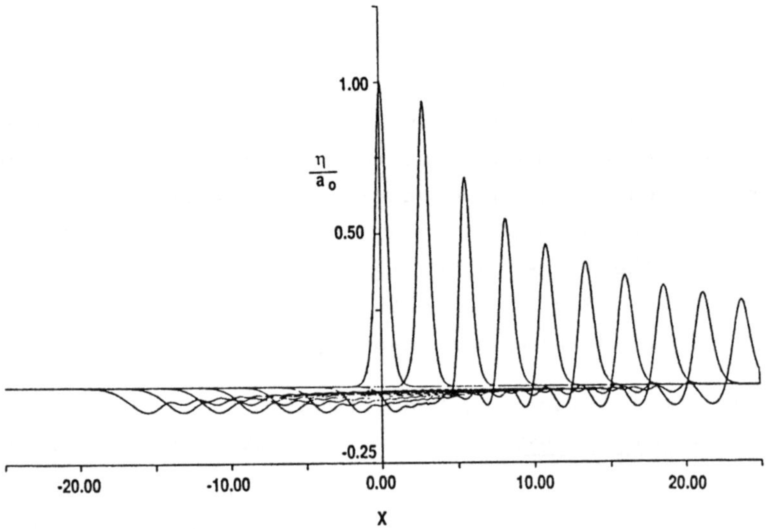


Fig. 2. Stages of wave evolution in diverging channel, including reflected component.

the form $\mu(b+2h)\eta$ to each equation. The value of $\mu$ was chosen based on fitting to one particular set of data in the diverging channel and then using the calibrated $\mu$ for the remaining conditions. The damped results are shown as solid lines in Fig. 1.

The effect of reflection has only a small effect on the evolution of the wave height in the present case. However, for the case of a diverging channel, the negative reflected wave is responsible for the depression of the region following the wave crest to below mean water level, as is evident in the wave profiles shown by Chang *et al*. An example of several stages in the time evolution of the surface displacement in the diverging channel is shown in Fig. 2.

## Diffraction of Solitary Waves by a Wall

The KP model is tested here in comparison with Melville's [7] measurements of Mach reflection of obliquely incident solitary surface waves by a vertical wall in water of constant depth. In this experiment, the action of nonlinearity is to reduce the amplitude of runup at the wall through the mechanism of formation of a stem wave with crest oriented perpendicular to the wall. Melville conducted a range of tests which included cases with $h$ = 0.3 m and relative incident wave amplitudes $a_i$ = 0.1 and 0.15, where $a_i = MAX|\eta_i|/h$. In Fig. 3, we show the predictions of relative runup $a/a_i$ along the wall for $a_i$ = 0.1 and a range of incidence angles, in comparison with Melville's data. Figure 4 shows plots of $a/a_i$ on a section perpendicular to the wall located at a distance $x/h$ = 16.7 along the wall, for $a_i$ = 0.15. The results indicate that the KP model is valid for oblique angles up to and including 35°, which is usually quoted as the limit of validity of the corresponding parabolic approximation for short waves. The only consistent discrepancy between data and prediction occurs near the leading edge of the wall, where there is some indication that the maximum numerical runup builds up too slowly. This effect is quickly corrected as the wave pattern develops.

## Evolution of Internal Solitary Waves in a Trapezoidal Channel

We now consider the evolution of an initial wave of depression in a two-layer channel with upper layer depth $h_1$ and lower layer depth $h_2(y)$ which varies linearly across the channel. We take $h_1$ = 0.1 m and $(h_2)_{MAX}$ = 0.15 m at y=0 and use a wave of depression

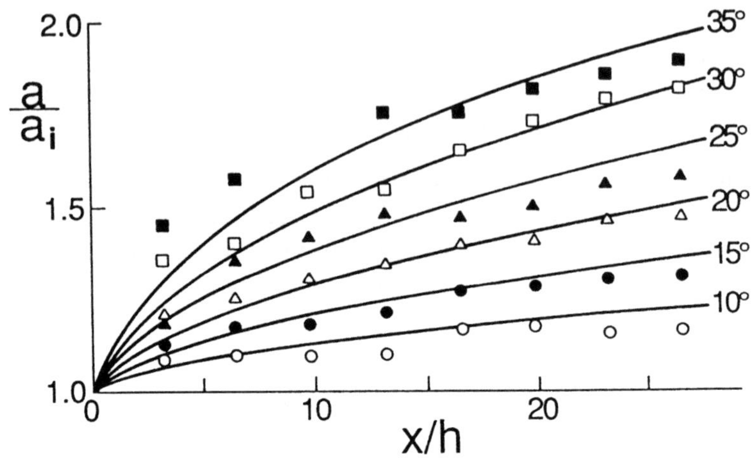


Fig. 3. Diffraction of solitary wave; maximum relative runup along wall. Data from Melville [7]. o,10°; •,15°; Δ,20°; ▲,25°; □,30°; ■,35°.

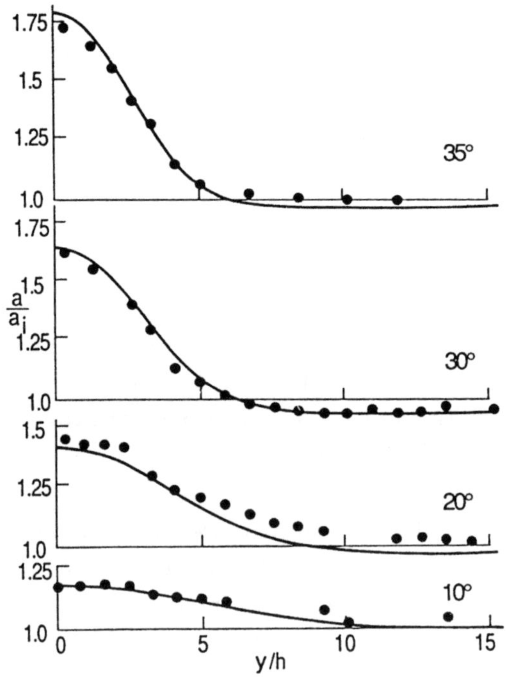


Fig. 4. Diffraction of solitary wave; maximum waveheight perpendicular to wall at x/h = 16.7. Data from Melville [7].

which has the characteristics of a permanent solitary wave in the deeper portion of the channel. In these examples the amplitude of the wave is $a_0 = 0.01$ m and the density ratio $\delta = 0.004$. We then take $(h_2)_{MIN} = 0.05$ m at y=20 m so that the lower layer is shallower than the upper layer along one boundary, with the transition occurring along the centerline. Figure 5a shows an early view of the evolution in which the curvature of the solitary wave crest (due to slowing at the shallower side) and the initial grazing incidence and reflection of the wave from the lateral boundary are apparent. Also apparent is the generation of short, steep waves along the shallow boundary, trailing the initial depression. This evolution is consistent with results of shoaling waves of depression in one-dimension, in which a train of solitary waves of elevation may evolve in the shoaled region with smaller lower-layer depth [4].

Figure 5b shows a view of the evolution at a later time and shows a complex wave pattern. The pattern is dominated by two wave-trains: one consisting of short-waves which have propagated back out into the channel and are dispersing in a nearly linear fashion due to the weak quadratic nonlinearity in this region, and a second consisting of trapped waves propagating along the channel boundary. These waves may develop into a train of edge waves in the absence of the leading long wave; however, further energy input into this portion of the wave train causes it to again separate from the wall, and later stages of evolution show a nearly linear long wave pattern followed by a confused pattern of dispersive short waves.

## References


1. Pierini, S.: Solitons in a channel emerging from a three-dimensional initial wave. Il Nuovo Cimento 9C (1986) 1045-1061.

2. Katsis, C.; Akylas, T.R.: On the excitation of long nonlinear water waves by a moving pressure distribution, Part 2. Three-dimensional effects. J. Fluid Mech. 177 (1987) 49-65.

3. Shuto, N.: Nonlinear waves in a channel of variable section. Coastal Eng. Japan 17 (1974) 1-12.

4. Helfrich, K.R., Melville, W.K.; Miles, J.W.: On interfacial solitary waves over slowly varying topography. J. Fluid Mech. 149 (1984) 305-317.

5. Eilbeck, J.C.; McGuire, G.R.: Numerical study of the regularized long-wave equation. I. Numerical methods. J. Comp. Phys. 19 (1975) 43-57.

6. Chang, P.; Melville, W.K.; Miles, J.W.: On the evolution of a solitary wave in a gradually varying channel. J. Fluid Mech. 95 (1979) 401-414.

7. Melville, W.K.: On the Mach reflection of a solitary wave. J. Fluid Mech. 98 (1980) 285-297.

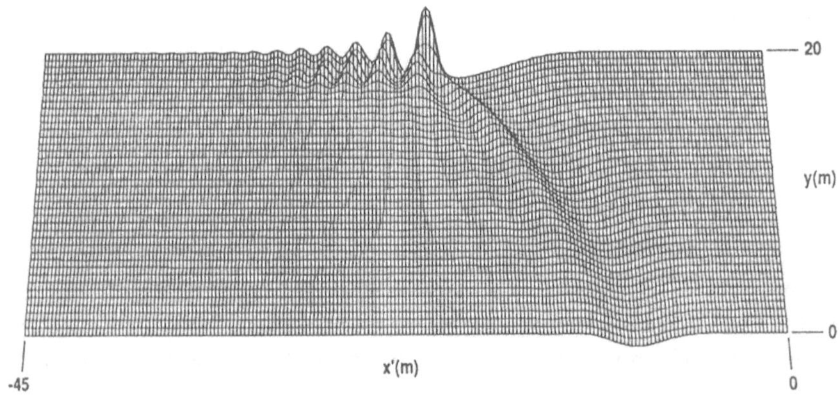


a)

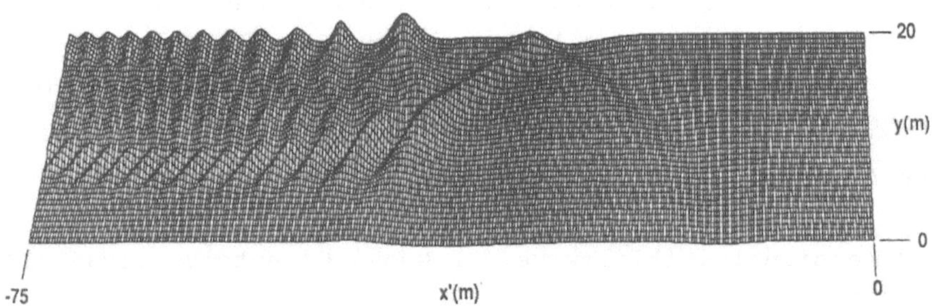


b)

Fig. 5. Evolution of an internal wave of depression in a trapezoidal channel. a) early stage of development b) evolved wave field.

# A Theoretical and Experimental Study of Precursor Solitary Waves Generated by Moving Disturbances

S. J. LEE[2], G. T. YATES and T. Y. WU
Engineering Science
California Institute of Technology
Pasadena, California USA

## Abstract

This is a joint theoretical, numerical and experimental study of the recently identified resonance phenomenon whereby a forcing disturbance moving steadily with a transcritical velocity in shallow water can generate, periodically, a succession of solitary waves, advancing upstream of the disturbance. A series of numerical and laboratory experiments was carried out with a cambered topography serving as the forcing disturbance moving over the floor of a water layer with velocity $U$, or Froude number $F = U/\sqrt{gh_o}$, where $g$ is the acceleration of gravity and $h_o$ the original uniform water depth. Within the transcritical range of about $0.4 < F < 1.2$ tested, and over a range of disturbance amplitudes, a series of upstream advancing solitary waves were generated, one after another, while a train of weakly nonlinear and weakly dispersive waves developed downstream of the disturbances immediately behind a nearly constant but elongating depressed water surface. For a review of the history and literature on this subject, we refer to the recent work by Wu (1987).

## Theoretical Models

Theoretical investigation of the phenomenon of upstream advancing solitary waves has been pursued using two mathematical models, one being the generalized Boussinesq (gB) model proposed by Wu (1979) and the other the forced Korteweg-de Vries (fKdV) equation developed here to include such forcing functions as moving surface pressures and submerged topographies. The fluid is assumed inviscid and incompressible and the motion is taken as irrotational. Laplace's equation in two-dimensional space $(x, z)$, with $z$ vertical, is solved for the velocity potential $\varphi(x, z, t)$ under appropriate initial and boundary conditions. The kinematic and

[1]This work was jointly sponsored by the U.S. Office of Naval Research and the National Science Foundation.

[2]Present address: Chungnam National University, Daeduk Science Town, KOREA.

K. Horikawa, H. Maruo (Eds.)
Nonlinear Water Waves
IUTAM Symposium Tokyo/Japan 1987

dynamic boundary conditions at the free surface are respectively

$$\varsigma_t + \varphi_x \varsigma_x - \varphi_z = 0, \qquad \text{on} \quad z = \varsigma(x,t), \tag{1}$$

$$\varphi_t + \frac{p_o}{\rho} + \frac{1}{2}(\varphi_x^2 + \varphi_z^2) + gz = 0, \qquad \text{on} \quad z = \varsigma(x,t). \tag{2}$$

where $\varsigma(x,t)$ is the free surface elevation, $p_o(x,t)$ is a prescribed distribution of pressure on the free surface, and the subscripts $x$, $z$ and $t$ stand for partial differentiation. The kinematic boundary condition on the bottom is

$$b_t + \varphi_x b_x - \varphi_z = 0 \qquad \text{on} \quad z = -h_o + b(x,t), \tag{3}$$

where $b(x,t)$ is a prescribed bottom topography. After suitably scaling the space and time, the nonlinear and dispersive effects are brought into a proper balance such that (Boussinesq class model),

$$\epsilon = (h_o/\lambda)^2 \ll 1, \qquad \alpha = a/h_o \ll 1, \qquad \alpha = O(\epsilon),$$

where $\lambda$ is the characteristic wavelength, and $a$ is a typical wave amplitude. By expanding the free surface elevation $\varsigma(x,t)$ and the layer mean velocity potential $\phi(x,t)$ in power series, the gB equations are derived as follows (see Lee 1985, Wu 1979),

$$\varsigma_t + [(1 - b + \varsigma)\phi_x]_x = b_t, \tag{4}$$

$$\phi_t + \varsigma + \frac{1}{2}\phi_x^2 - \frac{1}{3}\phi_{xxt} = -\frac{1}{2}b_{tt} - p_o, \tag{5}$$

where all lengths are referenced to $h_o$, $\phi$ to $c_o h_o$, $t$ to $h_o/c_o$ and

$$\phi(x,t) = \frac{1}{1 - b + \varsigma}\int_{-1+b}^{\varsigma} \varphi(x,z,t)dz .$$

Equation (4) is exact, and (5) has an error of $O(\epsilon^3)$. The disturbance strengths $p_o$ and $b$ are also taken $O(\epsilon)$. Thus, the generated wave system is weakly nonlinear and weakly dispersive.

It can be shown from these equations that whenever the disturbances are small, a surface pressure may produce a very similar solution as a bottom bump if both disturbances are prescribed by the same function. As the disturbance becomes larger, the $b_{tt}$ term in (5) will no doubt make a contribution in addition to the $b_t$ term in (4) and some differences between the two types of disturbances will become apparent.

Under the further assumptions that the motion is near-resonant, the disturbances are of $O(\epsilon^2)$, and that the waves are unidirectional (here, left going), the fKdV equation and the relation between the layer mean horizontal velocity $u(x,t)$ and $\varsigma(x,t)$ are obtained (Lee, 1985),

$$\varsigma_t - (1+\frac{3}{2}\varsigma)\varsigma_x - \frac{1}{6}\varsigma_{xxx} = \frac{1}{2}(p_o + b)_x \; , \tag{6}$$

$$u = -\varsigma + \frac{1}{4}\varsigma^2 - \frac{1}{6}\varsigma_{xx} + \frac{1}{2}(b - p_o). \tag{7}$$

Note that, in contrast to the gB model (4 – 5), the fKdV equation (6) does not differentiate the mode of the disturbance since $p_o$ and $b$ appear summed.

The wave resistance coefficient (referenced to $\rho g h_o^2$) is the streamwise component of the pressure acting on the body, which is

$$C_w = \int [-p_o\, \varsigma_x + p(x,t,-h_o+b)\; b_x\,]\; dx\; . \tag{8}$$

## Numerical Computations

Both models (gB and fKdV) were solved numerically using finite difference methods with a predictor-corrector scheme without iteration at the corrector stage (Wu & Wu 1982). Central differences for the space and forward difference for the time were used in discretization of the theoretical models. Figure 1 shows typical results for the wave elevation profiles at several time instants and the time variation of the resistance $C_w$ obtained numerically using the gB model with the surface pressure taken as

$$p_o(x,t) = \frac{1}{2}\, p_m \left[1 - \cos\frac{2\pi}{L}(x + F\, t)\right], \qquad 0 \le (x + F\, t) \le L,$$

$$= 0, \qquad\qquad \text{otherwise}\; ,$$

which starts to act from $t = 0$ on the water initially at rest. For all transcritical speeds, a series of precursor solitary waves are generated periodically and propagate upstream of the disturbance. During this motion, the pressure disturbance, while being kept steady, experiences a fluctuating resistance which becomes periodic shortly after its start. The relative size of precursor waves to the waves in the trailing region increases as $F$ increases.

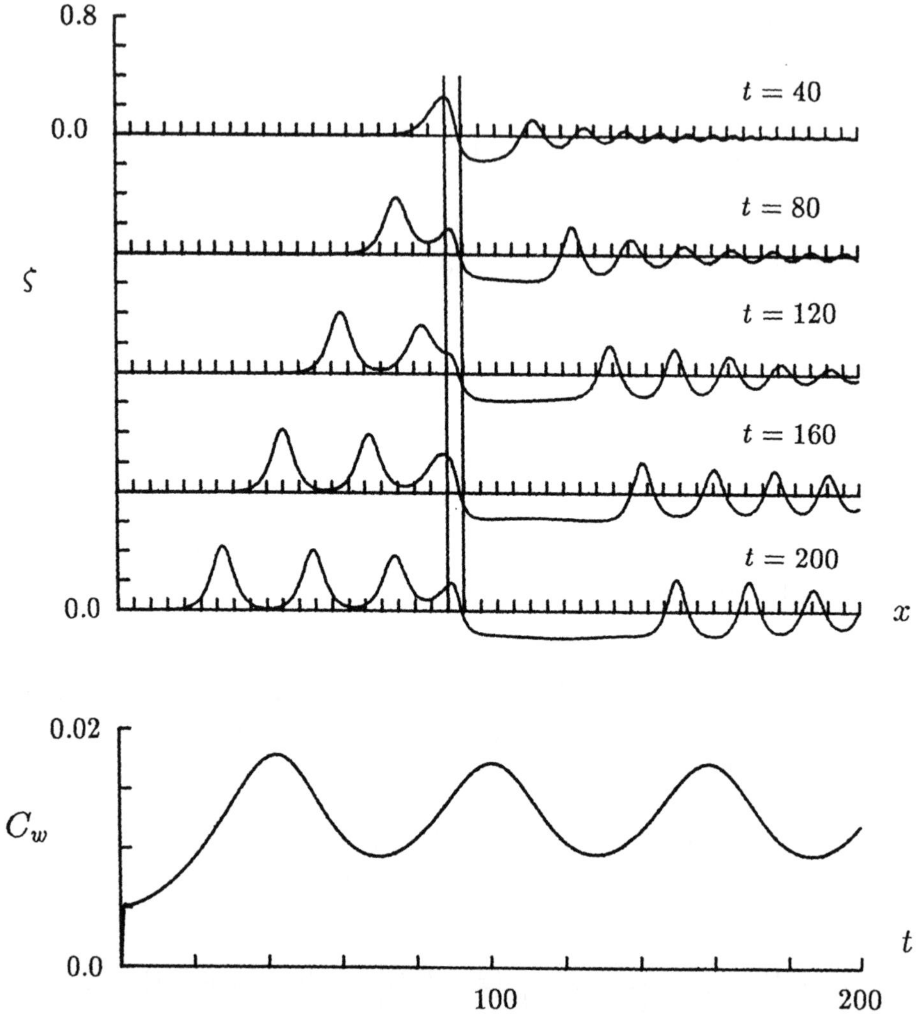


Figure 1. The free surface wave elevations $\varsigma$ are shown in the top plots versus $x$ at the specified times $t$, while the wave resistance coefficient $C_w$ is given as a function of time. All curves were computed using the gB model for a cosine distribution of surface pressure with $p_m = 0.1$, $h_o/L = 0.5$ and $F = 1.0$.

We find that the gB model predicts slightly greater wave heights and shorter periods of generation of precursor waves for a surface pressure than for a bottom topography of the same distribution. In general, the values of wave heights and periods of generation of precursor waves given by the fKdV model, which does not distinguish between the two kinds of forcing functions, lie somewhere between those given by the gB model for a surface pressure and for the bottom topography, their distribution being equal.

Although linear dispersive theory predicts unbounded surface elevations for a disturbance moving with $F$ which tends to 1 from below, it provides a finite value of the wave resistance in the same limit. Figure 2 shows the temporal mean values of $C_w$ predicted by the gB model as functions of $F$ for various $p_m$. A finite jump at $F = 1.0$ of $\overline{C}_w$ given by the linear theory is now replaced by continuous but rapid decrease in accordance with the gB model.

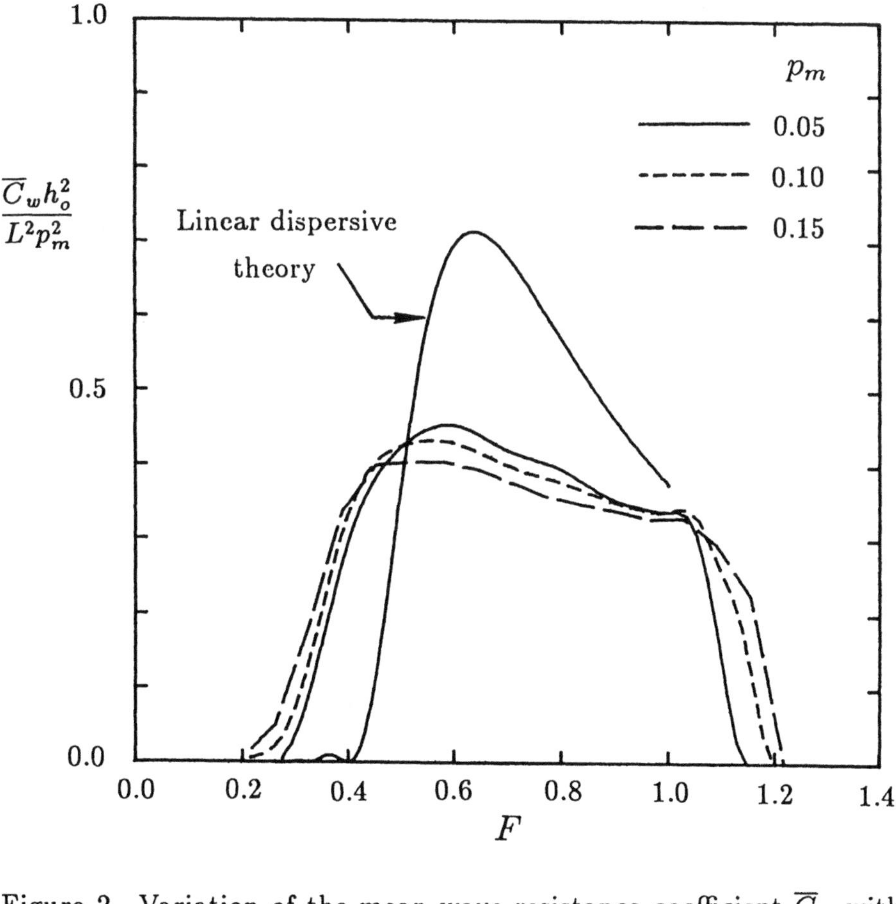


Figure 2. Variation of the mean wave resistance coefficient $\overline{C}_w$ with $F$ and comparison between the linear and nonlinear (gB) theories.

## Experimental Results and Comparisons

Experiments were conducted using a two-dimensional topography with a chord of $L = 4.9$ cm and height of 0.65 cm at its mid-chord. In cross section the 'bump' had a circular arc top and a flat base. While the bump was towed along the floor of a 7.5 m long tank, wave elevations were recorded at various locations along the center plane of the flume using both moving and fixed wave gages of resistance type (Lee, 1985). The ratio of the maximum height of the bump and the water depth ranged from 0.15 to 0.2, $h_o/L$ from 0.82 to 1.09, and $F$ from 0.7 to 1.2. Typical experimental data are shown in Figure 3 along with the corresponding numerical results given by the gB model for the corresponding topography. Dotted and solid curves represent experimental and numerical results, respectively.

All the main features predicted by the two theoretical models, evaluated by numerical methods, are seen also to appear in the experimental data, with some refined differences between the two models. The numerical results of the mean wave resistance (see Figure 2) indicate that the phenomenon of precursor solitary wave generation generally manifests over a quite broad transcritical range of Froude numbers ($0.2 < F < 1.2$) for forcing functions of small to moderate amplitude. This is corroborated by our experimental observations. Outside this speed range, the precursor waves were not found to appear (for two-dimensional disturbances) and a steady wave pattern relative to the disturbance quickly emerges after the motion is impulsively started.

For very weak disturbances the two numerical models give similar results and show a broad agreement with the experiments in terms of both the magnitude and the phase of the waves generated over the range of parameters covered in the current study. As the disturbance amplitude increases the fKdV model becomes slightly more effective in generating upstream advancing waves than the gB model. However, the differences between the results predicted by the two numerical models are about the same as that between the experimental data and the results predicted by either of the two models.

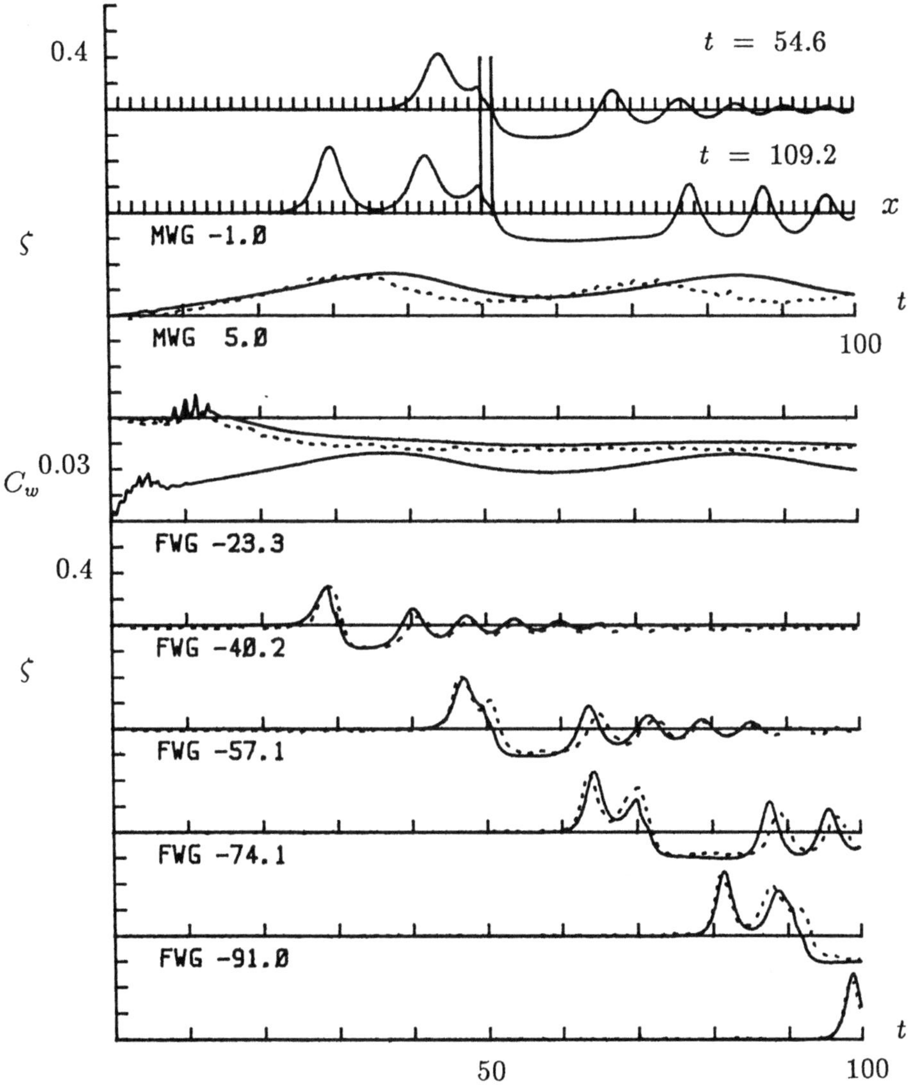


Figure 3. Comparison of the surface elevations between experimental data (dashed) and numerical solution (solid) of the gB model; $b_m = 0.2$, $h_o/L = 0.82$ and $F = 1.01$. The top two plots show $\varsigma$ versus $x$ for fixed time $t$. The next two plots show $\varsigma$ versus $t$ for wave gages moving with the disturbance. The fifth plot gives $C_w$ as a function of time, while the last 5 curves show $\varsigma$ versus $t$ for wave gages fixed with respect to the tank.

## References

Lee, S.J. 1985 Generation of long water waves by moving disturbances. Ph.D. Thesis, California Institute of Technology, Pasadena, CA. USA

Wu, T.Y. 1979 On tsunami propagation - evaluation of existing models. Tsunamis, Proc. of the the National Science Foundation Workshop organized and edited by L.S. Hwang & Y.K. Lee pp. 110-149.

Wu, D.M. & T.Y. Wu 1982 Three-dimensional nonlinear long waves due to moving surface pressure. Proc 14th Sym Naval Hydro. pp. 103-125.

Wu, T.Y. 1987 Generation of upstream-advancing solitons by moving disturbances. J. Fluid Mechanics (In Press)

# Instability and Breaking of a Solitary Wave

M. TANAKA**, J.W. DOLD*, M. LEWY* AND D.H. PEREGRINE*

** Department of Applied Mathematics, Faculty of Engineering Gifu University, JAPAN

* School of Mathematics, University of Bristol, U.K.

Abstract

The result of a linear stability calculation of solitary waves which propagate steadily along the free surface of a liquid layer of constant depth is examined numerically by employing a time-stepping scheme based on a boundary-integral method. The initial growth rate that is found for sufficiently small perturbations agrees with the growth rate expected from the linear stability calculation. In calculating the later 'nonlinear' stage of the instability, it is found that two distinct types of long-term evolution are possible. These depend only on the sign of the unstable normal-mode perturbation which is superimposed initially on the steady wave. The growth of the perturbation ultimately leads to breaking for one sign. Unexpectedly, for the opposite sign, there is a monotonic decrease in the total height of the wave. In this latter case there is a smooth evolution to a stable solitary wave of lesser amplitude but very nearly the same energy.

## Introduction

The linear stability problem for a solitary wave with an arbitrary amplitude which propagates steadily along the free surface of a liquid layer of constant depth was investigated by Tanaka [1]. He showed numerically that for solitary waves higher than that corresponding to the maximum total energy there is at least one growing (unstable) normal-mode. He also calculated the growth-rate eigenvalue and the eigenfunction of this mode. More recently this coincidence of the critical amplitude and the amplitude of energy maximum was confirmed analytically by Zufiria and Saffman [2] by using Zakharov's Hamiltonian formulation of water waves.

However all of this work is based on a linear stability theory with the assumption that the perturbation is infinitesimally small compared with the basic steady wave. It does not give any information about 'nonlinear' growth of the instabili-

K. Horikawa, H. Maruo (Eds.)
Nonlinear Water Waves
IUTAM Symposium Tokyo/Japan 1987

ty; for example one might ask if the instability leads eventually to breaking. In order to address this problem we have calculated the long-time evolution of the instability by using the very efficient time-stepping method developed by Dold and Peregrine [3]. In this work we have concentrated particularly on a solitary wave with the amplitude-to-depth ratio a=0.80374 which is well above the energy-maximum level $a_e$ (=0.78066).

## Initial Growth of Perturbation

The initial condition used for following the unstable evolution of a solitary wave is provided by

$$y = H + \varepsilon\, \tilde{\eta}\ ; \quad \varphi = \Phi + \varepsilon\, (\tilde{\varphi} + \Phi_y \tilde{\eta}) \tag{1}$$

where $H$ and $\Phi$ are the surface displacement and the velocity potential, respectively, of a steady solitary wave propagating in still water; $\tilde{\eta}$ and $\tilde{\varphi}$ are the unstable normal-mode perturbations to these quantities obtained by a linear stability calculation (Tanaka [1]). The small parameter $\varepsilon$ is used to select the size of the initial perturbation. We normalize space and time so that the undisturbed depth and the acceleration due to gravity are set equal to 1. With this normalization the basic solitary wave which was chosen for this study has an amplitude a=0.80374 and the corresponding phase speed c=1.29400.

In order to study the growth of perturbations, we evaluated a perturbation integral $P(t)$ defined by

$$P(t) = \frac{1}{2} \int_{-\infty}^{+\infty} \{y(x,t) - H(x-ct)\}^2\, dx. \tag{2}$$

According to the linear stability calculation, $P(t)$ should behave like $P(t) = P(0)\exp(2\lambda t)$ with $\lambda=0.0983$ for the basic steady wave under consideration. Tested on the unperturbed wave to check the accuracy of the numerical scheme, $P(t)$ remains less than $10^{-10}$ in value up to time $t=10$, when the wave has propagated about thirteen times the undisturbed depth. (Note, for comparison, that for the wave under consideration $1/2\int_{-\infty}^{+\infty} H^2 dx = 0.44900$.)

It is worth mentioning here the significance of the sign of the small parameter $\varepsilon$. As shown in Fig.1, the perturbation is not symmetric about $x=0$, the initial position of the crest of the solitary wave. This implies that the perturbed wave has unequal slope on each side of the crest. Throughout this work, we employ the convention that positive values of $\varepsilon$ always perturb the steady wave such that the rear face of the wave is steeper than the front face, while the negative values have the opposite effect. Although the sign of $\varepsilon$ does not have any relevance to the linear stability theory, the long-time evolution of the perturbation is found to be completely different as described in the next section.

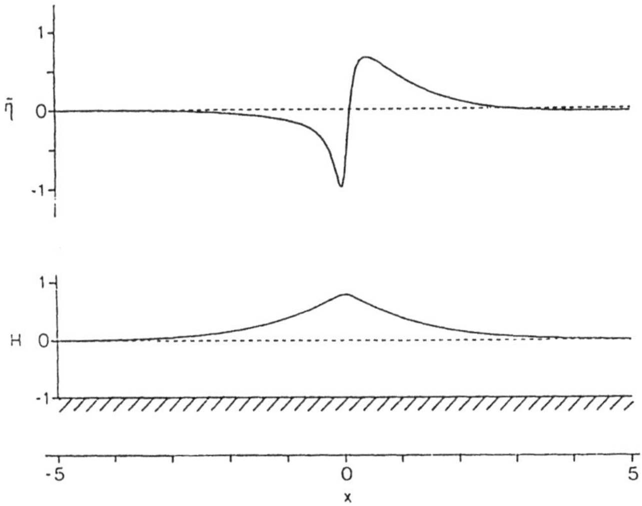


Fig.1. Unstable normal-mode $\tilde{\eta}(x)$ for a solitary wave (shown below to natural scale) propagating from right to left.

With $\varepsilon = \pm 0.001$, numerical result for the evolution of $P(t)$ is shown in Fig.2 which shows good agreement between the result of the present calculation and the prediction of the linear stability theory, providing an independent validation of the linear stability results.

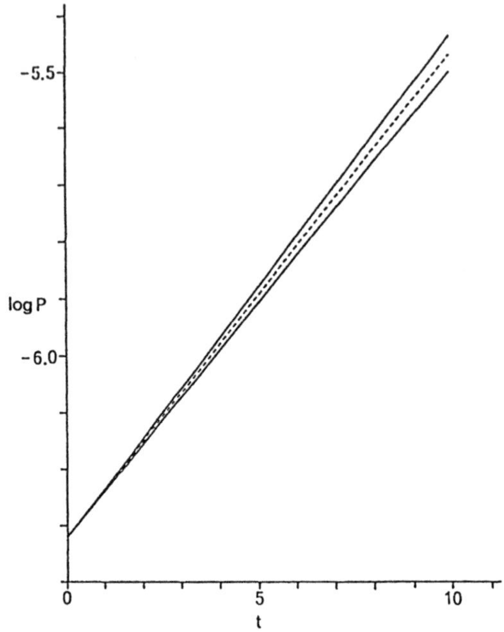


Fig.2. Development of the perturbation integral $P(t)$ with $\varepsilon$ =+0.001 (lower curve) and $\varepsilon$ =-0.001 (upper curve). The dashed line gives the exponential growth predicted by linear stability theory.

Long-Time Evolution

In order to investigate the nonlinear stage of the instability, we increased the values of $\varepsilon$ to +0.01. The behaviour is quite different for the two cases with different signs of $\varepsilon$ and we shall discuss them separately.

(1) Breaking Case ($\varepsilon$ = -0.01)

The front face of the wave which was initially only by a small amount steeper than the rear face gradually steepens further as the instability grows and small portion of the crest of the wave eventually overturns at $t \simeq 8.8$ as shown in Fig.3. It may also be noted that the time scale of the breaking event seems such shorter than the characteristic time scale of the basic steady wave $1/c$ ($\simeq 0.8$) or the growth of the instability $1/\lambda$ ($\simeq 10.2$). The local behaviour around the breaking crest shows some similarity between the present case for a solitary wave and that for a periodic wave on deep water investigated by Longuet-Higgins and Cokelet [4].

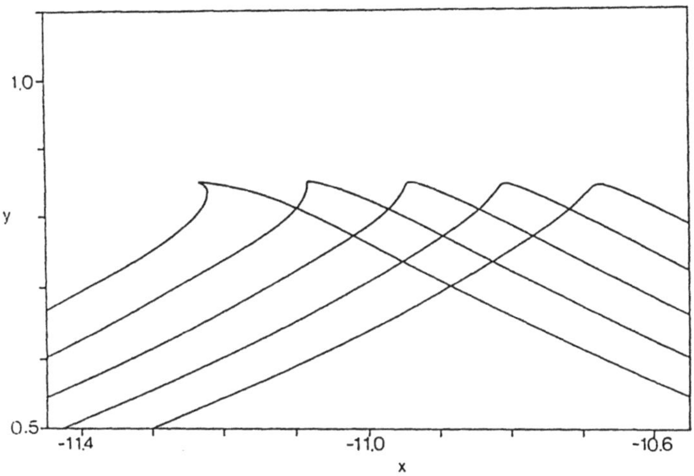


Fig. 3. Overturning of the breaking wave crest ($\varepsilon=-0.01$). Profiles are shown at times $t=8.5$, 8.6, 8.7, 8.8 and 8.9.

(2) Non-Breaking Case ($\varepsilon = +0.01$)

In Fig.4 we show the profile of the wave at $t=100$ for the case of $\varepsilon = +0.01$. The normal-mode which was superimposed on the steady solution is exactly the same as that used in the breaking case, the only difference being in the sign of $\varepsilon$. For this case the growth of the instability unexpectedly reduces the amplitude of the wave (see Fig.5). The wave profile becomes more gentle as the perturbation grows. The asymmetry in the initial condition does not grow as it does for the breaking case and the wave remains almost symmetric throughout the evolution. The profile at $t=100$ has been compared with that of the steady solitary wave of the same height; the two waves are graphically indistinguishable.

Since the steady wave we are employing as the basic wave

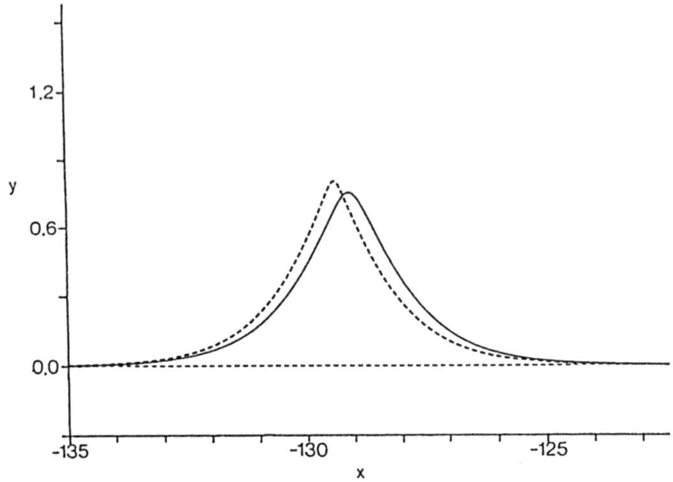


Fig. 4. Profile of the free surface at $t=100$ for the non-breaking case ($\varepsilon = 0.01$). The dashed line is the unperturbed case.

is beyond the energy maximum level, there exists another steady solitary wave with exactly the same amount of energy but with a smaller amplitude $\tilde{a}$. For the case under consideration $\tilde{a}=0.7539$. The evolution of the amplitude $a(t)$ shown in Fig.5 strongly suggests that the perturbed solitary wave is exponentially approaching another steady solitary wave with very nearly the same amount of total energy.

Unlike the total energy, the mass and momentum of the steady solitary wave with the amplitude $\tilde{a}$ are not the same as those of the initial condition. There is an increase in magnitude of both the mass and momentum. This increase of mass and momentum can be accounted for by the emission of two long waves behind the solitary wave. One is the radiated wave travelling in the opposite direction to the solitary wave and the other is a "trailing" wave following the solitary wave at the linear wave speed. It can be shown that mass, momentum and total energy are all conserved throughout the calculation within the numerical accuracy if the contributions of these emitted long waves are appropriately taken into account.

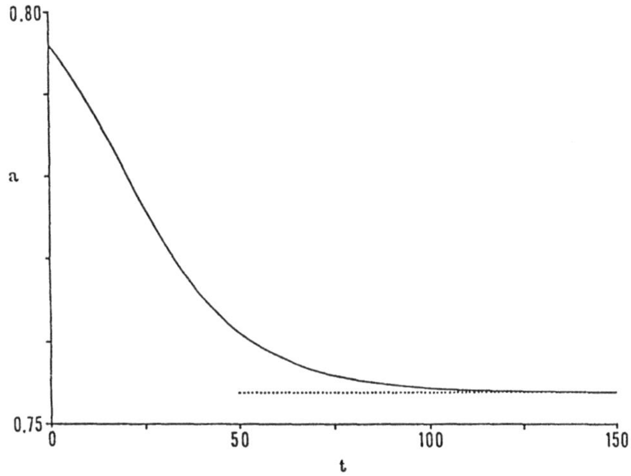


Fig. 5. Evolution of the amplitude $a(t)$. The dotted line shows the amplitude of equivalent energy as the initial condition $\tilde{a}$ (=0.7539).

It might be thought that the observed 'exponential' approach to the lower-amplitude solitary wave contradicts the result of linear stability analysis that all the solitary waves below the energy maximum level are neutrally stable. However, in the present case as $t \to \infty$ , the deviation of the wave from the final solitary wave does not necessarily satisfy the condition of going to zero as $x \to \infty$ which was a condition imposed on the perturbation when the linear stability calculation was carried out.

Further analysis of the profile of the trailing and radiated waves shows that these waves behave like exponential functions with respect to $x$ on their sides that face to the solitary wave. Moreover these waves propagate with speeds very close to that of linear long wave (i.e., one under our normalization of space and time) and do not propagate with the solitary wave. These two facts give intuitive explanation for the 'exponential' evolution of the wave to a neutrally stable solitary wave. This intuitive idea has also been confirmed quantitatively by comparing the exponential growth rate with $x$ of the trailing and radiated waves and those expected from the temporal decay rate of the amplitude of the solitary wave.

More detailed report on this work will appear soon in Journal of Fluid Mechanics (Tanaka et al. [5]).

## References


1. Tanaka, M.: The stability of solitary waves. Phys. of Fluids 29 (1986) 650-655.

2. Zufiria, J.A. and Saffman, P.G.: The superharmonic instability of finite-amplitude surface waves on water of finite depth. Stud. in Appl. Math. 74 (1986) 259-266.

3. Dold, J.W. and Peregrine, D.H.: An efficient boundary-integral method for steep unsteady water waves. in Numerical Methods for Fluid Dynamics II (eds. Morton, K.W. and Baines, M.J. Oxford, Clarendon Press 1986) 671-679.

4. Longuet-Higgins, M.S. and Cokelet, E.D.: The deformation of steep surface waves on water II. Growth of normal-mode instabilities. Proc. R. Soc. Lond. A364 (1978) 1-28.

5. Tanaka, M., Dold, J.W., Lewy, M. and Peregrine, D.H.: Instability and breaking of a solitary wave. to appear in J. Fluid Mech. 1987

# The Turbulent Flow Fields and Vortex Structures Inside Surf Zone

H. H. Hwung, C. Lin and C. C. Kao

Hydraulic and Ocean Engineering Department and Tainan Hydraulics Laboratory, National Cheng Kung University, Taiwan, China.

## Summary

In order to understand the characteristics of turbulence due to wave breaking, flow visualization techniques are used to observe the external and internal flow structures. Two typical spilling and plunging breakers are made in wave flume. 2W Argon-Ion laser, and 120W ultraviolet lamps are used to brighten the tracing particles respectively. In this situation, not only the external structures of wave breaking can be visualized but also the instantaneous velocity of particles in the internal flow field will be outlined. Moreover, it is valuable to point out that the three dimensional flow structures inside surf zone are also found in our elaborated experimental studies.

## Introduction

Since it is known that the characteristics of wave breaking are closely related to the physical properties in the shallow water area, recently a number of researchers paid much attention to this problem. Experimental studies on the generation mechanics and turbulent characteristics due to wave breaking have been made by Stive [1], Aono et al. [2], Sakai [4], Hino et al. [3], Basco [6], Nadaoka et al. [3], [5] and Jansen [8]. However due to the phenomena of wave motion after breaking are very complicated, the systematic results have not been reported completely till now.

One of principal objectives of the present research program is to examine the flow field structures; the other is to observe the formation of vortex structures and the mechanism of vortex stretching due to air entrainment by means of flow visualization.

## Experimental Method

The experiments are carried out in a laser-measuring wave flume (9.5m x 0.3m x 0.7m), Tainan Hydraulics Laboratory , National Cheng Kung University. In order to catch the flow structures clearly, the illuminating system is composed of 2W Argon laser, optical guiding device and a cylindrical lens. In this condition, a bright and monochromatic fan-

K. Horikawa, H. Maruo (Eds.)
Nonlinear Water Waves
IUTAM Symposium Tokyo/Japan 1987

shaped light sheet is produced to illuminate the motion of tracing particles on the vertical plane along the longitudinal direction and on the horizontal plane across the transverse direction shown in Fig-1 and Fig-2 respectively.

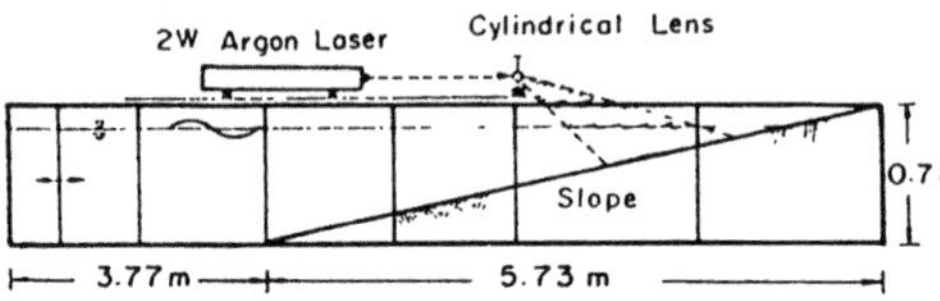


Fig-1. Installation of laser flow visualization

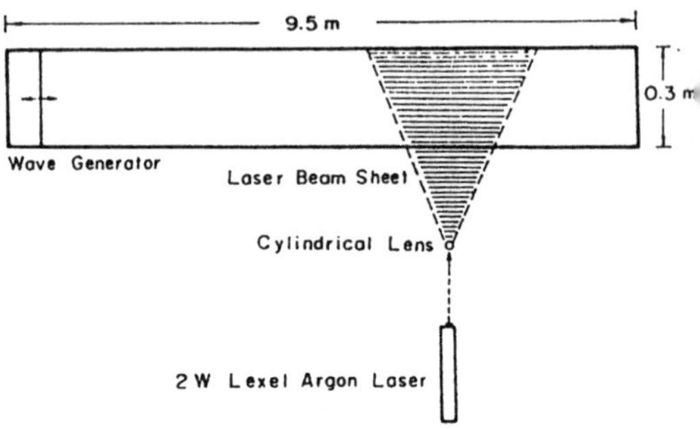


Fig-2. Installation of laser flow visualization

Furthermore, since it is very difficult to observe the flow structures of air bubbles which are caused by air entrainment inside surf zone, 120W ultraviolet lamps are taken to be the illuminating light and high speed photography is used to catch the flow fields in the testing area. Accordingly, the characteristics of flow structures inside air bubbles are able to be analysed.

For convenience to observe the complicated flow structures in surf zone, the tracers are made by the white polystyrene beads coated with three kinds of fluorescent liquid respectively (Rhodamine B, Rhodamine 6G and Fluorescein Sodium dissolved in Varnish). The different colored polystyrene beads (the specific gravity is 1.040) will be excited to scatter the red, yellow and green light in wave flume. In this situation, the external and internal flow structures can be visualized in the photographs and the instantaneous velocities are also calculated from high speed cine-film.

As for the experiment, its conditions are listed in the following table:

TABLE 1. Wave conditions

| case | slope | T(sec) | Hi(cm) | Ho/Lo | d(cm) | Hb/db | breaking type |
|---|---|---|---|---|---|---|---|
| 1 | 1/13.5 | 1.40 | 8.0 | 0.029 | 32.1 | 0.992 | plunging |
| 2 | 1/13.5 | 1.02 | 8.1 | 0.061 | 30.0 | 0.892 | spilling |
| 3 | 1/10.0 | 1.08 | 7.1 | 0.043 | 43.2 | 0.901 | plunging |
| 4 | 1/15.0 | 0.99 | 7.3 | 0.052 | 32.9 | 0.894 | spilling |

## Results and Discussions

According to the experimental results, Photo-1 and Photo-2 are spilling breakers. The formation of surface roller and internal instantaneous flow pattern can be seen in the photographs.

Photo-3 and Photo-4 are plunging breakers. It is obvious to see the plunging vortex is produced by the impinging jet of surface water and strong air entrainment. Further, between the tip of the plunging vortex and the sloping bottom, we find the "dividing region" is formed apparently due to the forward motion of the upstream flow and the rotating motion of the plunging vortex. Inside this "dividing region", the water particles are almost still; and it is valuable to notice that the sand ripples may be formed by the mechanism of this "dividing region".

Photo-5 and Photo-6 are the experimental results of flow structures on the horizontal plane of 2cm under still water level across the transverse direction as wave trough reaches. From the photographs, we find that the vertical vortex induced by air bubbles exists inside surf zone. Based on the above mentioned experimental results, we can conclude that the flow structures are three-dimensional motion within surf zone after wave breaking. And the results are similar to Nadaoka's [8] conclusion.

Moreover, the motions of air bubbles and internal flow structures of spilling and plunging breakers are caught by 120W ultraviolet lamps. The results of spilling breakers are shown in Fig-3. We can see the whitecap is generated at the crest and a clockwise vortex induced by the breaker is sliding down along the water surface. Afterwards, a splash motion takes place and the internal flow fields become turbulent due to the mixing of air bubbles and the interaction of backwash.

Figure-4 are the results of plunging breakers. It is obvious to see a surface roller is stirred by the strong air entrainment at the crest of plunging breaker. Then, the forward jet impinges on water surface. In this situation, it is apparent to see that two opposite horizontal eddies are produced at the point of jet impinging on the water surface. And the results are similar to those of Hino et al. [4] and Nadaoka [8].

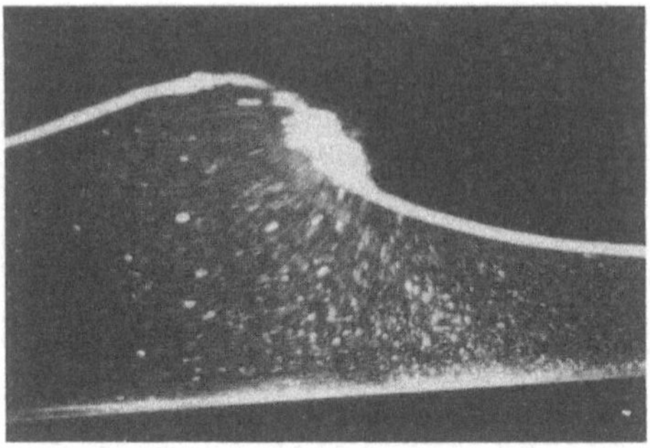

Photo-1. Spilling breaker (Case 2)

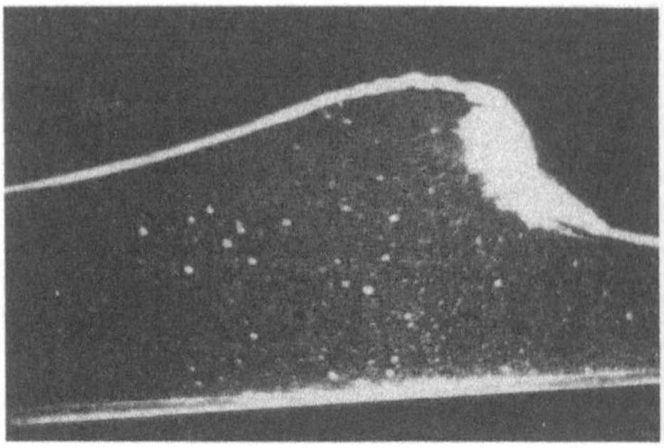

Photo-2. Spilling breaker (Case 2, continued)

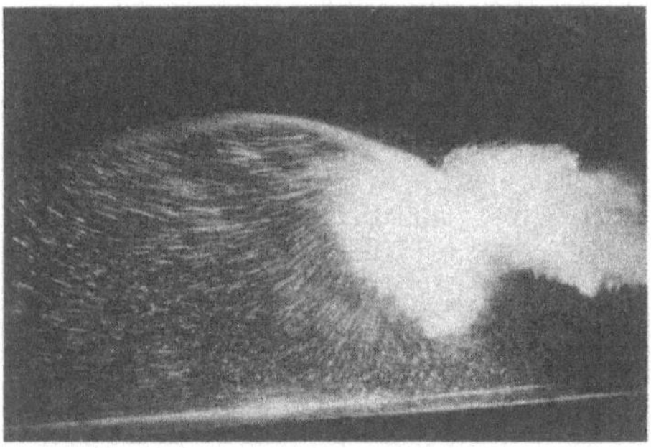

Photo-3. Plunging breaker (Case 1)

Photo-4. Plunging breaker (Case 1, continued)

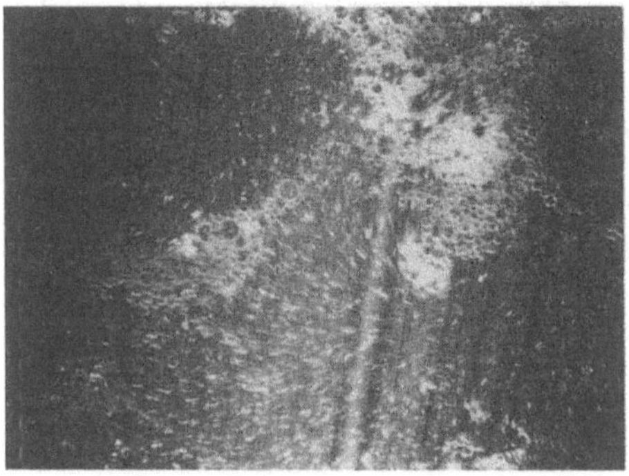

Photo-5. Vertical vortex is observed inside surf zone.

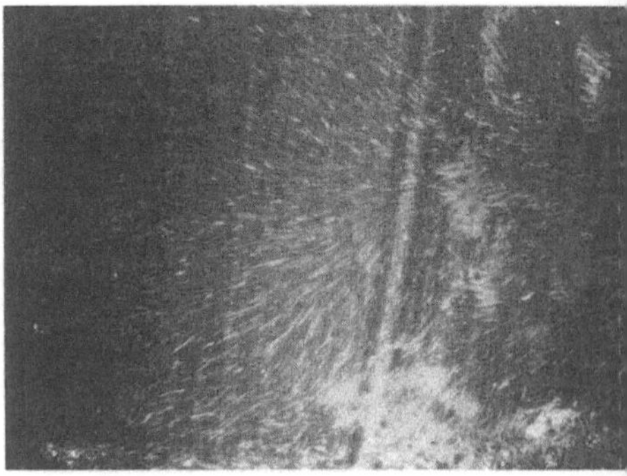

Photo-6. Vertical vortex is observed inside surf zone. (continued)

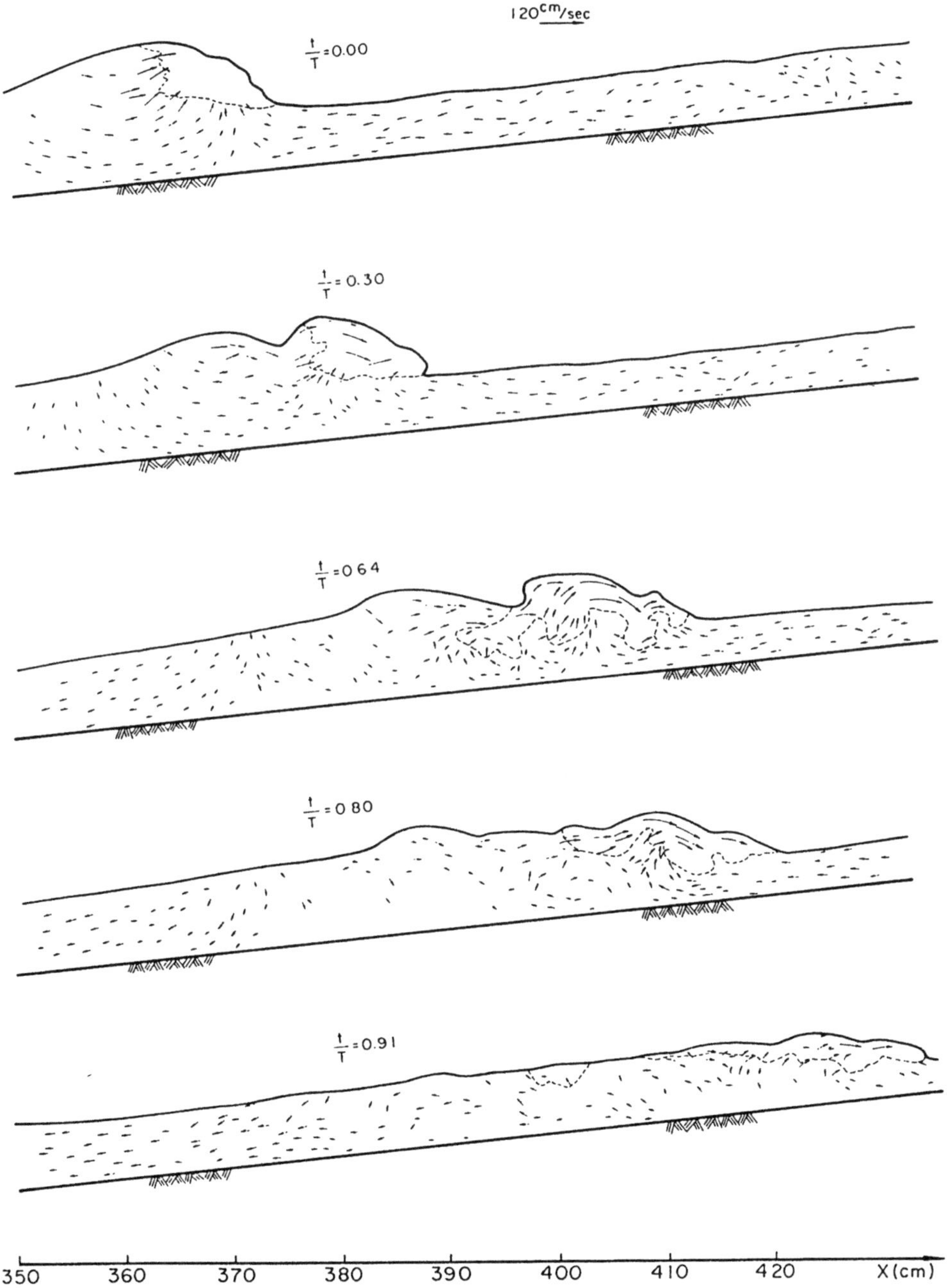


Fig-3. The evolutions and flow field structures of spilling breaker. (Case 4)

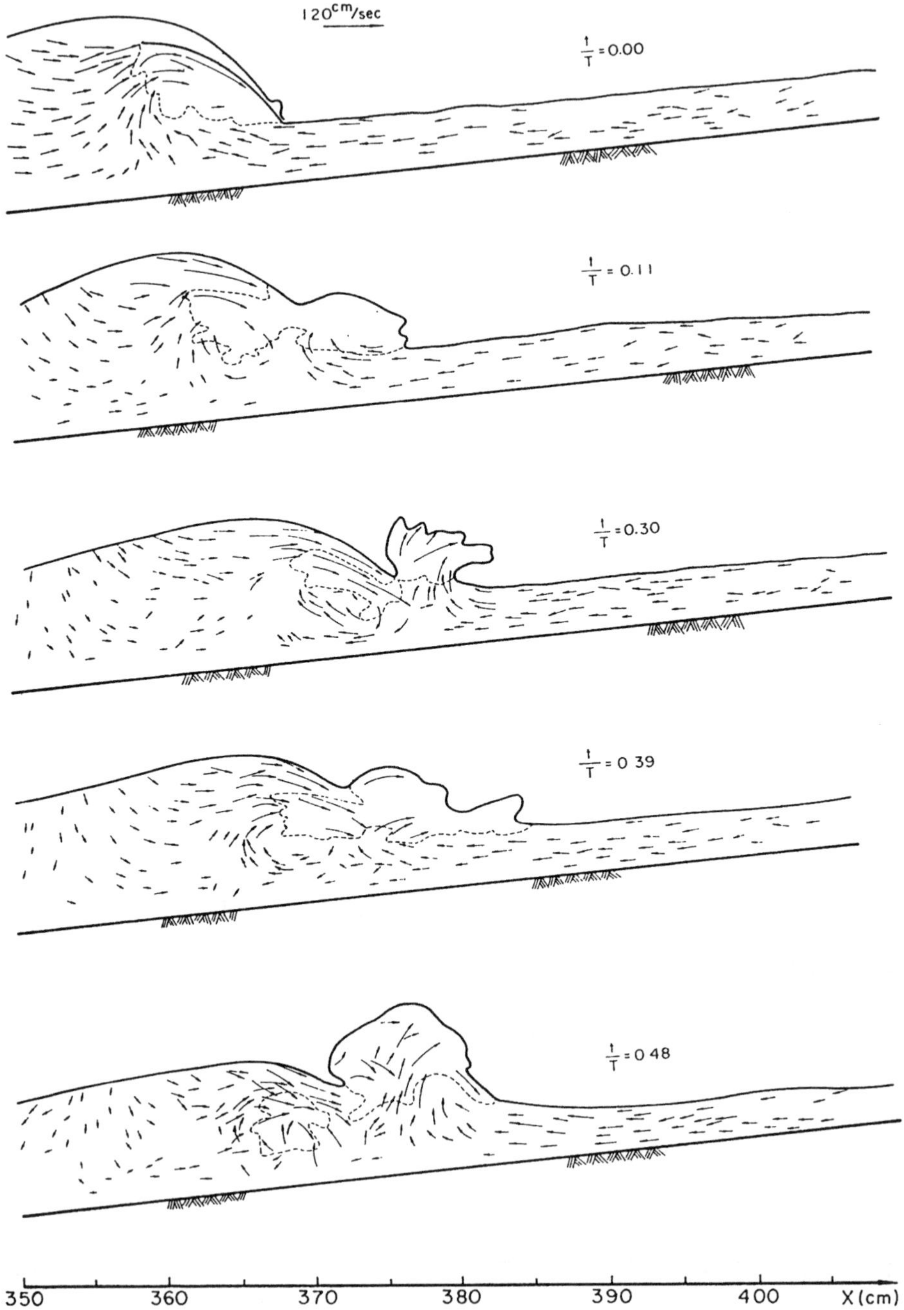


Fig-4. The evolutions and flow field structures of plunging breaker. (Case 3)

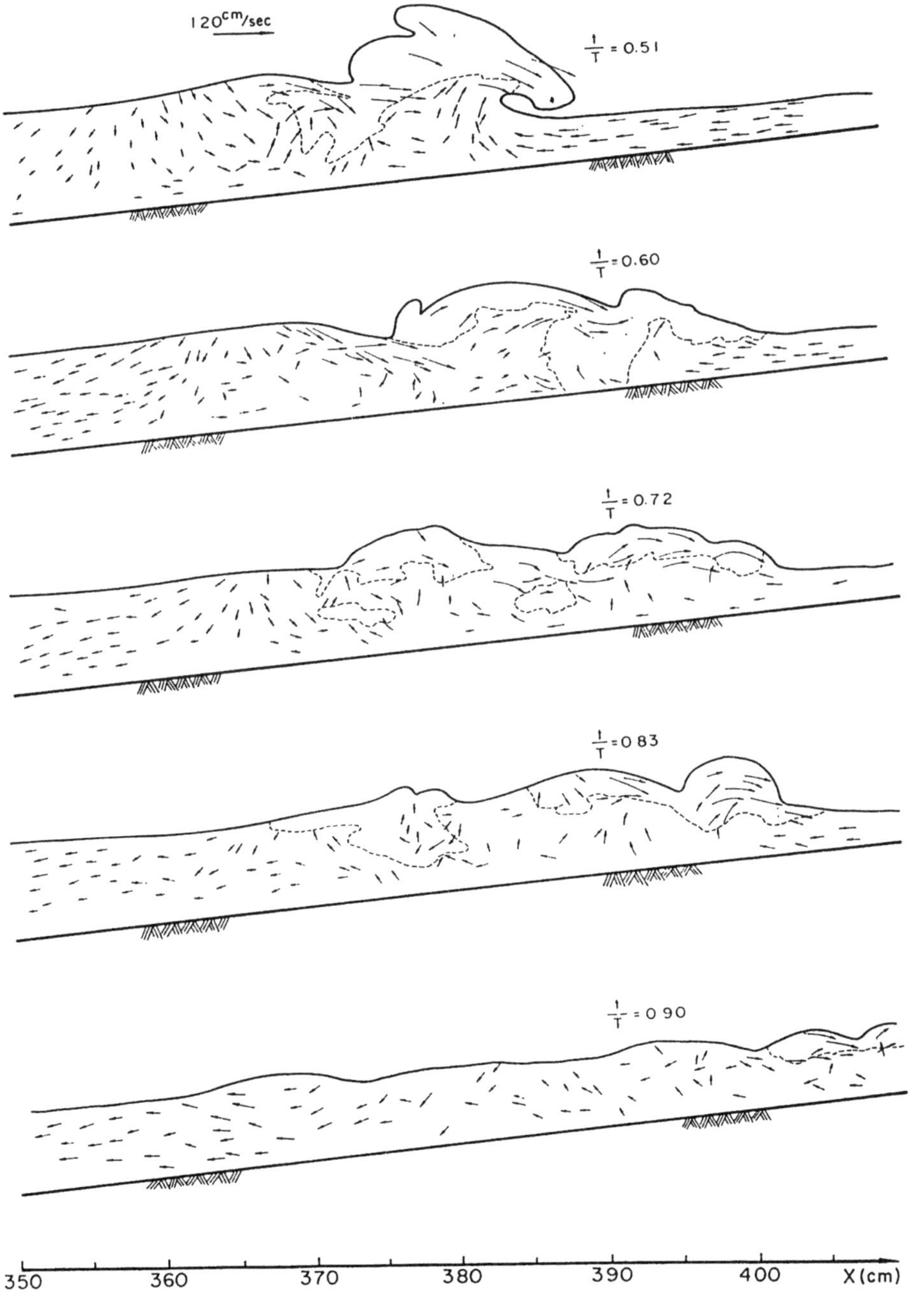


Fig-4. The evolutions and flow field structures of plunging breaker. (Case 3, continued)

## Conclusions

According to the experimental results, the noticeable phenomena of the turbulent flow structures inside surf zone after wave breaking are presented as follows: (1) Owing to the forward motion of the upstream flow and the rotating motion of the plunging vortex, between the tip of the vortex and the sloping bottom, where water particles are almost still is called "dividing region". (2) The flow structures inside surf zone are three-dimensional motions. (3) Two opposite horizontal eddies exist at the point of jet impinging on water surface.

This paper is only concerned with the characteristics of the turbulent flow structures with flow visualization techniques. As for the conditional sampling measurements by LDV are going to be presented in the future.

## REFERENCES


1. Stive, M.J.F. (1980): Velocity and pressure field in spilling breakers, Proc. of 17th Conf. on Coastal Eng. ASCE, PP.547-566.

2. Aono, T. and Hattori, K. (1984): Experiments on coherent structures of large scale turbulence under breaking waves, Proc. of 31st Japanese Conf. on Coastal Eng. PP.6-10.

3. Hino, M., Nadaoka, K. and Omata, A. (1984): Organized large scale structure of turbulence and related motion of water particle within the surf zone, Proc. 31st Japanese Conf. on Coastal Eng. PP.1-5 (in Japanese).

4. Sakai, T., Mizutani, T. and Tanaka, H. (1985): Vortex formation in plunging breaker, Proc. of 32nd Japanese Conf. on Coastal Eng. PP.70-74.

5. Nadaoka, K., Hino, M. and Koyano, Y. (1986): Turbulent flow field structure of breaking wave in the surf zone, Proc. 20th ICCE, Abstracts, PP.71-72.

6. Basco, D. R. (1985): A qualitative description of wave breaking, Journal of Waterway, Port, Coastal and Ocean Eng., ASCE, Vol.111(2), PP.171-188.

7. Nadaoka, K. (1986): A fundamental study on shoaling and velocity field structure of water waves in the nearshore zone, Technical Report. No.36, Department of Civil Eng., Tokyo Institute of Technology.

8. Jansen, P.C.M. (1986): Laboratory observations of the kinematics in the aerated region of breaking waves, Coastal Engineering, Vol.9, No.5, PP.453-478.

# Poster Session 2

# Nonlinear Wave-body Interaction

# Nonlinear Instability Waves in Channels and Jets

G. GRABITZ and G.E.A. MEIER

Max-Planck-Institut für Strömungsforschung, Göttingen,
F.R. Germany

## Summary

A theoretical and experimental investigation of the influence of travelling time effects on the stability of jets and channel flows is presented. Both the theoretical solution and experimental results show the development of an originally small velocity distortion in a water jet and an open channel flow. The growth of the waves is essentially determined by the velocity modulation at the origin of the jet. The two main differences of our non-linear theory compared to the classical linear stability theory are the peaky form of the wave ridges and the infinite amplitude (catastrophic amplification) for finite time and distance.

## Introduction

Essential aspects of the boundary wave amplification and the final disintegration of velocity modulated liquid jets can be explained by travelling time effects. Similar instabilities occur also in velocity modulated flow in open channels and ducts. For the explanation of this instability an exact solution of the non-linear unsteady one-dimensional basic equations is given. The experiments with velocity modulated water jets will prove the validity of this solution and demonstrate the mechanism of the disintegration of water jets. The experiments in an channel show that similar mechanisms occur also in other types of velocity modulated flows.

The travelling time instability mechanism explained by our solution of the non-linear equations of motion shows (in contrast to the classical linear theory of Lord Rayleigh [1]) for finite times and distances an infinite amplitude. In the usual linear stability theories which are concerned with velocity modulated liquid jets (Lee [2] Pimbley [3]) the travelling time effects lead to an exponential amplitude growth.

K. Horikawa, H. Maruo (Eds.)
Nonlinear Water Waves
IUTAM Symposium Tokyo/Japan 1987

## Theoretical Solution

We consider the incompressible fluid stream as an inviscid one-dimensional unsteady flow without surface tension. The pressure inside the jet is equal to the ambient pressure. Thus the flow is determined only by the velocity $w(x,t)$ and the cross-sectional area $F(x,t)$ of the jet. The symbols t and x denote time and space co-ordinate in the flow direction, respectively. The symbol $\gamma$ means $\pm g$ or zero according to the orientation of the jet in the gravitational field (g is the gravity acceleration).

The conservation equations of mass and momentum are

$$\frac{\partial F}{\partial t} + w \frac{\partial F}{\partial x} + F \frac{\partial w}{\partial x} = 0 \tag{1}$$

$$\frac{\partial w}{\partial t} + w \frac{\partial w}{\partial x} = \gamma \tag{2}$$

and the boundary conditions at x = 0 are

$$w(0,t) = w_a(t) \tag{3}$$

$$F(0,t) = F_a(t) \tag{4}$$

We introduce Lagrange-co-ordinates by the following transform:

$$t(\vartheta,\tau) = \vartheta \tag{5}$$

$$x(\vartheta,\tau) = w_a(\tau)(\vartheta-\tau) + \frac{\gamma}{2}(\vartheta-\tau)^2 \tag{6}$$

The co-ordinate $\vartheta$ stands for the actual time and $\tau$ means the starting time of the flow particles at x = 0.

With the help of this transform one gets (Grabitz and Meier [4]) the following solution for the velocity w and the cross section F:

$$w(\vartheta,\tau) = w_a(\tau) + \gamma\,(\vartheta - \tau). \tag{7}$$

$$F(\vartheta,\tau) = \frac{F_a(\tau)\ w_a(\tau)}{\left| w_a(\tau) - (\vartheta-\tau)\left[\frac{d\ w_a(\tau)}{d\tau} - \gamma\right] \right|} \tag{8}$$

For comparison with our experiments we consider the special case of an axisymmetric jet which originates in a nozzle of constant radius R with a perturbed velocity of the form

$$w_a(\tau) = U[1 + \varepsilon\sin(\omega\tau)] \tag{9}$$

with the frequency $\omega$ and the relative velocity pertubation $\varepsilon$. The radius R of the jet is then given as

$$\frac{\hat{R}(\hat{\vartheta},\tau)}{R_a} = \left|\frac{1+\varepsilon\sin(\Omega\hat{\tau})}{1+\varepsilon\sin(\Omega\hat{\tau})+[G-\varepsilon\Omega\cos(\Omega\hat{\tau})](\hat{\vartheta}-\hat{\tau})}\right| \tag{10}$$

with the non-dimensional times $\hat{\tau} = \tau\, U/R_a$, $\hat{\vartheta} = \vartheta\, U/R_a$, the non-dimensional frequency $\Omega = \omega R_a/U$, and the Froude-number $G = \gamma R_a/U^2$.

If the distortion is $\varepsilon > G/\Omega$ the solution has points of infinite radius. Beginning with the position x = 0 (i.e. $\hat{\vartheta} = \hat{\tau}$) points of maximal cross-section are generated in periodic travelling distances. Downstream of the first point of infinite radius the distribution of the flow material is no longer unique. For the case $\varepsilon < G/\Omega$ damping of the distortions of the stream is predicted.

For the special case G = 0 (no acceleration) from the equations (10) and (6) the following much simpler result is derived for the radius R and the peak position $x_p$:

$$\frac{R}{R_a} = \left|\frac{x_p(\hat{\tau})}{x_p(\hat{\tau})-x}\right| \tag{11}$$

with

$$\frac{x_p(\hat{\tau})}{R_a} = \frac{[1+\varepsilon\sin(\Omega\hat{\tau})]^2}{\varepsilon\Omega\cos(\Omega\hat{\tau})} \tag{12}$$

The starting time $\tau$ of a liquid element determines if there is a positive, finite position $x_p$ where the radius becomes infinite. The shortest distance is given by the following equation:

$$\frac{x_{p\,min}}{R_a} = \frac{1}{\varepsilon\Omega} \tag{13}$$

In contrast to the linear stability theory, where exponential growth of the amplitude is typical, in this exact solution of the nonlinear equation of motion the maximum radius goes to infinity like $(x_p-x)^{-1/2}$ for $(x_p-x) \to 0$.

For the two-dimensional case, corresponding to flow in a channel or duct, the deflection of the interface between the distorted and the undistorted stream is

$$\frac{h}{h_a} = \left|\frac{x_p(\hat{\tau})}{x_p(\hat{\tau})-x}\right| \tag{14}$$

In this case the amplification of the wave ridges of the interface goes with $(x_p-x)^{-1}$ for $(x_p-x)$ approaching 0. This means that in two dimensions the amplitudes go faster to infinity than in the case of rotational symmetry.

## Experiments with water jets

The experimental setup which was used for the observation of the velocity modulated water jets is shown in figure 1. The water is supplied through a porous resistive element to a plenum chamber where a porous piston is moving periodically. Thus, a velocity modulated jet is emanating from the nozzle. The velocity modulation at the mouth of the nozzle is measured by a hot film probe. The pressure is measured by a transducer inside the plenum chamber.

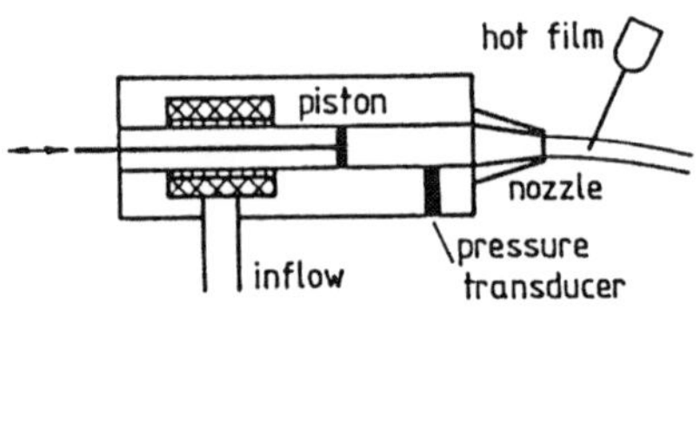


Fig. 1: Experimental setup

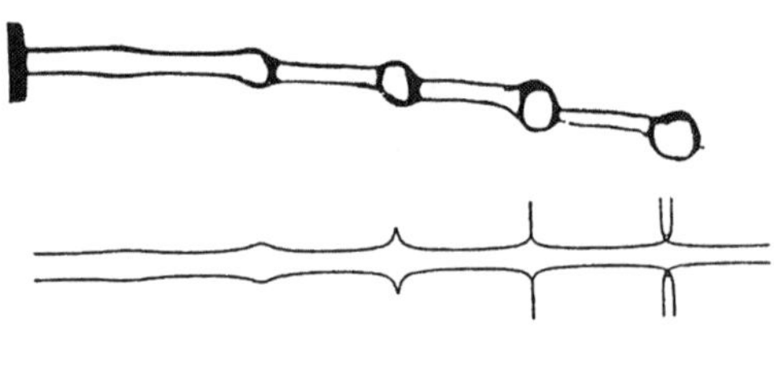

Fig. 2: Horizontal jet harmonically modulated at the nozzle ($\Omega$=0,6, $\varepsilon$=0,05)

Experimental results with water jets

In the upper part of figure 2 a photographic picture of a horizontal jet is shown. For comparison, we show the calculated shape of the jet at the same frequency ($\Omega$ = 0.6) and the same amplitude of the velocity modulation ($\varepsilon$ = 0.05) in the figure below the picture.

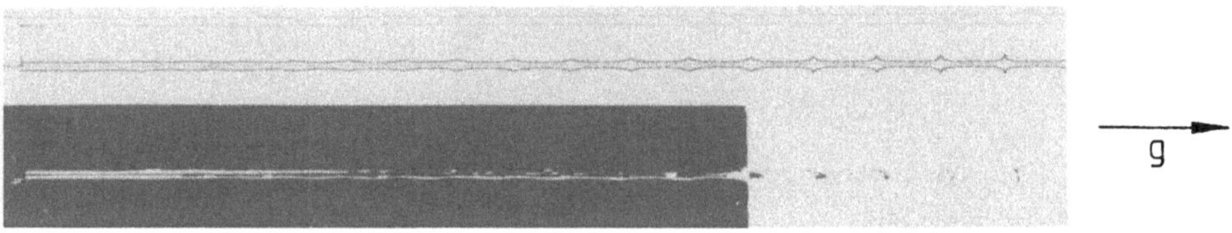


Fig. 3: Vertically downward flowing jet harmonically modulated at the nozzle ($\Omega$=0.54, $\varepsilon$=0.011).

Figure 3 shows a jet which is flowing vertically downward with very small velocity distortion ($\varepsilon$ = 0.011), together with the corresponding picture resulting from the calculation.

The distortion grows in both cases in good agreement with experiment and theory in the initial phase. Obviously the distribution of the fluid mass is determined in general by the travelling time effects. But the surface tension limits the amplification of the distortions and the peaky-structure of the wave ridges. Finally, the surface tension generates droplets by quenching the jet in the areas of smallest diameters. This influence of the surface tension $\sigma$ is more of a principal kind and not depending to much of the actual value. Figure 4 shows a comparision of two jets of different liquids with very different surface tensions. All other experimental parameters are comparable. Although the Weber number We = $R_a \rho U^2/\sigma$ of the two jets is different by a factor of more than $10^4$, both jets have roughly the same amplification of the distortion. This emphasizes that, for the mass-distribution in jets, the surface tension and the Rayleigh mechanism of jet instability are really important for Weber numbers smaller than 10 only.

a)
b)

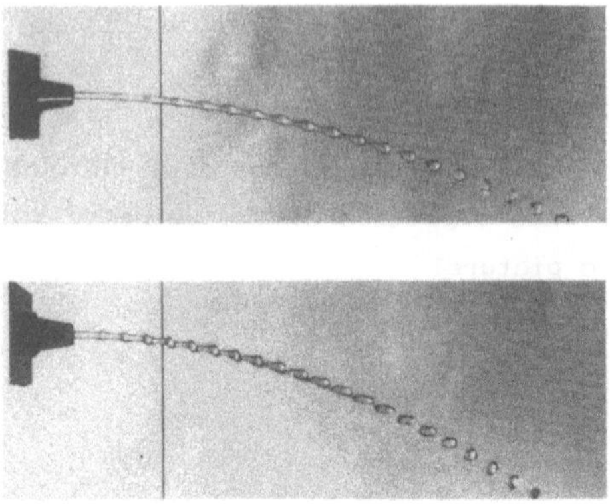

Fig. 4: Velocity distorted jets of different liquids with very different surface tensions

a) Water, $\sigma = 7.27 \cdot 10^4$ μN/m  b) IME 82, $\sigma = 2.49$ μN/m

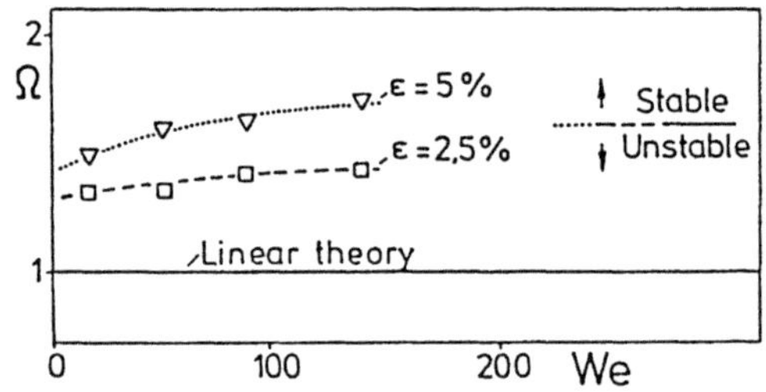


Fig. 5: Stability diagramm for velocity modulated liquid jets

A stability diagram which shows a comparison of the linear stability theory and the non-linear travelling wave amplification is given in figure 5. The non-linear travelling time effect displaces the stability limit (disintegration into droplets) depending on the amplitude of the distortion in the direction of higher excitation frequencies (Klöpper [5]).

## Experiments in the open channel

The experimental set-up is shown schematically in figure 6. Here the internal instability waves in the channel have been observed. From a large settling chamber the water runs into the channel through a mesh wire screen which is located at the intake. The channel has a widths of 5cm and a water height of 15cm. The length of the channel is about two meters.

With the help of a shaker, which is situated in the settling chamber, the lower part of the mesh wire is moved sinusoidally with an amplitude of about 3mm. The lower part of the duct flow is modulated by frictional forces according to equation (9) by this way. Two filaments of ink are supplied to the central part of the channel flow by an injection needle with holes.

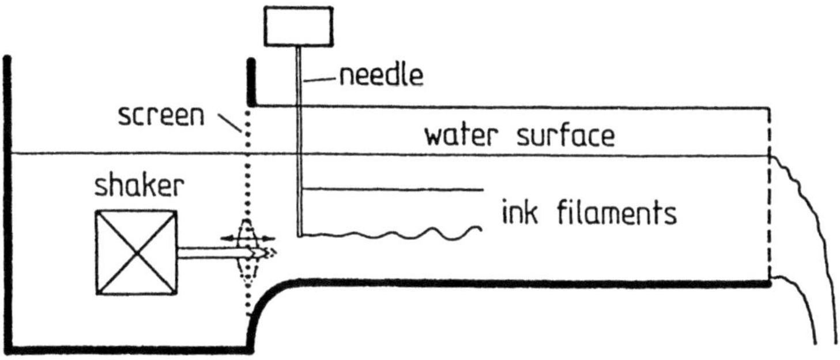


Fig. 6: Experimental setup

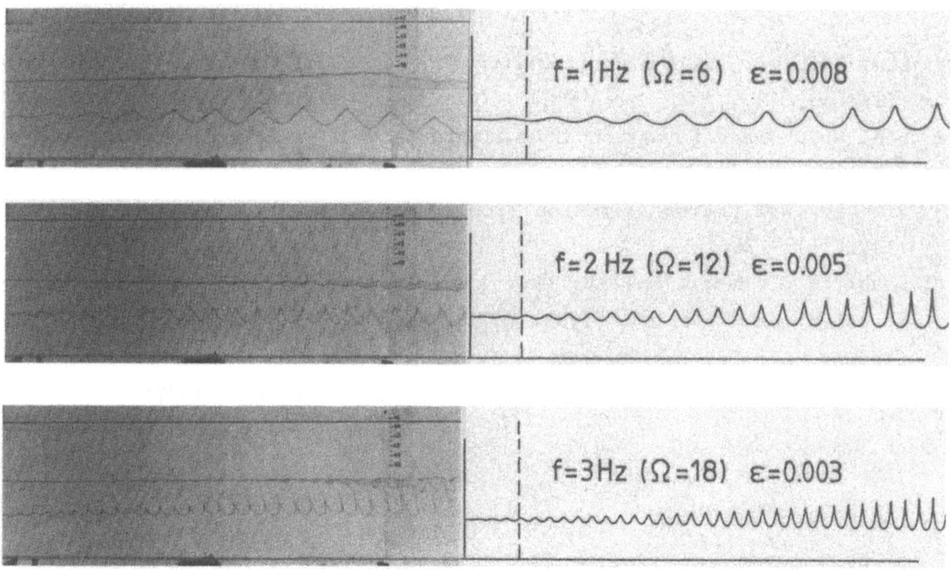


Fig. 7: Experimental ink filaments (left side) compared with calculated filaments (right side). The initial part of the filaments which is given by the calculated filaments up to the dahsed line is not to be seen on the photographs.

In figure 7 photographs of the ink filaments in the duct flow are compared with the theoretical solution. The average flow velocity in the channel was evaluated from the mass flow rate by U=4.5cm/sec, which corresponds to a Reynolds number of Re=Ub/ν=2250. The velocity modulation was performed with frequencies f = 1, 2, and 3Hz corresponding to reduced frequencies Ω = 6, 12, and 18. Because of inertia of the water the excitation of the flow oscillations by the mesh wire is less effective for higher frequencies. The theoretical curve has been calculated from equation (8), the amplitude of the distortion ε has been matched to the experimental value.

Our experiment in the open channel is a qualitative demonstration of the traveling time effect, which was quantitatively verified with the water jet. In general our opinion is that in many cases of "instability" in flows the linear or non-linear travelling time effect is an essential reason for the amplitude growth of the distortions in mass distribution introduced by initial velocity waves.

## References


1. Lord Rayleigh; On the instability of jets. Proc. London Math. Soc. X, 4 (1878).
2. Lee, H.C.; Drop formation in a liquid jet, IBM. J. Res. Develop. (1974) 364-369.
3. Pimbley, W.T.; Drop formation from a liquid jet. IBM. J. Res. Develop. (1976) 148-156.
4. Grabitz, G; Meier, G.E.A.; Über Laufzeitinstabilitäten in Flüssigkeitsstrahlen. ZAMM Vol. 63, (1983) T255-T257
5. Klöpper, A.; Der Einfluß von Laufzeit-Effekten auf den Zerfall von Flüssigkeitsstrahlen. Diplom-Arbeit, Universität Göttingen (1987).

# Long Internal Waves of Large Amplitude

MOTOYASU MIYATA

Geophysical Institute
University of Tokyo
Tokyo 113, Japan

## Summary

The results of an analytical investigation of the propagation of a strongly nonlinear, long internal gravity wave in a two-fluid system are presented. The governing equation is derived and shown to possess a steady-state solitary wavelike solution. The theory is tested experimentally by comparing measured and theoretical shapes of solitary waves.

## Introduction

Finite-amplitude effects on long internal waves have been extensively studied in recent years. Benny [1] showed that the nonlinear wave in two shallow-water layers are governed by the Korteweg-de Vries equation (see also [2]-[6]). The case where one of the layers has infinite extent was investigated by Benjamin [7] and Ono [8] (see also [9]-[13]). Joseph [14] derived the intermediate equation which connects the KdV and the Benjamin-Ono equations (see also [15]-[21]). All of these three evolution equations possess steady-state soliton solutions, which are based on the small expansion parameter $\varepsilon = A/h_1$, where $A$ is the wave amplitude and $h_1$ is the smaller of the two fluid depths. In this sense ($\varepsilon \ll 1$) these waves are all weakly nonlinear. On the other hand Miyata [22], using the three conservation laws of mass, energy and flow force, suggested the existence of strongly-nonlinear ($\varepsilon > 1$) internal solitary wave. The relationship among the various regimes is schematically shown in Fig. 1.

In this paper the author will first present the evolution equations for strongly-nonlinear internal waves in a two-fluid system. Then he will show that, when steady waveform of solitary type is assumed, the obtained equations reduce to the same differential equation as derived in his previous paper [22]. Finally the solitary wave solution will be tested experimentally by comparing the theoretical and measured waveforms.

K. Horikawa, H. Maruo (Eds.)
Nonlinear Water Waves
IUTAM Symposium Tokyo/Japan 1987

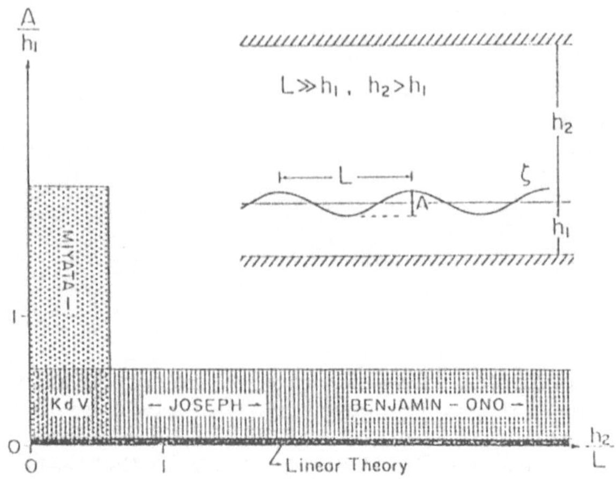


Fig. 1 Various regimes in two-fluid system

Evolution Equations for Long Internal Waves of Large Amplitude

Consider a two-fluid system, in which a layer of lighter fluid overlies a layer of heavier fluid, resting on a horizontal impermeable bed in a constant gravity field. For simplicity the free surface is replaced with a rigid lid, eliminating the surface wave (see Fig. 1). Both lower and upper fluids are assumed to be incompressible and irrotational with constant densities $\rho_1$ and $\rho_2$ respectively. This configuration excludes higher vertical modes and is the simplest model that supports internal gravity waves (see Fig. 1). The waves are assumed to be long, or more explicitly

$$\left(\frac{h_1+A}{L}\right)^4 \ll 1 \quad \text{and} \quad \left(\frac{h_2}{L}\right)^4 \ll 1 \,, \tag{1}$$

where $h_1$ and $h_2$ ($>h_1$) are the depths of the lower and upper layers respectively, L is the wave length, and the symbol $\ll$ should read as "can be neglected compared with". Note that the wave amplitude A may be larger than $h_1$, as long as the condition (1) is satisfied. Then starting with the Laplace equation for each layer, and applying the boundary conditions, the following equations can be derived.

$$\frac{\partial \eta_j}{\partial t} + \frac{\partial}{\partial x}(\eta_j U_j) = 0 \quad (j = 1 \text{ or } 2)\,, \tag{2}$$

$$||\rho_j[\frac{\partial U_j}{\partial t} + U_j\frac{\partial U_j}{\partial x} + g\frac{\partial \eta_j}{\partial x} + \frac{1}{6}\frac{\partial}{\partial t}(\eta_j^2\frac{\partial^2 U_j}{\partial x^2}) - \frac{1}{2}\frac{\partial}{\partial x}(\eta_j^2\frac{\partial^2 U_j}{\partial t\partial x})$$

$$+ \frac{\partial}{\partial x}\{\frac{1}{2}\eta_j^2(\frac{\partial U_j}{\partial x})^2 - \frac{1}{3}\eta_j^2 U_j\frac{\partial^2 U_j}{\partial x^2}\}]|| = 0 , \tag{3}$$

where $\eta_1 = \zeta+h_1$, $\eta_2 = \zeta-h_2$, $||f_j|| = f_1-f_2$ and $U_j$ is the depth-averaged horizontal velocity,

$$U_1 = \frac{1}{\eta_1}\int_{-h_1}^{\zeta} u_1 dz, \qquad U_2 = \frac{1}{\eta_2}\int_{\zeta}^{h_2} u_2 dz . \tag{4}$$

Other notations are conventional. The method of deriving these equations is essentially the same as described in Mei [23], in which equations (1.30) and (1.31) of Chapter 11 are the single layer version of Eq's (2) and (3).

## Solitary Wave Solution

Assume that the evolution equations derived above have a solution of solitary waveform. Then introducing the new variables $\xi = x-ct = h_1\xi'$, $\zeta = h_1\zeta'$, and applying the boundary conditions,

$$(\zeta, \frac{d\zeta}{d\xi}, \frac{d^2\zeta}{d\xi^2}, U_j) \to 0 \quad \text{as} \quad |\xi| \to \infty \tag{5}$$

we obtain, after lengthy but straightforward manipulation, the following differential equation.

$$\frac{F^2r^2(1+rs)}{3(r+s)}(\frac{d\zeta'}{d\xi'})^2 + \frac{E\zeta'^2+D\zeta'^3-\zeta'^4}{1+B\zeta'} = 0 , \tag{6}$$

where

$$B = \frac{r^2s - 1}{r(1+rs)} , \qquad D = \frac{r^2-s + r(F^2-1)(1-s)}{r + s} , \qquad E = -r(F^2 - 1) ,$$

$$r = \frac{h_2}{h_1} , \qquad s = \frac{\rho_2}{\rho_1} , \qquad F^2 = \frac{c^2(\rho_1 h_2+\rho_2 h_1)}{g(\rho_1-\rho_2)h_1 h_2} .$$

Equation (6) is exactly the same as derived in [22]. It can be analytically solved in terms of incomplete elliptic integrals of the first kind $F(\psi, k)$ and the third kind $(\psi, \alpha^2, k)$ (see e.g. Byrd & Friedman [24]). That is

$$\frac{1}{Fr}\sqrt{\frac{3(r+s)}{1+rs}}\,\xi' + \text{const.} = \frac{2B}{\sqrt{aB+1}}\{F(\psi, k) - \Pi(\psi, bB+1, k)$$

$$= \frac{2B}{\sqrt{aB+1}}\{sn^{-1}(\sin\psi, k) - \int_0^u \frac{dw}{1-(bB+1)sn^2 w}\} \tag{7}$$

where

$$\sin\psi = \text{sn } u = \sqrt{\frac{B\zeta'+1}{bB+1}}, \qquad k^2 = \frac{bB+1}{aB+1}, \quad a + b = -D, \quad ab = E$$

The solution (7) represents a solitary wave (for some examples of its waveshape and phase speed see [22]). To compare the characterictics of the waveform with other theories, the variation of wavelength against wave amplitude is shown in Fig.2.

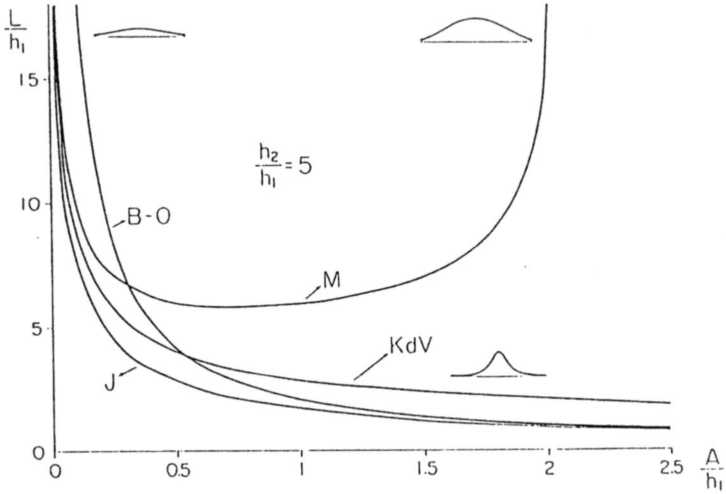


Fig. 2 Wavelength versus wave amplitude. L is equivalent-square-wave length defined by $2\Lambda L = \int_{-\infty}^{\infty} \zeta d\xi$. B-O: Benjamin-Ono, J: Joseph, M: Miyata.

The B-O curve in the figure decreases most rapidly as the wave amplitude increases. This implies that the Benjamin-Ono solitary wave is "more nonlinear" than any of the other three, because the effect of nonlinearity tends to steepen the wave. It is worth noting that the wavelength in Miyata's solitary wave, unlike the other three, increases for larger amplitude ranges. This corresponds to the physical mechanism that as the amplitude approaches its maximum wave height (see [22]), the effect of the upper fluid increases and tends to cancel the nonlinear motion of the lower fluid.

Experimental Results and Comparison with Theory

Experiments were carried out in a glass tank 180cm long, 30 cm high and 7cm wide, as sketched in Fig. 3. The two immiscible fluids chosen for use in this study were (i) aniline ( $\rho_1$=1.02) and water, (ii) water and oil

( $\rho_2$=0.87). The interfacial disturbances were generated by first pumping up the fluid at one end of the tank and then releasing it abruptly, thus imparting the mass and momentum at the interface. Photographs (see Fig. 4) were taken every second using two cameras and the wave profile was determined by reading the enlarged prints with a digitizer.

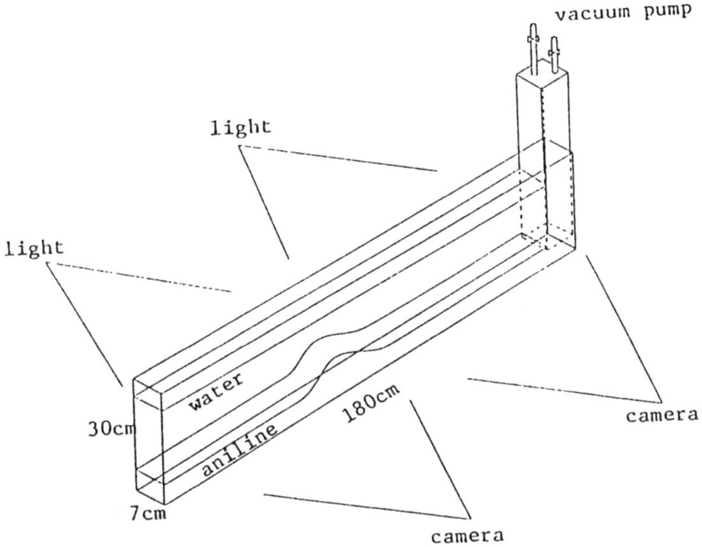


Fig. 3 Schematic diagram of experimental apparatus.

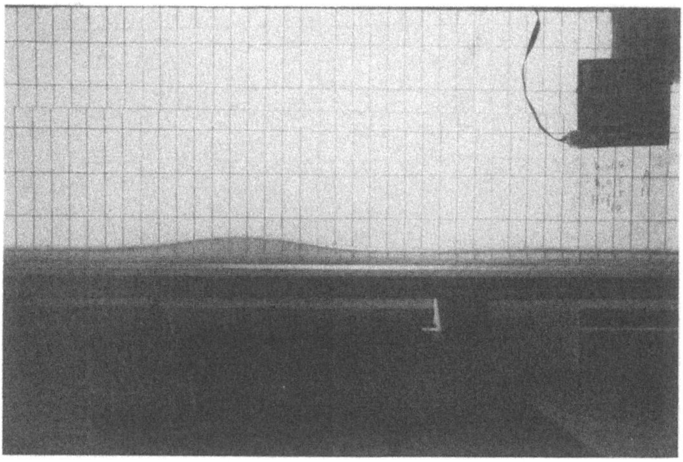

Fig. 4 Photograph of generated solitary wave.

Some measured waveforms for the case (i) are shown in Fig. 5. As can be seen in the figure the theoretical curve defined by (7) fairly well agrees with the experiments, although for some parameter ranges the KdV theory fits as well. It is not surprising that Benjamin-Ono's infinite-depth theory poorly describes the finite-depth experiments. The reason why Joseph's theory fails to predict experimental data is discussed by Segur & Hammack [25] (see also [26]). The results of the experiment for the case (ii) are not shown but they are not much different from the case (i). The more detailed analyses of the experiments will be published elsewhere.

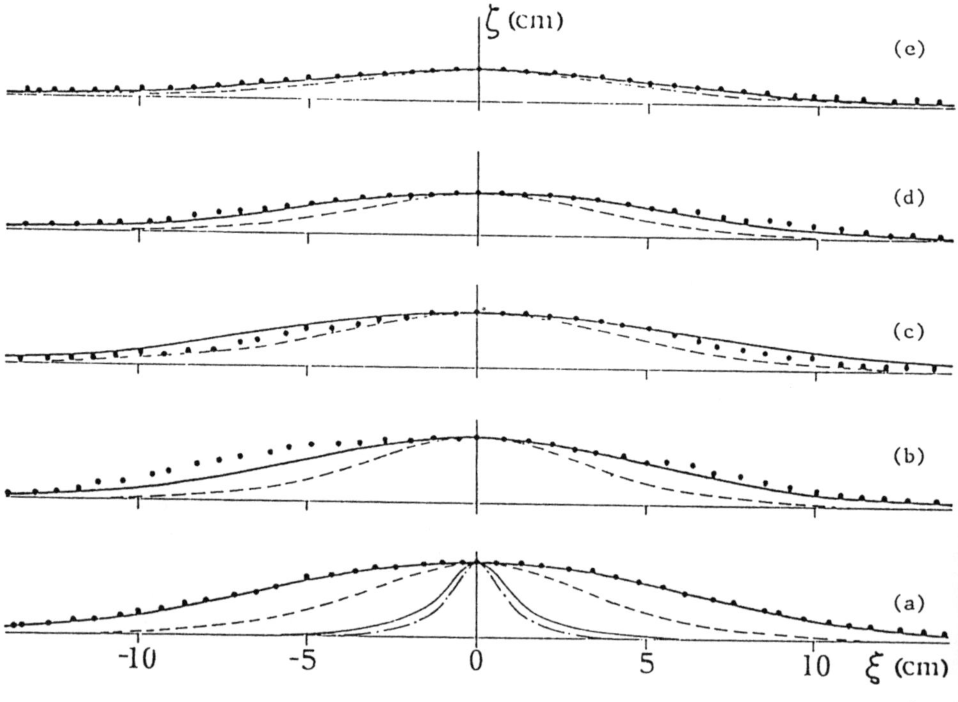


Fig. 5 Comparison of measured waves (dots) with theoretical shapes; Miyata: solid line, KdV: broken line, Benjamin-Ono: thin solid line, Joseph: dot-dashed line. The latter two theoretical curves are not shown for (b)-(e), but they are all below the KdV curve.

| | (a) | (b) | (c) | (d) | (e) | |
|---|---|---|---|---|---|---|
| $h_2$(cm) | 15 | 10 | 20 | 10 | 10 | $\rho_2 = 1$ |
| $h_1$(cm) | 1.4 | 1.5 | 1.4 | 1.3 | 1.4 | $\rho_1 = 1.02$ |
| A (cm) | 2.2 | 1.9 | 1.6 | 1.2 | 0.9 | |

## Concluding Remarks

It has been shown that the evolution equations ((2) and (3)) for strongly nonlinear internal wave possess a solitary wave solution. The theory is supported by experimental data at least for a certain parameter range. Whether the derived solitary wave behaves like a soliton, which resumes its shape after collision, is unknown. Extension of the theory to continuously stratified fluid and its application to geophysical nonlinear internal gravity waves observed in recent years (see [27] - [33]) remain to be studied.

## Acknowledgement

The author acknowledges his dept to Mr. N. Misawa and Dr. R. Kimura who assisted in performing the experiments at Ocean Research Institute, University of Tokyo. He wishes to thank Ms. T. Osada for drawing the figures and Ms. K. Kudo for typing the manuscript. Financial support was provided in part by Grant-in-Aid for Scientific Research from the Ministry of Education, Science and Culture, Japan (Project No. 62540300).

## References

1. Benny,D.J.: Long non-linear waves in fluid flows. J.Math. & Phys. 45 (1966) 52-63.
2. Keulegan, G.H.: Characteristics of internal solitary waves. J. Res. Nat. Bur. Stand. 51 (1953) 133-140.
3. Long, R.R.: Solitary waves in one and two-fluid system. Tellus 8 (1956) 460-471.
4. Peters, A.S.; Stoker, J.J.: Solitary waves in liquids having non-constant density. Comm. Pure Appl. Math. 13 (1960) 115-164.
5. Benjamin, T.B.: Internal waves of finite amplitude and permanent form. J. Fluid Mech. 25 (1966) 97-116.
6. Miles. J.W.: On internal solitary waves. Tellus 31 (1979) 456-462.
7. Benjamin, T.B.: Internal waves of permanent form in fluids of great depth. J. Fluid Mech. 29 (1967) 559-592.
8. Ono, H.: Algebraic solitary waves in stratified fluids. J. Phys. Soc. Japan 39 (1975) 1082-1091.
9. Case, K.M.: The N-soliton solution of the Benjamin-Ono equation. Proc. Natl. Acad. Sci. USA 75 (1978) 3562-3563.
10. Meiss, J.D.; Periera, N.R. : Internal wave solitons. Phys. Fluids. 21 (1978) 700-702
11. Satsuma, J.; Ishimori, Y.: Periodic wave and rational soliton solutions of the Benjamin-Ono equation. J. Phys. Soc. Japan 46 (1979) 681-687.
12. Matsuno, Y.: Exact multi-soliton solution of the Benjamin-Ono equation. J. Phys. A12 (1979) 619-621.
13. Chen, H.H.; Lee, Y.C.; Pereira, N.R.: Algebraic internal wave solitons and the integrable Calogero-Moser-Sutherland N-body problem. Phys. Fluids 22 (1979) 187-188.

14. Joseph, R.I.: Solitary waves in a finite depth fluid. J.Phys. A: Math. Gen. 10 (1977) L225-L227.
15. Kubota, T; Ko, D.R.S.; Dobbs, L.D.: Weakly nonlinear, long internal gravity waves in stratified fluids of finite depth. J. Hydronaut 12 (1978) 157-165.
16. Joseph, R.I.; Egri, R.: Multi-soliton solutions in a finite depth fluid. J. Phys. A: Math. Gen. 11 (1978) L97-L102.
17. Chen, H.H.; Lee, Y.C.: Internal wave solitons of fluids with finite depth. Phys. Rev. Lett. 43 (1979) 264-266.
18. Satsuma, J.; Ablowitz,M.J.; Kodama,Y.: On an internal wave equation describing a stratified fluid with finite depth. Phys. Lett. 73A (1979) 283-286.
19. Satsuma, J.; Ablowitz, J.: Solutions of an internal wave equation describing a stratified fluid with finite depth. Nonlinear Partial Diffential Equations in Engineering and Applied Science (eds). R.L. Sternberg, A.J. Kalinowki & J.S. Popadakis, New York; Marcel Dekker (1980) 397-414.
20. Heney, F.S. : Finite-depth and infinite-depth internal-wave solitons. Phys. Rev. A 21 (1980) 1054-1056.
21. Kodama, Y.; Satsuma, J.; Ablowitz, M.J.: Nonlinear intermediate long-wave equation: Analysis and method of solution. Phys. Rev. Lett. 46 (1981) 687-690.
22. Miyata, M.: An internal solitary wave of large amplitude. La mer. 23 (1985) 40-45.
23. Mei, C.C.: The Applied Dynamics of Ocean Surface Waves. Wiley-Interscience 1983.
24. Byrd, P.F.; Friedman, M.D.: Handbook of Elliptic Integrals for Engineers and Physicists. Springer-Verlag 1954.
25. Segur, H.; Hammack, J.L.: Soliton models of long internal waves. J. Fluid Mech. 118 (1982) 285-304.
26. Koop, C.G.; Butler, G.: An investigation of internal solitary waves in a two-fluid system. J. Fluid Mech. 112 (1981) 225-251.
27. Lee, O.S.: Observations on internal waves in shallow water. Oceanology and Limnology. 6 (1961) 313-321.
28. Christie, D.R.; Muirhead, K; Hales, A.: On solitary waves in the atmosphere. J. Atm. Sci. 35 (1978) 805.
29. Osborne, A.R.; Burch, T.L.: Internal solitons in the Andaman Sea. Science 208 (1980) 451-460.
30. Sandstrom, H.; Elliott, J.A.: Internal tide and solitons on the Scotian Shelf: A nutrient pump at work. J. Geophys. Res. 89 (1984) 6415-6426.
31. Apel, J.R.; Holdbrook, J.R.; Liu, A.K.; Tsai, J.: The Sulu Sea internal soliton experiment. J. Phys. Oceanogr. 15 (1985) 1525-1651.
32. Liu, A.K.; Holbrook, J.R.; Apel J.R.: Non-linear internal wave evolution in the Sulu Sea. J. Phys. Oceanogr. 15 (1985) 1614-1624.
33. Smith, R.K.; Coughlan, M.J.; Lopez, J.L.: Southerly nocturnal wind surges and bores in northeastern Australia. Monthly Weather Review 114 (1986) 1501-1518.

# Transmission and Reflection of a Planar Soliton

Y.Z.LIU, G.P.MIAO, Q.Q.LÜ, D.X.ZHU

Ship Hydrodynamics Laboratory
Shanghai Jiao Tong University, China

## Summary

The matched-asymptotic method is used to treat the problem of transmission and reflection of a planar solitary wave passing by a two-dimensional floating rectangular cylinder. The fluid motion in the outer region is governed by the Boussinesq equations which are solved numerically by the finite difference method. The inner region solutions, which satisfy the Laplace equation, can be expressed as eigenfunction expansions for the cylinder discussed. The outer and inner solutions are finally determined through matching and iteration process. As a numerical example, the time histories of transmitted and reflected waves as well as the forces and moment acting on the cylinder are given and discussed.

## Introduction

More interests are concentrated recently to the problem of interaction between structures and solitary wave or cnoidal waves, which are typical and simple in some senses as kinds of progressive weakly nonlinear shallow water waves. Some authors have investigated the interaction of a solitary wave ( to be called soliton hereafter) and a vertical cylinder (Isaacson [1], Li et.al.[2]) numerically. Because of the complexity of the 3-D character, it is very hard to understand the reaction behaviour and the change of wave pattern during their interaction. On the other hand, the reflection of a planar soliton at vertical wall (Power et.al.[3]) or inclined shoreline (Sugimoto et.al.[4]) has demonstrated the possibility to solve a more complicated problem: a planar soliton acting on a 2-D cylinder floating on the free surface. In this case, not only reflection but also transmission will occur.

We report on the computations for this problem by using matched-asymptotic method.

K. Horikawa, H. Maruo (Eds.)
Nonlinear Water Waves
IUTAM Symposium Tokyo/Japan 1987

## Formulation of the Problem and Its Solution

The transmission and reflection of a soliton on a 2-D rectangular floating body are discussed here. As indicated in Fig.1, a 2-D Cartesian coordinate system is adopted with $oy'$ axis vertically upwards and $ox'$ axis lying on the seabed. We denote the water depth as a constant h, free surface elevation $y'=h+\eta'$, the breadth of the cylinder $2x'_B$, and draught (h-a'). It is also shown in Fig.1 that the whole fluid domain is divided into five regions, each of which is denoted by $\Sigma_j$ (j=1,....5). $\Sigma_1$ and $\Sigma_5$ are regarded as outer fluid regions, where the fluid motion is governed by the Boussinesq equations. The regions $\Sigma_2$ and $\Sigma_4$ locate nearby the cylinder and the flow there is governed by the Laplace equation.

The following non-dimensional variables are used, i.e.

$$(\xi, y) = \frac{1}{h}(x', y') = \frac{k_0}{\beta^{1/2}}(x', y') \quad , \quad t = k_0 t' (gh)^{1/2}$$
$$p = p'/\rho g h \quad , \quad \eta = \eta'/A \quad , \quad \phi = k_0 h \phi'/[A(gh)^{1/2}]$$
$$\alpha = A/h \quad , \quad \beta = (k_0 h)^2 \quad , \quad \xi_b = x'_B/h$$

where A is the wave amplitude, $\lambda_0$ is the characteristic wave length, $k_0=2\pi/\lambda_0$ is the characteristic wave number, g is the gravitational acceleration. For weakly nonlinear waves, $\alpha \sim \beta \ll 1$ are required.

In the outer fluid region, we further take the outer horizontal variable as

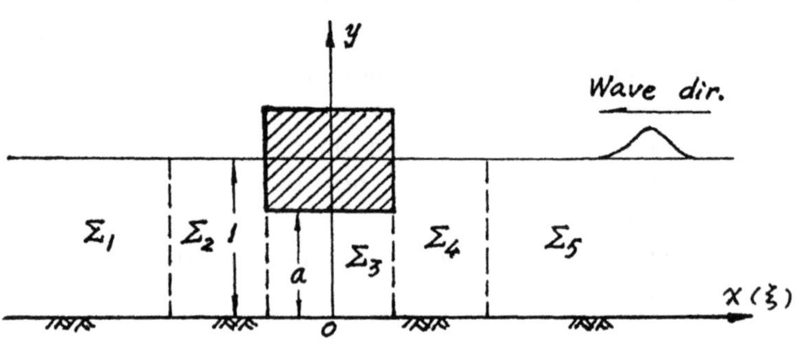


Fig.1. Geometrical configuration

$$x=\beta^{\frac{1}{2}}(\xi \mp \xi_b) \qquad \xi \gtrless 0 \tag{1}$$

The potential in that region can be formally expanded as

$$\phi(x,y,t)=\sum_{n=0}^{\infty}\left(-\beta y^2 \frac{\partial^2}{\partial x^2}\right)^n \frac{\phi_0(x,t)}{(2n)!} \tag{2}$$

With the usual derivation procedure [5], the following Boussinesq equations, which are accurate up to the order of $O(\alpha,\beta)$ and govern the fluid motion in the outer region, are finally obtained, i.e.

$$\eta_t + [(1+\alpha\eta)u]_x = 0 \tag{3a}$$
$$u_t + \alpha u u_x + \eta_x - \frac{\beta}{3}u_{xxt} = 0 \tag{3b}$$

where u is the horizontal velocity defined as depth averaged value, i.e.

$$u=\frac{1}{1+\alpha\eta}\int_0^{1+\alpha\eta}\phi_x\,dy \tag{4}$$

and it can also be written as

$$u=\phi_{0x}-\frac{1}{6}\beta\phi_{0xxx}+O(\beta^2) \tag{5}$$

The Boussinesq equations (3) may be numerically solved by means of an implicit finite difference scheme for given initial values and boundary conditions. The solutions at any time $t=(n+1)\Delta t$ may be obtained by solving the finite difference equation set with Thomas algorithm step by step. The boundary conditions needed for solving the Boussinesq equations are the hoizontal velocity u at $x=0^{\pm}$ (denoted as $u_0$ since $u_0^+=u_0^-$ can be proved) for $\Sigma_1$ and $\Sigma_5$ respectively.

We assume that the inner solutions in $\Sigma_2$ and $\Sigma_4$ may be written as

$$\phi_2(\xi,y,t)=\phi_\infty^- + \beta^{\frac{1}{2}}\psi^-(\xi,y,t) \tag{6a}$$
$$\phi_4(\xi,y,t)=\phi_\infty^+ + \beta^{\frac{1}{2}}\psi^+(\xi,y,t) \tag{6b}$$

where

$$\phi_\infty^{\pm}=f^{\pm}+\beta^{\frac{1}{2}}f_x^{\pm}(\xi\mp\xi_b)+\frac{1}{2}\beta f_{xx}^{\pm}[(\xi\mp\xi_b)^2-y^2]+O(\beta^{3/2}) \tag{7}$$

are the inner expansions of the outer solutions for regions $\Sigma_1$ and $\Sigma_5$ at $x=0^{\pm}$ (i.e. $\xi=\pm\xi_b$) respectively, in which $f^{\pm}=\phi_0(0^{\pm},t)$, $f_x^{\pm}=\phi_{0x}(0^{\pm},t)$ etc.

It can be proved that the functions $\psi^{\pm}$ should satisfy the Laplace equation with homogeneous Neumann conditions at the seabed and the undisturbed water surface. Further, $\psi^{\pm}\to 0$ as $\xi\to\pm\infty$ are required by the outer region matching conditions. The formal solutions of $\psi^{\pm}$ may be expressed as the following eigenfunction series, i.e.

$$\psi^{+}(\xi,y,t)=\sum_{m=1}^{\infty} b_m(t)\,e^{-m\pi(\xi-\xi_b)}\cos m\pi y \tag{8a}$$

$$\psi^{-}(\xi,y,t)=\sum_{n=1}^{\infty} a_n(t)\,e^{n\pi(\xi+\xi_b)}\cos n\pi y \tag{8b}$$

The formal solution in the fluid region $\Sigma_3$ under the cylinder may also be written as

$$\phi_3=\beta^{\pm}\sum_{q=1}^{\infty}(c_q e^{\delta\xi}+d_q e^{-\delta\xi})\cos\delta y+g_1\xi+g_2 \tag{9}$$

in which $\delta=q\pi/a$. $\phi_3$ should be matched with $\phi_2$ and $\phi_4$ on their common boundaries. The matching condition states that the potentials and their normal derivatives are equal across the common boundaries. For convenience, their time derivatives are in fact used for matching.

By matching, a set of relations connecting the coefficients in expressions (8) and (9) with $u_0$ and $\eta^{\pm}$ can be obtained. $u_0$ and $\eta^{\pm}$ are the averaged horizontal velocity and wave elevations at $x=0^{\pm}$ ($\xi=\pm\xi_b$) respectively.

With the above relations and the implicit finite difference scheme for the outer regions an iteration process may be taken to get the time domain solutions numerically. The following steps have been shown effective:

(a) Give the distribution of the initial values of wave elevation and the averaged horizontal velocity. For computation starting ($t=n\Delta t$, $n=1$), $(\eta^{\pm})^{n}=(\eta^{\pm})^{n-1}$ is chosen, and $(u_0)^{n+1}$ is assumed:

(b) $(\eta^-)^{n+1}$ and $(\eta^+)^{n+1}$ can be obtained from the finite difference equation set for the outer region with $(\eta^\pm)^n$ and $(u^\pm)^n$ at time $n\Delta t$ known;

(c) A set of coefficients $a_{nt}$, $b_{mt}$, $c_{qt}$ and $d_{qt}$ etc. (n, m, q=1, 2,...,K, K denote the number of truncated terms) can be obtained from the inner region matching relations with the known $u_0^{n+1}$, $(\eta^-)^{n+1}$ and $(\eta^+)^{n+1}$ substituted;

(d) A corrected $u_0^{n+1}$ can also be obtained by substituting $a_{nt}$ and $b_{mt}$ into one of the inner region matching relations. Repeat the iteration process from (b) to (d) until the error of $u_0^{n+1}$ between the two successive iterative steps is within the given tolerance. Then the computation enters the next time step and the above iteration process is carried out again.

The pressure distribution on the cylinder is given by the expression as follows, i.e.

$$p+\alpha\phi_t+y-1+\frac{1}{2}\frac{\alpha^2}{\beta}(\phi_\xi^2+\phi_y^2)=0 \tag{10}$$

Accurate up to $O(\alpha\beta^{\frac{1}{2}})$, the pressure distribution on the weather side of the cylinder $p^+$ is

$$p^+=1-y+\alpha\eta^+-\alpha\beta^{\frac{1}{2}}\sum_{m=1}^{\infty} b_{mt}\cos m\pi y \tag{11}$$

and that on the lee side of the cylinder $p^-$ is

$$p^-=1-y+\alpha\eta^--\alpha\beta^{\frac{1}{2}}\sum_{n=1}^{\infty} a_{nt}\cos n\pi y \tag{12}$$

The hydrodynamic pressure on the bottom of the cylinder $p_\xi$ is

$$p_\xi=-\alpha\beta^{\frac{1}{2}}\sum_{q=1}^{\infty}(c_{qt}e^{\delta\xi}+d_{qt}e^{-\delta\xi})\cos q\pi-\alpha\xi g_{1t}-\alpha g_2(t) \tag{13}$$

The coefficients needed in these expressions have been already determined by the above iteration process at any specified instant. Proper integrations of these pressure distributions along the cylinder surface will give the horizontal force $F_H$

the vertical force $F_V$ and the turning moment M acting on the cylinder. The integrals appeared may be evaluated by the trapezoidal rule.

## Computational Results

Here we show one of the computational examples. A soliton is assumed propagating from far right to the left with initial wave profile and horizontal velocity as

$$\eta = \mathrm{sech}^2\left\{\frac{\sqrt{3\alpha}}{2\beta^{1/2}}(x-x_0)\right\} \tag{14}$$

$$u=-(1+\alpha/2)\eta/(1+\alpha\eta) \tag{15}$$

where the parameters of the incident soliton are

$$\alpha=0.2,\quad \beta=0.1,\quad x_0=4.6$$

The time and space intervals in the computation are $\Delta t=0.04$ and $\Delta x=0.08$ respectively. The non-dimensional half breadth of the cylinder $\xi_b$ equals 1.0. The gap a between the cylinder and the seabed is taken as 0.56. All the series in the formal solutions of inner fluid regions are truncated up to 20 terms. Usually, three iterative steps are sufficient to ensure the convergence of the computations with the accuracy of 4 digits.

Figure 2 shows the time history of the transmitted and reflected waves of the soliton on the cylinder from t=3.0 till t=10.0. The horizontal axis is scaled with the outer variable x, 1 unit equals 0.46. At the initial time, the peak of the soliton locates at $x_0=4.6$ ( 10 units in the figure ). The soliton propagates to the left with time increasing. When t=3.0, the soliton moves to the near field of the cylinder. The transmitted wave obviously appears in the outer region of the lee side. The peak value of the wave amplitude at $x=0^+$ reaches its maximum $\eta^+_{max}=1.2665$ around t=4.2. However, the peak of the transmitted wave does not reach its maximum value until t=4.52, when $\eta^-_{max}=0.8604$ . After that time, the transmitted wave moves to the left with the peak value almost unchanged and returns to a perfect soliton profile gra-

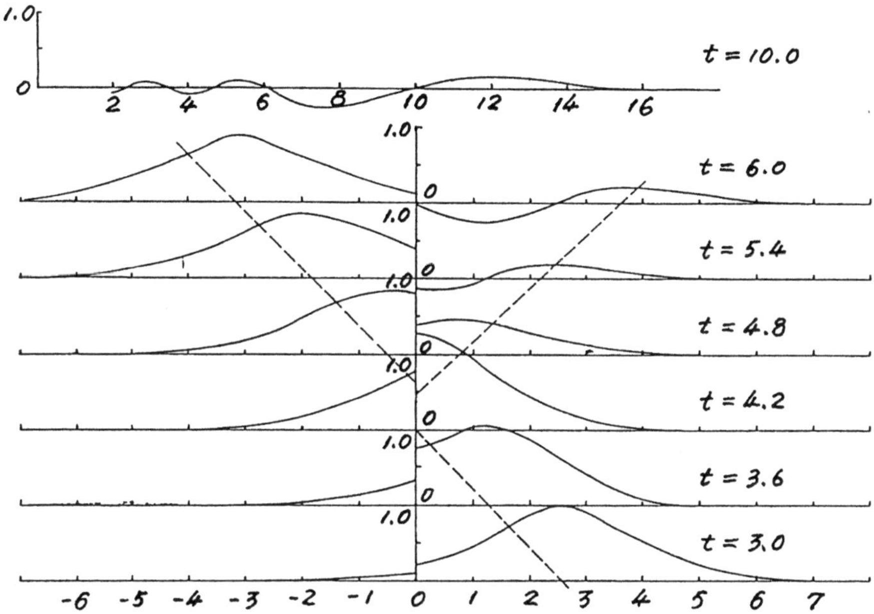


Fig.2 Reflection and transmission of a planar solitary wave due to a rectangular cylinder (outer region). "----" denotes the location of wave crest. The uppermost curve represents the reflected waves at t=10.

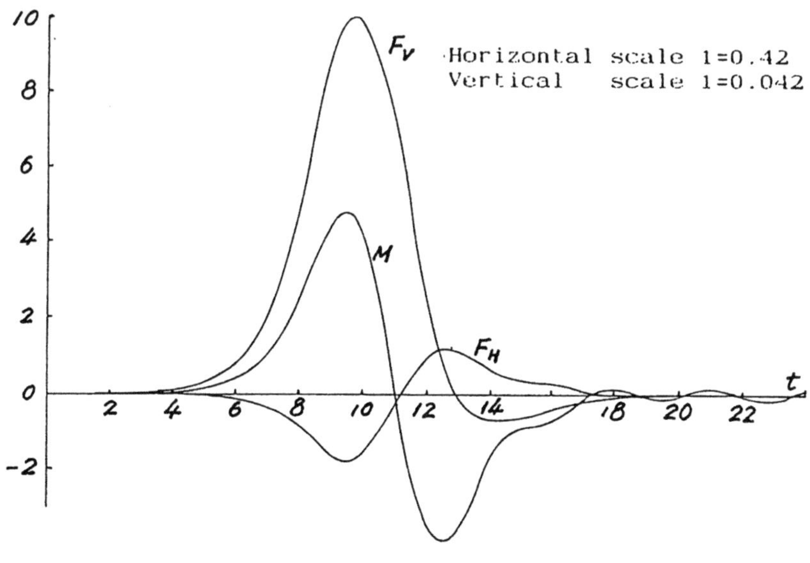


Fig.3 Time histories of the forces and moment on the cylinder

dually. On the contrary, the reflected wave moves to the right in a form of profile with a wave front of relatively large amplitude and an oscillating tail of decreasing amplitudes.

The time histories of the horizontal force $F_H$, the vertical force $F_V$ and the moment M acting on the cylinder when the soliton passing is shown in Fig.3. As can be observed in the figure, their maximum values are reached before $\eta^+$ reaches its maximum value (t=4.2). It is also observed that the horizontal force $F_H$ (<0) is in the same direction as that of the soliton propagating, the vertical force $F_V$ (>0) points upwards and the moment M tends to turn the cylinder to the left before t=4.2. With the time increasing to about t=4.62, the horizontal force and the moment change their directions. The vertical force changes its acting direction at about t= 5.46. From the present example, forces $F_H$ and $F_V$ on the cylinder tend to zero monotonously with the time further increasing, and the moment M tends to zero with oscillation.

## Reference


1. Isaacson, M. (1982), Nonlinear-wave effect on fixed and floating bodies, J. Fluid Mech., vol. 120, pp. 267-281.

2. Li, B. Y. and Lü, Y.L. (1986), A numerical model for nonlinear wave diffraction around large offshore structures, (to be published), Dalian Inst. of Technology, China.

3. Power, H. and Chwang, A. T. (1984), On reflection of a planar solitary wave at a vertical wall, Wave Motion, vol. 6, pp. 183-195.

4. Sugimoto, N. and Kakutani, T. (1984), Reflection of shallow water soliton, Part 1, Edge layer for shallow-water, J. Fluid Mech., vol. 146, pp. 369-382.

5. Whitham, G. B. (1974), Linear and Nonlinear Waves, Wiley-Interscience, New York.

# Nonlinear Interaction of Waves and Vertical Wall

**HONG GUANGWEN**

**East China Technical University of Water Resources**
**Nanjing, China**

## Abstract

A fourth order solution for nonlinear interactions among three-dimensional gravity waves on uniform current in water of constant depth is presented in this paper. The expressions to the same order for main kinematic and dynamic elements such as: velocity and acceleration components; dynamic pressure; total wave pressure per unit or given wave crest width and its moment around the toe, as well as the wave energy, the mass flux, momentum flux and energy flux per unit area from an arbitrary point in water to the free surface elevation are also given. The applications to the nonlinear interactions of wave with arbitrary incident angle and vertical wall or corner are discussed. These solutions can be generalized to cases of regular short-crested wave; short gravity wave and long wave; random sea waves.

## 1. Introduction

Nonlinear interaction of nonbreaking wave and vertical wall is of particular interest in the theory of coastal hydrodynamics, as well as in the design of breakwaters and seawalls. In solving this problem, factors such as propagation direction, height and length of incident wave, reflection coefficient and phase shift, water depth, nonlinearity, three-dimensional and random characters of wave motion and sectional length of wall should be considered. The effects of current and long wave may not be ignored in some situations. In recent years several analytic solutions for regular waves in water of constant depth have been developed. Based on the Eulerian system the analytic solutions of second to fourth order approximation for standing wave were derived by Biesel (1952); Tadjbakch and Keller (1960); Chermozhskaya (1975); and Goda (1966), respectively. By using the Lagrangian system the solutions of first to third order for standing wave were presented by Sainflou (1928); Miche (1944); and Sekerzh-Zenkovich (1951), respectively. The solution to second order for partial reflection of normal incident wave was given by Rundgren (1958). For full reflection of obliquely incident wave (short-crested wave) the solutions of second to fourth order approximation were presented by Fuchs (1952), Kuznetzov (1973); Hsu et al

K. Horikawa, H. Maruo (Eds.)
Nonlinear Water Waves
IUTAM Symposium Tokyo/Japan 1987

(1979), Fenton (1985); and Hong (1980), respectively. The numerical solution for highly nonlinear short-crested waves was given by Roberts (1983). For full and partial reflection of obliquely-incident wave the solution of second to fourth order approximation was given by the present author (1974, 1980), according to the nonlinear interaction among gravity waves. A third order explicit expression and corresponding calculating methods were also derived for main kinematic and dynamic elements, such as: velocity potential function, free surface elevation, wave pressure, total pressure on unit or given wave crest width from an arbitrary point in water to the free surface elevation and its moment around the toe, total pressure and its moment on any given rectangular area of vertical wall. In this paper a fourth order analytic solution for nonlinear interaction among three-dimensional gravity waves on uniform current in water of constant depth is presented, including the author's solution mentioned above in non-current case as a special one. The applications to nonlinear interactions of wave with arbitrary incident angle and vertical wall or corner are discussed. These solutions can be generalized to cases of regular short-crested wave; short gravity wave and long wave; random sea waves.

## 2. Mathematical Formulation and Theoretical Solution

The problem of gravity surface waves on uniform current in water of constant depth as an incompressible inviscid irrotational flow can be reduced to solving the velocity potential function, satisfying the Laplace's equation

$$\Delta\varphi = 0, \quad \tilde{\varphi} = \varphi^{\circ} + \varphi, \quad \varphi^{\circ} = u_x x + u_y y, \quad \tilde{V} = \nabla\tilde{\varphi} \tag{1}$$

with the boundary conditions at the free surface and at the horizontal bottom:

$$\tilde{\varphi}_{tt} + g\tilde{\varphi}_z + (\tilde{V}^2)_t + \nabla\tilde{\varphi}\cdot\nabla(\tilde{V}^2/2) = 0, \quad \text{for } z = \eta \tag{2}$$

$$\tilde{\varphi}_z = 0, \quad \text{for } z = -d \tag{3}$$

The free surface elevation is

$$\eta = -\frac{1}{g}\left\{\tilde{\varphi}_t + \frac{1}{2}\tilde{V}^2 - \tilde{c}(t)\right\} \tag{4}$$

$$\tilde{c}(t) = C(t) + C^{\circ}(t), \quad C^{\circ}(t) = \frac{1}{2}(u_x^2 + u_y^2)$$

where the x, y and z form a rectangular coordinate system with z measured vertical upwards from the still water level.

With the perturbation method, the solutions are expressed by the following series expansions with respect to a small parameter, related to the surface slope:

$$\varphi = \sum_{S=1}^{S}\varphi^{(S)}, \quad \eta = \sum_{S=1}^{S}\eta^{(S)}, \quad C(t) = \sum_{S=1}^{S} C^{(S)}(t)$$

and the absolute frequencies are: $\sigma = \sum_{S=0}^{S}\sigma^{(S)}$

Thus the free surface boundary conditions of the first to fourth order approximations at the still water level z=0 are as follows:

The first order approximation

$$I^{(1)} = 0 \tag{5}$$

$$g\eta^{(1)} = -Q^{(1)} + C^{(1)}(t)$$

Second order approximation

$$I^{(2)} = -\eta^{(1)} I_z^{(1)} - II^{(2)} \tag{6}$$

$$g\eta^{(2)} = -\left\{Q^{(2)} + \eta^{(1)} Q_z^{(1)}\right\} - \frac{1}{2}(V_{11}^2) + C^{(2)}(t)$$

Third order approximation

$$I^{(3)} = -\left\{\eta^{(1)} I_z^{(2)} + \eta^{(2)} I_z^{(1)} + \frac{1}{2}(\eta^{(1)})^2 I_{zz}^{(1)}\right\} - \left\{II^{(3)} + \eta^{(1)} II_z^{(2)}\right\} - III^{(3)} - I^{(1)*} \quad (7)$$

$$g\eta^{(3)} = -\left\{Q^{(3)} + \eta^{(1)} Q_z^{(2)} + \eta^{(2)} Q_z^{(1)} + \frac{1}{2}(\eta^{(1)})^2 Q_{zz}^{(1)}\right\} - \frac{1}{2}\left\{(V_{12}^2) + \eta^{(1)}(V_{11}^2)_z\right\} - Q^{(1)*} + C^{(3)}(t)$$

Fourth order approximation

$$I^{(4)} = -\left\{\eta^{(1)} I_z^{(3)} + \eta^{(3)} I_z^{(1)} + \eta^{(2)} I_z^{(2)} + \frac{1}{2}(\eta^{(1)})^2 I_{zz}^{(2)} + \eta^{(1)}\eta^{(2)} I_{zz}^{(1)} + \frac{1}{6}(\eta^{(1)})^3 I_{zzz}^{(1)}\right\} -$$

$$- \left\{II^{(4)} + \eta^{(1)} II_z^{(3)} + \eta^{(2)} II_z^{(2)} + \frac{1}{2}(\eta^{(1)})^2 II_{zz}^{(2)}\right\} - \left\{III^{(4)} + \eta^{(1)} III_z^{(3)}\right\} - II^{(2)*} \quad (8)$$

$$g\eta^{(4)} = -\left\{Q^{(4)} + \eta^{(1)} Q_z^{(3)} + \eta^{(3)} Q_z^{(1)} + \eta^{(2)} Q_z^{(2)} + \eta^{(1)}\eta^{(2)} Q_{zz}^{(1)} + \frac{1}{2}(\eta^{(1)})^2 Q_{zz}^{(2)} + \frac{1}{6}(\eta^{(1)})^3 Q_{zzz}^{(1)}\right\} -$$

$$- \frac{1}{2}\left\{(V_{13}^2) + (V_{22}^2) + \eta^{(1)}(V_{12}^2)_z + \eta^{(2)}(V_{11}^2)_z + \frac{1}{2}(\eta^{(1)})^2 (V_{11}^2)_{zz}\right\} - Q^{(2)*} + C^{(4)}(t)$$

where $I^{(S)} = \varphi_{tt}^{(S)} + g\varphi_z^{(S)} + (V_{0S}^2)_t + \frac{1}{2}\left[U_x (V_{0S}^2)_x + U_y (V_{0S}^2)_y\right]$, $S = 1, 2, 3, 4$

$$II^{(S)} = (V_S^2)_t + \frac{1}{2}\left[U_x (V_S^2)_x + U_y (V_S^2)_y\right] + \frac{1}{2}\sum_{p+q=S}\left[\varphi_x^{(p)}(V_{0q}^2)_x + \varphi_y^{(p)}(V_{0q}^2)_y + \varphi_z^{(p)}(V_{0q}^2)_z\right], \quad p, q \geq 1$$

$$III^{(S)} = \frac{1}{2}\sum_{p+q=S}\left[\varphi_x^{(p)}(V_q^2)_x + \varphi_y^{(p)}(V_q^2)_y + \varphi_z^{(p)}(V_q^2)_z\right], \quad p \geq 1, \ q \geq 2 \qquad Q^{(S)} = \varphi_t^{(S)} + \frac{1}{2}(V_{0S}^2),$$

$$V_{0S}^2 = 2\left[U_x \varphi_x^{(S)} + U_y \varphi_y^{(S)}\right], \quad V_S^2 = \sum_{p+q=S} V_{pq}^2, \quad p = q = \frac{S}{2} : V_{pq}^2 = [\varphi_x^{(p)}]^2 + [\varphi_y^{(p)}]^2 + [\varphi_z^{(p)}]^2,$$

$$p \neq q : \ V_{pq}^2 = 2\left[\varphi_x^{(p)}\varphi_x^{(q)} + \varphi_y^{(p)}\varphi_y^{(q)} + \varphi_z^{(p)}\varphi_z^{(q)}\right], \quad p \geq 1, \ q \geq 2$$

When current is absent, $I^{(S)} = \varphi_{tt}^{(S)} + g\varphi_z^{(S)}$, $II^{(S)} = (V_S^2)_t$, $Q^{(S)} = \varphi_t^{(S)}$. In the preceding equations the terms with asterisk* are concerned with higher order of frequencies. The fourth order solution of the nonlinear non-resonant interactions among gravity waves is found as follows:

$$\varphi = \Sigma_j a_j A_j \, ch\, k_j(d+z) \sin\theta_j + \sum_{S=2}^{4}\left\{\Sigma_{(S)}^{**} a_{(S)} A_{(S)} \, ch k_{(S)}(d+z) \sin\theta_{(S)}\right\} \quad (9)$$

$$\eta = \Sigma_j a_j \cos\theta_j + \sum_{S=2}^{4}\left\{\Sigma_{(S)} a_{(S)} \quad l_{(S)} \cos\theta_{(S)}\right\} \quad (10)$$

$$\sigma_j = (U_x \xi_j + U_y \zeta_j) + \mu_j^0 + \mu_j^{(2)}; \quad (\mu_j^0)^2 = g k_j \ th k_j d, \quad A_j = \frac{g}{\mu_j^0 \, ch k_j d}$$

$$\mu_j^{(2)} = \mu_j^0 \frac{a_j^2 k_j^2}{16 sh^4 k_j d}\left[8 sh^4 k_j d + 8 sh^2 k_j d + 9\right] + \sum_{n \neq j}\left\{\frac{1}{2}\mu_j^0 k_n^2 - \frac{1}{g}(C_{njn}^{+-} - C_{njn}^{-+})\right\} a_n^2 \quad (11)$$

where $\theta_j = \xi_j x + \zeta_j y - \sigma_j t + \varepsilon_j$, $\vec{k}_j = (\xi_j \ \zeta_j)$, $\theta_{(S)} = \xi_{(S)} x + \zeta_{(S)} y - \sigma_{(S)} t + \varepsilon_{(S)}$,

$\vec{k}_{(S)} = (\xi_{(S)}, \zeta_{(S)})$, $A_{(S)} = C_{(S)}/CH_{(S)}$, $CH_{(S)} = \left[\omega_{(S)}^2 - \mu_{(S)}^2\right] ch k_{(S)} d$, $\omega_{(S)}^2 = g k_{(S)} th k_{(S)} d$,

$l_{(S)} = B_{(S)} + F_{(S)} + C_{(S)}(t)/g$, $B_{(S)} = \frac{1}{g}\mu_{(S)} A_{(S)} \, ch k_{(S)} d$, $\sigma_{(S)} = \mu_{(S)} + (U_x \xi_{(S)} + U_y \zeta_{(S)})$

$S = 2$. $\Sigma_{(2)} = \Sigma_{\pm}\Sigma_{ij}$, $a_{(2)} = a_i a_j$, $\theta_{(2)} = \theta_{ij}^{\pm} = \theta_i \pm \theta_j$, $\mu_{(2)} = \mu_{ij}^{\pm} = \mu_i \pm \mu_j$

$\omega_{(2)}^2 = (\omega_{ij}^{\pm})^2 = g k_{ij}^{\pm} \ th k_{ij}^{\pm} d$, $A_{ij}^{\pm} = C_{ij}^{\pm} / CH_{ij}^{\pm}$

$$C_{(2)}(t) = \frac{g}{4}\left(\frac{k_i}{sh 2k_i d} + \frac{k_j}{sh 2k_j d}\right)\Delta_{ij}^{-}, \quad i = j, \ \Delta_{ij}^{-} = 1; \quad i \neq j, \ \Delta_{ij}^{-} = 0$$

$$C_{ij}^{\pm} = -\frac{1}{4}\left\{2\mu_{ij}^{0\pm}\mu_i^0\mu_j^0\left[cth k_i d \, cth k_j d \cos(\alpha_i - \alpha_j) \mp 1\right] + (\mu_i^0)^3 (cth^2 k_i d - 1) \pm (\mu_j^0)^3 (cth^2 k_j d - 1)\right\}$$

$$F_{(2)} = F_{ij}^{\pm} = \frac{1}{4g}\left\{(\mu_i^0)^2 + (\mu_j^0)^2 - \mu_i^0 \mu_j^0 HH_{ij}^{\pm}(0) / (k_i k_j \, sh k_j d \, sh k_j d)\right\}$$

$S = 3$, $\Sigma_{(3)} = \Sigma_{\hat{\pm}\pm}\Sigma_{nij}$, $a_{(3)} = a_n a_i a_j$, $\theta_{(3)} = \theta_n \hat{\pm} \theta_{ij}^{\pm} = \theta_{nij}^{\hat{\pm}\pm}$

$$C_{nij}^{\hat{\pm}\pm} = -\frac{1}{2}\left\{\hat{\pm} A_{ij} k_{ij}\left[g k_{ij} Ch k_{ij} d - (\mu_{ij}^0)^2 sh k_{ij} d\right] + (\mu_n^0)^3 l_{ij} / sh^2 k_n d\right.$$

$$+ \frac{1}{2} A_i A_j (\mu_n^0/2 \hat{\pm} \mu_{ij}^0) DH_{ij}(0) + 2 A_n A_{ij} \mu_{nij}^0 HH_{nij}(0)$$

$$\left.\hat{\mp}\frac{1}{4} A_n A_i A_j \, ch k_n d \, (\xi_n \xi_{ij} + \zeta_n \zeta_{ij}) HH_{ij}(0)\right\}$$

$$F_{nij}^{\hat{\pm}\pm} = \frac{1}{2g}\left\{A_{ij}\mu_{ij}^0 sh k_{ij} d + \mu_n^{02} l_{ij} + \frac{g}{4} k_n^2 - A_n A_{ij} HH_{nij}(0) - \frac{1}{4} A_i A_j DH_{ij}(0)\right\}$$

$$C_{nij}(t) = 0$$

$$S=4,\ \Sigma_{(4)} = \Sigma_{\hat{\hat{\pm}}\hat{\pm}\pm}\Sigma_{mnij}\ ,\ a_{(4)} = a_m a_n a_i a_j\ ,\ \theta_{(4)} = \theta_m \hat{\hat{\pm}}\theta_{nij}^{\hat{\pm}\pm} = \theta_{mnij}^{\hat{\hat{\pm}}\hat{\pm}\pm}$$

$$\begin{aligned}
-C_{mnij}^{\hat{\hat{\pm}}\hat{\pm}\pm} = {} & \hat{\hat{\pm}}\frac{1}{2}A_{nij}k_{nij}\left[gk_{nij}\,chk_{nij}d-(\mu_{nij}^{o})^{2}shk_{nij}d\right]+\frac{1}{2}k_{mn}A_{mn}l_{ij}\cdot \\
& \cdot\left[gk_{mn}\,chk_{mn}d-(\mu_{mn}^{o})^{2}shk_{mn}d\right]+\frac{1}{8}A_{mn}k_{mn}^{2}\left[gk_{mn}\,shk_{mn}d-\right. \\
& \left.-(\mu_{mn}^{o})^{2}chk_{mn}d\right]+\mu_{mnij}^{o}\left[A_m A_{nij}HH_{mnij}(0)+\frac{1}{2}A_{mn}A_{ij}GH_{mnij}(0)\right]+ \\
& +\frac{1}{4}(A_m k_m\, shk_m d\ \hat{\hat{\pm}}\ 2\mu_{nij}^{o})A_n A_{ij}DH_{nij}(0)+(l_{nij}+k_m^2/24)\cdot \\
& \cdot gk_m\mu_m^{o}/sh\,2k_m d+\frac{1}{4}A_m A_n(\mu_{mn}^{o}l_{ij}+k_{mn}k_{ij}A_{ij}\,shk_{ij}d)DH_{mn}(0)+ \\
& +\frac{1}{16}A_m A_n\mu_{mn}^{o}HG_{mn}(0)\ \hat{\hat{\mp}}\ \frac{1}{4}A_m A_n A_{ij}\,chk_n d(\xi_m\xi_{nij}+\zeta_m\zeta_{nij})\cdot \\
& \cdot HH_{nij}(0)\ \hat{\mp}\ \frac{1}{4}A_m A_n A_{ij}\,chk_{ij}d(\xi_{mn}\xi_{ij}+\zeta_{mn}\zeta_{ij})HH_{mn}(0) \\
& \hat{\hat{\pm}}\ \frac{1}{16}A_n A_i A_j\left\{chk_n d\left[k_n^2\ \hat{\mp}\ (\xi_n\xi_{ij}+\zeta_n\zeta_{ij})\right]DH_{ij}(0)+k_n\,shk_n d\ \cdot\right. \\
& \left.\cdot\left[HG_{mn}(0)\ \hat{\hat{\mp}}\ (\xi_n\xi_{ij}+\zeta_n\zeta_{ij})HH_{ij}(0)\right]\right\}
\end{aligned}$$

$$\begin{aligned}
F_{mnij}^{\hat{\hat{\pm}}\hat{\pm}\pm} = {} & \frac{1}{2g}\left\{\mu_{nij}^{o}A_{nij}k_{nij}\,shk_{nij}d+\mu_{ij}^{o}A_{ij}k_{ij}l_{mn}\,shk_{ij}d+\mu_m^{o}A_m k_m\,shk_m d\,(l_{nij}+\right. \\
& +k_m^2/24)+\frac{1}{4}A_{ij}k_{ij}^{2}\,chk_{ij}d+\frac{1}{2}\mu_n^{o}A_n k_n^2 l_{ij}\,chk_n d-A_m A_{nij}HH_{mnij}(0)- \\
& \left.-\frac{1}{2}A_{ij}\left[A_{mn}GH_{mnij}(0)+A_n DH_{nij}(0)\right]-\frac{1}{4}A_i A_j\left[2l_{mn}DH_{ij}(0)+HG_{ij}(0)\right]\right\}
\end{aligned}$$

$$HH_{(pq)}(z)=\left\{(\xi_{(p)}\xi_{(q)}+\zeta_{(p)}\zeta_{(q)})chk_{(p)}(d+z)chk_{(q)}(d+z)\mp k_{(p)}k_{(q)}shk_{(p)}(d+z)shk_{(q)}(d+z)\right\}$$

$$DH_{(pq)}(z)=\frac{\partial}{\partial z}HH_{(pq)}(z)\ ,\quad HG_{(pq)}(z)=\frac{\partial^2}{\partial z^2}HH_{(pq)}(z)\ ,\quad p+q=S$$

$$HH_{(11)}=HH_{ij}\ ,\ HH_{(12)}=HH_{nij}\ ,\ HH_{(13)}=HH_{mnij}\ ,\ HH_{(22)}=GH_{mnij}$$

$$C_{mnij}(t)=-g\Delta_{mnij}F_{mnij}\ ,\quad \theta_{mnij}=0\ ,\quad \Delta_{mnij}=1\ ,\quad \theta_{mnij}\neq 0\ ,\quad \Delta_{mnij}=0$$

The summations with double asterisks ** are taken with respect to subscripts m,n,i,j which do not satisfying $\theta_{(S)}=\theta_j$ or $\theta_{(S)}=0$ . When $\omega_{(S)}=\mu_{(S)}$, zero divisors for $A_{(S)}$ appear, and the harmonic resonance may occur.

Once the $\varphi$ , $\eta$ , C(t) and $\sigma$ are determined, the other kinematic and dynamic elements can be derived, such as:

wave pressure
$$\frac{p(z)}{\rho g}=-z-\frac{1}{g}\left\{Q+\frac{1}{2}V^2-c(t)\right\} \tag{12}$$

components of water particle acceleration
$$w_x=(Q+\frac{1}{2}V^2)_x\ ;\quad w_y=(Q+\frac{1}{2}V^2)_y\ ;\quad w_z=(Q+\frac{1}{2}V^2)_z \tag{13}$$

where
$$Q=\varphi_t+\frac{1}{2}V_*^2=\sum_{S=1}^{4}Q^{(S)}\ ,\quad Q^{(S)}=\Sigma_{(S)}a_{(S)}\left[-\mu_{(S)}\right]A_{(S)}chk_{(S)}(d+z)\cos\theta_{(S)}$$
$$\frac{1}{2}V_*^2=u_x\varphi_x+u_y\varphi_y=\sum_{S=1}^{4}V_{0S}^2\ ,\quad V_{0S}^2=Q^{(S)}\left[\sigma_{(S)}-\mu_{(S)}\right]/\left[-\mu_{(S)}\right]$$
$$\begin{aligned}\frac{1}{2}V^2=\frac{1}{2}\sum_{S=2}^{4}V_S^2=\frac{1}{4}\Big\{&\Sigma a_i a_j A_i A_j HH_{ij}(z)\cos\theta_{ij}+2\Sigma a_n a_i a_j A_n A_{ij}HH_{nij}(z) \\ &\cos\theta_{nij}+\Sigma a_m a_n a_i a_j(2A_m A_{nij}HH_{mnij}(z)+A_{mn}A_{ij}GH_{mnij}(z))\cos\theta_{mnij}\Big\}\end{aligned}$$

total pressure per unit area from Z to $\eta$
$$R(z)=\frac{1}{\rho g}\int_z^{\eta}p\,dz=\frac{1}{2}z^2+\frac{1}{g}(\eta-z)C(t)-\frac{1}{g}\int_z^{\eta}Q\,dz-E(z)/\rho g \tag{14}$$

total energy per unit area from Z to $\eta$
$$\widetilde{E}(z)=KE^*(z)+KE(z)+PE=KE^*(z)+E(z) \tag{15}$$

where

$$PE = \frac{1}{2}\rho g \eta^2$$

$$KE^*(z)/\rho g = \frac{1}{2g}\int_z^\eta V_*^2 dz = \frac{1}{2g}\sum_{S=1}^{4}\Sigma_{(S)} a_{(S)} [IV_*^2]_{(S)} \cos\theta_{(S)}$$

$$KE(z)/\rho g = \frac{1}{2g}\int_z^\eta V^2 dz = \frac{1}{2g}\sum_{S=2}^{4}\Sigma_{(S)} a_{(S)} [IV^2]_{(S)} \cos\theta_{(S)}$$

$$-\int_z^\eta Q\,dz = \sum_{S=1}^{4}\Sigma_{(S)} a_{(S)} [IQ]_{(S)} \cos\theta_{(S)}$$

total pressure moment from Z to $\eta$ around Z

$$M(z) = \frac{1}{\rho g}\int_z^\eta (u-z)\,p(u)\,du = -zR(z) + \frac{1}{3}z^3 - \frac{1}{2g}z^2 C(t) + \frac{1}{g}\sum_{S=1}^{4}\Sigma_{(S)} a_{(S)} [IM]_{(S)} \cos\theta_{(S)} + \frac{1}{4g}\Sigma a_i a_j C^{(2)}(t)\cos\theta_{ij} \quad (16)$$

mass flux per unit area from Z to $\eta$

$$q_x(z)/\rho = U_x(\eta - z) + \int_z^\eta \varphi_x dz \; ; \quad q_y(z)/\rho = U_y(\eta - z) + \int_z^\eta \varphi_y dz \quad (17)$$

energy flux per unit area from Z to $\eta$

$$F_x(z)/\rho = \tilde{C}(t) q_x(z)/\rho - \int_z^\eta \varphi_t (U_x + \varphi_x) dz \quad (18)$$

momentum flux per unit area from Z to $\eta$

$$S_{xx}(z)/\rho = gR(z) + U_x q_x(z)/\rho + \int_z^\eta \varphi_x (U_x + \varphi_x)\, dz \quad (19)$$

$$S_{xy}(z)/\rho = U_y q_x(z)/\rho + \int_z^\eta \varphi_y (U_x + \varphi_x) dz = U_x q_y(z)/\rho + \int_z^\eta \varphi_x (U_y + \varphi_y) dz = S_{yx}(z)/\rho$$

$$(F_x + U_x S_{xx})/\rho = gU_x R(z) + (\tilde{C}(t) + U_x^2)\, q_x(z)/\rho - \int_z^\eta (\varphi_t - U_x \varphi_x)(U_x + \varphi_x) dz \quad (20)$$

The expressions for $[IQ]_{(S)}$ and $[IV^2]_{(S)}$ are given in appendix I. The formulas for $[IV_*^2]_{(S)}$ , $[I\varphi_x]_{(S)}$ , $[I\varphi_y]_{(S)}$ and $[-I\varphi_t]_{(S)}$ may be found by substituting $[\sigma_{(S)} - \mu_{(S)}]$, $[\xi_{(S)}]$, $[\zeta_{(S)}]$ and $[\sigma_{(S)}]$ instead of $[\mu_{(S)}]$ in formulas for $[IQ]_{(S)}$, respectively. Similarly, the expressions for $[-I\varphi_t \varphi_x]_{(S)}$ , $[-I\varphi_t \varphi_y]_{(S)}$ and $[I\varphi_x \varphi_y]_{(S)}$ may be found by substituting $[\sigma_{(S)}\xi_{(S)}]$, $[\sigma_{(S)}\zeta_{(S)}]$ and $[\xi_{(S)}\zeta_{(S)}]$ instead of $[\xi_{(p)}\xi_{(q)} + \zeta_{(p)}\zeta_{(q)} \pm (\mp) k_{(p)} k_{(q)}]$ in formulas for $[IV^2]_{(S)}$ , respectively.

## 3. Nonlinear Interaction of Waves and Vertical Wall

### 3.1 Interaction of Oblique Regular Wave and Vertical Wall

Let's consider the case of that a vertical wall is set along the x-axis and the incident wave makes an angle- $\alpha$ with the wall. The nonlinear interaction of oblique wave and vertical wall can be considered as a particular nonlinear interaction between incident and reflected waves, whose linear components of free surface elevations are given by

$$\eta^{(1)} = \eta_I^{(1)} + \eta_R^{(1)} = a_2 \cos\theta_2 + a_1 \cos\theta_1$$

in which $a_1 = Ka_2$ , K=reflection coefficient, $\varepsilon_r = \varepsilon_1 - \varepsilon_2$ = phase shift; $\xi_1 = \xi_2 = \xi = k\cos\alpha$ , $\zeta_1 = -\zeta_2 = \zeta = k\sin\alpha$ ; $U_y = 0$, $\mu_1^0 = \mu_2^0$ , $\sigma_2 = \sigma^0 + \mu_2^{(2)}$ , $\sigma_1 = \sigma^0 + \mu_1^{(2)}$ , $\mu_r^{(2)} = \sigma_2 - \sigma_1 = \mu_2^{(2)} - \mu_1^{(2)}$ , $\sigma^0 = \mu^0 + \xi U_x$ . Based on the preceding general solutions, an explicit analytic expression of the third order approximation had been presented. For example

$$\varphi = A\,ch\,k(d+z)(a_2 \sin\theta_2 + a_1 \sin\theta_1) + (\tfrac{3}{8}\mu^0/sh^4kd)\,ch2k(d+z)(a_2^2 \sin 2\theta_2 + a_1^2 \sin 2\theta_1) + (\tfrac{1}{2}a_2 a_1 \mu^0 A_0/ch2\xi d)\,ch2\xi(d+z)\sin(\theta_2+\theta_1) + \alpha_4\,ch3k(d+z)(a_2^3 \sin 3\theta_2 + a_1^3 \sin 3\theta_1) + a_2 a_1 \alpha_5\,ch k_{112}^{++}(d+z)[a_2 \sin(2\theta_2+\theta_1) + a_1 \sin(2\theta_1+\theta_2)] + a_2 a_1 \alpha_6\,ch k_{112}^{+-}(d+z)[a_2 \sin(2\theta_2-\theta_1) + a_1 \sin(2\theta_1-\theta_2)] + \varphi^{(4)} \tag{21}$$

$$\eta = (a_2\cos\theta_2 + a_1\cos\theta_1) + [(\beta_7)_2 a_2^3 \cos\theta_2 + (\beta_7)_1 a_1^3 \cos\theta_1] + \beta_1[a_2^2 \cos 2\theta_2 + a_1^2\cos 2\theta_1] + a_2 a_1[\beta_2 \cos(\theta_2+\theta_1) + \beta_3\cos(\theta_2-\theta_1)] + a_2 a_1 \Delta k/sh2kd + a_2 a_1 \beta_5[a_2\cos(2\theta_2+\theta_1) + a_1\cos(2\theta_1+\theta_2)] + a_2 a_1 \beta_6[a_2\cos(2\theta_2-\theta_1) + a_1\cos(2\theta_1-\theta_2)] + \eta^{(4)} \tag{22}$$

$$c(t) = (a_2^2 + a_1^2 + 2\Delta a_2 a_1)gk/(2sh2kd) + c^{(4)}(t), \quad \alpha = 0,\ \Delta = 1; \quad \alpha \neq 0,\ \Delta = 0$$

The expressions of velocity potential, surface elevation, pressure, total pressure and its moment can be rewritten in the following general form:

$$Y = A\cos\psi + B\cos 2\psi + C\cos 3\psi + A'\sin\psi + B'\sin 2\psi + C'\sin 3\psi + D, \quad \psi = \xi x - \sigma_2 t + \varepsilon_2$$

when $\mu_r \equiv 0$ or $\mu_r \simeq 0$ (e.g.k $\simeq$ 1) the extreme values and corresponding phases of wave processes can be determined analytically. As special cases, this solution can be reduced to progressive wave ($a_1 = 0$), short-crested wave (K=1, $\varepsilon_r = 0$) and standing wave ($\alpha = \pi/2$). In Eq.(22) the component with factor of $\cos(\theta_2 + \theta_1)$ is similar to Mach wave.

For partial reflection the solution (21) is satisfying the inhomogeneous mixed boundary condition on the wall

$$\varphi_y^{(S)} + \beta^{(S)}\varphi^{(S)} = G^{(S)}(x, y, z, t) \tag{23}$$

where $\quad \frac{1}{S}\beta^{(S)} = i\zeta\frac{1-R}{1+R}, \quad R = Ke^{i\varepsilon_r}, \quad i = \sqrt{-1}$

$$G^{(1)} = (-ia_2)R[(i\zeta + \beta^{(1)})(exp(-i\mu_r t) - 1)]\,A\,ch\,k(d+z)\,e^{i\psi}$$

The calculating results for full reflection may be found in Ref.(13). The agreement between computed and some laboratory data was good.

### 3.2 Interaction of Regular Short-crested Wave and Vertical Wall

In this special case, $\eta_I^{(1)}$ and $\eta_R^{(1)}$ are expressed as follows:

$$\eta_I^{(1)} = a_2\cos\theta_2 + a_4\cos\theta_4, \quad \eta_R^{(1)} = a_1\cos\theta_1 + a_3\cos\theta_3,$$

in which $\xi_1 = \xi_2 = k\cos(\alpha - \alpha_0)$, $\xi_3 = \xi_4 = k\cos(\alpha + \alpha_0)$, $\zeta_1 = -\zeta_2 = k\sin(\alpha - \alpha_0)$, $\zeta_3 = -\zeta_4 = k\sin(\alpha + \alpha_0)$; $a_2 = a_4 = a$, $a_1 = a_3 = Ka$, $\varepsilon_2 = \varepsilon_4 = \varepsilon$, $\varepsilon_1 = \varepsilon_3 = \varepsilon + \varepsilon_r$; $U_y = 0$, $\mu_j^0 = \mu^0$; ctg $\alpha_0$=three-dimensional index of wave, $-\alpha$ =incident angle of short-crested wave.

### 3.3 Interaction of Short Gravity Wave, Long Wave and Current in Front of Vertical Wall

In this special case, $\eta^{(1)}$ is expressed as follows:

$$\eta^{(1)} = \sum_{j=1}^{4} a_j \cos\theta_j$$

in which $a_1 = K_1 a_2$, $a_3 = K_2 a_4$; $\xi_1 = \xi_2$, $\zeta_1 = -\zeta_2$; $\xi_3 = \xi_4$, $\zeta_3 = -\zeta_4$; $\varepsilon_1 = \varepsilon_2 + \varepsilon_{r1}$

$$\varepsilon_3 = \varepsilon_4 + \varepsilon_{r3}; \quad \sigma_1^0 = \sigma_2^0, \quad \sigma_3^0 = \sigma_4^0, \quad U_y = 0$$

### 3.4 Interaction of Irregular Waves and Vertical Wall

In this special case, $\eta^{(1)}$ is expressed as follows:

$$\eta^{(1)} = \sum_{j=1}^{2N} a_j\cos\theta_j, \quad \eta_I^{(1)} = \sum_{j=1}^{N} a_j\cos\theta_j, \quad \eta_R^{(1)} = \sum_{N+1}^{2N} a_j\cos\theta_j$$

in which $a_{j+N}=a_j$, $\xi_{j+N}=\xi_j$, $\zeta_{j+N}=-\zeta_j$, $\mu^o_{j+N}=\mu^o_j$, $\sigma_j=\sigma^o_j+\mu^{(2)}_j$; $\varepsilon_j$ =random phase, uniformly distributed in interval of $0\sim 2\pi$; and j=1,2,...,N. The Edgeworths series (Ref.15) and "Generalized Normal Distribution" (Ref.14) may be used to describe the statistical property of nonlinear random waves.

3.5 Interaction of Waves and Corner

Consider a regular wave train approaching a reflecting corner with inner angle of $\gamma\pi$ ($\gamma\leq 1$), one of vertical wall is set alone the x-axis (polar angle $\theta_0=0$), the other one is on $\theta_0=\gamma\pi$. The incident wave makes an angle $\alpha$ with positive x. The linear solution of wave diffraction by vertical wedge and corner was given by author as follows

$$\varphi^{(1)}_j=Re\left\{ia_jA_j chk_j(d+z)f_\gamma(\rho,\theta_0)e^{i(\mu_j t+\varepsilon_j)}\right\} \tag{24}$$

$$\eta^{(1)}_j=Re\left\{a_j f_\gamma(\rho,\theta_0)e^{i(\mu_j t+\varepsilon_j)}\right\} \tag{25}$$

$$f_\gamma(\rho,\theta_0)=f^-_\gamma(\rho,\theta_0-\alpha)+Rf^+_\gamma(\rho,\theta_0+\alpha),\quad 0<\gamma\leq 2 \tag{26}$$

$$f^\pm_\gamma(\rho,\theta_0\pm\alpha)=\sum_{n=0}^{\infty}\frac{\delta_n}{\gamma}e^{i\frac{n\pi}{2\gamma}}J_{n/\gamma}(k\rho)\cos\frac{n}{\gamma}(\theta_0\pm\alpha)$$
$$=f^\pm_{2\gamma}(\rho,\theta_0\pm\alpha)+f^\pm_{2\gamma}(\rho,\theta_0\pm\alpha+2\gamma\pi) \tag{27}$$

in which $x=\rho\cos\theta_0$, $y=\rho\sin\theta_0$; n=0, $\delta_n=1$, $n\neq 0$, $\delta_n=2$; $J_{n/\gamma}$ =Bessel function of order $n/\gamma$, $u_x=u_y=0$.

For special cases of $\gamma=1/N$ (N=1,2,3,...), the solution (25) is reduced to the following expression

$$\eta^{(1)}_j=\sum_{p=1}^{N}a_j\left\{exp\left[i\left(k_j\rho\cos\left(\theta_0-\alpha_j+(p-1)\frac{2\pi}{N}\right)+\mu_p t+\varepsilon_j\right)\right]+R\,exp\left[i\left(k_j\rho\cdot\cos\left(\theta_0+\alpha_j+(p-1)\frac{2\pi}{N}\right)+\tilde{\mu}_p t+\varepsilon_j\right)\right]\right\} \tag{28}$$

in which $\eta^{(1)}_I=a_j\,exp\left[i(k_j\rho\cos(\theta_0-\alpha_j)+\mu_j t+\varepsilon_j\right]$, $\mu^o_p=\tilde{\mu}^o_p=\mu^o_j$, $\mu^{o2}_j=gk_j thk_j d$

The boundary conditions on the corner are as follows

$$\theta_0=0,\quad \frac{1}{\rho}\varphi^{(1)}_{\theta_0}+B^{(1)}_0\varphi^{(1)}=G^{(1)}_0,\quad B^{(1)}_0=-ik_j\sin\alpha_j\frac{1-R}{1+R} \tag{29}$$

$$G^{(1)}_0=-ik_j\sum_{p=2}^{N}\left\{\left[\frac{1-R}{1+R}\sin\alpha_j-\sin\left(\alpha_j-(p-1)\frac{2\pi}{N}\right)\right]exp\left[ik_j\rho\cos\left(\alpha_j-(p-1)\frac{2\pi}{N}\right)\right]\right.$$
$$\left.+R\left[\frac{1-R}{1+R}\sin\alpha_j+\sin\left(\alpha_j+(p-1)\frac{2\pi}{N}\right)\right]exp\left[ik_j\rho\cos\left(\alpha_j+(p-1)\frac{2\pi}{N}\right)\right]\right\}\left[iA_j chk_j(d+z)\right]$$

$$\theta_0=\gamma\pi,\quad -\frac{1}{\rho}\varphi^{(1)}_{\theta_0}+B^{(1)}_\gamma=G^{(1)}_\gamma;\quad B^{(1)}_\gamma=B^{(1)}_0\Big|_{\alpha_j=\gamma\pi-\alpha_j},\quad G^{(1)}_\gamma=G^{(1)}_0\Big|_{\alpha_j=\gamma\pi-\alpha_j} \tag{30}$$

The nonlinear interaction of waves and corner ($\gamma=1/N$) is a special case of nonlinear interactions among wave trains, whose linear components are given by Eq. (28).

Appendix I. —— Expressions for $[IQ]_{(S)}$, $[IV^2]_{(S)}$ and $[IM]_{(S)}$

$$[IQ]_j=[\mu_j]A_j/k_j\,(shk_jd-shk_j(d+z))$$

$$[IQ]_{ij}=[\mu_{ij}]A_{ij}/k_{ij}\,(shk_{ij}d-shk_{ij}(d+z))+[\mu_j]g/(2\mu^o_j)$$

$$[IQ]_{nij}=[\mu_{nij}]A_{nij}/k_{nij}(shk_{nij}d-shk_{nij}(d+z))+[\mu_j]\left(\frac{g}{2\mu^o_j}l_{ni}+\frac{\mu^o_j}{8}\right)+\frac{1}{2}[\mu_{ij}]A_{ij}chk_{ij}d$$

$$[IQ]_{mnij}=[\mu_{mnij}]A_{mnij}/k_{mnij}\left(shk_{mnij}d-shk_{mnij}(d+z)\right)+[\mu_j]\left(\frac{g}{2\mu^o_j}\left(l_{mni}+\frac{1}{24}k^2_j\right)+\frac{1}{4}\mu^o_j\right.$$
$$\left.l_{ni}\right)+\frac{1}{8}[\mu_{ij}]A_{ij}\left(4l_{mn}chk_{ij}d+k_{ij}shk_{ij}d\right)+\frac{1}{2}[\mu_{nij}]A_{nij}chk_{nij}d$$

$$[IV^2]_{ij}=\frac{1}{4}A_iA_j\left\{[\xi_i\xi_j+\zeta_i\zeta_j\mp k_ik_j](sh(k_i+k_j)d-sh(k_i+k_j)(d+z))/(k_i+k_j)+\right.$$
$$\left.+[\xi_i\xi_j+\zeta_i\zeta_j\pm k_ik_j](sh(k_i-k_j)d-sh(k_i-k_j)(d+z))/(k_i-k_j)\right\}$$

$$[IV^2]_{nij}=\frac{1}{2}A_nA_{ij}\left\{[\xi_n\xi_{ij}+\zeta_n\zeta_{ij}\mp k_nk_{ij}](sh(k_n+k_{ij})d-sh(k_n+k_{ij})(d+z))/(k_n+k_{ij})\right.$$

$$+\left[\xi_n\xi_{ij}+\zeta_n\zeta_{ij}\hat{\pm}k_nk_{ij}\right](sh(k_n+k_{ij})d-sh(k_n-k_{ij})(d+z))/(k_n-k_{ij})\Big\}$$
$$+\frac{1}{8}A_iA_j\Big\{\left[\xi_i\xi_j+\zeta_i\zeta_j\mp k_ik_j\right]ch(k_i+k_j)d+\left[\xi_i\xi_j+\zeta_i\zeta_j\pm k_ik_j\right]ch(k_i-k_j)d\Big\}$$

$$[IV^2]_{mnij}=\frac{1}{2}A_mA_{nij}\Big\{\left[\xi_m\xi_{nij}+\zeta_m\zeta_{nij}\hat{\hat{\mp}}k_mk_{nij}\right](sh(k_m+k_{nij})d-sh(k_m+k_{nij})(d+z))/(k_m+k_{nij})$$
$$+\left[\xi_m\xi_{nij}+\zeta_m\zeta_{nij}\hat{\hat{\pm}}k_mk_{nij}\right](sh(k_m-k_{nij})d-sh(k_m-k_{nij})(d+z))/(k_m-k_{nij})\Big\}$$
$$+\frac{1}{4}A_{mn}A_{ij}\Big\{\left[\xi_{mn}\xi_{ij}+\zeta_{mn}\zeta_{ij}\hat{\hat{\mp}}k_{mn}k_{ij}\right](sh(k_{mn}+k_{ij})d-sh(k_{mn}+k_{ij})(d+z))/(k_{mn}+k_{ij})$$
$$+\left[\xi_{mn}\xi_{ij}+\zeta_{mn}\zeta_{ij}\hat{\hat{\pm}}k_{mn}k_{ij}\right](sh(k_{mn}-k_{ij})d-sh(k_{mn}-k_{ij})(d+z))/(k_{mn}-k_{ij})\Big\}$$
$$+\frac{1}{32}A_iA_j\Big\{\left[\xi_i\xi_j+\zeta_i\zeta_j\mp k_ik_j\right](4l_{mn}ch(k_i+k_j)d+(k_i+k_j)sh(k_i+k_j)d)$$
$$+\left[\xi_i\xi_j+\zeta_i\zeta_j\pm k_ik_j\right](4l_{mn}ch(k_i-k_j)d+(k_i-k_j)sh(k_i-k_j)d\Big\}+\frac{1}{4}A_nA_{ij}$$
$$\Big\{\left[\xi_n\xi_{ij}+\zeta_n\zeta_{ij}\hat{\mp}k_nk_{ij}\right]ch(k_n+k_{ij})d+\left[\xi_n\xi_{ij}+\zeta_n\zeta_{ij}\hat{\pm}k_nk_{ij}\right]ch(k_i-k_j)d\Big\}$$

$$[IM]_j=-\mu_jA_jJ(k_j)/k_j$$

$$g[IM]_{ij}=-\mu_{ij}A_{ij}J(k_{ij})/k_{ij}+A_iA_j\Big\{\left[\xi_i\xi_j+\zeta_i\zeta_j\mp k_ik_j\right]J(k_i+k_j)/(k_i+k_j)+$$
$$+\left[\xi_i\xi_j+\zeta_i\zeta_j\pm k_ik_j\right]J(k_i-k_j)/(k_i-k_j)\Big\}$$

$$g[IM]_{nij}=-\mu_{nij}A_{nij}J(k_{nij})/k_{nij}+g/24+\frac{1}{4}A_nA_{ij}\Big\{\left[\xi_n\xi_{ij}+\zeta_n\zeta_{ij}\hat{\mp}k_nk_{ij}\right]$$
$$J(k_n+k_{ij})/(k_n+k_{ij})+\left[\xi_n\xi_{ij}+\zeta_n\zeta_{ij}\hat{\pm}k_nk_{ij}\right]J(k_n-k_{ij})/(k_n-k_{ij})\Big\}$$

$$g[IM]_{mnij}=-\mu_{mnij}A_{mnij}J(k_{mnij})/k_{mnij}-g(2l_{ij}+l_{mn})/12+\frac{1}{8}A_{ij}\mu_{ij}ch\,k_{ij}d$$
$$+\mu_m^{o2}/12+\frac{1}{8}A_{mn}A_{ij}\Big\{\left[\xi_{mn}\xi_{ij}+\zeta_{mn}\zeta_{ij}\hat{\hat{\mp}}k_{mn}k_{ij}\right]J(k_{mn}+k_{ij})/(k_{mn}+k_{ij})$$
$$+\left[\xi_{mn}\xi_{ij}+\zeta_{mn}\zeta_{ij}\hat{\hat{\pm}}k_{mn}k_{ij}\right]J(k_{mn}-k_{ij})/(k_{mn}-k_{ij})\Big\}+\frac{1}{4}A_mA_{nij}$$
$$\Big\{\left[\xi_m\xi_{nij}+\zeta_m\zeta_{nij}\hat{\hat{\mp}}k_mk_{nij}\right]J(k_m+k_{nij})/(k_m+k_{nij})+\left[\xi_m\xi_{nij}+\zeta_m\zeta_{nij}\hat{\hat{\pm}}k_mk_{nij}\right]$$
$$J(k_m-k_{nij})/(k_m-k_{nij})-\frac{1}{64}A_iA_j\Big\{\left[\xi_i\xi_j+\zeta_i\zeta_j\mp k_ik_j\right]ch(k_i+k_j)d+$$
$$+\left[\xi_i\xi_j+\zeta_i\zeta_j\pm k_ik_j\right]ch(k_i-k_j)d\Big\}$$

$$J(u)=\left[ch\,ud+uz\,sh\,u(d+z)-ch\,u(d+z)\right]/u,\qquad u=k_j,\ k_i+k_j,\cdots$$

Appendix II. —— References


1. Biesel, F.: La Houille Blanche, No.3 (1952) 372–376.
2. Tadjbakhsh, I. & Keller, J.B.: Standing surface waves of finite amplitude, J. Fluid Mech. 8(1960) 442–451.
3. Goda, Y.: The fourth order approximation to the pressure of standing waves, Coastal Engineering in Japan, Vol. 10 (1967) 1–11.
4. Секерж-Зенькович, Я. И.: АНСССР, серияГеогр. И Геофиз., No. 1(1951) 57–73.
5. Rundgren, L.: Water wave forces, Bulletin No.54 (1958), The Institution of Hydraulics, Inst. of Technology, Stockholm.
6. Fuchs, R.A.: On the theory of short-crested oscillatory waves., U. S. Nat. Bur. Stand. Circular 521 (1952) 187–200.
7. Кузнецов, А. И.: Труды Союзморнии проекта, 34(40) (1973) 3–9.
8. Hsu, J.R.C., Tsuchiya, Y., and Silvester, R.: Third-order approximation to short-crested waves, J. Fluid Mech. 90(1977) 179–196.
9. Fenton, J. D.: Wave forces on vertical walls, J. of waterway, Port, Coastal and Ocean Div., A S C E, Vol. 111, No. 2 (1985) 216–234.
10. Roberts, A. J.: Highly nonlinear short-crested water waves. J. Fluid Mech. 135(1983)301-321.
11. Hong, G. W.: On the nonlinear interactions among gravity surface waves, A C T A OCEAN-OLOGICA SINICA, Vol. 2, No. 2 (1980) 158–180 (in Chinese). Vol,No.1(1982),144-145(in English).
12. Hong, G. W.: Nonlinear interactions among gravity surface waves, Proc. SYMPOSIUM UBER . HYDROLOGIE UND KUSTENINGENIEURWESEN, Nanjing, China, October, 1984.
13. Hong, G. W.: Nonlinear interaction between wave in oblique advance and vertical wall. Proc. of Second International Conf. on Coastal and Port Engineering in Developing countries, Beijing, 1987.
14. Hong, G. W.: Generalized Normal Distribution, China Ocean Engineering, No. 2 (1983) 18–37 (in Chinese), No. 1 (1987) 59–82 (in English).
15. Longuet-Higgins, M.S.: The effect of nonlinearities on statistical distributions in the theory of sea waves, J.Fluid Mech. 17 (1963) 459–479.

# Superharmonic Waves Generated by the Large Amplitude Heaving Motion of a Submerged Body

A. CLEMENT, P. FERRANT

Laboratoire d'Hydrodynamique Navale.
UA 1217 C.N.R.S.
E.N.S.M. Nantes - France.

Summary.

A novel method for the computation of free surface flows generated by submerged bodies with a large amplitude of the motion is described.

A submerged three-dimensional body is subject to a periodic motion of translation under the free-surface of an incompressible, inviscid heavy fluid of infinite depth. The amplitude of the motion is not necessarily small with respect to a typical body dimension, and the body boundary condition has to be satisfied at the exact time-depending position of the hull. The other parameters such as frequency and submergence are supposed to remain in such a range that the wave steepness remains small, so that the linearized free-surface condition is still valid. Assuming the steady state to be established, the velocity potential is expressed as a Fourier series of the fundamental frequency . The resulting set of coupled integral equations for the potential and its normal derivative is then approximated by a discretization of the surface integrals which are restricted to the mean body position by using the proper Kelvin operators. The solutions of the resulting linear algebraic systems give access to the hydrodynamic forces and to the radiated wave field. Some results are presented for an immersed heaving sphere. It is shown that even for moderate amplitudes, the delinearization of the body boundary condition has very significant effects, especially an important dispersion of the radiated energy on the harmonic components of the wave field. The results are compared with those of a time-domain method (Pot [1]) for the force computation, and with experimental data (Dassonville [2]) for the wave field. The agreement is good in both cases.

## The mathematical problem.

We consider the irrotational flow generated by the forced motion of a rigid body under the free surface of an incompressible, inviscid heavy fluid of infinite depth. The fluid flow can thus be represented by a velocity potential $\Phi(M,t)$, the velocity being given by $\vec{V} = \nabla\Phi$. The potential satisfies Laplace's equation in the fluid domain $D(t)$:

$$\Delta\Phi(M,t) = 0 \qquad M \in D(t) \tag{1}$$

The vertical coordinate is positive up, and the undisturbed free surface is represented by the plane $z = 0$. The wave steepness is assumed to remain small, and $\Phi$ is subject to the classical linearized free surface condition:

K. Horikawa, H. Maruo (Eds.)
Nonlinear Water Waves
IUTAM Symposium Tokyo/Japan 1987

$$\frac{\partial^2 \Phi}{\partial t^2} + g \frac{\partial \Phi}{\partial z} = 0 \quad \text{for } z = 0 \tag{2}$$

where g is the gravity acceleration.

On the body surface S(t), the normal fluid velocity is given by the normal component of the prescribed body velocity $\vec{V}_B(t)$.

$$\left.\frac{\partial \Phi(M,t)}{\partial n}\right|_{M \in S(t)} = \vec{V}_B \cdot \vec{n} \tag{3}$$

where $\vec{n}$ is the unit normal exterior to the body. A far field condition has also to be imposed:

$$\Phi(M,t) \longrightarrow 0 \quad \text{for } R \longrightarrow \infty \tag{4}$$

where R is the distance from a fixed reference point to the point M.

The body is subject to a periodic motion, with frequency ω. The steady state fluid response is then also periodic, and the associated physical quantities will be expressed as Fourier series, with fundamental frequency ω. for example, the velocity potential will be written as:

$$\Phi(M,t) = \sum_{n=0}^{\infty} \phi_n^{*}(M)\cos(n\omega t)+\phi_n^{**}(M)\sin(n\omega t) = \sum_{n=0}^{\infty} \mathrm{Re}(\Phi_n\ e^{-in\omega t}) \tag{5}$$

According to the preceding considerations, the velocity potential is solution of the following problem:

$$\left[\begin{array}{l}
\Delta \phi = 0 \qquad \text{or} \qquad \Delta \phi_p = 0 \qquad p = 0,1,\ldots\ldots. \\[2mm]
\dfrac{\partial \phi_p}{\partial z} = K_p\, \phi_p \qquad ; \qquad K_p = \dfrac{p^2\omega^2}{g} \quad ; (z=0) \\[2mm]
\phi_p \xrightarrow[R \longrightarrow \infty]{} 0 \\[2mm]
\displaystyle\sum_p \frac{\partial \phi_p}{\partial n}\, e^{-ip\omega t} = \vec{V}_B \cdot \vec{n} \qquad M \in S(t) \\[2mm]
+ \text{ radiation condition}
\end{array}\right. \tag{6}$$

<u>Integral equations</u>

We further introduce a fictitious surface Σ enclosing the body and limited by the free surface (for example an hemisphere). Applying Green's identity to the fluid domain D'(t) enclosed by S(t), F' and Σ, we have from (1):

$$\iint_{\partial D(t)=S(t)+F'+\Sigma} \left[\phi \frac{\partial}{\partial n}\left(\frac{1}{r}\right) - \frac{1}{r}\frac{\partial}{\partial n}(\phi)\right] d\,S_{M'} = \left[\begin{array}{ll} 0 & M \in CD'(t) \\ -2\pi\ \phi\ (M) & M \in \partial D'(t) \\ -4\pi\ \phi\ (M) & M \in D'(t) \end{array}\right. \tag{7}$$

According to (5), with M on ∂D', (7) gives:

$$\sum_{p=0}^{\infty} \iint_{F'+\Sigma+S(t)} \left[ \phi_p^*(M') \frac{\partial}{\partial n'} \left(\frac{\cos p\omega t}{R}\right) + \phi_p^{**}(M') \frac{\partial}{\partial n'} \left(\frac{\sin p\omega t}{R}\right) - \frac{\cos p\omega t}{R} \frac{\partial}{\partial n'} (\phi_p^*(M')) - \frac{\sin p\omega t}{R} \frac{\partial}{\partial n'} (\phi_p^{**}(M')) \right] ds_{M'} = \begin{cases} 0 & M \in C_{D'} \\ -2\pi\, \phi(M,t) & M \in \delta D' \\ -4\pi\, \phi(M,t) & M \in D' \end{cases} \tag{8}$$

with R a function of time when M' is on S(t).

Various integral methods may be deduced from this basic integral equation, whether Rankine or Kelvin operators are used. In all cases, the computations will have to account for the singularity displacement, for the body boundary condition to be satisfied at the time depending body surface S(t). A method based on the use of Kelvin operators has been developed, for which, as in the linearized radiation problem, the integral equations are reduced to the body surface only ([3]). The difficulty lies then in the determination of the Green functions, $G_{1p}$ and $G_{2p}$, solutions of the following problem:

$$\begin{cases} \Delta G_{ip}(M,M',t) = \delta(M,M') \cdot \begin{bmatrix} \cos(p\omega t) \\ \sin(p\omega t) \end{bmatrix} \begin{matrix} i=1 \\ i=2 \end{matrix} \\ \left[ \frac{\partial^2}{\partial t^2} + 2\epsilon \frac{\partial}{\partial t} + g \frac{\partial}{\partial z} \right] G_{ip}(M,M',t) = 0 \qquad ,i=1,2 \;,\; M \in D \\ G_{ip}(M,M',t) \longrightarrow 0 \quad \text{for } R \longrightarrow \infty \end{cases} \tag{9}$$

with: $M'(x',y',z'=z_o+a\cos\omega t)$ $\qquad z_o + a < 0$ (prescribed motion)

$M\,(x,y,z)$ $\qquad z < 0$

$R = |MM'|$

and

$$G_{1p}(M,M',t) = \frac{-\cos(p\omega t)}{4\pi R(t)} + g_{1p}(M,M',t)$$

$$G_{2p}(m,m',t) = \frac{-\sin(p\omega t)}{4\pi R(t)} + g_{2p}(M,M',t)$$

where $g_{1p}$ and $g_{2p}$ are regular harmonic functions in the lower half space. Assuming that such functions are available, they can be substituted to the Rankine operators in (8). An infinite set of coupled integral equations for the harmonic components of the velocity potential on the body surface is then derived:

$$\begin{cases} \displaystyle\sum_{p=0}^{\infty} \iint_{S(t)} \left[ \phi_p^*(M') \frac{\partial}{\partial n'} G_{1p}^{n*} + \phi_p^{**} \frac{\partial}{\partial n'} G_{2p}^{n*} - G_{1p}^{n*} \frac{\partial}{\partial n'} \phi_p^*(M') - G_{2p}^{n*} \frac{\partial}{\partial n'} \phi_p^{**}(M') \right] ds_{M'} = \frac{1}{2} \phi_n^*(M) \\ \text{for } 0 \leqslant n \leqslant \infty \end{cases} \tag{10}$$

with:

$$G_{ip}(M,M',t) = \sum_{n=0}^{\infty} \left[ G_{ip}^{n*} \cos(n\omega t) + G_{ip}^{n**} \sin(n\omega t) \right] \qquad (11)$$

and $\Phi_n^{**}$ by substituting $G_{ip}^{n*}$ with $G_{ip}^{n**}$ in (10).

The singularity distributions equivalent to the body (sources and dipoles or sources only) will be obtained as solutions of integral equations deduced from (10). We shall return to that after some comments on the Green functions.

The Green functions

The solutions of (9) are obtained by a Fourier transform method which naturally gives them as Fourier series of the potential for a fixed field point. The details of the calculations being omitted, $G_{1n}$ is given by:

$$G_{1n}(M,M',t) = -\frac{\cos n\,\omega t}{4\pi}\left[\frac{1}{MM'(t)} + \frac{1}{MM_1'(t)}\right]$$
$$-\sum_{p=1}^{\infty}\left[\frac{2K_p}{\pi^3}\int_0^{\pi} \cos n\sigma \cos p\sigma \int_0^{\pi/2} R_e\left[e^{K_p\zeta} E_1(K_p\zeta)\right] d\theta\, d\sigma \cos p\omega t\right.$$
$$\left. + K_p\, C_{np}(aK_p)\, e^{K_p(Z+Z_o')}\left[H_o(K_p r)\cos p\omega t - J_o(K_p r)\sin p\omega t\right]\right] \qquad (12)$$

with the following notations:

| | |
|---|---|
| $M'$ | source point. coordinates: $x', y', z' = z'_o + a \cos\omega t$ |
| $M'_1$ | symmetric point with respect to the plane $z = 0$ |
| $M$ | field point. coordinates $x, y, z$ |
| $H_o, J_o$ | Struve and Bessel functions of order 0 |
| $r$ | horizontal distance between M and M' $\quad r = [(x-x')^2+(y-y')^2]^{1/2}$ |
| $\zeta = Z + Z'_o + a \cos\sigma + i\, r \cos\theta$ | |
| $C_{np} = I_{n+p} + I_{n-p}$ | with $I_k$ modified Bessel function of order k |
| $S_{np} = -I_{n+p} + I_{n-p}$ | |
| $E_1$ | exponential integral function |
| $K_p = p^2\omega^2/g$ | wavenumber of order p |

$G_{2n}$ is obtained by inverting sin and cos, substituting $S_{np}$ to $C_{np}$ and the sign before $J_o$.

These expressions have been used for a detailed study of $G_{11}$ ([4]). They are particularly useful for the computation of the wave field, which is a superposition of regular waves of celerity $C_n = g/n\omega$. For $a/z_o = .5$, $K_1 z_o = .25$ and $t = 0$. (or an integer number of periods), we give in figure (1) a radial cut of the free surface, obtained from the five first harmonics of $G_{11}$ together with the wave field radiated by a moving singularity of same strength starting from rest, computed by a time domain simulation. The results illustrate the convergence of the transient wave field towards the periodic results issued from (12).

Solution of the integral problem.

After truncation of the Fourier series, and discretization of the body surface by plane panels, the system (10) leads to a finite linear system which is directly solved for the harmonic components of the potential on the body. The involved influence coefficients are computed from new expressions of the Green functions, giving the harmonics of the potential when both source and field points are subject to the heaving motion. These expressions involve series of the amplitude parameter $a/z_0$, the coefficients of which are pre-integrated over the panels. This feature allows a low cost investigation of the amplitude influence of the nonlinearities. The complete new formulas may be found in [3]. After solution of the linear system, the forces are computed by pressure integration, the dynamic pressure on the body being written in the following form:

$$P_d = -\rho \left[ \frac{d\Phi}{dt} - \vec{V} \cdot \overrightarrow{grad}\, \Phi + \frac{1}{2} \left| \overrightarrow{grad}\, \Phi \right|^2 \right] \qquad (13)$$

The wave field is then computed from a set of specific influence coefficients based on the expressions (12). As for the isolated source, this wave field is a superposition of regular waves of celerity $g/n\omega$, $n=1,2,\ldots$, the amplitude of the harmonics being a non-linear function of the amplitude and frequency parameters.

Results and comments

A sample of the results is given thereafter, for a sphere whose mean depth of submergence is equal to the diameter (figure 2). In figure (3), we give a plot of the hydrodynamic force exerted on the sphere, for a wave number $K_1R = \omega^2R/g = 1.0$ and for an amplitude $A/R = 0.5$. The super-harmonic components are about 15% of the total. In figure (4), a comparison is made between the super-harmonics components of the force given by our frequency-domain method, and the result of a time-domain simulation for a sphere starting from rest, with the same assumptions regarding the boundary conditions [1]. Although the steady-state has clearly not been reached in the time domain, the two results tend to a good agreement. For the same amplitude, we give in figure (5) a plot of the first four oscillating components of the force, versus the wavenumber $K_1R$. The figure (6) gives a waterfall plot of a radial cut of the free surface, over a distance of about three wavelengths and a time interval of two periods. In this case, the amplitude of the second harmonic is about the same as the first. This result may seem unrealistic, for an amplitude which is not so high. This phenomenon is nevertheless confirmed by comparison with experimental data, for the same amplitude and for various values of the wavenumber (figure 7). A sphere of radius $R = 0.105$m in a 17x9m tank has been used, and we give the signal of a wave probe located at a distance of 17R from the center of the sphere [2], together with the results of our computations. The agreement is quite good over the whole frequency range, even when the computed amplitude of the second harmonic is greater than the first ($K_1R = 0.16$). The computed characteristics of the radiated wave field are summarized in figures (8) and (9), with plots of the amplitudes of the first three superharmonic components of the wave field, and of the rate of energy radiated by the first four harmonics, relative to the total power input by the body. A strong dispersion of the energy on the wave field harmonics is observed. For $K_1R = 0.18$ and $A/R = 0.5$, only 55% of the power input by the body, which is only due to the first harmonic of the force, is recovered on the fundamental component of the wave field. These numerical and experimental results confirm the phenomenon of nonlinear dispersion of energy in the higher harmonics which was first pointed out in the study of the associated Green function [4].

## References

[1] G. Pot: *Etude Théorique et Numérique des Mouvements Libres de Corps Flottants et Immergés. Extension au Cas des Grands Mouvements de Corps Immergés.* Thèse de Docteur-Ingénieur, ENSTA, Sept. 1986.

[2] B. Dassonville: *Etude Expérimentale des Effets Non-Linéaires du Mouvement de Grande Amplitude d'un Corps Faiblement Immergé.* Thèse de Docteur-Ingénieur, ENSM, 1987.

[3] P. Ferrant: *Sphère Immergée en Mouvement de Pilonnement de Grande Amplitude.* Premières Journées de l'Hydrodynamique, Nantes, Fev. 1987.

[4] A. Clément, P. Ferrant: *Free Surface Potential of a Pulsating Singularity in Harmonic Heave Motion* 4th Int. Conf. Num. Ship Hydro. Washington, Sept. 1985.

**Acknowledgments**: Thanks are due to Institut de Recherche de la Construction Navale (PARIS) for partial financial support, and to B. Dassonville for the experimental results.

-000-

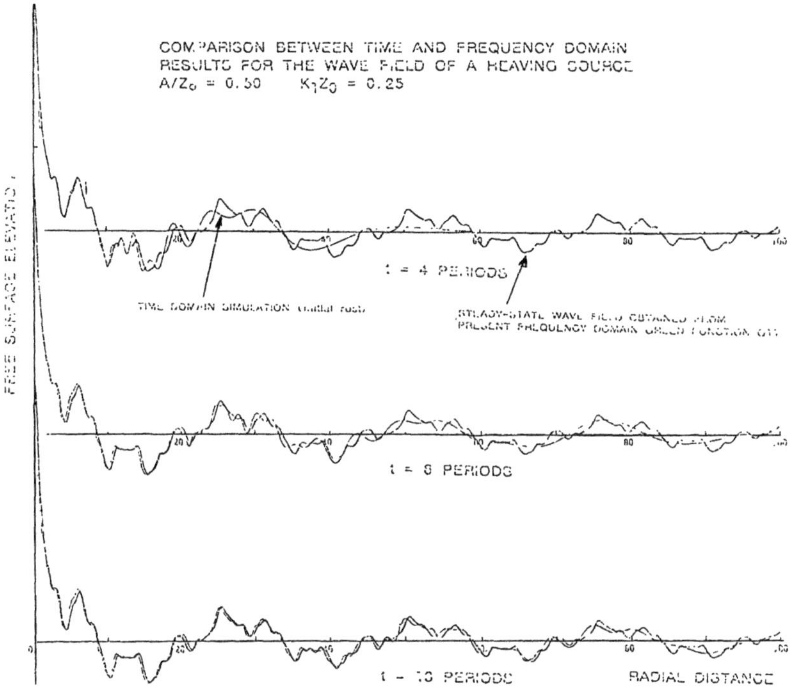


**FIGURE 1**

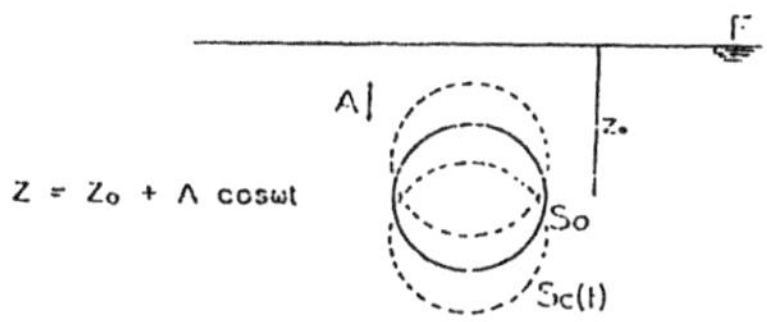


**FIGURE 2**

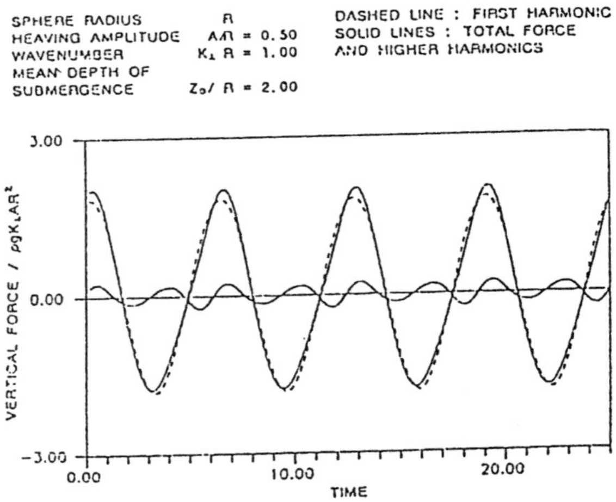


FIGURE 3

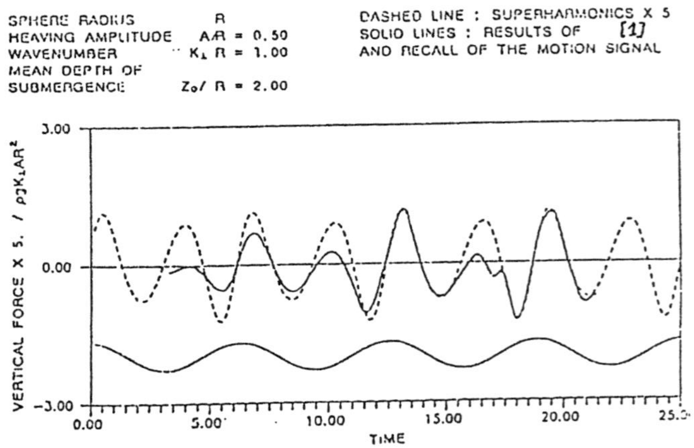


FIGURE 4

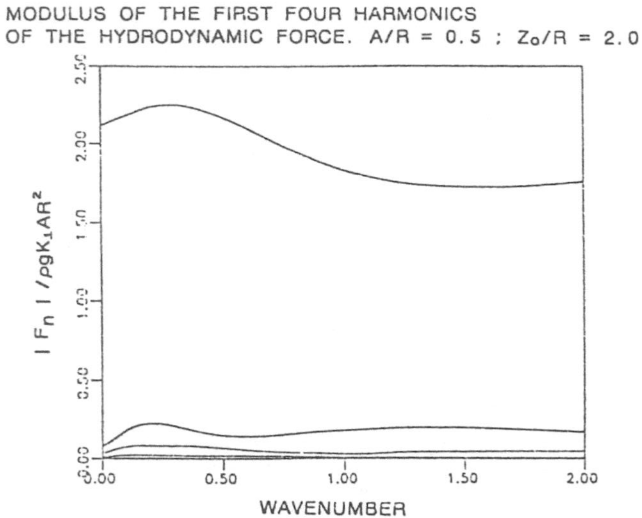


FIGURE 5

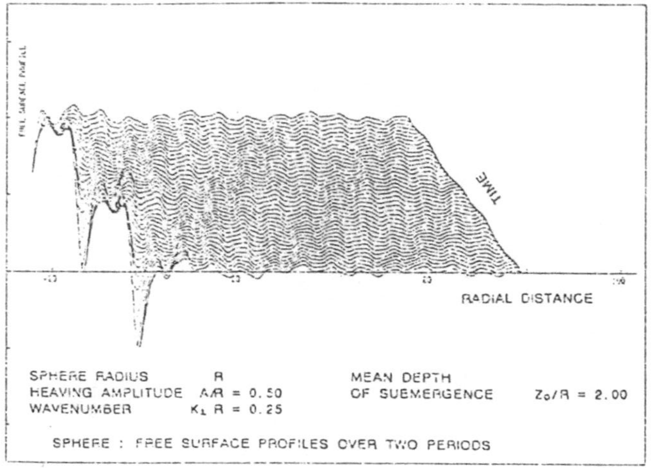


FIGURE 6

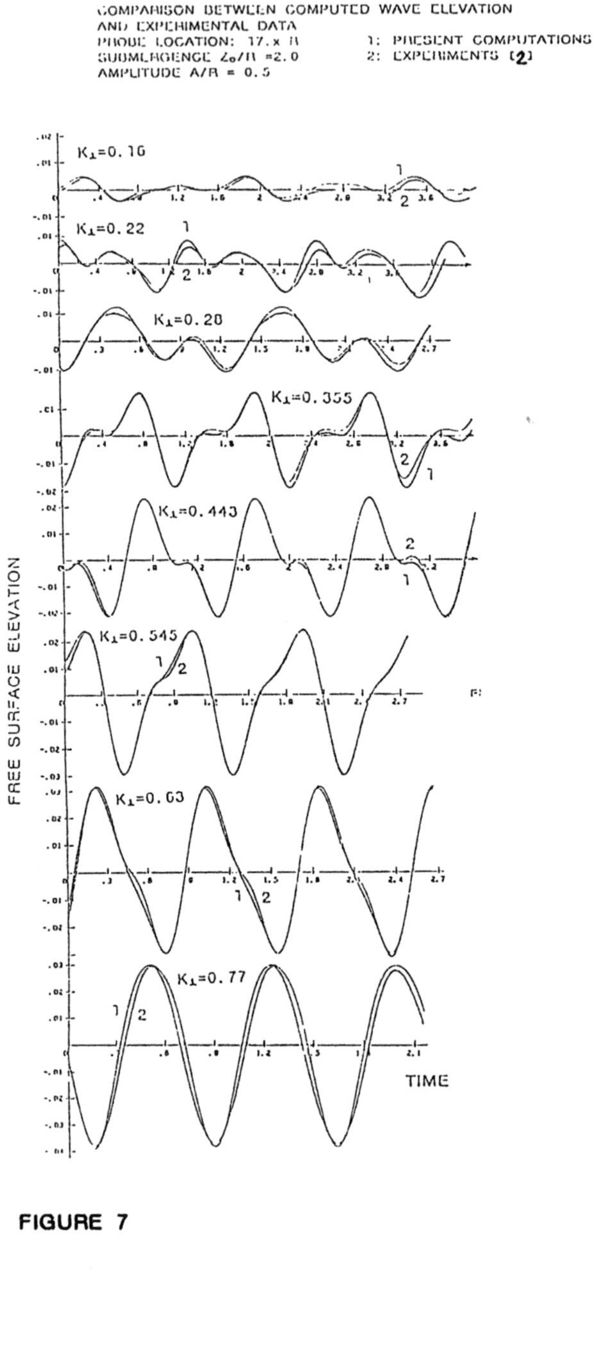
COMPARISON BETWEEN COMPUTED WAVE ELEVATION AND EXPERIMENTAL DATA
PROBE LOCATION: 17.x R
SUBMERGENCE Z₀/R =2.0
AMPLITUDE A/R = 0.5
1: PRESENT COMPUTATIONS
2: EXPERIMENTS [2]
K⊥=0.16
K⊥=0.22
K⊥=0.28
K⊥=0.355
K⊥=0.443
K⊥=0.545
K⊥=0.63
K⊥=0.77
FREE SURFACE ELEVATION
TIME

**FIGURE 7**

COMPUTED CHARACTERISTICS OF THE RADIATED WAVE FIELD VERSUS WAVENUMBER $K_{\perp}R = \omega^2 R/g$

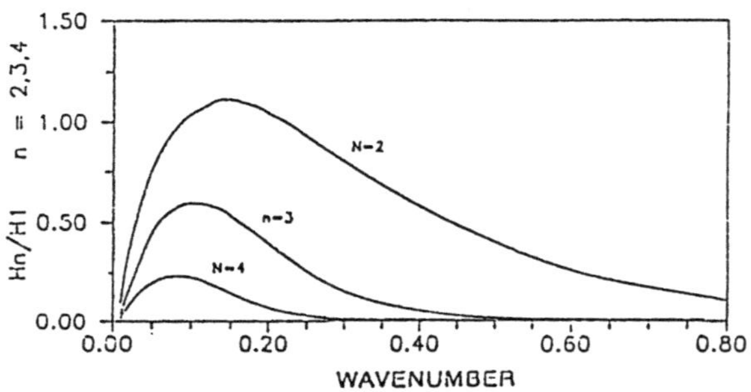


AMPLITUDE OF THE 2nd, 3rd AND 4th HARMONICS OF THE WAVE FIELD, RELATIVE TO THE AMPLITUDE OF THE 1st
$A/R = 0.5$; $Z_o/R = 2.0$

**FIGURE 8**

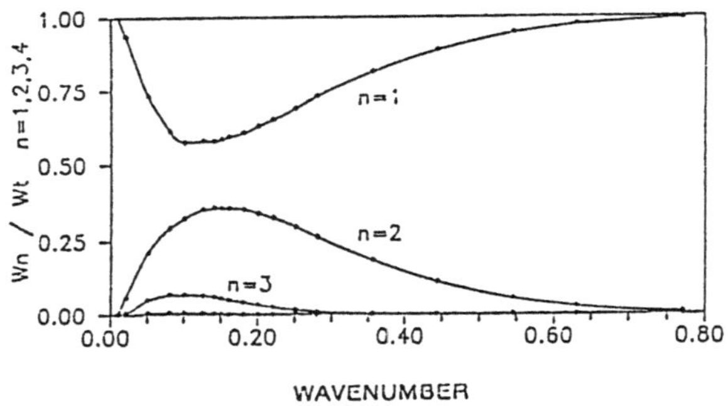


ENERGY RADIATED BY EACH OF THE FIRST FOUR COMPONENTS OF THE WAVE FIELD, RELATIVE TO THE TOTAL POWER INPUT. $A/R = 0.5$; $Z_o/R = 2.0$

**FIGURE 9**

# A New Approach to Nonlinear Waves Generated by a Body Moving Steadily at a Free Surface

V. Bertram, G. Jensen

Institut für Schiffbau
Universität Hamburg
Lämmersieth 90
2000 Hamburg 60, Germany

Summary

The steady wave resistance problem is formulated in a differential equation with non-linear boundary conditions. For non-breaking waves it is solved iteratively using a panel method. The radiation condition is enforced by a new numerical technique which is in the authors' opinion superior to the traditional four-point finite difference operator introduced by Dawson [1]. Good results are obtained for a submerged body and a Series 60 hull.

List of used symbols

| | |
|---|---|
| $E$ | error term in Dawson's free surface formulation |
| $g$ | gravity acceleration |
| $\vec{n}$ | unit normal vector |
| $x, y, z$ | coordinates fixed to the ship, $x$ points upstream, $z$ vertically downward |
| $U$ | speed of body |
| $\varepsilon$ | error in the non-linear free surface boundary condition |
| $\lambda$ | wave length |
| $\phi$ | potential |
| $\Phi$ | approximation for potential |
| $\phi_N$ | approximation for $\phi$, result of iterative step |
| $\Phi_N$ | approximation for $\phi$, result of weighted average |
| $\zeta$ | $z$-coordinate of free surface |
| $\bar{\zeta}$ | approximation for $\zeta$ |

Indices denote partial derivatives.

K. Horikawa, H. Maruo (Eds.)
Nonlinear Water Waves
IUTAM Symposium Tokyo/Japan 1987

## 1. Introduction

The steady potential flow past a body in or near the free surface of an ideal fluid is of importance for many engineering applications. It is often solved by superimposing singularities satisfying Laplace's equation, the condition of vanishing disturbance far below the free surface and a linearized condition at the undisturbed free surface. If, however, one tries to satisfy the non-linear free surface condition, we prefer to use the simple Rankine source potential, and compute the distribution of the sources so that the boundary conditions at the free surface (constant pressure, no flow across boundary and waves only in a sector downstream) are satisfied in an iteration. For this purpose a boundary condition linearizing the difference between any known approximation and the unknown potential and wave elevation is used.

A major problem is convergence of the iteration for nearly or actually breaking waves. Söding and Jensen [2] used for the two-dimensional problem an iterative scheme that was capable of handling cases quite close to breaking. They also introduced a simple, accurate technique for the numerical treatment of the radiation condition. We extended these methods to the three-dimensional problem.

## 2. Problem

We want to determine the steady, symmetrical, vortex-free potential flow of an incompressible fluid around a body at or near a free surface. The potential $\phi$ and the position of the free surface $\zeta$ are required to meet the following conditions:

$$\Delta\phi = 0 \quad \text{for } z > \zeta \tag{1}$$

$$\vec{n}\nabla\phi = 0 \qquad \text{on the wetted body surface} \tag{2}$$

$$\frac{1}{2}(\nabla\phi)^2 - g\zeta = \frac{1}{2}U^2 \quad \text{on } z = \zeta \tag{3}$$

$$\nabla\phi\nabla\zeta = \phi_z \quad \text{on } z = \zeta \tag{4}$$

(For simplification, $\zeta(x, y, z)$ with $\zeta_z = 0$ is assumed here.)

$$\lim_{x^2+y^2+z^2\to\infty} \nabla\phi = (-U, 0, 0)\ . \tag{5}$$

Far-field waves may appear only behind the body (radiation condition).

Eliminating $\zeta$ from (3) and (4) results in the free-surface condition

$$\frac{1}{2}\nabla\phi\nabla(\nabla\phi)^2 - g\phi_z = 0 \quad \text{on } z = \zeta \tag{6}$$

## 3. Linearisation

In the problem constituted by (1), (2), (5) and (6), only (6) is non-linear. It is to be satisfied iteratively by linearizing (6) around approximations $\Phi$ and $\bar{\zeta}$ of $\phi$ and $\zeta$ resp. For this purpose (6) is developed in a Taylor expansion for $\phi - \Phi$ about $\zeta$, and quadratic terms are neglected. This yields - details of the derivation can be found in [2] - the condition

$$\nabla\Phi\nabla\left[-(\nabla\Phi)^2 + \nabla\Phi\nabla\phi\right] + \frac{1}{2}\nabla\phi\nabla(\nabla\Phi)^2 - g\phi_z \tag{7}$$

$$+\left[\frac{1}{2}\nabla\Phi\nabla(\nabla\Phi)^2 - g\Phi_z\right]_z \frac{\frac{1}{2}\left[-(\nabla\Phi)^2 + 2\nabla\Phi\nabla\phi - U^2\right] - g\bar{\zeta}}{g - \nabla\Phi\nabla\Phi_z} = 0$$

on $z = \bar{\zeta}$. The denominator in the last term is zero if the vertical particle acceleration $\nabla\Phi\nabla\Phi_z$ is equal to $g$, i. e. if the approximated flow becomes unstable. Actually wave breaking occurs already at somewhat smaller accelerations.

## 4. Comparison with other linearized free-surface conditions

Using the potential of the parallel stream and the undisturbed free surface as an approximation simplifies (7) to the classical Kelvin condition.

$$U^2\phi_{xx} - g\phi_z = 0 \tag{8}$$

Dawson [1] proposed a linearisation of the free-surface condition around the approximations $\bar{\zeta} = 0$ and $\Phi$ = double-body potential. He introduced the index $l$ to denote a derivative in the streamline direction of the double-body flow. Using this notation and approximating $\nabla(\Phi_l)$ by $(\nabla\Phi)_l$ (7) transforms to

$$\Phi_l\left(-\Phi_l^2 + \Phi_l\phi_l\right)_l + \Phi_{ll}\Phi_l\phi_l - g\phi_z - E = 0, \tag{9}$$

with

$$E = \frac{1}{2}\Phi_{zz}\left(-(\nabla\Phi)^2 + 2\nabla\Phi\nabla\phi - U^2\right) \quad .$$

This expression still differs from Dawson's condition

$$\left(\Phi_l^2\phi_l\right)_l - g\phi_z = 2\Phi_l^2\Phi_{ll}. \tag{10}$$

by the Term $E$.

5. Numerical solution

The potential $\phi$ is approximated by the potential of the parallel stream plus simple point sources above a local part of the free surface plus triangular source panels with linearly varying strength [3] distributed within the body. The unknown source strengths are determined from linearized boundary conditions at collocation points on the free surface and on the body.

Also the radiation condition has to be treated numerically. For this purpose Dawson [1] introduced a method using a special four point upstream finite difference operator along the double body streamlines. He determined the operator from numerical experiments for a two-dimensional free surface flow, mainly from the condition that the computed wave amplitude should be almost constant over several wave lengths at some distance behind the disturbance. This method can only be used with a formulation in terms of numerical, streamwise derivatives. Drawbacks of this method are that it requires streamline tracing and produces wave lengths which are too short by about 5 % for usual discretizations.

Another method for satisfying the radiation condition was first described in [2] for the two-dimensional case. Söding and Jensen called it a backward shifting of source points above the free surface relative to the collocation points. We extended this method to the three-dimensional case. As in the two-dimensional case trial computations were performed using the linearized free surface condition (8), simple point sources above the free surface and a submerged dipole as the disturbance. For an equidistant mesh the sources were shifted downstream by an amount $\Delta\xi$ and also in direction of the y-axis by an amount increasing linearly from zero at $y = 0$ to $\Delta\eta$ (Fig.1). As in [2] different values for $\Delta\xi$ and $\Delta\eta$ were tried to find maxima for the determinant modulus of the system of equations. They were found at $\Delta\xi = \Delta\eta = 0$ and at $\Delta\xi = \Delta\Xi$ and $\Delta\eta = 0.25\Delta H$, where $\Delta\Xi$ and $\Delta H$ are the spacing of the mesh in $x$ and $y$ direction. However, the maximum in y-direction was extremely shallow. Using the first maximum waves are produced both upstream and downstream, while the latter satisfies the radiation condition with transverse wave lengths very exactly equal to $2\pi U^2/g$ and the disturbance of the free surface is limited to the Kelvin angle. For an equidistant mesh shifting the sources by $\Delta\xi = \Delta\Xi$ is equivalent to adding a row of collocation points at the front end of the source mesh and deleting one at the back end. This is what we do in the case of local mesh refinements.

## 6. Numerical results

Equation (7) is used iteratively to obtain a solution for the flow with the non-linear boundary condition (6). For $\Phi$ we begin with $-Ux$. In the following iteration cycles the approximation $\phi_N$ of the previous cycle is used as $\Phi$. $\zeta$ is determined from (3) applied to $\Phi$. For moderate wave steepness, this scheme converges nicely. For steeper waves, however, the changes of $\phi$ computed from (7) are excessive. Therefore, a weighted average $\Phi_N$ between $\Phi$ and $\phi_N$ is better used as approximate potential in the next iteration cycle. In order to determine the weighing factor the error of the exact boundary condition (6) is defined:

$$\varepsilon(\phi) = \max\left|(\frac{1}{2}\nabla\phi\nabla(\nabla\phi)^2 - g\phi_z)\right| \tag{11}$$

If in one iteration step the error $\varepsilon$ increases, the weighted average $\Phi_N$ is used instead of $\phi_N$:

$$\Phi_N = \frac{\varepsilon(\Phi)\phi_N + \varepsilon(\phi_N)\Phi}{\varepsilon(\Phi) + \varepsilon(\phi_N)} \tag{12}$$

If necessary this weighing process is repeated.

## 7. Results

For the simple submerged test body shown in fig.2 the error (11) was reduced by a factor of 45 within 8 iteration steps. The maximum vertical component of the particle acceleration was about 0.65 g, the maximum wave steepness $\lambda_o/\zeta_{max} = 15.4$ with $\lambda_o$ $2\pi U^2/g$. Figure 3 shows a perspective view of the free surface for the final non-linear solution. For clarity the wave elevation is enlarged by factor 3. Figure 4 shows contour wave resistance coefficient (figs. 5 and 6). The error (11) was reduced by several orders half the final solution. Dotted lines are crests, solid lines troughs. Comparing the two solutions it can be seen that as expected the crests are higher, the troughs longer and the wave length shorter in the non-linear solution.

For a surface piercing body the iteration is far more difficult because there are almost always parts close to the bow where wave breaking occurs even at rather low speeds. Even for the Kelvin condition (8) we get strong oscillations due to the large velocity gradients near stagnation points. To avoid this we place the innermost line of collocation points a small distance to the side of the water line. Thus these local effects do not disturb the overall solution. Preliminary computations for a Series 60 ship with block coefficient 0.6 show satisfactory non-linear results for the wave elevation and the wave resistance coefficient (fig.5 and 6). The error (11) was reduced by several orders of magnitude. The solution depends, however, on the gap between the hull and the innermost line of collocation points, which has to be chosen from experience.

References

1. Dawson, C.W.
A Practical Computer Method for Solving Ship-Wave Problems
Second International Conference on Naval Hydrodynamics, Berkeley, 1977
2. Jensen,G., Z.-X. Mi, H. Söding
Rankine Source Methods for Numerical Solutions of the Steady Wave Resistance Problem
16th Symposium on Naval Hydrodynamics, Berkeley, 1986
3. Webster, W.C.
The Flow about Arbitrary Three-Dimensional Smooth Bodies
Journal of Ship Research Vol. 19, No. 4, 1975
4. Proceedings of the Workshop on Ship Wave-Resistance Computations, Vol.1
DTNSRDC, Bethesda, 1979

Figures

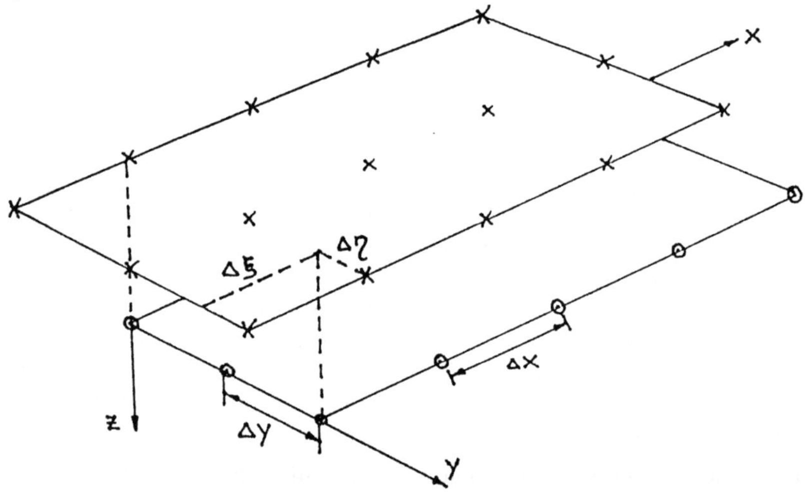


Fig. 1: Position of sources (×) relative to collocation points (⊙)

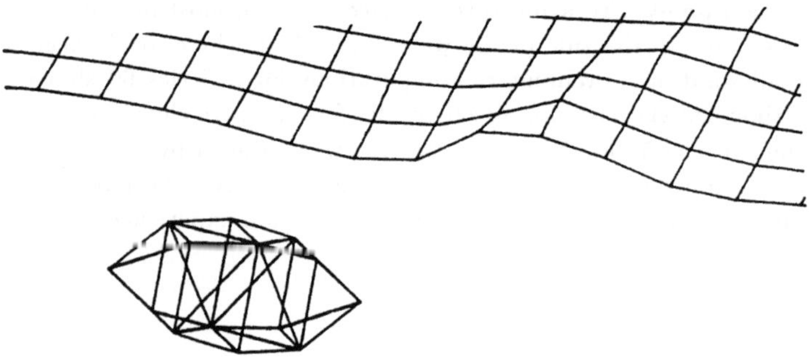

Fig. 2: Submerged test body and part of the free surface mesh

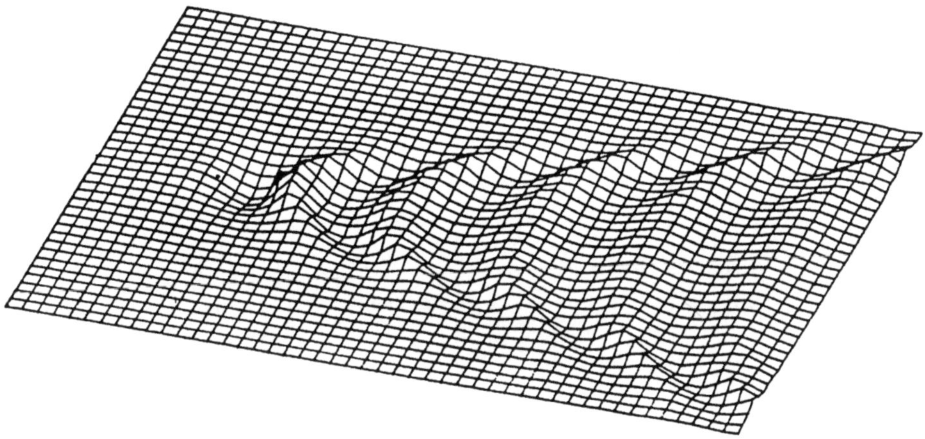

Fig. 3: Wave elevation for submerged body

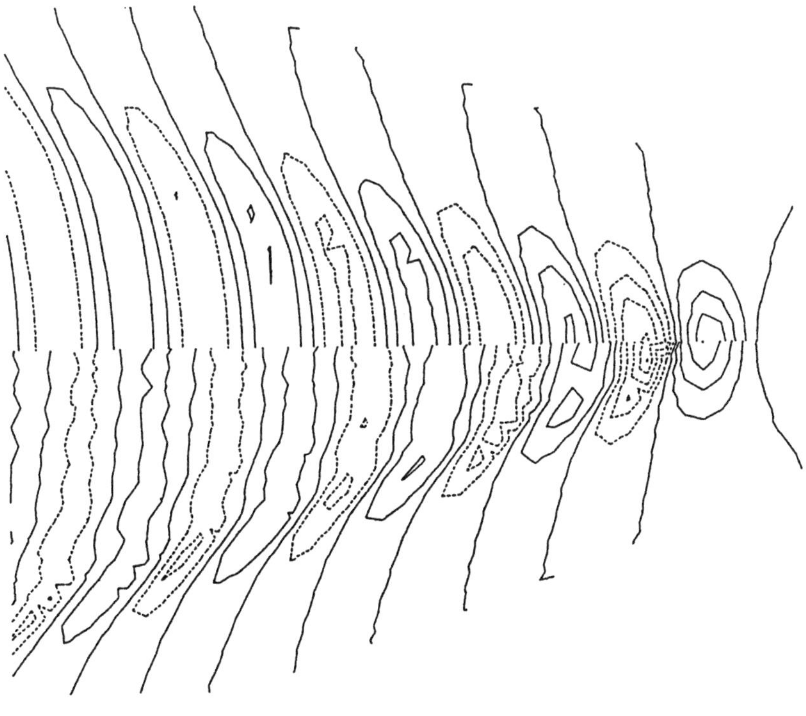

Fig. 4: Contour lines for waves generated by submerged body
(upper half linearized solution, lower half non-linear solution)

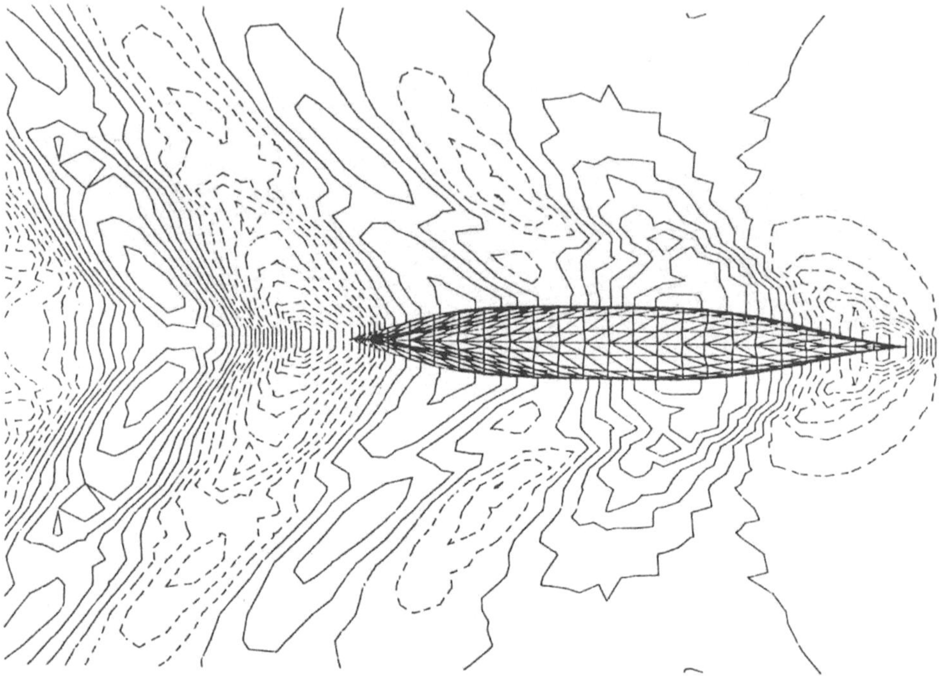

Fig. 5: Contour lines of wave elevation for Series 60 ship at $F_n = 0.3$ (dotted lines are crests, solid lines troughs)

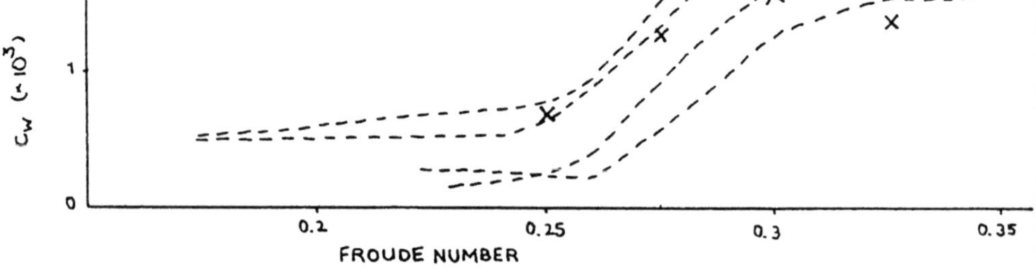


Fig. 6: Wave resistance coefficients for Series 60 ship

# Extreme Wave Forces Acting on a Floating Structure

S. NAITO , K. TAKAGI and S. NAKAMURA

Dept. of Naval Architecture
Osaka University, Japan

Summary

Extreme wave forces acting on a floating structure are calculated in the following three cases. The first case is wave forces acting on a column with lowerhull. The second one is wave forces acting on a moving column. The last one is wave forces acting on a fixed column at the condition in which the concentrated impulsive wave and the large swell coexist. The velocity potential of extreme wave is calculated by BEM and the forces acting on a column and a lower-hull are estimated by Morison's formula. Calculated results are compared with experimental ones in the first and the second cases. Effects of the existence of swell on the extreme wave forces are shown in the last.

## 1. Intoroduction

Usually semi-submersible oil rig is situated at rough sea, where sometimes in winter season, large amplitude waves are generated and attack the oil rig. It is not so easy to generate similar waves in experimental tank. Meanwhile impulsive wave can be generated easier , so we use it to simulate the wave condition of rough sea. Impulsive wave has strong nonlinearity near the free surface like the extreme large amplitude waves. The orbital velocity of large amplitude waves near the free surface has different characters compared with that of small amplitude waves defined by linear theory. It is necessary for the estimation of impulsive wave forces to use nonlinear wave theory.
A semi-sub consists of three parts, talking loosely, deck, column and lowerhull. The column is damageable by waves because it passes through the free surface. The lowerhull also is damageable at shallow draft condition. So it is important to estimate the wave forces acting on a column with lowerhull in severe wave condition.

K. Horikawa, H. Maruo (Eds.)
Nonlinear Water Waves
IUTAM Symposium Tokyo/Japan 1987

In the previous paper [1] authors showed one method to calculate the wave forces acting on a column only. In that case the column without a lowerhull did not move. The agreement between theoretical results and experimental ones was good. The velocity of wave orbit at the center of column was used to estimate the force acting on the column. Making use of the same hypotheses, wave forces are calculated in the following three cases. The first case is the wave force acting on a column with lowerhull. The second one is the wave force acting on a moving column. The last one is the wave force acting on a fixed column at the condition in which the concentrated wave and the large swell coexist.

## 2.Wave forces acting on a column with lowerhull

The velocity potential of concentrated wave is calculated by Boundary Element Method (BEM) [2] [3] and the forces acting on a column and a lowerhull are estimated by Morison's formula. The calculation method is described in detail in the previous paper [1]. Figure 1 and Fig.2 show the time histories of forces acting on a column with lowerhull. Dimensions of the model are shown in Fig.3. When a lowerhull has deep draft (Fig.1), the agreement between the calculated force and the measured one is good. On the contrary, when a lowerhull is near the free surface (Fig.2), since Morison's hypothesis for a lowerhull is not valid, the agreement between them is poor.

## 3.Wave forces acting on a moving column

In case of a column with surging motion, the computational method of wave forces is the same as that of the fixed column, provided that the relative velocity between column and wave orbit is used. The surging motion of column is expressed by

$$x(t) = A\cos\{\,\omega(t-t_0) + \alpha\,\} \qquad (1)$$

Where, A is the amplitude of surging motion, $\omega$ is the angular frequency and $\alpha$ is the phase difference between column and impulsive wave. It is defined that the maximum force is obtained when $t=t_0$. A typical time history of impact wave force acting on a moving column is shown in Fig.4. It is defined that impact wave force is obtained after subtracting the sinusoidal force, which is caused by forced oscillation test in still water, from the measured force. Comparison of maximum forces related to $\alpha$ between calculation and measurement is shown in Fig.4. Agreements between them are fairly good. The maximum force on the moving column is 1.6 times larger than that on the fixed column.

## 4.Wave forces acting on a fixed column in coexisting of large swell and impulsive wave

Properties of the concentrated impulsive wave are changed by the existence of large swell in actual sea. Therefore, wave profiles and wave forces are calculated in the following three cases.

(a) The convergent point of the impulsive wave is chosen at a crest of swell.

(b) The swell does not exist.

(c) The convergent point of the impulsive wave is chosen at a trough of swell.

On the open boundary of the wave field, it is assumed that the velocity potentials are shown as the following equations

$$\Phi = \Phi_0 + \Phi_1 \tag{2}$$

$$\Phi_0 = \sum_n \sqrt{2S(\omega_n)\Delta\omega}\ \frac{g \cosh K_n(y+h)}{\omega_n \cosh(K_n h)} \sin\{K_n(x-x_0)-\omega_n(t-t_0)\} \tag{3}$$

$$\Phi_1 = \frac{\delta}{2} H_I \frac{g \cosh K_0(y+h)}{\omega_0 \cosh(K_0 h)} \sin\{K_0(x-x_0)-\omega_0(t-t_0)\} \tag{4}$$

where $H_I = \sum_n \sqrt{2S(\omega_n)\Delta\omega}$

$$K_n \tanh(K_n h) = \frac{\omega_n^2}{g}$$

$\Phi$ : Total velocity potential

$\Phi_0$ : Velocity potential of the impulsive wave

$\Phi_1$ : Velocity potential of the swell

$S(\omega)$ : Spectrum of the impulsive wave

$h$ : Water depth

$x_0, t_0$ : Convergent point and time

$$\delta = \begin{cases} 1 & \text{case (a)} \\ 0 & \text{case (b)} \\ -1 & \text{case (c)} \end{cases} \quad \text{mentioned above}$$

It is chosen that $H_I = 0.07m$, $h = 0.5m$ and the wavelength of swell is 3.5m in model scale.
Calculated wave profiles are shown in Fig.6. Calculated time histories of impact wave forces are shown in Fig.7. The existence of swell strongly affects the maximum value of impact forces.
The propagative direction of impulsive wave is the same as that of swell in the previous cases. Then, other cases in which the direction of impulsive wave is opposite to that of swell are calculated. The convergent point of the impulsive wave is chosen at a crest of the swell in the case (d). And, it is chosen at a trough of the swell in the case (e). Calculated wave profiles in these cases are shown in Fig.8. Calculated time histories of impact wave forces are shown in Fig.9. The magnitude of impact wave forces depends on the steepness of wave and the velocity of fluid particle. In the case (d), the steepness of wave is large but the velocity of fluid particle is small. On the other hand, in the case (e), the steepness of wave is small but the velocity of fluid particle is large. Therefore, in these cases, effects of the swell are small when these are compared with the cases (a) and (c).

References

1. Takagi, K., Naito, S. and Nakamura, S. : The Impact Load and Pressure Acting on the Vertical Circular Cylinder. Fifth International Offshore Mechanics and Arctic Engineering Symposium (1986) Vol.1 p.242.

2. Vinje, T. and Brevig, P. : Numerical Calculations of Forces from Breaking Waves. International Symposium on Hydrodynamics in Ocean Engineering, The Norwegian Institute of Technology (1981) p.547.

3. Takagi, K., Naito, S. and Nakamura, S. : Computation of Nonlinear Hydrodynamic Forces on Two-Dimensional Body by Boundary Element Method. Journal of the Kansai Society of Naval Architects, Japan (1985) No.197 p.31.

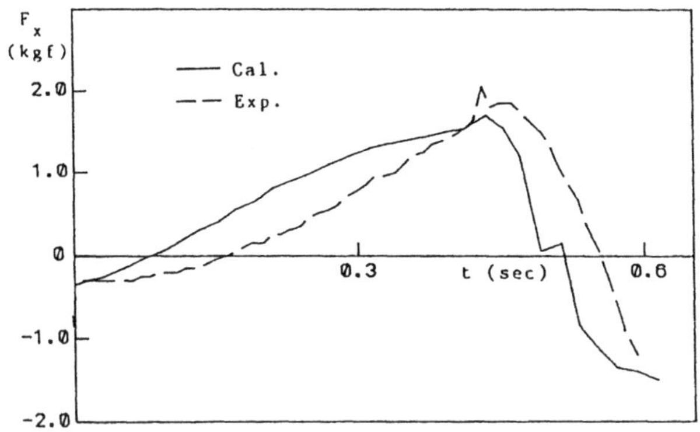


Fig.1. Comparison of wave forces acting on a column with a lowerhull between calculation and experiment. (Deep draft)

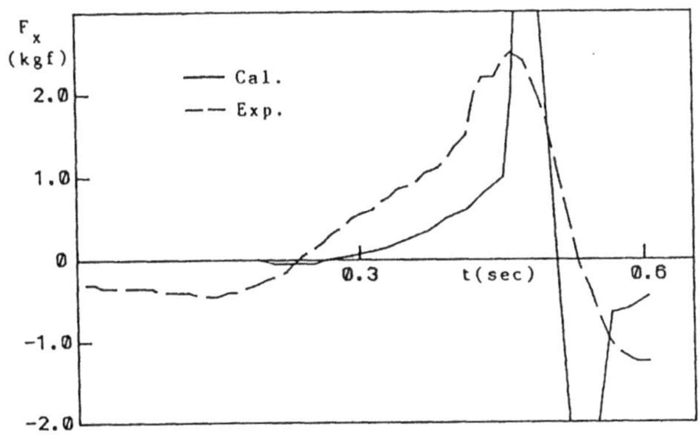


Fig.2. Comparison of wave forces acting on a column with a lowerhull between calculation and experiment. (Shallow draft)

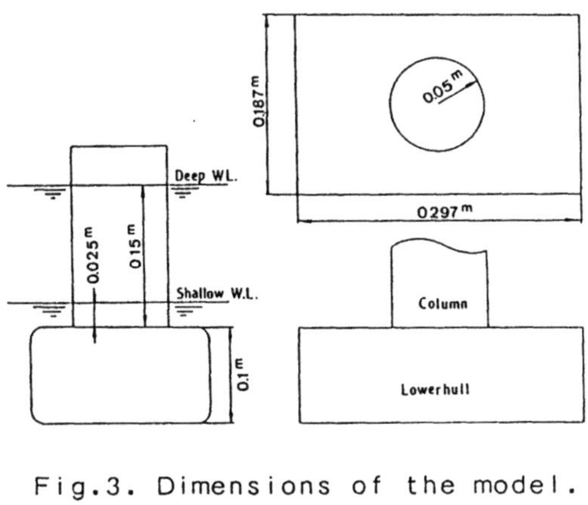


Fig.3. Dimensions of the model.

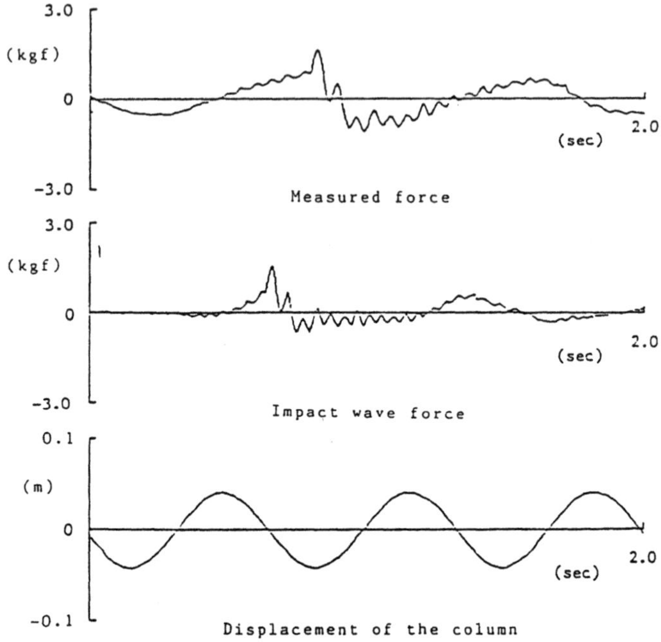


Fig.4. Typical time histories of impact wave forces acting on a moving column.

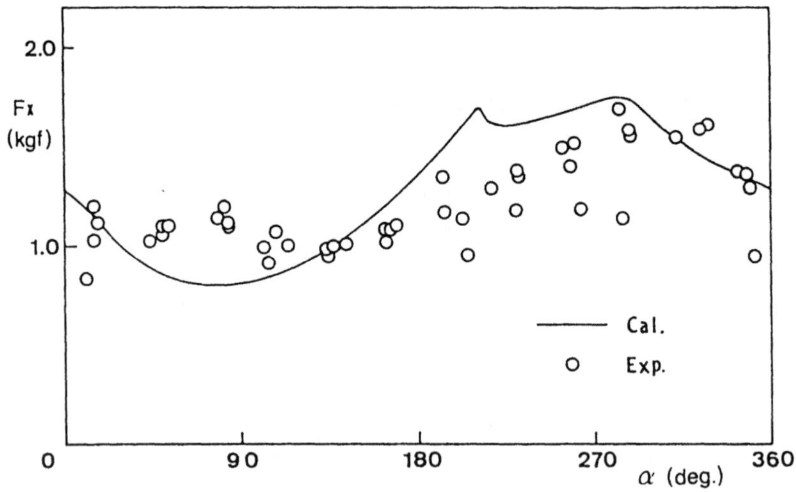


Fig.5. Comparison of the peak force between calculation and experiment.

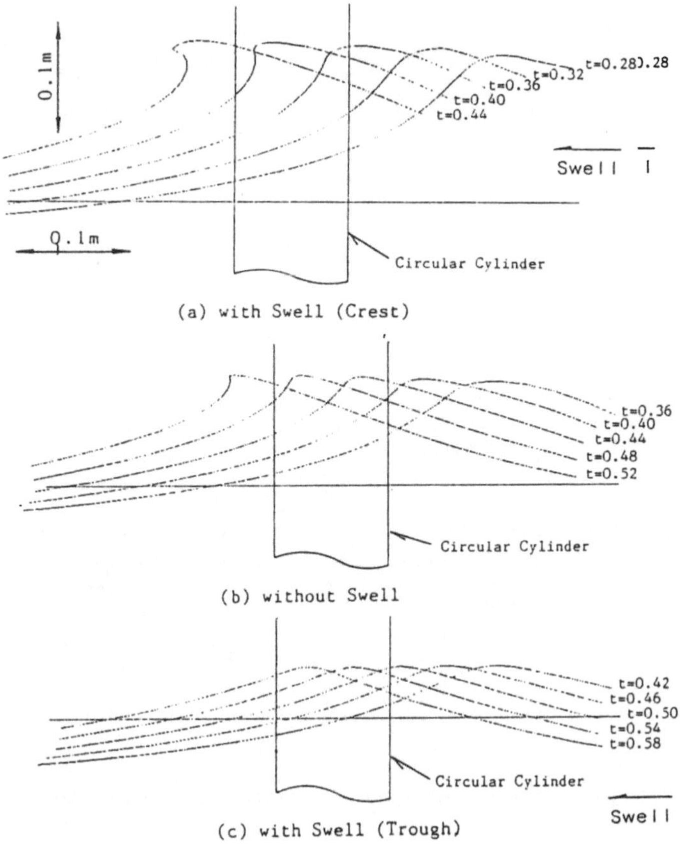


Fig.6. Calculated results of the concentrated wave profile.

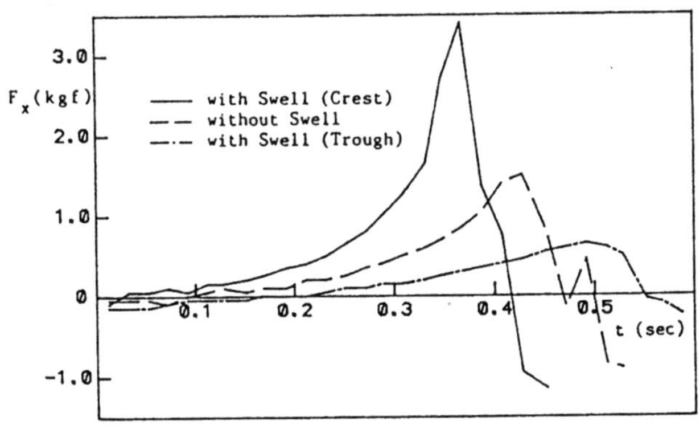


Fig.7. Calculated time histories of impact wave forces acting on a column in swell and impulsive wave.

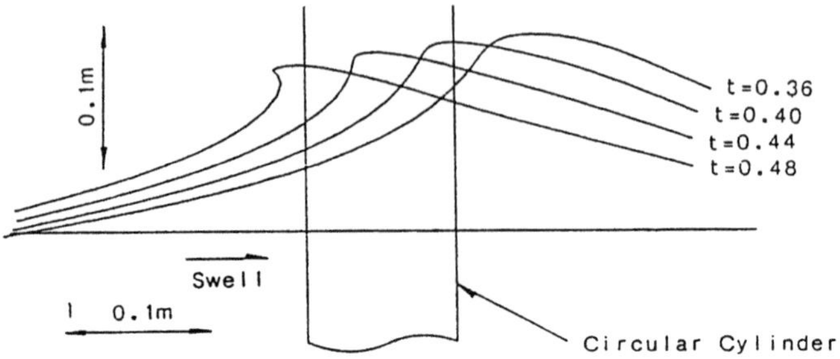


(d) Crest

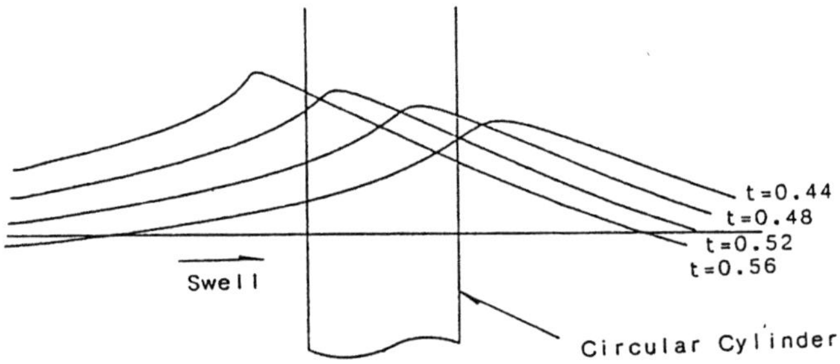


(e) Trough

Fig.8. Calculated results of the concentrated wave profile.

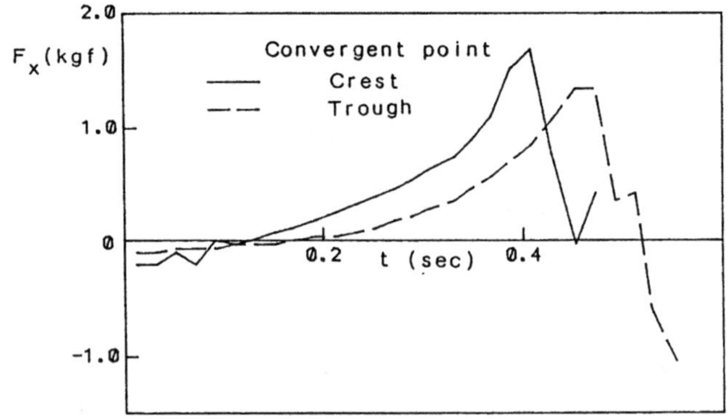


Fig.9. Calculated time histories of impact wave forces acting on a column in swell and impulsive wave.

# Large Slow Drift of a Ship in Slightly Modulated Beam Seas

H. S. CHOI and D. J. KIM

Department of Naval Architecture
Seoul National University, Korea

Summary

The slow sway motion of a moored fore-and-aft symmetric ship in slightly modulated beam seas is discussed. The slow motion is associated with long waves. Depending on the speed of propagation, they may be decomposed into the locked, free and radiated long waves. Numerical examples are given for a Series 60 ($C_B$= 0.6) ship in some different wave packets. Because of the restoring force developed by the mooring line, resonance may occur and the magnitude of the slow sway can be large. Real fluid effects are not included.

## Introduction

A moored vessel experiences slow drift motions on the horizontal plane as well as fast oscillations. In seas with narrow-banded spectra waves from neighboring frequencies interact due to the inherent nonlinearity on the free surface. If the low frequency excitation is caused by difference-frequency waves, the frequency can be as low as the natural frequency of the moored vessel. In consequence the magnitude of the drift motion can be large and it may cause significant strains in mooring cables.

An empirical method of calculating the slow drift motion was proposed and used widely in practice [1]. In irregular waves the slow drift force can be expressed in terms of quadratic transfer function of wave components. Newman [2] suggests that for small frequency differences the coefficients of the function can be approximated by those at zero difference, i.e the steady time-mean values. In many other papers, the slow motion is found as a part of the complete and complicated second order theory, for example see Pinkster [3]. Triantafyllou [4] considered the large

K. Horikawa, H. Maruo (Eds.)
Nonlinear Water Waves
IUTAM Symposium Tokyo/Japan 1987

slow motion of a vessel by introducing multiple scale only in time. The applicability of all the approaches above is limited to deep water problems.

While in water of finite depth sea bottom topography affects the formation and propagation of waves. The contrasting temporal and spatial scales involved in the wave propagation suggest the use of the perturbation method of multiple scales in both time and space. Steps have been made in this direction by Agnon and Mei [5]. They studied a sliding block in beam seas and clearly showed that the slow motion is associated with long waves. This work was extended to a floating long cylinder, in which the transient drift motion is solved analytically without solving any second-order potential explicitly [6]. They extensively examined the effects of the blockage coefficient and the mooring stiffness, and clarified the condition under which the slow sway can be large.

In this paper we apply this technique to a weakly moored fore-and-aft symmetric ship in slightly modulated beam seas. Only the slow sway is considered and real fluid effects are not included. Numerical examples are given for a Series 60 ($C_B$=0.60) in some different wave packets.

## Assumptions and Formulations

For simplicity we choose that the modulation ratio has the same order of magnitude as the steepness of carrier waves, $\Omega/\omega=O(kA)$ and $\varepsilon=kA=o(1)$, where k, $\omega$ and A denote the wave number, frequency and amplitude of carrier waves in this order. It implies that the nonlinearity and dispersion of waves are supposed to be comparable each other. The water depth is finite, kh=O(1). Since it belongs to the most interesting class of problem, the ship's draft is assumed to be comparable to the water depth so that the blockage coefficients have values in the range of C/h=O(1). The ship's motion shall be constrained by a single-point mooring, of which stiffness is weak compared to the ship's inertia, $K/\omega^2M=o(\varepsilon^2)$, where M is the ship's mass. In the light that the wave envelope has typical temporal scales of $O(\varepsilon\omega)^{-1}$ and spatial scales of $O(\varepsilon k)^{-1}$, we define the slow coordinates $(y_1,t_1)=\varepsilon(y,t)$ in order to separate the slow motion from the fast ones.

Under the usual assumptions of potential theory the fluid motion is described by the gradient of a velocity potential, which is the solution of the Laplace equation satisfying necessary boundary conditions. With the reference frame of moving coordinate system which follows the slow sway, $Y_o(t_1)$, the potential and the surface elevation are expanded asymptotically and decomposed into harmonics with respect to time t as like

$$\phi=\varepsilon[\phi_{10}+(\phi_{11}e^{-i\omega t}+*)]+o(\varepsilon^2) \tag{1}$$

where * means complex conjugate.

Formally the problem for $\phi_{11}$ is identical to the well-known boundary-value formulation of the linear, time harmonic problem of a ship without forward speed in regular beam waves. Several solution algorithms are available and the procedure is routine. Taking terms up to second order, the lateral drift force is given by [7]

$$\bar{F}y(y_1,t_1)=[\frac{\rho k^2}{8\pi}(2\frac{Cg}{Cp}\tanh kh)^{-1}\int_0^{2\pi}|H(\theta)|^2\sin\theta d\theta + \frac{1}{2}\rho\omega\coth kh\,\mathrm{Im}\{H(\tfrac{3}{2}\pi)\}]\,|A(t_1-y_1/Cg)|^2 \tag{2}$$

$$H(\theta)=\iint_{Sm}(\partial\phi_{11}/\partial n-\phi_{11}\,\partial/\partial n)\,(\cosh k\zeta+\sinh k\zeta\tanh kh) \times e^{ik(\xi\cos\theta+\eta\sin\theta)}ds \tag{3}$$

where Cg and Cp denote the group and phase velocities of carrier waves, respectively.

## Slow Sway Motion

Since the second-order potentials are bounded, it is assumed that the flow associated with slow motions is reflected by the leading-order slow potential, $\phi_{10}$. The flow has a quite different physics in the fluid region far from the ship compared to that near to it. It is therefore relevant to treat the problem separately in each fluid region and then to match two solutions in the sense of matched asymptotic expansions. The wave equation in the far field is readily derived as follows [5,6] :

$$\phi_{10\,t_1 t_1} - gh\phi_{10\,y_1 y_1} = g(\nabla\phi_{11}\zeta_{11}{}^* + *)_{y_1} -$$

$$-[\nabla\phi_{11}\cdot\nabla\phi_{11}{}^* - \sigma(\phi_{11}\phi_{11z}^* + *)]_{t_1} \qquad z=o \qquad (4)$$

with $\sigma = \omega^2/g$

The inhomogeneous solution corresponds to group-locked waves propagating at the group velocity of carrier waves, while the homogeneous solution represents free waves propagating at the celerity of shallow water waves. In addition to these, long waves are radiated away from the ship caused by the slow sway.

In the near field the second-order boundary condition on the ship's surface yields to the equation of the slow sway

$$MY_{o\,t_1 t_1} + K_2 Y_o = \rho\int \Delta\phi_{10\,t_1}\, dS + \bar{F}_y \qquad (y_1=o) \qquad (5)$$

where $\varepsilon^2 K_2 = K$ and $\Delta\phi_{10}$ stands for the potential jump between the lee and weather sides.

The jump in potential is related to pressure difference, which compels fluid going around the ship instead of passing directly underneath the keel. For a slender ship the potential jump is the solution of an integral equation of the form [8]

$$-\frac{i}{4}\int_{-L/2}^{L/2}\Delta\phi_{10}(\xi)\, Ho^{(1)}(k_1|x-\xi|)\,d\xi = -k_1^{-2} - k_1^{-1}\int_o^x \frac{\Delta\phi_{10}(\xi)}{2C(\xi)}\cdot$$

$$\cdot \sin k_1(x-\xi)\,d\xi + A\cos k_1 x + B\sin k_1 x \qquad (6)$$

where $Ho^{(1)}$ is the Hankel function of first kind and

the integral constants A and B are ultimately determined by forcing the potential jump to vanish at the ends $x=\pm L/2$.

The blockage coefficient can be easily calculated by a simple algorithm, for example see Taylor [9]. Once the potential jump is found it is straightforward to solve the equation of slow sway.

Numerical Results

As a numerical example we first consider a Series 60 ($C_B=0.60$) in a pure sinusoidal modulation. The dimensions of the ship are L/B=7.50, B/T=2.50. In this particular case the equation of slow sway takes simply the form of linear one.

If we choose H/h=0.5 and $kL=4\pi$, the lateral drift force reduces to $\bar{F}y=0.0584\ \rho g B^2 L (kA)^2$. In Figure 1 the normalized slow sway amplitudes are illustrated versus the modulation ratio for three different values of mooring stiffness. Because of the restoring force developed by the mooring line, resonance may occur. However, the peak value at resonance is of qualitative one only, as flow separation must be more important.

The second example is the transient response of the slow sway under the action of a wave packet

$$A=H(t_1)\ H(\pi/\Omega_1-t_1)\ A_o \sin \Omega_1 t_1 \quad (\varepsilon\Omega_1=\Omega) \quad o\le t_1\le \pi/\Omega_1 \qquad (7)$$

where H is Heaviside's step function.

The analytical solution is to be found in [6]. Figure 2 shows the normalized transient sway displacement versus the dimensionless time $(\Omega/\pi)t_1$ for $K_2=0.5\rho ghL$ and $\Omega_1/\omega=1.0$. The transient motion decays faster than that for a two-dimensional floating cylinder.

It is clearly shown that the slow sway motion of a ship in slightly modulated beam seas can be analysed by using the perturbation technique of multiple scales in both time and space, exactly in the same manner as for a two-dimensional floating cylinder, without solving any second-order potential explicitly. When the mooring stiffness is weak compared to the body inertia, resonance may occur and the slow sway motion can in principle be large.

We thank the financial support by the Korea Science and Engineering Foundation under the contract number 871-0316-001-2.

References

1. Hsu, F.H. ; Blenkarn, K.A. : Analysis of peak mooring forces caused by slow vessel drift oscillations in random waves. Proc. 2nd Offshore Tech. Conf., Houston, Vol.1, (1970)

2. Newman, J.N. : Second order, slowly varying forces on vessels in irregular waves. Proc. Intern, Symp. Dynamics Marine Vehicles and Offshore Structures in Waves. London (1974) 182-186.

3. Pinkster, J.A. : Low frequency second order wave exciting forces on floating structures. NSMB publ. No.650, 1976.

4. Triantafyllou, M.S. : A consistent hydrodynamic theory for moored and positioned vessel. J. Ship Res. 26 (1982) 97-105.

5. Agnon, Y. ; Mei, C.C. : Slow drift motion of a two dimensional block in beam seas. J. Fluid Mech. 151 (1985) 279-294.

6. Agnon, Y. ; Choi, H.S. ; Mei, C.C. : Slow drift of a floating cylinder in narrow-banded beam seas. (submitted to J. Fluid Mech. for possible publication).

7. Choi. H.S. ; The drift force and moment on a floating body in water of finite depth. (in Korean) J. Soc. Naval Arch. Korea 24 (1987) 9-16.

8. Tuck. E.O. : Ship motions in shallow water. J. Ship Res. 14 (1970) 317-328.

9. Taylor, P.J. : The blockage coefficient for flow about an arbitrary body immersed in a channel. J. Ship Res. 17 (1973) 97-105.

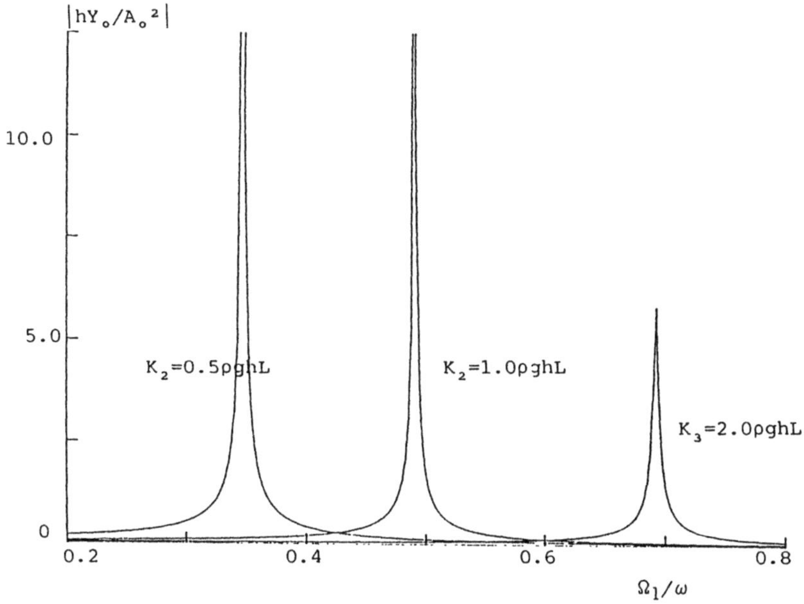


Figure 1. Normalized amplitudes of the slow sway of a ship (Series 60, $C_B$=0.6) versus the modulation ratio in a sinusoidal envelope for 3 different mooring stiffnesses

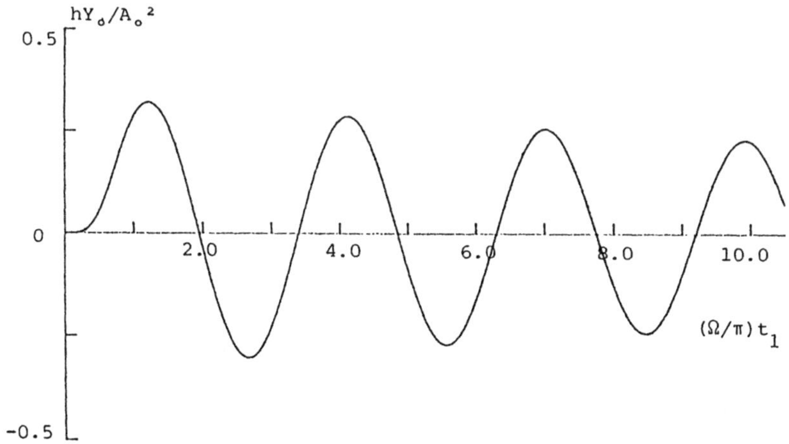


Figure 2. Transient slow sway of a ship (Series 60, $C_B$=0.6) in a wave packet

# The Statistics of Slow Drift Oscillations with Nonlinear Restoring Forces

TOM MARTHINSEN

Saga Petroleum a.s., Oslo, Norway.

## 1. INTRODUCTION

The theory for slowdrift oscillations has received quite some attention in the literature, reflecting the fact that the problem is rather complicated. This is due to several nonlinearities in the mathematical formulation. The oscillations are excited by slowly varying wind forces which are normally taken to be Gaussian and wave drift forces which are nonlinear and non-Gaussian wave forces. In the present study we are concerned with the statistics of slowdrift oscillations of a moored floating structure. We shall focus on the nonlinearities and we consider the wave drift forces as the only exciting sources. It is believed that the excitation due to wind can be included in the present theory at a later stage without major difficulties.

The equation of motion for the slowly varying motion of a moored floating structure in one degree of freedom may be written

$$m\ddot{X} + c(X,\dot{X}) + k(X) = F(t) \tag{1}$$

Here $X$ denotes one of the horizontal motion components, e. g. the surge motion, $m$ is the total mass, $c(X,\dot{X})$ represent a damping term usually due to viscous drag and $k(X)$ the restoring force from the mooring system. For simplicity we have assumed that the different degrees of freedom decouple. This is not a serious assumption for a semisubmersible, but may be highly questionable for a ship. The theory developed in section 2 may be carried through with full coupling, but the resulting numerical problem must probably wait for the next generation of supercomputers.

The functions $c(X,\dot{X})$ and $k(X)$ are in general nonlinear. When the mooring is very tight, as for example for a tension leg platform, the restoring force is nearly linear and the stochastic linearization technique may be used. However, for a catenary moored semisubmersible this procedure is not applicable as $k(X)$ is in general strongly nonlinear.

When the equations of motion are linear the probability density of $X$ has been derived by Næss [1],[2]. His work is based on the Kac and Siegert [3] expansion of a stochastic process described by a Volterra series carried out to second order. As our equations of motion may be strongly nonlinear this expansion procedure is not applicable to the motion variable $X$. Instead we shall approximate the drift force with a Markov process and thereby find a Fokker-Plank-Kolmogorov (FPK) equation for the joint probability of $X,\dot{X}$ and $F$. This equation must be solved numerically and given the solution, although it may be difficult to obtain, the statistics of $X(t)$ is determined.

K. Horikawa, H. Maruo (Eds.)
Nonlinear Water Waves
IUTAM Symposium Tokyo/Japan 1987

The FPK equation was first applied to this problem by Roberts [4]. He used, however, a rather simplified expression for the wave drift force. In the present paper we use the Kac and Siegert expansion of the complete slowly varying drift force and retain only the first two terms which turn out to be dominant, at least in our practical examples.

## 2. THEORY

Let $E(t)$ denote the wave process which is assumed stationary Gaussian. $E(t)$ may be considered as transformed white noise $N(t)$, that is

$$E(t) = \int_{-\infty}^{\infty} q(\tau)\, N(t-\tau)\, d\tau \tag{2}$$

where

$$\langle N(\tau)\, N(\tau - t)\rangle = \delta(t)$$

and $\langle\,\rangle$ denotes expected value.
We shall be interested in the second order process $F(t)$

$$\begin{aligned} F(t) &= \int_{-\infty}^{\infty}\int_{-\infty}^{\infty} h(\tau_1,\tau_2)\, E(t-\tau_1)\, E(t-\tau_2)\, d\tau_1\, d\tau_2 \\ &= \int_{-\infty}^{\infty}\int_{-\infty}^{\infty} K(\tau_1,\tau_2)\, N(t-\tau_1)\, N(t-\tau_2)\, d\tau_1\, d\tau_2 \end{aligned} \tag{3}$$

where the filter function $K$ is given by

$$K(t_1,t_2) = \int_{-\infty}^{\infty}\int_{-\infty}^{\infty} h(\tau_1,\tau_2)\, q(t_1-\tau_1)\, q(t_2-\tau_2)\, d\tau_1\, d\tau_2$$

More commonly the drift force is expressed as

$$F(t) = 4\pi^2 \int_{-\infty}^{\infty}\int_{-\infty}^{\infty} \tilde{h}(\omega_1,\omega_2)\, \tilde{E}(\omega_1)\, \tilde{E}(\omega_2)\, e^{i\omega_1 t}\, e^{i\omega_2 t}\, d\omega_1\, d\omega_2 \tag{4}$$

where $\tilde{}$ denotes Fourier transforms.
The Kac and Siegert expansion of $F(t)$ is written

$$F(t) = \sum_{j=1}^{\infty} \lambda_j\, W_j(t)^2 \tag{5}$$

Here $\lambda_j$ are the ordered eigenvalues of the integral equation

$$\int_{-\infty}^{\infty} \Gamma(\omega_1,\omega_2)\, \psi(\omega_2)\, d\omega_2 = \mu\, \psi(\omega_1) \tag{6}$$

where

$$\Gamma(\omega_1,\omega_2) = \frac{1}{2}(S_E(|\omega_1|)S_E(|\omega_2|))^{1/2}\tilde{h}(\omega_1,-\omega_2) \tag{7}$$

and $S_E$ is the one-sided wave spectrum. The eigenfunctions $\psi_j(\omega)$ are orthogonal. The $W_j(t)$ are Gaussian with zero mean and two-sided spectral density function $|\psi_j(\omega)|^2$.

When only the slowly varying part of the second order force $F(t)$ is considered, Næss [2] has shown that $\lambda_{2j-1} = \lambda_{2j}$ and $W_{2j-1}(t)$ and $W_{2j}(t)$ are Hilbert transform pairs. Thus we may write

$$F(t) = \sum_{j=1}^{\infty} \mu_j Z_j(t)^2 \tag{8}$$

where

$$Z_j(t) = [W_{2j-1}(t)^2 + W_{2j}(t)^2]^{1/2} \tag{9}$$

is a Rayleigh process, $\mu_j = \lambda_{2j-1} = \lambda_{2j}$.

The eigenvalue problem (6) has been solved, using the eight times eight transfer function matrix obtained by Faltinsen and Løken [5] for the slowly varying drift forces on an infinitely long circular cylinder with a draft of 10m and a beam of 20m. The eigenvalues are given in the table below.

| Eigenvalues $\mu_j$ in N/m. | |
|---|---|
| j | $\mu_j$ |
| 1 | 480.6 |
| 2 | -35.2 |
| 3 | 23.8 |
| 4 | 14.9 |
| 5 | 2.0 |
| 6 | 0.3 |
| 7 | -0.2 |
| 8 | 0.2 |

In this example we see that it is a good approximation to use only the first double set of eigenvalues. Thus we have

$$F(t) = \mu Z(t)^2 = \mu(W_1(t)^2+W_2(t)^2) \tag{10}$$

where the subscript 1 has been deleted for $\mu_1$ and $Z_1$.

We note at this point that if the so-called Newman's approximation is used, Newman [6], there will be only two pairs of eigenvalues, $\mu_1$ and $\mu_2$, one of which, $\mu_2$, is negative, Vinje[7]. The ratio $|\mu_2|/\mu_1$ is equal to the ratio between the probability of the drift force being negative to the probability of being positive. With negative force we mean a force acting opposite to the wave direction. Thus or physical understanding implies that $|\mu_2| << \mu_1$. This is also confirmed by calculations.

We shall now follow a procedure described by Stratonovich [8]. Define functions $A$ and $B$ and a frequency $\omega_0$ such that

$$W_1(t) = A(t)\sin\omega_0 t + B(t)\cos\omega_0 t \tag{11a}$$

$$W_2(t) = A(t)\cos\omega_0 t - B(t)\sin\omega_0 t \tag{11b}$$

The correlation functions for $A$ and $B$ are

$$R_s(t) = \langle A(t)A(t+\tau)\rangle = \langle B(t)B(t+\tau)\rangle \tag{12a}$$

$$R_a(t) = \langle A(t)B(t+\tau)\rangle = -\langle B(t)A(t+\tau)\rangle \tag{12b}$$

It can be shown that

$$\tilde{R}_s(\omega) = S_W(\omega+\omega_0) + S_W(\omega-\omega_0)$$

$$\tilde{R}_a(\omega) = S_W(\omega+\omega_0) - S_W(\omega-\omega_0)$$

where

$$S_W(\omega) = 2|\psi(\omega)|^2 \tag{13}$$

where $\psi(\omega)$ is one of the (normalized) eigenfunctions, corresponding to the eigenvalue $\mu$, of the integral equation (9).

An explicit assumption is now made about the functional form of the correlation functions $R_s(t)$ and $R_a(t)$. The spectrum $S_W(\omega)$ for the floating cylinder has been calculated and it is found that the spectrum is well approximated by

$$S_W(\omega) \approx \frac{1}{\pi}\frac{\epsilon}{\epsilon^2 + (\omega-\omega_0)^2} \tag{14}$$

with $\epsilon = 0.187\,s^{-1}$, see fig 1. With this approximation we have

$$R_s(t) = e^{-\epsilon|t|} \tag{15a}$$

$$R_a(t) = 0 \tag{15b}$$

As stated above $Z(t)$ is a Rayleigh process. The probability density function is

$$p(z) = z\exp(-\frac{z^2}{2}) \tag{16}$$

and when $R_s(t)$ is given by (15a), Z is also a Markov process. The governing stochastic differential equation (Langevin equation) for Z is

$$\dot{Z} = -\epsilon(Z - \frac{1}{Z}) + (2\epsilon)^{1/2}N(t) \tag{17}$$

and the Fokker-Planck-Kolmogorov (FPK) equation governing the transition probability $\hat{p}(z,t\,|\,z_0,t_0)$ is

$$\frac{1}{\epsilon}\frac{\partial\hat{p}}{\partial t} = \frac{\partial}{\partial z}((z-\frac{1}{z})\hat{p}) + \frac{\partial^2\hat{p}}{\partial z^2} \tag{18}$$

The probability density (16) is the stationary solution of (18).

By writing the equation of motion (1) as two first order equations

$$\dot{X}(t) = Y \tag{19a}$$

$$\dot{Y}(t) = -c(X,Y) - k(X) + \mu Z^2 \tag{19b}$$

and using (17) to determine the slowly varying drift force we have a system of stochastic differential equations for the Markov vector process $(X,Y,Z)$. The associated FPK-equation for the transition probability $\hat{p}(x,y,z,t \mid x_0,y_0,z_0,t_0)$ is

$$\frac{\partial \hat{p}}{\partial t} = -\frac{\partial}{\partial x}(y\hat{p}) + \frac{\partial}{\partial y}[(c+k-\mu z^2)\hat{p}] + \epsilon\frac{\partial}{\partial z}[(z-\frac{1}{2z})\hat{p}] + \epsilon\frac{\partial^2 \hat{p}}{\partial z^2} \tag{20}$$

The stationary solution of (20) gives the joint probability density $p(x,y,z)$. This solution must be obtained numerically.

By transforming $U = Z^2/2$ we obtain an FPK-equation that may be easier to handle numerically. The stochastic differential equation becomes

$$\dot{U}(t) = -2\epsilon(U-1) + 2(\epsilon U)^{1/2}N(t)$$

and the FPK-equation

$$\frac{\partial \hat{p}}{\partial t} = -\frac{\partial}{\partial x}(y\hat{p}) + \frac{\partial}{\partial y}[(c+k-2\mu u)\hat{p}] + 2\epsilon\frac{\partial}{\partial u}[(u-1)\hat{p}] + 2\epsilon\frac{\partial^2}{\partial u^2}[u\hat{p}] \tag{21}$$

## 3. NUMERICAL TEST CASE

The stationary solution of the FPK-equation (21) is found by a Galerkin method. $p(x,y,u)$ is sought on the form

$$\frac{1}{2\pi}\exp(-\frac{x^2}{2}-\frac{y^2}{2}-u)f(x,y,u) \tag{22}$$

and $f(x,y,u)$ is expanded

$$f(x,y,u) = \sum_{i,j,k} A_{ijk}\, He_i(x) He_j(y) L_k(u) \tag{23}$$

$He_i$ is the i-th order Hermite polynomial and $L_k$ is the k-th order Laguerre polynomial.

As a numerical test case the circular cylinder has been moored as a tension leg platform at 50 m water depth. The equations of motion have been approximated by

$$m\ddot{X} + b\dot{X} + k_1 X + k_3 X^3 = F(t) \tag{24}$$

The values chosen for the parameters are $k_1/m = 0.0025\ \mathrm{s}^{-1}$, $k_3/m = 0.0002\ \mathrm{m}^{-2}\mathrm{s}^{-2}$ and $b = 0.03\sqrt{k_1/m}$.

Figure 2 shows the computed probability for X (minus the static offset). Also shown is the Gaussian distribution with the same mean value and variance. For this lightly damped system the motion is nearly Gaussian in spite of the non-Gaussian exciting force and the nonlinear restoring force. The numerical method becomes unstable for the present problem as the importance of the nonlinear terms increase. Work with finite element methods is in progress.

## ACKNOWLEDGEMENT

The present study has been carried out as a part of the research program "Reliability of Marine Structures" sponsored by Det norske Veritas and Saga Petroleum a. s.

## REFERENCES

[1] Næss, A. "The statistical distribution of second-order slowly-varying forces and motions", Appl. Ocean Res., Vol. 8, 1986, pp. 110-118.

[2] Næss, A. "Statistical response of second-order response of marine structures", J. Ship Res., Vol. 29, 1985, pp. 270-284.

[3] Kac, M. and Siegert, A. J. F. "On the theory of noise in radio receivers with square law detectors", J. Appl. Phys., Vol. 18, 1947, pp. 383-397.

[4] Roberts, J. B. "Nonlinear analysis of slow drift oscillations of moored vessels in random seas", J. Ship Res., Vol. 25, 1981, pp. 130-140.

[5] Faltinsen, O. M. and Løken, A. E. "Slow drift oscillations of a ship in irregular waves", Appl. Ocean Res., Vol. 1, 1979, pp.21-31.

[6] Vinje, T. "On the statistical distribution of second order forces and motion", Int. Shipbuilding Prog., Vol. 30, 1983, pp. 58-68.

[7] Newman, J. N. "Second-order slowly varying forces on vessels in irregular waves", Int. Symp. on the Dynamics of Marine Vehicles and Structures in Waves", University College, London 1974.

[8] Stratonovich, R. L. "Topics in the theory of random noise", Gordon and Breach, New York, Vol. 1, 1963.

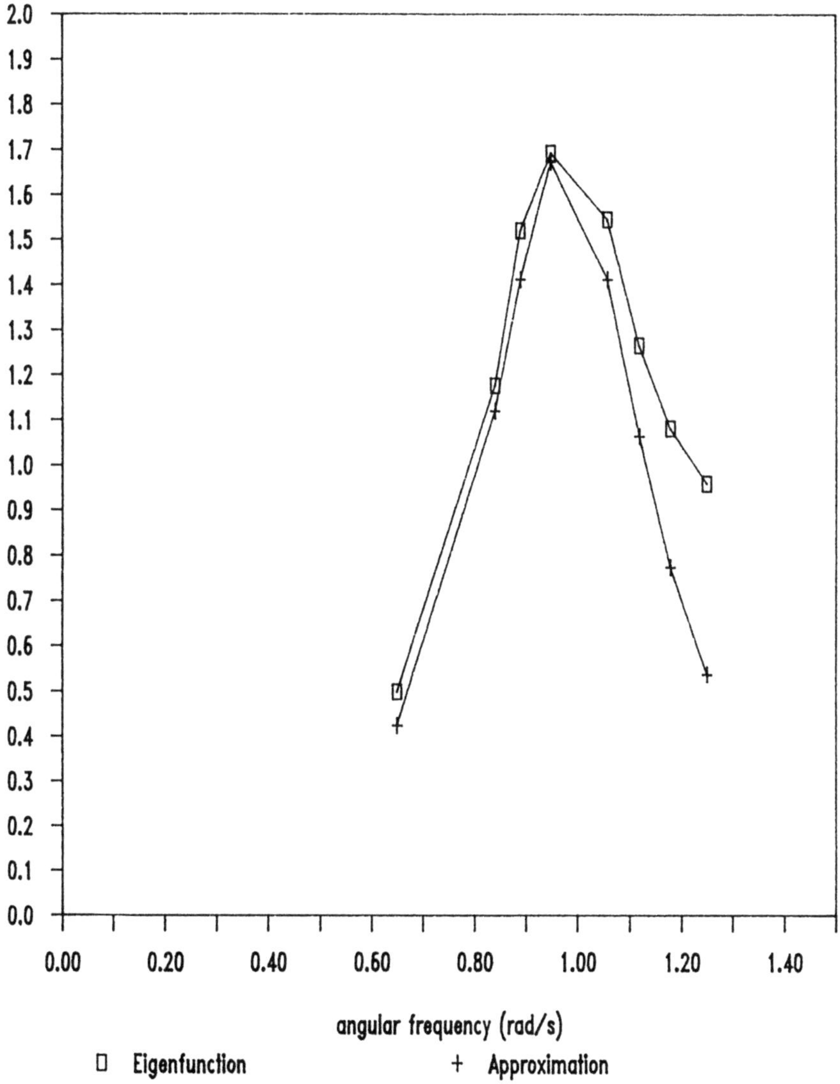

2.0
1.9
1.8
1.7
1.6
1.5
1.4
1.3
1.2
1.1
1.0
0.9
0.8
0.7
0.6
0.5
0.4
0.3
0.2
0.1
0.0
0.00
0.20
0.40
0.60
0.80
1.00
1.20
1.40
angular frequency (rad/s)
□ Eigenfunction
+ Approximation


Fig. 1

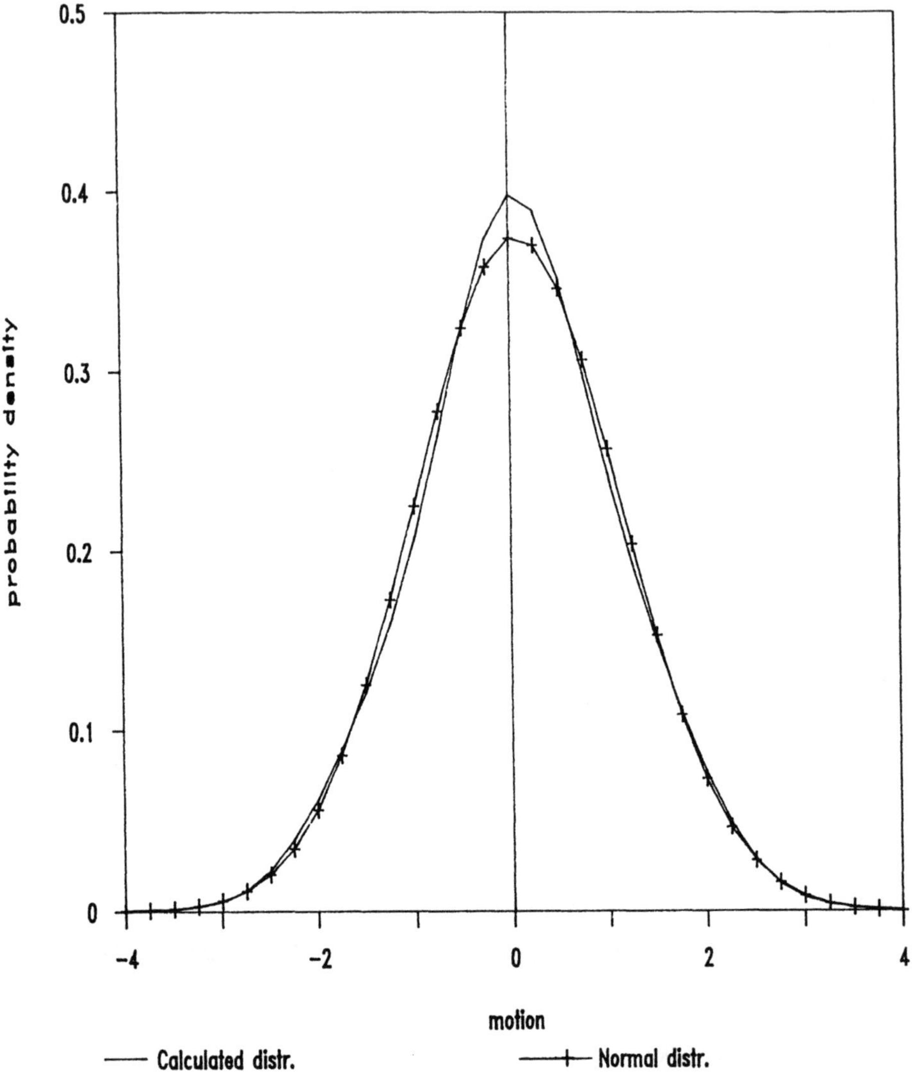
0.5
0.4
0.3
0.2
0.1
0
probability density
-4
-2
0
2
4
motion
Calculated distr.
Normal distr.

Fig. 2

MIX
Papier aus verantwortungsvollen Quellen
Paper from responsible sources
FSC® C105338

If you have any concerns about our products,
you can contact us on
**ProductSafety@springernature.com**

In case Publisher is established outside the EU,
the EU authorized representative is:
**Springer Nature Customer Service Center GmbH**
**Europaplatz 3, 69115 Heidelberg, Germany**

Printed by Libri Plureos GmbH
in Hamburg, Germany